Contracts Department

Spon's Mechanical and Electrical Services Price Book

Edited by
DAVIS, BELFIELD and EVEREST
Chartered Quantity Surveyors

1981

Twelfth Edition

LONDON
E. & F. N. SPON LTD

First published 1968
Twelfth edition 1980
E. & F. N. Spon Ltd
11 New Fetter Lane, London, EC4P 4EE
© 1980 E. & F. N. Spon Ltd
Printed in Great Britain by
Richard Clay (The Chaucer Press), Ltd,
Bungay, Suffolk

ISBN 0 419 12080 7
ISSN 0305 4543

All rights reserved. No part of this book
may be reprinted, or reproduced or utilized in
any form or by any electronic, mechanical or
other means, known now or hereafter invented
including photocopying and recording, or in
any information storage and retrieval system,
without permission in writing from the
Publisher.

British Library Cataloguing in Publication Data

Spon's mechanical and electrical services price book.
   1881: 12th ed.
   1. Heating – Estimates – Great Britain – Periodicals
   2. Air conditioning – Estimates – Great Britain – Periodicals
   I. Davis, Belfield and Everest (*Firm*)
   697   TH7335   80–41276

ISBN 0–419–12080–7
ISSN 0305–4543

Preface

The twelfth edition of Spon's Mechanical and Electrical Services Price Book has been revised to take account of the changes in labour and material costs which have occurred in the past year. In addition some changes have been made to more accurately reflect prevailing conditions within the mechanical and electrical installation industries.

In Part One, Mechanical Installations, the general revision takes account of labour rates in the Heating and Ventilating Industry and Plumbing Mechanical Engineering Services Industry which came into force in April 1980 and material costs which were current in March 1980.

Also in this section the basis for calculating labour costs has been changed in two ways; firstly the composition of the 'gang' has been varied and, secondly, the 'Constant' of labour used in the preparation of 'Prices for Measured Work' has been changed from 'pair' to 'man' hours.

In Part Two, Electrical Installations, the general revision takes account of labour rates in the Electrical Contracting Industry which will come into force in January 1981 and material costs which were current at March 1980. In addition this section has been revised throughout to provide a more comprehensive coverage.

No allowance has been made in any of the sections for Value Added Tax.

Major changes that may have occurred since the preparation of this edition are shown in the 'Stop Press'.

New readers may find the following guide to the book useful.

Part I: Mechanical Installations; contains Rates of Wages, Market Prices of Materials, Labour Constants and Prices for Measured Work.

Part II: Electrical Installations; contains Rates of Wages, Market Prices of Materials, Labour Constants and Prices for Measured Work.

Part III: Approximate Estimating; contains average rates for mechanical and electrical services installations in typical types of buildings applicable on a rate per square metre basis or in pricing approximate quantities, tables of cost indices and an elemental cost plan.

Part IV: Daywork; contains definitions of prime cost for Heating and Ventilating and Electrical Industries.

Part V: Fees; contains extracts from fee scales for Consulting Engineers.

Part VI: Large Industrial Projects; contains details of the types of contract normally used, details of site and national agreements, methods of estimating, costing guidelines and a table of cost indices.

Before referring to prices or other information in the book readers are advised to study the instructions or notes which precede each section.

The prime purpose of the book is to provide average prices for mechanical and electrical work to enable a Bill of Quantities to be priced to provide a reasonably accurate indication of the likely cost of a project; supplementary information is given which will enable the reader to make adjustments to suit his own requirements. It cannot be too strongly emphasized that it is not intended the prices are used in the preparation of an actual tender without adjustment for the circumstances of the particular project such as locality, size and current market conditions, this is particularly important at times of a high rate of inflation.

Part VI: Large Industrial Projects, covers a field that has limited common ground with mechanical and electrical services in buildings. This is why it is a self-contained section and its content, coverage and arrangement differ from the preceding sections. It has been assumed that not every reader will have a working knowledge of this part of the engineering industry and for this reason sub-sections have been included covering contracts and labour agreements.

The construction of large industrial engineering projects is not generally orientated towards the bill of quantities. There is therefore no convenient widely accepted standard method of measurement on which to base the order of the estimating sub-section and for this reason, a functional cost code framework has been incorporated. A typical cost analysis has been included to provide a practical link between the cost code list and the cost guide.

The diversity and complexity of work in the industry has made it impractical to treat each element in the cost guide in the same way. Each element is treated on its merits and whilst one element lends itself to reasonably detailed measurement and pricing, another can only be estimated in outline terms proportional to other elements.

While every effort is made to ensure the accuracy of the information given in this publication neither the Editors nor Publishers in any way accept liability for loss of any kind resulting from the use made by any person of such information.

In conclusion, the Editors record their appreciation of the indispensable assistance received from individuals and organizations including the Association of Cost Engineers, the Engineering Employers' Federation and the Oil and Chemical Plant Constructors' Association.

Davis, Belfield and Everest
Chartered Quantity Surveyors,
5 Golden Square,
London W1R 3AE

STOP PRESS

Rates of Wages

Plumbing Mechanical Engineering Services Industry (page 19)
Weekly holiday credit and sickness benefit stamps with effect from 13 October 1980 will be:

	£
Technical Plumber	10.53
Advanced Plumber	9.33
Trained Plumber, Government Trainee and Apprentice in last year of training	8.67
Working Principals	9.49
All Plumbers over 65 years of age	9.49
Apprentice Plumbers, 1st to 3rd year of training	5.14

Market Prices of Materials

Readers are advised to check the validity of material prices during the currency of this edition.

Daywork

Electrical Contracting Industry (page 337)
The definition of prime cost of daywork carried out under an electrical contract is being revised and is expected to be published by the end of the year. No date for its implementation has yet been set.

E & F N Spon

The technical division of Associated Book Publishers Ltd,
11 New Fetter Lane, London EC4P 4EE.

Make sure you always use the current editions of Spon... Inflation has made last year's editions obsolete.

Spon's Price Books are

Spon's Architects' and Builders' Price Book

Spon's Mechanical and Electrical Services Price Book

EDITED BY DAVIS, BELFIELD AND EVEREST

Spon's Price Books are

- a comprehensive pricing guide to building
- a management tool that can save you money
- completely updated with each new edition to incorporate the latest cost data
- constantly revised and expanded to meet the latest needs of users

Spon's Price Books allow you to

- save your company from potential – and possible expensive – errors by providing comparisons for all cost calculations
- consider the cost advantages of different building solutions
- analyse the effects of price and wage increases on overall costs
- aquaint yourself with building prices, conditions and procedures outside the U.K.
- compare the practices of your company to a national norm

Avoid Disappointment.
Get your copies on publication every year.

To gain maximum benefit from the Price Books it is important to start using them as soon as they are published. By placing a standing order this way you avoid the risk of the Price Books being unavailable when you need them. They have frequently sold out in the past despite increased print-runs.

For further information contact
Lyndsey Williams, E & F N Spon, North Way, Andover, Hants SP10 5BE.

Contents

page

PREFACE	v
STOP PRESS	vii
INDEX	xi
INDEX TO ADVERTISERS	xiv

PART ONE: MECHANICAL INSTALLATIONS

Directions	3
Rates of Wages and Working Rules	7
Market Prices of Materials	
Tubing and fittings	21
Flanges	46
Expansion joints	53
Pipe fixings	55
Boilers	57
Cisterns, tanks and cylinders	61
Calorifiers	68
Pumps, circulators and accelerators	69
Air distribution equipment	73
Fans	79
External louvres	81
Radiators	83
Convector heating units	85
Unit heaters	86
Fire fighting appliances	88
Valves, traps, regulators and gauges	90
Insulation	99
Pipes and fittings	100
Constants of Labour	105
Prices for Measured Work	
Tubing and fittings	120
Expansion joints	145
Pipe fixings	147
Ductwork	152
Boilers	177
Cisterns, tanks and cylinders	180
Calorifiers	184
Pumps, circulators and accelerators	186
Fans	188
Radiators	189
Convector heating units	191

Unit heaters	193
Fire fighting appliances	194
Valves, regulators, traps and gauges	196
Thermal insulation	205
Pipes and fittings	212

PART TWO: ELECTRICAL INSTALLATIONS

Directions	221
Rates of Wages and Working Rules	225
Market Prices of Materials	
Switchgear and distribution	235
Conduit and fittings	237
Cable trunking and fittings	243
Cable trays and fittings	246
Ladder rack and fittings	247
Cable and cords	248
Fittings and accessories	253
Constants of Labour	259
Prices for Measured Work	
Switchgear and distribution	272
Control gear	276
Conduit and fittings	277
Cable trunking and fittings	281
Cable trays and fittings	284
Ladder racks and fittings	285
Cables	286
Fittings and accessories	291
Light fittings	297

PART THREE: APPROXIMATE ESTIMATING

Directions	301
Mechanical installations	303
Lift and escalator installations	315
Electrical installations	317
Cost indices	324
Elemental cost plan	325

PART FOUR: DAYWORK

Heating and Ventilating Industry	333
Electrical Industry	337

PART FIVE: FEES FOR PROFESSIONAL SERVICES

Consulting Engineers' Fees	339

PART SIX: LARGE INDUSTRIAL PROJECTS

Directions	371
The Contract	373
Labour	383
Estimating	437

Index

A.B.S. tubing, 44
— fittings, 45
Accelerators, 71, 113, 187
Acoustic lining, 176
— louvre, 81
Air compressors, 454
— conditioning, approximate estimating, 303, 312
— distribution equipment, 73
— valves, 96, 115, 202
Altitude gauge, 98, 204
Anti-vibration mountings, 80
Approximate estimating, 301
— —, air conditioning installations, 303
— —, boiler plant, 307
— —, electrical installations, 316
— —, fire fighting installations, 313
— —, lift and escalator installations, 315
— —, mechanical installations, 303
Asbestos joint rings, 51
Association of Cost Engineers, 486
Attenuators, 78

Ball valves, 94, 115, 201
Boilers,
—domestic, 57, 111, 177
—**commercial packaged**, 58, 111, 178
— commercial sectional, 59, 111, 178
— commercial steam, 60, 111, 179
— approximate estimating, 307
Bolts, 50
Brass joint rings, 51
Bulk storage tanks, 447
Bus bar chambers, 235, 260, 272

Cable, 248, 266, 286, 472
— armoured, 248, 266, 287
— markers, 476
— trays, 246, 264, 284, 476
— trunking fittings, 244, 263, 282
— — p.v.c., 245, 263, 283
— — steel, 243, 263, 281
Calorifiers, 68, 112, 184, 311
Capillary type fittings, 36, 38, 41, 108, 137, 140
Cast iron pipes and fittings, 100, 116, 212
Ceiling diffuser, 73, 176
Central heating, 314
Check valves, 94, 115, 200, 469
Circulators, 71, 113, 187

Cisterns, 61, 112, 180
Clock circuits, 322
Cocks, 90, 115, 196, 310
Cold water service, 311
Columns, 449
Compression type fittings, 39, 42, 44, 108, 138
Conduit, aluminium, 237
— boxes, 239, 262, 278
— clips, 242
—, fittings, 238, 262, 277
—, p.v.c., 237
— saddles, 242
—, steel, 237, 262, 277
Consumer units, 256, 269, 276, 296, 317
Control gear, 276
— valves, 97, 203, 479
Convector units, 85, 114, 192
Cooker control units, 255, 268, 295, 317
Copper calorifiers, 68, 112, 184
— cylinders, 63, 112, 181
— saddles, 55, 109, 140
— tubing, 37, 107, 136, 311
— fittings capillary, 38, 41, 108, 137, 140
— — compression, 39, 108, 138
— — weldable, 40, 141
Cords, 248
Cost analysis, 443
— code list, 439
— guide, 446
— indices, 324, 486
Cover plates, 56
Cylinders, 63, 112, 181

Dampers, 77, 80, 153
Daywork, 333
Diffusers, 73, 176
Direct labour costs, 483
Distribution boards, 235, 260, 272
Ductwork, 152

Earthing, 472
Electric heating, 318
— motors, 456
Electrical Contracting Industry, 225
— conduit, 237, 262, 277
— fittings, 238, 262, 277
— installations, approximate estimating, 316
Elemental cost plan, 325
Emergency lighting circuits, 321

Escalators, 315
Expansion joints, 53, 110, 145
External lighting, 323

Fans, 79, 113, 188
Fees for professional services, 339
Fibreglass mattress, 205
— sectional insulation, 206
— slabs, 205
Field construction costs, 483
Filters, 75
Fire dampers, 77, 153
— extinguishers, 89, 195
— fighting appliances, 88, 194
— — installations, approximate estimating, 313
Flanges bronze, 49, 142
— steel, 46, 124, 130, 134, 464
Flow indicators, 480
Fluorescent fittings, 256, 270, 297, 321
Foam inlet equipment, 88, 194
Furnaces, 456
Fuse switches, 236, 261, 275

Gas valves, 92, 198
Gate valves, 92, 115, 198, 467
Glass tubing and fittings, 466
— valves, 471
Globe valves, 93, 200, 470
Grilles, 74, 176

Head office costs, 482
Heat exchangers, 452
Heating, cost of, 303
Heavy power installations, 319
Hose reels, 88, 195
Hot water service, 311

Immersion heaters, 269, 296
Instrument cables, 472
Instruments, 479
Insulation, 205, 312, 481
— jackets, 99
Isolators, 236, 260, 274

Joint rings, 51, 124, 130, 134, 143, 146, 465
Junction boxes, 241, 262, 280, 473

Labour constants, 105, 259
— element, 4, 119, 222, 271
— rates, 4, 7, 222, 225, 392, 433
Ladder rack, 247, 265, 285
Large industrial projects, 369
Lift and escalator installations, approximate estimating, 315
Lighting circuits, 317
— points, 317

Lighting switches, 253, 268, 291
Lightning protection, 323
Louvres, external, 81

Malleable fittings, 27, 122, 128
Manifolds, 42
Mechanical installations, approximate estimating, 303
Microbore copper tubing, 37
— fittings capillary, 41
— — compression, 42
Mineral insulated cable, 250, 266, 288
Motor control centres, 477
Motors, electric, 456

Orifice plates, 479
Overheads, 3, 4, 119, 221, 271, 485

Painting, 481
Pipe fixings and supports, 55, 109, 147, 465
— hangers, 151
— sleeves, 151
Pipework, 309
Polythene tubing, 43, 144
— —, fittings, 44, 144
Power cables, 474
— points, 317
Preformed insulation, 206
Preliminaries, 3, 119, 221, 271
Pressure indicators, 480
— vessels, 451
Profit, 3, 119, 221, 271, 485
Pumps, 69, 113, 186, 311, 455
P.V.C. tubing, 43
— —, fittings, 45

Radiator valves, 95, 115, 201
Radiators, 83, 113, 189, 311
Relief valves, 96, 202, 480
Roof extract units, 79, 113, 188
Rubber joint rings, 52

Safety valves, 96, 115, 202
Sectional tanks, 183
Services costs, 303, 316
Shaver sockets, 255, 268, 294
Sight glasses, 98
Sill line convectors, 83, 191
Silos, 448
Site agreement, 384, 404
Skirting radiators, 84
Socket outlets, 254, 268, 292, 317
Stainless steel tube and fittings, 36
Standard form of building contract, 3, 221
Standby generating set, 323
Starters, 69, 80, 188, 477

Index

Steam strainers, 98, 204
— traps, 98, 204
Steel boxes, 239, 262, 279
— tubing, 21, 106, 120, 131, 309, 457
— fittings malleable, 27, 122, 128
— — screwed and socketed, 21, 106, 120, 460
— — welding, 32, 107, 132, 461
— — wrought, 25, 120, 127
Sub-contractors, 4, 119
Substation, 319
Sump pump, 72, 113, 187
Switches, 236, 253, 260, 268, 274, 291
Switch fuses, 236, 260, 274
Switchgear and distribution, 235, 260, 272

Tanks, 62, 67, 112, 180, 183, 312
Temperature indicator, 480
Thermal insulation, 99, 205, 312, 481
Thermocouple cables, 473
Thermometers, 98, 204
Thermoplastic tubing, 43, 144
Thermostatic control valves, 97, 203
Thermostats, 97, 203
Transformers, 478

Traps, 103, 118, 204, 217
Trays, 450
Trunking, 243, 263, 281

Underfloor heating, approximate estimating, 318
Unit heaters, 86, 114, 192
UPVC pipes and fittings, 101, 117, 214

Valves, 90, 115, 196, 310, 467

Wage rates:
  electrical contracting industry, 225
  heating, ventilating, air conditioning, piping and domestic engineering industry, 7
  plumbing mechanical engineering services industry, 19
Washers, 50
Weather louvres, 82
Weldable fittings, 32, 40, 45, 107, 132, 141, 461
Welding, 131
Working rules:
  electrical contracting industry, 226
  heating, ventilating, air conditioning, piping and domestic engineering industry, 10
Wrought fittings, 25, 120, 127

INDEX TO ADVERTISERS

Cape Building Services Limited	*facing page* 206
J. Gardener & Co. Ltd	*facing page* 121
The Graham Group	*facing page* 249
Hayward Technical Services Ltd	*facing page* 120
Pirelli General Cable Works Ltd	*facing page* 248
Pullen Pumps Ltd	*facing page* 70
E. &. F.N. Spon Ltd	*facing pages* 71 *and* 207

Advertising agent:
T. G. Scott & Son Ltd
30–32 Southampton Street
London WC2E 7HR

PART ONE

Mechanical Installations

Directions, *page* 3
Rates of Wages and Working Rules, *page* 7
Market Prices of Materials, *page* 21
Labour Constants, *page* 105
Prices for Measured Work, *page* 119

MECHANICAL INSTALLATIONS

Directions

RATES OF WAGES AND WORKING RULES

This section gives rates of wages and extracts from the working rules of the Heating, Air Conditioning, Piping and Domestic Engineering Industry and for the Plumbing Mechanical Engineering Services Industry which were operative at April 1980.

MARKET PRICES OF MATERIALS

The prices given, unless otherwise stated, include for delivery to sites in the London area at March 1980 and represent the prices paid by contractors after the deduction of all trade discounts but exclude any charges in respect of V.A.T.

LABOUR CONSTANTS

Labour 'constants' are given for the major items of work for which prices are given in 'Prices for Measured Work'.

PRICES FOR MEASURED WORK

These prices are intended to apply to new work in the London area and include allowances for all overhead charges, preliminary items and profit. The prices are for reasonable quantities of work and the user should make suitable adjustments if the quantities are especially small or especially large. Adjustments may also be required for locality (e.g. outside London) and for the market conditions (e.g. volume of work on hand or on offer) at the time of use.

MECHANICAL INSTALLATIONS

The labour rate on which these prices have been based is £3·55 per man hour which is the London rate at April 1980 plus allowances for all other emoluments and expenses. To this rate has been added 40% to cover site and head office overheads and preliminary items together with a further 5% for profit, resulting in an inclusive rate of £5·22 per man hour. The rate of £3·55 per man hour has been calculated on a working year of 2043 hours; a detailed build-up of the rate is given at the end of these Directions.

PLUMBING INSTALLATIONS

The labour rate on which these prices have been based is £3·52 per man hour which is the rate at April 1980 plus allowances for all other emoluments and expenses. To this rate has been added similar percentages as for MECHANICAL INSTALLATIONS resulting in an inclusive rate of £5·17 per man hour.

In calculating 'Prices for Measured Work' the following assumptions have been made:

   (*a*) That the work is carried out as a sub-contract under the Standard Form of Building Contract and that such facilities as are usual would be afforded by the main contractor.

DIRECTIONS

(b) That, unless otherwise stated, the work is being carried out in open areas at a height which would not require more than simple scaffolding.

(c) That the building in which the work is being carried out is no more than six storeys high.

Where these assumptions are not valid, as for example where work is carried out in ducts and similar confined spaces or in multi-storey structures when additional time is needed to get to and from upper floors, then an appropriate adjustment must be made to the prices. Such adjustment will normally be to the labour element only.

No allowance has been made in the prices for any cash discount to the main contractor.

The labour element, inclusive of overheads and profit, in any particular item can be ascertained by deducting the amount which appears in italics below the measured price from the measured price. This amount shown in italics represents the value of material content contained in the measured price including relevant allowances for waste, overheads and profit.

The prices of materials upon which 'Prices for Measured Work' are based are as shown in 'Market Prices of Materials', or as indicated, with the addition of 10% to cover overheads and a further 5% for profit. Allowance has been made for waste where necessary.

SUB-CONTRACTORS

Where work would normally be carried out by sub-contractors, such as ductwork and insulation, prices have been based on sub-contractors' estimates with the addition of 15% to cover overheads, profits and any attendance necessary.

LABOUR RATE

The following detail shows how the labour rate for mechanical installation work of £3·55 per man hour has been calculated.

Total annual cost of notional eleven man gang

Hourly rate effective from February 1980

Comprising:

	£
1 Foreman	2·67
1 Advanced Fitter/Welder (gas/arc)	2·46
2 Advanced Fitter/Welders (gas or arc)	2·36
3 Advanced Fitters	2·26
2 Fitters	2·05
1 Mate	1·64
1 18-year-old Apprentice	1·33

Hours actually worked

		£	£
1021¼ hours Foreman	@	2·67	2727·41
2043 hours Advanced Fitter/Welder (gas/arc)	@	2·46	5025·78
4086 hours Advanced Fitter/Welder (gas or arc)	@	2·36	9642·96
6129 hours Advanced Fitters	@	2·26	13851·54
4086 hours Fitters	@	2·05	8376·30
2043 hours Mates	@	1·64	3350·52
1708¼ hours 18-year-old Apprentice	@	1·33	2272·31

C/f 45246·82

Mechanical Installations 5

DIRECTIONS

		£	£
			B/f 45246·82

Non-productive overtime
112½ hours Foreman . . . . @ 2·67   300·38
112¼ hours Advanced Fitter/Welder (gas/arc) . . . . @ 2·46   276·75
225 hours Advanced Fitter/Welder (gas or arc) . . . @ 2·36   531·00
337½ hours Advanced Fitters . . . . @ 2·26   762·75
225 hours Fitters . . . . @ 2·05   461·25
112¼ hours Mates . . . . @ 1·64   184·50
112½ hours 18-year-old Apprentice . . . . . . @ 1·33   149·63

Daily travelling allowance
2088 days Craftsmen & Improvers . . . . . . @ 1·43   2985·84
424 days Assistants, Mates & Apprentice . . . . @ 1·23   521·52

Daily fares
2512 days . . . . @ 1·90   4772·80

Trade supervision
1021¼ hours Foreman . . . . @ 2·67   2727·41

Apprentice day release
0334¼ hours 18-year-old Apprentice . . . . . . @ 1·33   444·89

National Insurance 'Non-contracted out contributions' based on weekly gross pay shown in brackets.
48 weeks Foreman (£133·50) . . . . @ 18·32   879·36
48 weeks Advanced Fitter/Welder (gas/arc) (123·50) . . @ 16·95   813·60
96 weeks Advanced Fitter/Welder (gas or arc) (£118·50) . @ 16·27   1561·92
144 weeks Advanced Fitters (£114·00) . . . . . @ 15·65   2253·60
96 weeks Fitters (£104·00) . . . . @ 14·28   1370·88
48 weeks Mates (£83·50) . . . . @ 11·17   536·16
48 weeks 18-year-old Apprentice (£68·50) . . . . @ 9·42   452·16

Weekly holiday credit/welfare stamp
48 weeks Foreman . . . . @ 10·54   505·92
288 weeks Advanced Fitter/Welder, etc.. . . . . @ 9·83   2831·04
96 weeks Fitter/Welder (gas or arc), etc. . . . . @ 9·13   876·48
96 weeks Assistant, Mates, etc. . . . . . . @ 8·08   775·68

Recognized holidays with pay
62 hours Foreman . . . . @ 2·67   165·54
62 hours Advanced Fitter/Welder (gas/arc) . . . . @ 2·46   152·52
124 hours Advanced Fitter/Welder (gas or arc) . . . @ 2·36   292·64
186 hours Advanced Fitters . . . . @ 2·26   420·36
124 hours Fitters . . . . @ 2·05   254·20
62 hours Mates . . . . @ 1·64   101·68
62 hours 18-year-old Apprentice . . . . . . @ 1·33   82·46

Training
C.I.T.B. Levy
10 Operatives . . . . @ 60·00   600·00

                                                                                 74291·74

Severance pay and sundry costs
Add 1% . . . .   742·92

                                                                            C/f 75034·66

Mechanical Installations
DIRECTIONS

	£	£
		B/f 75034·66
Employer's liability and third party insurance		
Add say £1·60%		1200·56
Annual cost of notional 11 man gang (10½ men actual working) .		76235·22
Average annual cost per working man		7260·50
All-in man hour cost		3·55

Notes:
The following assumptions have been made in the above calculations.

(1) The working week of 38 hours is made up of 8 hours Monday and 7½ hours Tuesday to Friday.

(2) Actual working week of 45 hours made up of 5 days at 9 hours per day.

MECHANICAL INSTALLATIONS

Rates of Wages and Working Rules

HEATING, VENTILATING, AIR CONDITIONING, PIPING AND DOMESTIC ENGINEERING INDUSTRY

Extracts from National Agreement made between:

Heating and Ventilating and The National Union of Sheet Metal
 Contractors' Association   Workers, Coppersmiths and
ESCA House,   Heating and Domestic Engineers
34 Palace Court,   75/77 West Heath Road,
Bayswater   Hampstead,
London W2 4JG   London, NW3 7TL
 Telephone: 01–229 2488   *Telephone:* 01–455 0053

RATES OF WAGES

EFFECTIVE FROM 4 FEBRUARY 1980

Hourly rates of Wages, all districts of the United Kingdom.

	Hourly rate £
Foreman	2·67
Chargehand	2·56
Advanced Fitter	2·26
Fitter	2·05
Welding supplement gas/arc	0·20
Welding supplement gas or arc	0·10
Improver	1·95
Assistant	1·85
Mate over 18	1·64
Craft apprentice (during four year apprenticeship)	
Up to 17	0·72
17–18	1·03
18–19	1·33
19–20	1·64
(for late entrants)	
20–21	1·74
21–22	1·85

Junior mates receive Apprenticeship rates

RATES OF WAGES (HEVAC)

Daily travelling allowance
C = Craftsmen including Improvers
M & A = Assistant, Mate, Junior Mate and Craft Apprentice
Direct distance from centre to job

Over	Not exceeding	C £	M & A £
0 miles	2 miles	0·54	0·46
2 miles	5 miles	1·07	0·92
5 miles	10 miles	1·43	1·23
10 miles	15 miles	1·79	1·54
15 miles	20 miles	2·33	2·00
20 miles	25 miles	2·87	2·47
25 miles	30 miles	3·22	2·77
30 miles	35 miles	3·58	3·08
35 miles	40 miles	3·94	3·39
40 miles	45 miles	4·30	3·70
45 miles	50 miles	4·66	4·01

Weekly holiday credit/welfare stamp
Effective from 3 September 1979

Foreman, Chargehand	10·54
Advanced Fitter/Welder (gas/arc),	
Advanced Fitter/Welder (gas or arc),	
Advanced Fitter,	
Fitter/Welder (gas/arc)	9·83
Fitter/Welder (gas or arc),	
Fitter,	
Improver,	
Apprentice aged 19–20	9·13
Assistant,	
Mate,	
Apprentice aged 18–19	8·08
Apprentice and Junior Mate aged 16–18	5·28

Abnormal conditions
Exceptionally dirty work or work under
abnormal conditions of such a character as
to be equally onerous.   £1·03 extra per day or part of a day

Exposed work at heights
Over 125 ft., not exceeding 250 ft.   £0·41 extra per day or part of a day

Swings, cradles and ladders
Agreed sum of £0·10 per hour as follows: 'An Operative working in swings or cradles shall be paid £0·10 per hour extra for the time actually worked in those conditions. An Operative working on ladders shall be paid an extra £0·10 per hour for the time actually worked at a height of 20 ft. and an additional £0·10 per hour for each additional 10 ft. The height shall be measured from the nearest fixed flooring or fixed scaffolding to the actual work.'
                                              £0·10 per hour extra as defined

RATES OF WAGES (HEVAC)

Lodging allowance
Effective from 4 February 1980
 £6·50 per night, including night of day of return plus reimbursement of any V.A.T. subject to provision by Operative of valid tax invoice.

Sickness and accident benefit
Effective from 3 September 1979

Grade	Weeks 1–2 £	Weeks 3–28 £	Weeks 29–52 £
(a)	33·04	22·26	20·23
(b)	27·93	18·06	15·26
(c)	22·61	13·51	10·08
(d)	15·12	10·01	2·73
(e)	10·01	10·01	N/A

Notes:
 (i) benefit is payable from the fourth day of an incapacity;
 (ii) the flat rate amounts in respect of weeks 1–28 will be reduced by £2·75 where industrial injury benefit may be claimed;
 (iii) second or subsequent periods of incapacity separated by less than 13 weeks from the original incapacity will be treated as 'continuous' claims;
 (iv) benefit for grade (e) is payable up to 26 weeks only.
 (v) the grades of Operatives entitled to the different rates of benefit are the same as the grades for weekly credit/welfare values.

Death benefit; Accidental dismemberment/total disability benefit

	£
Death Benefit – any cause	7500
Accidental Dismemberment/Total Disability Benefit	5000
Loss of four fingers or a thumb	1000
Loss of index finger	600
Loss of any other finger	100
Loss of big toe	200
Loss of any other toe	50

WORKING RULES (HEVAC)

Extracts from National Working Rules

HOURS OF WORK (CLAUSE 3)

(*a*) The normal working week shall consist of 38 hours to be worked in five days from Monday to Friday inclusive. The length of each normal working day shall be determined by the Employer but shall not be less than six hours or more than eight hours.

(*b*) The Employer and the Operative concerned may agree to extend the working hours to more than 38 hours per week for particular jobs, provided that overtime shall be paid in accordance with Clause 9.

MEAL AND TEA BREAKS (CLAUSE 4)

(*a*) The normal break for lunch shall be one hour except when such a break would make it impossible for the normal working day to be worked, in which case the break may be reduced to not less than half an hour.

(*b*) An Operative directed to start work before his normal starting time or to continue work after his normal finishing time shall be entitled to a quarter of an hour meal interval with pay at the appropriate overtime rate for each two hours of working (or part thereof exceeding one hour) in excess of the normal working day, which on Saturdays, Sundays and other holiday days shall mean eight hours. Where an Operative is entitled to a morning and/or evening meal interval under this clause, the meal interval shall replace the morning and/or afternoon tea break referred to in (*c*).

(*c*) A tea break shall, subject to (*b*), be allowed in the morning and in the afternoon without loss of pay, provided that Operatives co-operate with the Employer in minimizing the interruption to production.

GUARANTEED WEEK (CLAUSE 5)

(*a*) An Operative who has been continuously employed by the same Employer for not less than two weeks is guaranteed wages equivalent to his inclusive hourly normal time earnings for 38 hours in any normal working week; provided that during working hours he is capable of, available for and willing to perform satisfactorily the work associated with his usual occupation, or reasonable alternative work if his usual work is not available.

(*b*) In the case of a week in which holidays recognized by agreement, custom or practice occur, the guaranteed week shall be reduced for each day of holiday by the normal working day as determined in Clause 3 (a).

(*c*) In the event of a dislocation of production as a result of industrial action the guarantee shall be automatically suspended in respect of Operatives affected on the site or sites where the industrial action is taking place. In the event of such dislocation being caused by Operatives working under other Agreements and the Operatives covered by this Agreement not being Parties to the dislocation, the Employers will endeavour to provide other work or if not able to do so will provide for the return of the Operatives to the shop or office from which they were sent.

WAGES AND ALLOWANCES (CLAUSE 8)

(*d*) A Fitter or an Advanced Fitter who holds one or both current certificates of Competency issued by the Heating, Ventilating and Domestic Engineers' National Joint Indus-

WORKING RULES (HEVAC)

trial Council, in oxy-acetylene welding and/or metal arc and who is competent in such welding to the standard(s) required by such certificate(s) shall receive a Welding Supplement for one welding skill or for both welding skills as appropriate, the amounts of which shall be enumerated in an Appendix to this Agreement.

(e) Payment of merit money to an Operative may be made at the option of the Employer for mobility, loyalty, long service etc. etc. and for special skill over and above that detailed in definition of the Operative's grade.

(f) Wages shall be paid on Thursday, where practicable, or at times determined by mutual arrangement, but the Employer shall not hold more than four days' wages of the operative on any pay day.

(g) A Chargehand shall receive extra remuneration. The amount shall be dependent on the character of the charge. The extra remuneration shall be agreed between the Chargehand and the Employer.

(h) Junior mates under the age of 18 shall be paid the same rates as Craft Apprentices.

(i) Operatives engaged on exceptionally dirty work, or work under abnormal conditions, of such a character as to be equally onerous, shall receive an agreed sum extra per day or part of a day.

(j) An Operative working in swings or cradles shall be paid an agreed sum extra per hour for the time actually worked in those conditions. An Operative working on ladders shall be paid an agreed sum extra per hour for the time actually worked at a height of 20 ft. and an additional agreed sum per hour for each additional 10 feet.

Target incentive schemes may be introduced where they are mutually agreed by the Employer and the majority of the workforce. Agreed guidelines and criteria have been established by the Association and the Union for their operation.

OVERTIME (CLAUSE 9)

(c) Subject to the adjustment in Clause 9 (d) (iv) overtime rates shall be paid for all hours worked in excess of the normal working day.

(d) Time worked in excess of the normal working day on Monday to Friday inclusive shall be paid for as follows:

(i) First four hours, time and a half subject to the adjustment in Clause 9 (d) (iv); thereafter, until normal starting time next morning, double time.

(ii) If time is lost through the fault of the Operative, overtime rate shall not be paid until the full normal hours for the day have been worked.

(iii) An Operative directed to start work before the normal starting time shall be paid the appropriate overtime rates for all hours worked before the normal starting time, but if through the action of the Operative the normal working day is not worked, ordinary hourly rates shall be paid for all hours worked.

(iv) Each week the number of hours of overtime payable at time and a half shall be reduced and the number of hours payable at normal hourly rate shall be correspondingly increased so that up to 40 hours are paid at the normal hourly rate provided that the 40 hours shall be reduced by any hours of absence during the normal working day arising from

any day or days of recognized holiday in accordance with Clause 18a or a day or days in lieu thereof in accordance with Clause 18c
certified sickness
absence with the concurrence of the Employer

WORKING RULES (HEVAC)

absence for which the Operative can produce evidence to the satisfaction of the Employer that his absence was due to causes beyond his control

(v) An Operative called back to work at any time between the period commencing two hours after the normal finishing time and until two hours before normal starting time shall be paid such overtime rates as would apply had work been continuous from normal finishing time and shall be paid a minimum of two hours at the appropriate rate.

(e) Time worked on Saturday and Sunday shall be paid for as follows:
(i) Saturday – first five hours, time and a half; after the first five hours, double time but if time is lost through the fault of the Operative the double time rate shall not apply until time lost has been made up.
(ii) Sunday – double time for all hours worked until starting time on Monday morning.

PAYMENT FOR HOLIDAYS WORKED (CLAUSE 10)

(a) This clause applies to all days recognized as a holiday in Clause 18, and in England and Wales – Christmas Day and Boxing Day; in Scotland – three days (see Clause 10c) of Winter Holiday including New Year's Day.

(b) An Operative who works on any of the days in Clause 10a shall be paid a minimum of two hours at the appropriate rate.

(c) Time worked on such days shall be paid as follows:

In England and Wales
Christmas Day, Boxing Day, New Year's Day, Good Friday, Easter Monday, Spring Bank Holiday, May Day, Late Summer Bank Holiday.
                                        Double time for all hours worked.

In Scotland
Boxing Day, three consecutive days of Winter Holiday which shall include New Year's Day and the one or two holiday days which immediately follow it (if any), Spring Holiday and the Friday before the Spring Holiday, May Day, Autumn Holiday (one day).
                                        Double time for all hours worked.

Christmas Day and the one day of recognized holiday to be agreed locally.
   The normal working day as determined in Clause 3a time and a half; thereafter double time.
Friday before Autumn Holiday.
   The normal working day as determined in Clause 3a, normal hourly rates; thereafter overtime rates in accordance with Clause 9.

(d) The general conditions of the Agreement shall apply to men called back to work on these holidays.

NIGHT SHIFTS AND NIGHT WORK (CLAUSE 11)

(a) For an Operative who works for at least five consecutive nights
(i) The basic rate, called the night shift rate, shall be one and a third times the normal rate.
(ii) Overtime rates and conditions shall be as for normal working days but the basic rate shall be the night shift rate.

WORKING RULES (HEVAC)

(*b*) An Operative who works for less than five nights and does not work during the day shall be paid at overtime rates as if the normal day had already been worked.

CONTINUOUS SHIFT WORK (CLAUSE 12)

Where jobs have to be continuously operated the work shall be carried out in two or three shifts of eight hours each according to requirements. The Operatives concerned shall be paid time and a third in cases where a six day shift is worked and time and a half in cases where a seven day shift is worked, overtime and night shift rates being compounded in these rates. Arrangements shall be made to change the shifts worked by each Operative.

ALLOWANCES TO MEN WHO TRAVEL DAILY (CLAUSE 15)

(*a*) Except where his centre is the job, an Operative who is required by his Employer to travel daily up to 50 miles to the job shall be paid fares and travelling time as stated in (*i*) and (*ii*) below.

(*i*) return daily travelling fares from his centre to the job. Where cheap daily or period fares or other cheap travel arrangements by public transport are available the Employer may pay fares on that basis. Where, however, a change in such travel arrangements results from a change in the working arrangements the Employer must pay the Operative for any additional cost. The Employer at his option may provide suitable conveyance for the Operative to and from the job in which case fares shall not be paid.

(*ii*) allowances for travelling time, provided that the normal hours are worked on the job. The allowances for travelling time shall be agreed from time to time by the Association and the Union and shall be enumerated in an Appendix to this Agreement. When a reasonably direct journey is not possible, a claim for special consideration may be made by the Operative and in case of dispute the matter shall be referred to the Chief Officials of the parties, whose decision shall be final.

(*b*) Except where his centre is the job, payment to the Operative of allowances for travelling time and fares for journeys beyond fifty miles daily from his centre to the job will be for agreement between the Employer and the Operative concerned.

ALLOWANCES TO MEN WHO LODGE (CLAUSE 16)

(*a*) Where an Operative is sent to a job to which it is impracticable to travel daily and where the Operative lodges away from his place of residence he shall (except if he is engaged at the job or if his centre is the job) be paid the items in (*i*) to (*v*) below where appropriate:

(*i*) a nightly lodging allowance including the night of the day of return and when on week-end leaves in accordance with Clause 17*a*. The nightly lodging allowance shall be agreed from time to time by the Association and the Union and shall be enumerated in an Appendix to this Agreement. The lodging allowance shall not be paid when an Operative is absent from work without the concurrence of the Employer, nor when suitable lodging is arranged by the Employer at no expense to the Operative, nor during the annual holidays defined in Clause 20 including the new week of Winter Holiday. The Operative shall provide the Employer with a statement signed by himself to the effect that he is in lodgings for the period of payment of lodging allowance under this clause. Without such evidence, the Employer shall deduct tax on lodging allowance paid.

(*ii*) when suitable lodging are not available within two miles from the job, daily return fares from lodging to job. The Employer at his option may provide suitable conveyance for the Operative between the lodgings and the job, in which case fares shall not be paid.

(*iii*) travelling time for the time spent in travelling from the centre at the commencement and completion of the job at the normal time rate but when an excessive number of hours of travelling is necessarily incurred, a claim for special consideration may be made by the Operative to the Employer or by the Employer to the Operative and in case of dispute the matter shall be referred to the Chief Officials of the parties, whose decision shall be final.

(*iv*) fares between his centre and the job at the commencement and the completion of the job. Return fares shall be used when available.

(*v*) week-end leaves in accordance with Clause 17*a*.

(*b*) An Operative whose employment is terminated by proper notice on either side during the course of a job, shall be entitled to travelling time back to his centre and a single fare for the journey from the job to his centre. This condition shall not apply to an Operative who is discharged for misconduct or who leaves the job without the concurrence of his Employer.

WEEK-END LEAVES (CLAUSE 17)

(*a*) An Operative who is in receipt of lodging allowance in accordance with Clause 16 shall be allowed a week-end leave every two weeks. Such Operative shall be entitled to return to his respective centre for the recognized holidays prescribed in Clause 18 and to facilitate this, the nearest normal week-end leave shall, where necessary, be deferred or brought forward to coincide with the holiday.

(*b*) Unless the Employer and the Operative agree otherwise the week-end leave shall be from normal finishing time on Friday to normal starting time on Monday.

(*c*) An Operative shall not be required to start his return journey before 6.00 a.m. on the appropriate day of return to the job but shall, where the return journey makes it impossible to commence work at the normal starting time, agree with his Employer the working arrangements for the day.

(*d*) Week-end return fares shall be paid for week-end leaves. If an Operative does not elect to return to his centre a single fare from the job to his centre shall be paid.

(*e*) An Operative on a week-end leave whose work is up to 150 miles from his centre shall travel home in his own time, but travelling time from the centre to the job shall be paid at the normal time rate. An Operative whose work is 150 miles or more from his centre shall be paid travelling time of four hours at normal rate from the job to the centre; travelling time for the journey back to the job to be paid at normal time rate. If an Operative elects to stay at the job travelling time shall not be paid.

(*f*) When a reasonably direct journey is not possible or when an excessive number of hours travelling is necessarily incurred on jobs more than 150 miles from an Operative's centre, a claim for special consideration in respect of travelling time may be made by the Operative to the Employer or by the Employer to the Operative and in case of dispute the matter shall be referred to the Chief Officials of the parties, whose decision shall be final.

WORKING RULES (HEVAC)

(g) An Operative on week-end leaves (including holidays provided under Clause 18), shall be paid the nightly lodging allowance, provided that the leave is within this Agreement or is agreed with the Employer.

RECOGNIZED HOLIDAYS (CLAUSE 18)

Junior Mates under the age of 18 years and Craft Apprentices shall have recognized holidays with pay in accordance with the terms of the Agreement of Service issued by the Heating, Ventilating and Domestic Engineers' National Joint Industrial Council. Adult Operatives aged 18 years and over (other than Craft Apprentices) shall take recognized holidays as provided hereinafter.

(a) The following days shall be holidays with pay at the normal hourly rate for the normal working day as determined in Clause 3a.

In England and Wales
New Year's Day; Good Friday; Easter Monday; Spring Bank Holiday; May Day; Late Summer Bank Holiday; Christmas Day; Boxing Day.

In Scotland
Spring Holiday (two days viz. Spring Holiday and the Friday before); May Day; Autumn Holiday (two days); Christmas Day; Boxing Day plus one other day to be agreed locally.

If any of these days comes within the annual holidays as provided in Clause 20, mutual arrangements shall be made to substitute some other day for the day or days included.

(b) Operatives who fail to report for work on the working day preceding and the working day following a holiday shall not qualify for payment for these holidays unless their absence was due to having worked continuously until after midnight the previous night in accordance with Clause 9f or they can produce evidence to the satisfaction of the Employer that their absence was due to causes beyond their control. In case of dispute the matter shall be referred to the Chief Officials of the Parties whose decision shall be final.

(c) Operatives who work on a recognized holiday as set out in Clause 18a shall be paid overtime in accordance with Clause 10c and shall be granted a day's holiday with pay for each holiday day worked, at a mutually agreed time.

(d) The general conditions of the Agreement shall apply to men called back to work on these holidays.

ANNUAL HOLIDAY CREDITS AND WELFARE PREMIUM (CLAUSE 19)

(a) All Operatives aged 18 years and over shall be provided with a card covering annual holidays and welfare insurance. The card of each such Operative shall, subject to the provisions below, be stamped by the Employer first employing him in any week in an accounting period (i.e. the Employer whose obligation it is to stamp the National Insurance card for that week) with a stamp covering a weekly credit in respect of annual holiday and a weekly premium in respect of welfare insurance.

(b) Values of the annual holiday credit shall be agreed from time to time between the Association and the Union.

(c) The weekly premium in respect of welfare insurance shall provide for payment to the operative of sickness and accident benefit, death benefit, accidental dismemberment and total disability benefit. Rates of benefits shall be agreed from time to time between the Association and the Union.

WORKING RULES (HEVAC)

(*f*) Each accounting period shall start from the commencement of the week in which 1 September falls and shall continue until the commencement of the next accounting period.

(*g*) No stamp shall be affixed in respect of the four weeks of the annual holiday but in respect of these weeks the four appropriate spaces on the card shall be marked with an 'H'. When the week or Winter Holiday and/or Autumn Holiday in England and Wales does not fall in one complete calendar week no stamp shall be affixed in respect of the week in which the major part of the holiday falls.

(*h*) An Operative who qualifies for payment of sickness or accident benefit under Clause 22 shall be entitled to have a stamp affixed to his holiday card by his Employer during the period of his disability and up to a maximum of four stamps in respect of any one period of disability and up to a maximum of eight stamps in any one accounting period. Where the provisions in the previous sentence do not apply no stamp shall be affixed in respect of any complete week or weeks during which an Operative is absent due to sickness or accident, but in respect of such week or weeks the appropriate space or spaces on the card shall be marked with an 'S'.

(*i*) No stamp shall be affixed in respect of any week in which the Operative:
    (*i*) is available for work less than 32 normal working hours unless he can produce evidence to the satisfaction of the Employer that his absence was due to causes beyond his control
    (*ii*) terminates his employment without giving the notice required in Clause 2.

(*k*) Not more than 48 stamps shall be affixed to an Operative's card in any accounting period.

ANNUAL HOLIDAYS (CLAUSE 20)

All Operatives aged 18 years and over shall have annual holidays with pay in accordance with this clause. The pay shall consist of the appropriate annual holiday credits standing to the credit of the Operative.

Annual holidays shall consist of:
(*i*) four days of Spring Holiday.
(*ii*) two weeks of Summer Holiday.
(*iii*) Six days of Winter Holiday.

(*a*) **Winter Holiday**
(*i*) in England and Wales the Winter Holiday shall consist of five working days taken in conjunction with Christmas Day, Boxing Day and New Year's Day (or working days declared Bank Holidays in lieu thereof when Christmas Day and/or Boxing Day fall on Saturday and/or Sunday).
(*ii*) in Scotland the Winter Holiday shall consist of five consecutive working days in one complete calendar week. Such working days to include New Year's Day, or to follow New Year's Day when the latter falls on Saturday or Sunday.

(*b*) **Autumn Holiday**
(*i*) The Autumn Holiday shall commence in 1978 and shall consist of five consecutive working days to be taken where practicable in the period between 7 August and 31 October, at such time as the Employer shall determine and shall normally be at the Late Summer Holiday period in England and Wales and the Autumn Holiday period in Scotland.

WORKING RULES (HEVAC)

(c) **Other Annual Holiday**

(i) The other Annual Holiday shall consist of two normal working weeks as defined in this Agreement;

(ii) where practicable the Other Annual Holiday shall be taken as a whole. When this is impracticable, one week shall be taken as a whole and the remaining week shall be taken as a whole or in parts. The period of the whole holiday or the first week if the holiday is not taken as a whole shall be fixed by mutual consent. When the holiday is taken in parts the second week whether taken as a whole or in parts shall be taken at such time or times as the Employer may determine;

(iii) the Other Annual Holiday shall where practicable be taken in the period between 1 June and 31 October and the Operative not taking his holiday as a whole shall have the right to have his first week during this period. The whole of the holiday must be taken by 31 May immediately following the accounting period in which the appropriate holiday credits accrue.

(d) **Deferment of holiday**

Any Operative who is unable to take the Winter Holiday at the designated time or the Autumn Holiday or the Other Annual Holiday at the time decided by the Employer, either because of accident or sickness or because he is required by the Employer to work during that period, shall be granted his holiday by the Employer as soon thereafter as is reasonably convenient.

Where any Operative is required by his Employer to work during such holiday period he shall be entitled to be paid in respect of work done during that period only at ordinary rates of pay, provided nevertheless that, in the event of the holiday period coinciding with one or more of the holidays defined in Clause 10a and the Operative being required to work on such a day, he shall be entitled to payment for work done on that day or days at the rate set out in Clause 10c.

ANNUAL HOLIDAY CREDITS – PAYMENT (CLAUSE 21)

The sum standing to the credit of each Operative, being the sum of the weekly credits less any administrative charge approved by the Parties to the Agreement, shall be paid to the Operative by the Employer on taking his holidays in the following manner:

(i) in relation to the one week of Autumn Holiday, the total amount of holiday credits on Part 1 of the card for the current accounting period

(ii) in relation to the first week of Other Annual Holiday, the total amount of holiday credits on Part 2 of the card for the current accounting period

(iii) in relation to the second week of Other Annual Holiday, the total amount of holiday credits on Part 3 of the card for the current accounting period

(iv) In relation to the one week of Winter Holiday, the total amount of holiday credits on Part IV of the card for the previous period

(v) the Operative shall receive on the last customary pay-day prior to the commencement of the holiday in question a sum equivalent to the total amount of the appropriate credits

(vi) if any week of annual holiday is not taken as a whole but in days, then the Employer shall pay to the Operative in those pay-weeks in which the day/s holiday is/are taken a sum equivalent to one-fifth of the total amount of the appropriate credits for the week of holiday in question, for each holiday day.

WORKING RULES (HEVAC)

WELFARE BENEFITS – PAYMENT (CLAUSE 22)

(a) All Operatives aged 18 years and over shall be entitled to sickness and accident benefit, death benefit and accidental dismemberment and total disablement benefit. H. & V. Welfare Ltd have arranged policies of insurance for these welfare benefits but by arrangement with the insurers all benefits falling due under the policies shall be paid on behalf of the insurers by H. & V. Welfare Ltd to the Operative or his estate.

Mechanical Installations – Rates of Wages and Working Rules

PLUMBING MECHANICAL ENGINEERING SERVICES

Authorised rates of wages agreed by the Joint Industry Board for the Plumbing Mechanical Engineering Services Industry in England and Wales

The Joint Industry Board for Plumbing Mechanical
  Engineering Services in England and Wales,
Brook House, Brook Street,
St Neots,
Huntingdon, Cambs. PE19 2HW
  Telephone: 0480 76925

RATES OF WAGES

EFFECTIVE FROM 4 FEBRUARY 1980

	Hourly rate £
Technical plumber	2·55
Advanced plumber	2·24
Trained plumber	2·04
Apprentices	
1st year of training	0·68
2nd year of training	1·02
3rd year of training	1·28
4th year of training	1·63
An Apprentice in his third or fourth year who gains his City & Guilds Advanced Craft Certificate in accordance with the requirements of the Board	1·84

Daily travelling allowance

Over	Not exceeding	Operatives	Apprentices and Government trainees
0 miles	2 miles	51p	41p
2 miles	5 miles	102p	82p
5 miles	10 miles	136p	109p
10 miles	15 miles	170p	136p
15 miles	20 miles	204p	163p
20 miles	25 miles	238p	190p
25 miles	30 miles	272p	217p
30 miles	35 miles	306p	245p
35 miles	40 miles	340p	272p
40 miles	45 miles	374p	299p
45 miles	50 miles	408p	326p

RATES OF WAGES (PLUMBING)

Abnormal conditions	89p per day
Lodging allowance	£6·75 per night
Responsibility money	15p per hour
Plumbers welding supplement	
Possession of Gas or Arc Certificate	10p per hour
Possession of Gas and Arc Certificate	20p per hour
Tool allowance	50p per week

Weekly holiday credit/Sick benefit stamp
Effective from 1 October 1979

	£
Technical Plumber	8·92
Advanced Plumber	8·11
Trained Plumber, Government Trainee and Apprentices in last year of training	7·36
All Plumbers over 65	8·27
Working Principals	8·27
Apprentice Plumbers 1st to 3rd year of training	3·96

MECHANICAL INSTALLATIONS

Market prices of materials

In the section which follows the prices given, unless otherwise stated, include the cost of delivery to sites in the London area at March 1980 and represent the prices paid by contractors after the deduction of all trade discounts. Prices do not allow for any charges in respect of V.A.T.

STEEL TUBING AND FITTINGS

BLACK MEDIUM WEIGHT SCREWED AND SOCKETED STEEL TUBING AND TUBULARS
(to B.S. 1387 ex works in 1 tonne lots pieces and longscrews in assorted standard lengths and standard length nipples)

	Unit	Nominal bore					
		15 mm £	20 mm £	25 mm £	32 mm £	40 mm £	50 mm £
Tubing, random lengths 5400 to 6400 mm	100 metres	49·90	60·23	87·20	112·20	132·85	188·70
Piece 300 mm to under 600 mm long	No.	0·31	0·35	0·51	0·56	0·68	1·02
Piece under 300 mm long	,,	0·14	0·17	0·26	0·28	0·34	0·46
Longscrew 300 mm to under 600 mm long	,,	0·35	0·41	0·58	0·62	0·77	1·15
Longscrew under 300 mm long	,,	0·20	0·24	0·34	0·37	0·45	0·62
Double longscrew 300 mm to under 600 mm long	,,	0·41	0·47	0·64	0·69	0·86	1·28
Double longscrew under 300 mm long	,,	0·26	0·31	0·41	0·45	0·55	0·76
Running nipple	,,	0·07	0·08	0·10	0·14	0·18	0·28
Barrel nipple	,,	0·10	0·11	0·15	0·19	0·23	0·33
Bends and springs	,,	0·20	0·26	0·40	0·68	0·74	1·30
Double bends	,,	0·91	1·07	1·33	1·91	2·29	3·34

		Nominal bore				
		65 mm £	80 mm £	100 mm £	125 mm £	150 mm £
Tubing, random lengths 5400 to 6400 mm	100 metres	249·30	324·06	461·06	595·20	772·04
Piece 300 mm to under 600 mm long	No.	1·48	1·96	3·07	5·66	6·84
Piece under 300 mm long	,,	0·65	0·92	1·47	3·03	4·63
Longscrew 300 mm to under 600 mm long	,,	1·67	2·26	3·30	5·93	7·07
Longscrew under 300 mm long	,,	0·94	1·31	1·94	3·63	5·58
Double longscrew 300 mm to under 600 mm long	,,	1·92	2·60	3·70	6·39	7·57
Double longscrew under 300 mm long	,,	1·20	1·66	2·36	4·12	4·38
Running nipple	,,	0·60	0·94	1·48	5·37	7·79
Barrel nipple	,,	0·59	0·81	1·46	2·94	4·63
Bends and springs	,,	2·54	4·03	6·95	29·67	41·97
Double bends	,,	12·45	15·99	24·96		

Note: Tubulars (except nipples) are normally supplied with sockets which will be charged extra at the prices shown on page 25. For tubulars to exact lengths add 20%.

21

Mechanical Installations – Market Prices of Materials
STEEL TUBING AND FITTINGS

GALVANIZED MEDIUM WEIGHT SCREWED AND SOCKETED STEEL TUBING AND TUBULARS
(to B.S.1387 ex works in 1 tonne lots pieces and longscrews in assorted standard lengths and standard length nipples)

	Unit	15 mm £	20 mm £	Nominal bore 25 mm £	32 mm £	40 mm £	50 mm £
Tubing, random lengths, 5400 to under 6400 mm	metres	66·49	80·26	116·19	149·51	177·03	251·40
Piece 300 mm to under 600 mm long	No.	0·42	0·46	0·68	0·75	0·90	1·37
Piece under 300 mm long	,,	0·18	0·23	0·34	0·38	0·45	0·62
Longscrew 300 mm to under 600 mm long	,,	0·47	0·54	0·77	0·82	1·02	1·54
Longscrew under 300 mm long	,,	0·27	0·32	0·45	0·50	0·60	0·82
Double longscrew 300 mm to under 600 mm long	,,	0·55	0·63	0·86	0·92	1·14	1·71
Double longscrew under 300 mm long	,,	0·34	0·41	0·54	0·60	0·73	1·01
Running nipple	,,	0·09	0·11	0·13	0·18	0·24	0·37
Barrel nipple	,,	0·13	0·15	0·20	0·25	0·31	0·43
Bends and springs	,,	0·27	0·35	0·53	0·91	0·99	1·73
Double bends	,,	1·22	1·42	1·77	2·55	3·05	4·46

		65 mm £	80 mm £	Nominal bore 100 mm £	125 mm £	150 mm £
Tubing, random lengths, 5400 to under 6400 mm	100 metres	332·19	431·82	614·37	793·10	1028·74
Piece 300 mm to under 600 mm long	No.	1·96	2·60	4·06	7·48	9·05
Piece under 300 mm long	,,	0·85	1·22	1·95	4·01	6·13
Longscrew 300 mm to under 600 mm long	,,	2·21	2·99	4·37	7·85	9·35
Longscrew under 300 mm long	,,	1·25	1·73	2·57	4·80	7·37
Double longscrew 300 mm to under 600 mm long	,,	2·54	3·44	4·90	8·46	10·01
Double longscrew under 300 mm long	,,	1·60	2·19	3·13	5·45	5·79
Running nipple	,,	0·80	1·25	1·96	7·10	10·30
Barrel nipple	,,	0·78	1·07	1·94	3·90	6·13
Bends and springs	,,	3·36	5·33	9·19	39·24	55·51
Double bends	,,	16·47	21·15	33·01		

Note: *Tubulars (except nipples) are normally supplied with sockets which will be charged extra at the prices shown on page 26. For tubulars to exact lengths add 28%.*

Mechanical Installations – Market Prices of Materials
STEEL TUBING AND FITTINGS

BLACK HEAVY WEIGHT SCREWED AND SOCKETED STEEL TUBING AND TUBULARS
(to B.S. 1387 ex works in 1 tonne lots pieces and longscrews in assorted standard lengths and standard length nipples)

	Unit	15 mm £	20 mm £	*Nominal bore* 25 mm £	32 mm £	40 mm £	50 mm £
Tubing, random lengths, 5400 to under 6400 mm	100 metres	58·12	70·17	101·59	130·71	154·77	219·80
Piece 300 mm to under 600 mm long	No.	0·34	0·38	0·55	0·60	0·73	1·10
Piece under 300 mm long	„	0·15	0·18	0·28	0·31	0·37	0·50
Longscrew 300 mm to under 600 mm long	„	0·38	0·44	0·62	0·67	0·83	1·24
Longscrew under 300 mm long	„	0·22	0·26	0·36	0·40	0·49	0·67
Double longscrew 300 mm to under 600 mm long	„	0·44	0·51	0·69	0·75	0·92	1·38
Double longscrew under 300 mm long	„	0·28	0·33	0·44	0·48	0·59	0·82
Running nipple	„	0·07	0·09	0·10	0·15	0·20	0·30
Barrel nipple	„	0·11	0·12	0·16	0·20	0·25	0·35
Bends and springs	„	0·22	0·29	0·43	0·73	0·80	1·40
Double bends	„	0·98	1·15	1·43	2·06	2·47	3·60

		65 mm £	80 mm £	*Nominal bore* 100 mm £	125 mm £	150 mm £
Tubing, random lengths, 5400 to under 6400 mm	100 metres	290·44	377·54	537·14	693·41	899·43
Piece 300 mm to under 600 mm long	No.	1·60	2·11	3·30	6·08	7·36
Piece under 300 mm long	„	0·69	0·99	1·58	3·26	4·98
Longscrew 300 mm to under 600 mm long	„	1·79	2·43	3·55	6·38	7·60
Longscrew under 300 mm long	„	1·01	1·41	2·09	3·91	5·60
Double longscrew 300 mm to under 600 mm long	„	2·07	2·79	3·98	6·88	8·14
Double longscrew under 300 mm long	„	1·30	1·78	2·54	4·43	4·71
Running nipple	„	0·65	1·01	1·60	5·78	8·37
Barrel nipple	„	0·64	0·87	1·57	3·17	4·98
Bends and springs	„	2·73	4·33	7·47	31·90	45·13
Double bends	„	13·38	17·19	26·84		

Note: *Tubulars (except nipples) are normally supplied with sockets which will be charged extra at the prices shown on page 25. For tubulars to exact lengths add 20%.*

STEEL TUBING AND FITTINGS

GALVANIZED HEAVY WEIGHT SCREWED AND SOCKETED STEEL TUBING AND TUBULARS
(to B.S. 1387 ex works in 1 tonne lots pieces and longscrews in assorted standard lengths and standard length nipples)

	Unit	\multicolumn{5}{c}{Nominal bore}
		15 mm £	20 mm £	25 mm £	32 mm £	40 mm £	50 mm £
Tubing, random lengths, 5400 to under 6400 mm	100 metres	77·45	93·49	135·36	174·17	206·23	292·89
Piece 300 mm to under 600 mm long	No.	0·44	0·49	0·72	0·79	0·95	1·45
Piece under 300 mm long	,,	0·19	0·24	0·36	0·40	0·48	0·65
Longscrew 300 mm to under 600 mm long	,,	0·50	0·58	0·81	0·87	1·08	1·63
Longscrew under 300 mm long	,,	0·28	0·34	0·47	0·52	0·64	0·87
Double longscrew 300 mm to under 600 mm long	,,	0·58	0·66	0·91	0·98	1·21	1·81
Double longscrew under 300 mm long	,,	0·36	0·43	0·57	0·63	0·73	1·07
Running nipple	,,	0·09	0·11	0·13	0·19	0·26	0·39
Barrel nipple	,,	0·14	0·15	0·21	0·26	0·32	0·46
Bends and springs	,,	0·28	0·37	0·56	0·96	1·05	1·83
Double bends	,,	1·29	1·50	1·88	2·69	3·33	4·72

		\multicolumn{5}{c}{Nominal bore}
		65 mm £	80 mm £	100 mm £	125 mm £	150 mm £
Tubing, random lengths, 5400 to under 6400 mm	100 metres	387·01	503·07	715·74	923·97	1198·48
Piece 300 mm to under 600 mm long	No.	2·07	2·75	4·29	7·91	9·57
Piece under 300 mm long	,,	0·90	1·29	2·06	4·23	6·48
Longscrew 300 mm to under 600 mm long	,,	2·33	3·16	4·62	8·29	9·88
Longscrew under 300 mm long	,,	1·32	1·83	2·72	5·08	7·79
Double longscrew 300 mm to under 600 mm long	,,	2·69	3·63	5·18	8·94	10·58
Double longscrew under 300 mm long	,,	1·69	2·32	3·30	5·76	6·12
Running nipple	,,	0·84	1·32	2·07	7·51	10·88
Barrel nipple	,,	0·83	1·13	2·05	3·37	6·48
Bends and springs	,,	3·55	5·63	9·71	41·47	58·67
Double bends	,,	17·40	22·35	34·89		

Note: Tubulars (except nipples) are normally supplied with sockets which will be charged extra at the prices shown on page 26. For tubulars to exact lengths add 28%.

STEEL TUBING AND FITTINGS

BLACK WROUGHT FITTINGS
(to B.S. 1740)

	Unit	Nominal bore					
		15 mm £	20 mm £	25 mm £	32 mm £	40 mm £	50 mm £
Elbow, male and female	No.	0·81	1·07	1·62	2·14	2·76	5·22
Elbow female, equal and reducing.	,,	0·66	0·81	1·07	1·62	2·14	2·76
Tee equal or reducing	,,	0·69	0·85	1·34	1·93	2·44	3·57
Cross	,,	1·37	1·63	2·05	2·82	3·42	5·30
Socket, equal	,,	0·12	0·14	0·17	0·24	0·31	0·46
Socket, diminished	,,	0·31	0·35	0·48	0·67	0·86	1·20
Cap	,,	0·28	0·34	0·54	0·74	0·91	1·26
Plug	,,	0·19	0·23	0·28	0·38	0·45	0·65
Backnut	,,	0·12	0·13	0·20	0·26	0·31	0·54

		Nominal bore				
		65 mm £	80 mm £	100 mm £	125 mm £	150 mm £
Elbow, male and female	,,	6·36	9·36	17·60	55·47	87·66
Elbow female, equal and reducing.	,,	5·79	8·51	16·07	50·45	79·66
Tee equal or reducing	,,	6·16	9·96	17·22	53·23	85·61
Cross	,,	11·91	21·74	35·90	122·82	188·94
Socket, equal	,,	0·84	1·17	2·22	4·44	6·66
Socket, diminished	,,	2·81	3·86	6·06	18·35	28·78
Cap	,,	2·40	3·26	5·73	17·91	26·86
Plug	,,	1·34	2·38	5·09	15·88	25·41
Backnut	,,	1·07	1·72	2·70	9·15	13·21

STEEL TUBING AND FITTINGS

GALVANIZED WROUGHT FITTINGS
(to B.S. 1740)

	Unit	*Nominal bore* 15 mm £	20 mm £	25 mm £	32 mm £	40 mm £	50 mm £
Elbow, male and female	No.	0·93	1·24	1·87	2·47	3·19	6·03
Elbow female, equal and reducing	,,	0·76	0·93	1·24	1·87	2·47	3·19
Tee	,,	0·79	0·98	1·54	2·23	2·81	4·12
Cross	,,	1·58	1·88	2·37	3·26	3·95	6·12
Socket, equal	,,	0·13	0·15	0·19	0·28	0·36	0·53
Socket, diminished	,,	0·36	0·40	0·56	0·77	1·00	1·38
Cap	,,	0·33	0·40	0·62	0·85	1·05	1·45
Plug	,,	0·22	0·27	0·33	0·44	0·51	0·75
Backnut	,,	0·14	0·15	0·23	0·30	0·36	0·63

		Nominal bore 65 mm £	80 mm £	100 mm £	125 mm £	150 mm £
Elbow, male and female	,,	7·31	10·77	20·24	63·80	100·81
Elbow female, equal and reducing	,,	6·65	9·79	18·48	58·01	91·61
Tee	,,	7·08	11·46	19·80	61·21	98·46
Cross	,,	13·70	25·00	41·29	141·24	217·28
Socket, equal	,,	0·96	1·34	2·56	5·11	7·65
Socket, diminished	,,	3·23	4·44	6·97	21·10	33·09
Cap	,,	2·76	3·74	6·59	20·59	30·89
Plug	,,	1·54	2·73	5·86	18·27	29·22
Backnut	,,	1·23	1·97	3·10	10·53	15·19

STEEL TUBING AND FITTINGS

BLACK BEADED MALLEABLE FITTINGS

	Unit	Nominal bore					
		15 mm £	20 mm £	25 mm £	32 mm £	40 mm £	50 mm £
Elbow	No.	0·19	0·26	0·40	0·64	0·83	1·19
Elbow, male and female	,,	0·22	0·29	0·43	0·69	0·95	1·31
45 degree elbow	,,	0·36	0·41	0·59	0·83	1·07	1·54
Tee	,,	0·29	0·38	0·52	0·83	1·07	1·66
Cross	,,	0·48	0·72	0·83	1·31	2·02	3·33
Concentric reducing socket	,,	0·17	0·23	0·31	0·48	0·57	0·83
Eccentric reducing socket	,,		0·46	0·55	0·83	1·02	1·24
Equal socket	,,	0·16	0·19	0·26	0·38	0·53	0·95
Plug, hollow	,,	0·12	0·15	0·21	0·29	0·38	0·48
Cap	,,	0·15	0·18	0·24	0·40	0·48	0·67
Male and female bend	,,	0·30	0·41	0·62	0·95	1·19	1·90
Female bend	,,	0·30	0·41	0·57	1·14	1·17	1·90
Male bend	,,	0·36	0·50	0·76	1·54	1·95	2·38
Twin elbow	,,	0·60	0·95	1·13	2·38	3·33	5·23
Pitcher tee	,,	0·57	0·72	1·02	1·54	2·26	3·68
Return bend	,,	0·71	1·07	1·54	2·85	3·45	4·16
Hexagon bush	,,	0·12	0·17	0·19	0·36	0·43	0·69
Hexagon nipple	,,	0·17	0·20	0·26	0·38	0·48	0·74
Hexagon reducing nipple	,,	0·20	0·29	0·36	0·59	0·67	0·83
Plug, solid	,,	0·13	0·15	0·23	0·32	0·42	0·52
Backnut	,,	0·17	0·18	0·21	0·24	0·33	0·48
Union, female	,,	1·01	1·14	1·49	2·06	2·50	3·16
Union, male and female	,,	1·27	1·71	2·15	2·98	3·55	4·73
Union elbow, male and female	,,	0·77	0·97	1·38	1·89	2·50	3·58
Tee reducing	,,	0·29	0·38	0·52	0·83	1·07	1·66

	Unit	Nominal bore				
		65 mm £	80 mm £	100 mm £	125 mm £	150 mm £
Elbow	,,	2·26	3·33	5·70	11·88	22·10
Elbow, male and female	,,	2·61	3·56	6·65		
45 degree elbow	,,	2·61	3·92	7·60		22·33
Tee	,,	3·33	3·80	6·89	16·40	26·14
Cross	,,	3·92	5·23	9·50		
Concentric reducing socket	,,	1·54	1·90	3·45		9·03
Eccentric reducing socket	,,	1·85	3·02			
Equal socket	,,	1·43	2·02	3·33		12·36
Plug, hollow	,,	0·90	1·35	2·50	6·53	7·60
Cap	,,	1·31	1·43	2·97		
Male and female bend	,,	4·04	5·47	12·83		37·07
Female bend	,,	4·04	5·47	11·41	26·14	39·92
Twin elbow	,,	6·42	10·93			
Pitcher tee	,,	4·75	6·53	16·16		
Hexagon bush	,,	1·26	1·90	3·33	6·89	8·55
Hexagon nipple	,,	1·30	2·02	3·21		8·79
Hexagon reducing nipple	,,	1·43	2·38			
Backnut	,,	1·00	1·14			
Union, female	,,	6·44	10·51	20·60		
Plug solid	,,	1·00	1·50			
Tee reducing	,,	3·32	3·80	6·89	16·39	26·14

STEEL TUBING AND FITTINGS

GALVANIZED BEADED MALLEABLE FITTINGS

	Unit	Nominal bore 15 mm £	20 mm £	25 mm £	32 mm £	40 mm £	50 mm £
Elbow female	No.	0·26	0·35	0·54	0·86	1·11	1·58
Elbow, male and female	,,	0·29	0·38	0·57	0·92	1·27	1·74
45 degrees elbow	,,	0·48	0·54	0·79	1·11	1·43	2·06
Tee equal or reducing	,,	0·38	0·51	0·70	1·11	1·43	2·22
Cross	,,	0·64	0·95	1·11	1·74	2·69	4·44
Concentric reducing socket	,,	0·23	0·30	0·41	0·63	0·76	1·11
Eccentric reducing socket	,,	0·61	0·73	1·11	1·36	1·65	2·47
Equal socket	,,	0·21	0·26	0·35	0·51	0·70	1·27
Plug, hollow	,,	0·16	0·19	0·27	0·38	0·51	0·63
Cap	,,	0·19	0·24	0·32	0·54	0·64	0·89
Male and female bend	,,	0·40	0·54	0·83	1·27	1·58	2·53
Female bend	,,	0·40	0·54	0·76	1·52	1·55	2·53
Male bend	,,	0·48	0·67	1·01	2·06	2·60	3·17
Twin elbow	,,	0·79	1·27	1·51	3·17	4·44	6·97
Pitcher tee	,,	0·76	0·95	1·36	2·06	3·01	4·91
Return bend	,,	0·95	1·43	2·06	3·80	4·59	5·54
Hexagon bush	,,	0·16	0·22	0·25	0·48	0·57	0·92
Nipple	,,	0·23	0·27	0·35	0·51	0·63	0·98
Hexagon reducing nipple	,,	0·27	0·38	0·48	0·79	0·89	1·11
Plug, solid	,,	0·18	0·21	0·30	0·43	0·55	0·70
Backnut	,,	0·22	0·24	0·27	0·32	0·44	0·63
Union, female	,,	1·34	1·52	1·99	2·75	3·33	4·21
Union, male and female	,,	1·69	2·28	2·87	3·98	4·74	6·32
Union elbow, male and female	,,	1·03	1·30	1·84	2·52	3·33	4·77

		Nominal bore 65 mm £	80 mm £	100 mm £	125 mm £	150 mm £
Elbow	,,	3·01	4·44	7·60	15·84	29·46
Elbow, male and female	,,	3·48	4·75	8·87		
45 degrees elbow	,,	3·48	5·23	10·14		29·78
Tee equal or reducing	,,	4·44	5·07	9·19	21·86	34·85
Cross	,,	5·23	6·97	12·67		
Concentric reducing socket	,,	2·06	2·53	4·59		12·04
Eccentric reducing socket	,,	4·02				
Equal socket	,,	1·90	2·69	4·44		16·47
Plug, hollow	,,	1·20	1·81	3·33	8·71	10·14
Cap	,,	1·74	1·90	3·96		
Male and female bend	,,	5·39	7·29	17·10		49·42
Female bend	,,	5·39	7·29	15·20	34·84	53·22
Twin elbow	,,	8·55	14·57			
Pitcher tee	,,	6·34	8·71	21·54		
Hexagon bush	,,	1·68	2·53	4·44	9·19	11·41
Nipple	,,	1·74	2·69	4·28		11·72
Hexagon reducing nipple	,,	1·90	3·17			
Backnut	,,	1·33	1·52			
Union, female	,,	8·60	14·04	27·51		

STEEL TUBING AND FITTINGS

HOT FINISHED SEAMLESS CARBON STEEL TUBING
(with plain ends to B.S. 3601 ex works in random lengths)

For quantities 1 tonne to under 6 tonnes Thickness	Unit	32 mm £	40 mm £	Nominal bore 50 mm £	65 mm £	80 mm £	100 mm £	150 mm £
3·2 mm	100 metres	210·68	242·72					
3·6 mm	,,		270·70					
4·0 mm	,,	246·10	283·77					
4·5 mm	,,			388·13	498·40	572·79		1069·72
5·0 mm	,,		340·23	419·44		593·81	768·49	1157·89
5·6 mm	,,	320·14	400·65	473·35	590·57	646·22	844·20	1275·47
6·3 mm	,,			536·54	690·64	745·17	978·06	1481·20
7·1 mm	,,							1657·51
8·0 mm	,,							1857·36
10·0 mm	,,							2082·41

	Unit	200 mm £	250 mm £	Nominal bore 300 mm £	350 mm £	400 mm £	450 mm £
4·9 mm	,,	1476·70					
5·6 mm	,,	1624·95					
6·3 mm	,,	1887·25	2366·90	2818·54			
7·1 mm	,,		2664·20	3173·01			
8·0 mm	,,	2371·86	2990·05	3561·75	4003·63		
10·0 mm	,,	2666·16	3527·44	4213·53	4733·16	5538·36	6253·00
12·5 mm	,,		4682·59	5598·14	6186·38	7205·84	8158·68

Extra on above tubing for bevelled end for welding *Add* 2½%

STEEL TUBING AND FITTINGS

HOT FINISHED SEAMLESS CARBON STEEL TUBING
(with plain ends to B.S. 3602 ex works in random lengths)

			32	40	Nominal bore				
					50	65	80	100	150
For quantities 1 tonne to under 6 tonnes		Unit	mm	mm	mm	mm	mm	mm	mm
Thickness		100	£	£	£	£	£	£	£
3·2 mm	.	. metres	229·64	264·56					
4·0 mm	. .	,,	268·25	309·31					
4·5 mm	. .	,,			423·07	543·26	624·33		1165·99
5·0 mm	. .	,,		370·86	457·19		647·25	837·66	1262·11
5·6 mm	. .	,,	348·95	436·71	515·96	643·73	704·39	920·17	1390·76
6·3 mm	. .	,,			584·83	752·80	812·24	1066·09	1614·50
7·1 mm	. .	,,							1806·69
8·0 mm	. .	,,							2024·52
10·0 mm	. .	,,							2269·83

			200	250	Nominal bore			
					300	350	400	450
			mm	mm	mm	mm	mm	mm
			£	£	£	£	£	£
5·0 mm	. .	,,	1609·60					
5·6 mm	. .	,,	1771·19					
6·3 mm	. .	,,	2057·09	2579·91	2520·96			
7·1 mm	. .	,,		2903·99	3458·58			
8·0 mm	. .	,,	2585·33	3259·16	3882·31	4363·95		
10·0 mm	. .	,,	2906·12	3844·91	4592·76	5159·14	6036·81	6815·77
12·5 mm	. .	,,		5104·02	6101·97	6743·16	7854·37	8892·96

Extra on above tubing for bevelled end for welding *Add* 2½%

STEEL TUBING AND FITTINGS

HOT FINISHED SEAMLESS CARBON STEEL TUBING
(with bevelled ends to API 5L grades A and B ex works in long random lengths)

For quantities 1 tonne to under 6 tonnes Thickness	Unit	32 mm	40 mm	50 mm	Nominal bore 65 mm	80 mm	100 mm	150 mm
	100	£	£	£	£	£	£	£
3·56 mm	metres	236·24						
3·68 mm	,,		283·05					
3·91 mm	,,				367·08			
4·85 mm	,,	297·49						
5·08 mm	,,			360·08		553·96		
5·49 mm	,,						673·73	
5·54 mm	,,			480·70				
6·02 mm	,,							958·95
7·01 mm	,,					747·90		
7·11 mm	,,							1702·57
7·62 mm	,,						911·18	
8·56 mm	,,							1357·94
10·97 mm	,,							2490·79

STEEL TUBING AND FITTINGS

STEEL BUTT WELDING FITTINGS
(to B.S. 1965 – where applicable – for use with tubes corresponding in thickness to B.S. 1387 medium and heavy weight)

	Unit	\multicolumn{6}{c}{Nominal bore}
		15 mm £	20 mm £	25 mm £	32 mm £	40 mm £	50 mm £
Medium weight							
45 degree elbow, long radius	No.	1·21	1·21	1·40	1·40	1·40	1·40
90 degree elbow, long radius	,,	1·21	1·21	1·54	1·54	1·54	1·54
Branch bend	,,	2·93	2·93	3·70	3·70	3·70	3·70
180 degree bend	,,	3·50	3·50	4·42	4·42	4·42	4·42
Heavy weight							
45 degree elbow, long radius	,,			1·59	1·59	1·59	1·59
90 degree elbow, long radius	,,			1·84	1·84	1·84	1·84
Short radius elbow	,,			4·29	4·29	4·29	4·29
Equal tee	,,			5·32	5·32	5·54	5·74
Reducing tee	,,			5·52	5·52	6·06	7·22
Cap	,,			3·13	3·13	3·13	3·40
Branch bend	,,			4·42	4·42	4·42	4·42
180 degree bend	,,			5·13	5·13	5·13	5·13
Medium and heavy weight reducer							
Concentric							
20 mm ×	,,	1·71					
25 mm ×	,,	1·90	1·80				
32 mm ×	,,	2·98	2·42	2·24			
40 mm ×	,,	3·55	3·11	2·42	2·42		
50 mm ×	,,	4·45	4·11	2·71	2·71	2·58	
65 mm ×	,,	4·92	4·74	3·35	3·21	2·86	
80 mm ×	,,			5·90	5·90	3·62	2·93
100 mm ×	,,				20·77	6·95	4·96
125 mm ×	,,					7·61	
150 mm ×	,,					14·08	
Eccentric							
20 mm ×	,,	2·55					
25 mm ×	,,	3·21	3·03				
32 mm ×	,,	4·45	3·56	3·31			
40 mm ×	,,	5·51	5·41	3·74	3·74		
50 mm ×	,,	6·85	5·90	4·51	3·74	3·74	
65 mm ×	,,		10·96	7·65	5·41	4·36	3·92
80 mm ×	,,			8·20	8·20	6·29	4·55
100 mm ×	,,					10·73	8·18
125 mm ×	,,						16·08
150 mm ×	,,						33·86

STEEL TUBING AND FITTINGS

STEEL BUTT WELDING FITTINGS
(to B.S. 1965 – where applicable – for use with tubes corresponding in thickness to B.S. 1387 medium and heavy weight)

	Unit	Nominal bore				
		65 mm £	80 mm £	100 mm £	125 mm £	150 mm £
Medium weight						
45 degree elbow, long radius	No.	1·77	2·12	3·49	6·57	9·03
90 degree elbow, long radius	,,	2·21	2·40	4·23	7·54	10·91
Branch bend	,,	5·34	5·77	10·28	18·42	26·67
180 degree bend	,,	5·61	6·89	12·25	21·61	29·35
Heavy weight						
45 degree elbow, long radius	,,	2·08	2·59	4·20	7·40	11·04
90 degree elbow, long radius	,,	2·55	2·99	4·95	8·64	12·57
Short radius elbow	,,	4·29	5·43	8·96	14·58	21·27
Equal tee	,,	9·07	10·58	14·33	28·22	31·53
Reducing tee	,,	11·29	13·24	17·94	35·32	39·47
Cap	,,	4·02	4·16	5·37	6·64	8·23
Branch bend	,,	6·12	7·17	11·64	20·47	29·78
180 degree bend	,,	6·36	8·45	14·08	24·68	34·86
Medium and heavy weight reducers						
Concentric						
80 mm ×	,,	2·93				
100 mm ×	,,	4·11	3·34			
125 mm ×	,,	7·23	6·10	5·79		
150 mm ×	,,	10·96	7·51	6·29	6·29	
Eccentric						
80 mm ×	,,	4·55				
100 mm ×	,,	7·13	6·42			
125 mm ×	,,	14·73	12·51	9·73		
150 mm ×	,,	18·82	14·09	11·76	11·76	

STEEL TUBING AND FITTINGS

STEEL BUTT WELDING FITTINGS
(to B.S. 1965 – where applicable – for use with tubes corresponding in thickness to B.S. 3601)

	Unit	6·3 mm £	7·1 mm £	Thickness 8·0 mm £	9·5 mm £	11·0 mm £	12·5 mm £
45 degree elbow, long radius							
Nominal bore							
150 mm	No.	11·83	12·65				
200 mm	,,	19·58		20·32			
250 mm	,,	37·00		38·82	39·00		
300 mm	,,		55·59	57·62	59·39		
350 mm	,,				92·55	117·91	
400 mm	,,				118·20		147·86
90 degree elbow, long radius							
Nominal bore							
150 mm	,,	13·88	14·85				
200 mm	,,	23·03		23·92			
250 mm	,,	43·54		45·71	45·84		
300 mm	,,		65·32	67·75	69·86		
350 mm	,,				108·89	138·71	
400 mm	,,				139·08		173·98
90 degree elbow, short radius							
Nominal bore							
150 mm	,,		24·43				
200 mm	,,			33·31			
250 mm	,,			60·48	62·17		
300 mm	,,			88·91	94·73		
350 mm	,,				128·83	164·05	
400 mm	,,				164·54		205·79
Branch bend							
Nominal bore							
150 mm	,,	33·32	35·62				
200 mm	,,	55·26		57·39			
250 mm	,,	105·16		110·33	113·40		
300 mm	,,		156·76	162·61	172·85		
350 mm	,,				261·35	332·80	
400 mm	,,				333·78		417·50
180 degree bend							
Nominal bore							
150 mm	,,	35·44	36·64				
Equal tee							
Nominal bore							
150 mm	,,		37·23				
200 mm	,,			50·11			
250 mm	,,				82·61		
300 mm	,,				119·60		
350 mm	,,				140·45	178·90	
400 mm	,,				240·26		292·43
Reducing tee							
Nominal bore							
200 mm	,,			62·65			
250 mm	,,			103·33			
300 mm	,,			149·03			
350 mm	,,			175·61			
400 mm	,,			240·29			

STEEL BUTT WELDING FITTINGS
(to B.S. – where applicable – for use with tubes corresponding in thickness to B.S. 3601) *continued*

Cap

Nominal bore	Unit	6·3 mm £	7·1 mm £	8·0 mm £	9·5 mm £	11·0 mm £	12·5 mm £
150 mm	No.		8·23				
200 mm	,,			12·77			
250 mm	,,				19·44		
300 mm	,,				22·01		
350 mm	,,				26·68	33·96	
400 mm	,,				31·34		39·01

Reducer, concentric

	Unit	100 mm £	125 mm £	150 mm £	200 mm £	250 mm £	300 mm £	350 mm £
200 mm ×	No.	13·13	12·73	9·03				
250 mm ×	,,	24·77		17·58	15·02			
300 mm ×	,,			27·70	25·24			
350 mm ×	,,				47·17	43·73	43·73	
400 mm ×	,,				89·13	88·03	73·39	73·39

Reducer, eccentric

	Unit	100 mm £	125 mm £	150 mm £	200 mm £	250 mm £	300 mm £	350 mm £
200 mm ×	,,	20·62	20·34	15·72				
250 mm ×	,,	44·13		29·82	26·90			
300 mm ×	,,			56·63	53·82	45·53		
350 mm ×	,,				78·40	65·98	65·98	
400 mm ×	,,				148·25	132·31	109·76	109·76

STEEL TUBING AND FITTINGS

STAINLESS STEEL TUBE
(to B.S. 4127: Part 2 1972 suitable for domestic water services (in 1 tonne lots))

	Unit	15 mm £	22 mm £	Nominal bore 28 mm £	35 mm £	42 mm £
Tubing	100 metres	50·25	76·53	101·58	145·72	189·70

STAINLESS STEEL FITTINGS
(capillary solder ring type)

Coupling female	No.	0·79	1·60	2·21	3·56	4·69
45 degree bend	,,	1·87	2·53	3·54	10·82	17·04
90 degree bend	,,	1·60	2·08	2·98	13·46	18·72
Tee equal	,,	2·50	3·50	5·04	15·12	20·71
Tee reducing	,,		7·40	9·12	15·55	20·00
Reducer	,,		5·36	7·22	16·01	20·50
Union coupling (cone seat)	,,	18·79	23·16	25·51	33·65	44·06
Union coupling (flat seat)	,,	18·79	23·16	25·51	33·65	44·06
Tap connector	,,	9·89	11·71	23·54		

COPPER TUBING AND FITTINGS

COPPER TUBING
(in lots of 2500 metres and over)

	Unit	12 mm £	15 mm £	Nominal size 22 mm £	28 mm £	35 mm £	42 mm £
B.S. 2871 Table X	100 metres	49·58	56·84	103·88	131·32	236·18	317·52

		54 mm £	67 mm £	Nominal size 76 mm £	108 mm £	133 mm £	159 mm £
B.S. 2871 Table X	,,	409·64	596·82	754·60	1058·40	1274·00	2048·20

	Unit		12 mm £	Nominal size 15 mm £	22 mm £	28 mm £	35 mm £
B.S. 2871 Table Y	100 metres		61·83	88·21	154·84	197·96	313·60

		42 mm £	54 mm £	Nominal size 67 mm £	76 mm £	108 mm £	
B.S. 2871 Table Y		383·18	662·48	837·90	967·26	1744·40	

	Unit		15 mm £	Nominal size 22 mm £	28 mm £	35 mm £	42 mm £
B.S. 2871 Table Z	100 metres		49·00	83·01	103·88	152·88	232·26

	Unit		54 mm £	Nominal size 67 mm £	76 mm £	108 mm £	
B.S. 2871 Table Z	100 metres		382·20	509·60	650·72	916·30	

	Unit			Nominal size 6 mm £	8 mm £	10 mm £	
B.S. 2871 Table W	100 metres			25·77	33·81	47·23	

COPPER TUBING AND FITTINGS

CAPILLARY TYPE FITTINGS
(to B.S. 864 – where applicable)

	Unit	15 mm £	22 mm £	Nominal size 28 mm £	35 mm £	42 mm £	54 mm £	67 mm £
Straight coupling copper to copper	No.	0·14	0·21	0·30	0·65	0·93	1·88	4·67
Straight coupling metric copper to imperial copper	,,	0·24	0·31	0·45	1·02	1·32		
Straight connector copper to female iron	,,		0·32	0·42	0·96	1·58	2·30	
Straight connector copper to male iron.	,,	0·51	0·75	1·09	1·71	2·07	3·44	
Reducing connector copper to female iron	,,	0·39	0·68	0·92	1·64	2·02	3·00	5·45
Reducing connector copper to male iron	,,	1·04	1·46	1·09				
Reducing connector copper to MI	,,	0·54	0·99	1·41				
Straight connector copper to lead	,,	0·34	0·48	0·68				
Flanged connector	,,	6·59	8·34	9·46	14·31	17·67		
Tank connector .	,,	0·81	1·21	1·66	2·23	2·83	4·21	
Tank connector with long thread.	,,	1·17	1·62	2·27				
Reducer	,,		0·26	0·41	0·76	1·03	1·86	3·52
Adaptor copper to female iron	,,	0·61	0·92	1·41	2·03	2·83	3·57	
Adaptor copper to male iron	,,	0·61	0·92	1·41	2·03	2·83	3·57	
Adaptor imperial female copper to metric male copper .	,,	0·22	0·27	0·43	0·53	0·68	1·40	
Union coupling .	,,	1·15	1·76	2·42	3·46	4·92	7·99	12·92
Elbow .	,,	0·23	0·39	0·63	1·42	2·15	4·91	10·94
Backplate elbow	,,	0·92	1·74					
Overflow bend .	,,	1·80	1·87					
Return bend	,,	1·58	1·87	2·76				
Obtuse elbow	,,	0·42	0·68	1·05	2·63	4·81		9·65
Equal tee .	,,	0·42	0·65	1·20	2·22	3·29	5·94	13·42
Reducing tee – reduced branch	,,		0·63	1·16	2·09	3·84	6·06	10·57
Reducing tee – one end and branch reduced .	,,		0·66	1·27	2·74	3·84	6·06	
Reducing tee – both ends reduced	,,		0·73	1·47	2·33	3·84	6·06	
Reducing tee – one end reduced .	,,		0·66	1·27	2·19	3·84	6·06	10·57
Back plate tee	,,	1·79						
Heater tee.	,,	1·61						
Union heater tee	,,	2·17						
Air vent tee	,,	1·62						
Sweep tee-equal .	,,	0·92	1·34	2·53	4·08	5·74	7·12	11·49
Sweep tee-reducing	,,		1·41	2·53	4·08			
Sweep tee-double	,,	1·47	2·01	3·05				
Cross	,,	1·68	2·18	3·11				
Stop end .	,,	0·33	0·48	0·82	1·27	1·66	2·45	
Straight tap connector	,,	0·52	0·76					
Bent tap connector	,,	0·62	0·96					
Bent male union connector .	,,	1·44	1·96	2·88	4·47	7·41	11·68	
Bent female union connector	,,	1·44	1·96	2·88	4·47	7·41	11·68	
Straight union adaptor	,,	0·69	1·04	1·44	1·91	2·68	3·49	6·40
Straight union connector male	,,	1·33	1·78	2·49	3·41	5·57	7·35	10·92
Straight union connector female .	,,	1·33	1·78	2·49	3·41	5·57	7·35	
Male nipple .	,,	0·64	0·74	1·04	1·51	2·90	3·87	4·06
Female nipple .	,,	0·64	0·74	1·04	1·51	2·90	3·87	

Mechanical Installations – Market Prices of Materials

COPPER TUBING AND FITTINGS

COMPRESSION TYPE FITTINGS
(to B.S. 864 – where applicable)

Nominal size

Brass	Unit	15 mm £	22 mm £	28 mm £	35 mm £	42 mm £	54 mm £
Straight coupling copper to copper.	No.	0·58	0·77	1·27	2·09	2·80	4·40
Straight connector copper to imperial copper	,,		0·81				
Male coupling copper to male iron.	,,	0·46	0·66	0·92			
Male coupling with long thread and backnut	,,	0·85	1·16	1·54			
Female coupling copper to female iron	,,	0·50	0·62	0·92	1·62	2·20	3·43
Lead coupling copper to lead.	,,	0·58	0·81				
Elbow.	,,	0·66	1·00	1·46	2·55	3·93	6·36
Male elbow.	,,	0·62	0·85	1·35	2·59	4·09	
Female elbow	,,	0·62	0·89	1·35	2·20	3·81	
Backplate elbow	,,	0·89	1·58				
Tee equal or reducing	,,	0·93	1·27	2·12	3·54	5·63	9·04
Tee with male iron branch	,,	1·16	1·46				
Straight tap connector	,,	0·66	0·89				
Backplate tee	,,	1·27					
Reducing set	,,	0·35	0·31	0·50	0·92	1·47	3·43
Tank coupling	,,	0·73	0·89	1·35	2·42		
Tank coupling – long thread.	,,		1·20				

Dezincifiable

Straight coupling copper to copper.	,,	0·70	0·96	1·48	2·83	3·58	
Straight connector copper to imperial copper	,,	1·15					
Male coupling copper to male iron.	,,	0·58	0·80	1·42	2·46	3·31	
Male coupling with long thread and backnut	,,		1·64	2·18			
Female coupling copper to female iron	,,	0·64	0·82	1·31	2·28	2·70	
Elbow.	,,	0·80	1·11	1·94	3·41	4·56	
Male elbow.	,,	0·96	1·16	1·91			
Female elbow	,,	0·82	1·19	1·84			
Backplate elbow	,,	1·17	2·04				
Tee equal	,,	1·19	1·71	2·82	4·37	5·90	
Tee reducing	,,	1·26	1·71	2·82			
Straight tap connector	,,	0·96	1·43				
Reducing set	,,	0·49	0·44	0·71	1·30	1·90	
Blanking plug	,,	0·23	0·27				

COPPER TUBING AND FITTINGS

WELDABLE TYPE FITTINGS
(to B.S. 864 – where applicable)

	Unit	15 mm £	22 mm £	28 mm £	35 mm £	42 mm £	54 mm £
Straight coupling, copper to copper	No.	0·57	0·66	0·76	1·13	1·60	2·26
Straight connector, copper to iron	,,	0·66	0·94	1·38	1·97	3·27	4·49
Straight union, copper to copper	,,	2·26	3·07	4·31	5·33	7·37	10·66
Elbow, ditto	,,	0·57	0·76	1·23	1·77	2·36	3·74
Elbow, copper to male iron	,,	1·23	1·59	2·50	3·63	5·11	8·50
Return bend, copper to copper	,,	1·34	2·14	2·90	5·23	7·01	10·37
Obtuse elbow, ditto	,,	0·60	0·76	1·23	1·77	2·36	3·66
Tee, all ends copper	,,	0·66	1·13	1·49	2·83	3·59	5·11
Sweep tee, ditto	,,	0·87	1·59	2·26	3·84	5·23	6·81
Concentric reducing socket	,,		0·66	0·86	1·34	1·70	2·36
Eccentric reducing socket	,,		0·94	1·38	1·87	2·36	3·27
Cap	,,	0·57	0·66	0·76	0·86	0·94	1·59

	Unit	67 mm £	76 mm £	108 mm £	133 mm £	159 mm £
Straight coupling, copper to copper	,,	2·90	4·10	6·18	7.73	11.04
Straight connector, copper to iron	,,	7·00	8·58	12·43		
Straight union, copper to copper	,,	18·01	23·15	48·39		
Elbow, ditto	,,	5·11	10·66	21·28	39·85	53·29
Elbow, copper to male iron	,,	14·94	22·24	44·65		
Return bend, copper to copper	,,	17·42	27·09	57·97		
Obtuse elbow, ditto	,,	4·97	10·66	21·28	39·85	50·25
Tee, all ends copper	,,	7·36	11·78	23·17	46·40	57·97
Sweep tee, ditto	,,	10·66	19·33	30·91	63·37	81·13
Concentric reducing socket	,,	3·27	4·67	8·69	10·66	12·64
Eccentric reducing socket	,,	4·86	7·00	12·60	16·82	20·27
Cap	,,	2·36	2·90	4·86		

Mechanical Installations – Market Prices of Materials
COPPER TUBING AND FITTINGS

CAPILLARY TYPE HIGH DUTY FITTINGS

	Unit	15 mm £	22 mm £	28 mm £	35 mm £	42 mm £	54 mm £
Straight coupling copper to copper	No.	0·68	1·09	1·56	2·71	2·98	4·36
Reducing coupling	,,	1·22	1·49	2·04			
Straight female connector	,,	1·60	1·84	2·71			
Straight male connector	,,	1·62	1·84	2·71		5·43	8·83
Reducer	,,	0·82	0·82	1·49	1·90	2·45	3·94
Adaptor male copper to female iron	,,	2·31	2·58				
Union coupling	,,	3·06	3·94	5·43	9·52	11·20	
Elbow	,,	1·97	2·10	3·14	4·90	6·12	10·40
Return band	,,			6·46	7·48		
Equal tee	,,	2·30	2·85	3·75	6·46	8·17	12·90
Reducing tee	,,	3·11	3·67	5·24	8·35	10·61	16·89
Tee made iron branch	,,	4·60	5·43				
Stop end	,,	1·49					
Straight union adaptor	,,	1·36	1·84	2·45	4·42	5·59	
Best union adaptor	,,	3·54	4·76	6·46			
Bent male union connector	,,	4·55	6·12	11·23			
Composite flange	,,		7·48	8·43	10·75	12·58	17·66

CAPILLARY TYPE FITTINGS SUITABLE FOR MICROBORE TUBES
(to B.S. 864)

	Unit	6 mm £	8 mm £	10 mm £	12 mm £
Straight coupling copper to copper	No.	0·39	0·26	0·26	0·25
Elbow	,,	0·61	0·61	0·55	0·49
Reducer	,,		0·47	0·36	0·58
Tee equal	,,	0·94	0·94	0·79	0·82
Tee reducing 15 mm ×	,,		0·89	0·89	0·93
Tee reducing 22 mm ×	,,		1·09	1·09	1·15
Stop end	,,	0·32	0·32	0·32	0·42
Straight connector copper to iron	,,	0·55	0·39	0·39	0·60
Bent radiator union-brass	,,	1·09	1·09	1·09	1·09
Ditto chromium plated	,,	1·36	1·36	1·36	1·36
Reducing coupling 8 mm ×	,,	0·56			
10 mm ×	,,		0·41		
15 mm ×	,,		0·45		
22 mm ×	,,		0·53	0·53	

COPPER TUBING AND FITTINGS

COMPRESSION TYPE FITTINGS SUITABLE FOR MICROBORE TUBES
(to B.S. 864)

	Unit	Nominal size 6 mm £	8 mm £	10 mm £	12 mm £
Straight coupling copper to copper	No.	0·54	0·54	0·58	0·58
Reducing coupling	,,		0·54	0·58	
Elbow	,,	0·67	0·67	0·75	0·70
Reducing set	,,		0·21	0·25	0·29
Equal tee	,,	0·96	0·96	1·08	0·96
Reducing tee 15 mm ×	,,				1·00
22 mm ×	,,				1·37
Stop end	,,	0·50	0·50	0·50	0·50
Straight connector copper to iron	,,	0·42	0·42	0·42	0·42

MANIFOLDS

		Connections 4 × 8 mm £	6 × 8 mm £	2 × 10 mm £	4 × 10 mm £
Manifold side entry one way flow 22 mm body with the connections stated	Unit No.	2·76	3·31	1·81	2·90
Manifold – linear flow with the connections stated					
22 mm body	,,	2·39			3·07
28 mm body	,,		3·41		

		12 × 10 mm £	18 × 10 mm £
Manifold – two way flow and return 28 mm body with the connections stated	Unit No.	7·34	10·34

		8 × 10 mm £
Ditto but 22 mm body	Unit No.	5·26

THERMOPLASTIC TUBING AND FITTINGS

LOW DENSITY POLYTHENE TUBING
(to B.S. 1972)

	Unit	$\frac{1}{2}"$	$\frac{3}{4}"$	1"	$1\frac{1}{4}"$	$1\frac{1}{2}"$	2"
	100	£	£	£	£	£	£
Class B	metres			39·39	60·35	94·01	123·96	190·64
Class C	,,	33·70	52·26	81·21	129·08	168·14	262·54
Class D	,,	41·58	62·44	98·80	159·09	207·72	

HIGH DENSITY POLYTHENE TUBING
(to B.S. 3284)

	Unit	$\frac{1}{2}"$	$\frac{3}{4}"$	1"	$1\frac{1}{4}"$	$1\frac{1}{2}"$	2"
	100	£	£	£	£	£	£
Class C	metres	28·40	44·30	68·94	110·96	144·29	223·77
Class D	,,	35·19	54·43	87·49	136·22	178·48	279·68

UNPLASTICIZED PVC TUBING
(to B.S. 3505/6 in 9 metre lengths with spigot and socket)

	Unit	$\frac{1}{2}"$	$\frac{3}{4}"$	1"	$1\frac{1}{4}"$	$1\frac{1}{2}"$	2"	3"	4"	6"
	100	£	£	£	£	£	£	£	£	£
Class B	metres							224·00	323·00	607·00
Class C	,,						146·00	257·00	393·00	734·00
Class D	,,				70·00	88·00	186·00	329·00	480·00	922·00
Class E	,,	30·00	43·00	59·00	88·00	116·00	211·00	368·00	561·00	1129·00

HIGH IMPACT PVC TUBING
(for working pressures shown)

	Unit	$\frac{1}{2}"$	$\frac{3}{4}"$	1"	$1\frac{1}{4}"$	$1\frac{1}{2}"$	2"	3"	4"	6"
	100									
Class C (130 p.s.i.)	metres						140·00	372·00	532·00	1032·00
Class D (173 p.s.i.)	,,				92·00	115·00				
Class E (217 p.s.i.)	,,	40·00	55·00	74·00	108·00	147·00	228·00	468·00	680·00	1272·00
Class T (173 p.s.i.)	,,	66·00	80·00	107·00	134·00	168·00	264·00			

THERMOPLASTIC TUBING AND FITTINGS

ABS (ACRYLONITRILE BUTADIENE STYRENE) TUBING
(for the working pressures shown in 500 metre lots)

Nominal size

	Unit	1/2" £	3/4" £	1" £	1 1/4" £	1 1/2" £	2" £	3" £	4" £	6" £
Class C (9 bar)	100			51·00	82·00	117·00	153·00	336·00	520·00	1048·0
Class D (12 bar)	metres							424·00	624·00	1328·0
Class E (15 bar)	„	40·00	60·00	81·00	114·00	176·00	216·00	460·00	752·00	
Class T (12 bar)	„	60·00	68·00	101·00	180·00	196·00	260·00			

COMPRESSION TYPE METAL FITTINGS
(for polythene tube to B.S. 1972 and 3284)

Nominal size

Coupling	Unit	1/2" £	3/4" £	1" £	1 1/4" £	1 1/2" £	2" £
Polyethylene to polyethylene	No.	1·08	1·54	2·73	4·12	6·82	8·78
Polyethylene to copper (B.S. 2871 Table X)	„	1·08	1·58	2·62			
Polyethylene to male iron	„	1·04	1·46	2·04	3·81	5·12	6·82
Polyethylene to female iron	„	1·04	1·46	1·93	3·85	5·12	6·82
Tank coupling including backnut	„			2·04			
Elbow polyethylene to polyethylene	„	1·35	2·04	3·43	6·39	7·97	9·59
Male elbow	„	1·31	2·04	2·27			
Female elbow	„	1·27	1·89	2·97			
Female wall elbow	„	1·85	2·97				
Tee, all ends to polyethylene	„	1·81	2·89	4·24	8·78	10·47	15·71
Female tee with female iron branch	„	1·85	2·89	2·54			
Reducing set	„	0·55	0·69	1·31	1·93	2·39	5·12
Straight tap connector	„	1·58	2·73				
Bent tap connector	„	1·93	3·08				
Stop end	„	1·16	1·58				
Blank cap	„	0·27	0·39	0·66	1·04		
Copper liners for B.S. 1972							
Class B	„		0·14	0·16	0·24	0·39	0·48
Class C	„	0·12	0·14	0·16	0·24	0·39	0·48
Class D	„	0·12	0·14				
Copper liners for B.S. 3284							
Class B	„		0·16	0·24	0·39	0·48	
Class C	„	0·12	0·14	0·16	0·24	0·33	0·48
Class D	„	0·12	0·14	0·16	0·24	0·33	0·48

THERMOPLASTIC TUBING AND FITTINGS

HIGH IMPACT RIGID PVC AND ABS FITTINGS
(with plain ends for solvent cement joints)

	Unit	$\frac{1}{2}''$ £	$\frac{3}{4}''$ £	$1''$ £	$1\frac{1}{4}''$ £	$1\frac{1}{2}''$ £	$2''$ £	$3''$ £	$4''$ £	$6''$ £
Socket	No.	0·22	0·24	0·29	0·40	0·54	0·78	2·76	3·92	11·92
Reducing socket	,,	0·17	0·21	0·29	0·34	0·42	0·77	2·40	3·80	13·68
Elbow 90 degree	,,	0·28	0·32	0·42	0·67	0·97	1·46	3·80	6·24	23·52
Elbow 45 degree.	,,	0·42	0·46	0·61	0·78	1·01	1·41	4·04	7·52	18·40
Bend 90 degree .	,,	0·96	1·26	1·48	2·44	2·64	4·48	11·04	22·00	46·80
Bend 45 degree .	,,	0·96	1·26	1·46	2·24	2·48	4·08	8·56	17·04	36·20
Union	,,	0·90	0·98	1·35	1·58	2·32	3·28			
Saddle	,,						2·08	2·88	3·20	3·72
Equal tee .	,,	0·31	0·41	0·61	0·80	1·10	1·76	4·68	7·60	28·24
End cap	,,	0·20	0·22	0·26	0·31	0·62	0·82	2·20		
Straight tap connector	,,	0·54	0·59							
Bent tap connector	,,	0·80	0·88							
Tank connector .	,,	0·50	0·55							
Stub flange drilled to B.S. 10 table E .	,,	0·93	0·96	1·02	1·18	1·39	1·84	3·48	4·56	8·24
Cast iron backing ring to flange	,,	1·20	1·24	1·29	1·32	1·50	2·88	4·48	5·68	9·68
Neoprene gasket	,,	0·25	0·27	0·32	0·35	0·42	0·58	0·82	1·22	2·08

FLANGES TO B.S. 10

SOLID FORGED MILD STEEL BLANK FLANGES

	Unit	15 mm £	20 mm £	Nominal size 25 mm £	32 mm £	40 mm £	50 mm £
Table D							
Black	No.	0·74	0·74	0·86	0·96	1·08	1·26
Galvanized	,,	0·96	0·96	1·11	1·25	1·41	1·64
Table E							
Black	,,	0·74	0·74	0·86	0·96	1·08	1·26
Galvanized	,,	0·96	0·96	1·11	1·25	1·41	1·64
Table F							
Black	,,	0·90	0·90	1·07	1·20	1·33	2·09
Galvanized	,,	1·18	1·18	1·39	1·56	1·73	2·72
Table H							
Black	,,	1·06	1·06	1·20	1·46	1·64	2·44
Galvanized	,,	1·39	1·39	1·56	1·90	2·13	3·17

		65 mm £	Nominal size 80 mm £	100 mm £	125 mm £	150 mm £
Table D						
Black	,,	1·51	1·93	2·31	3·80	4·76
Galvanized	,,	1·97	2·51	3·00	4·93	6·19
Table E						
Black	,,	1·51	1·93	2·79	4·11	5·93
Galvanized	,,	1·97	2·51	3·62	5·34	7·70
Table F						
Black	,,	2·51	3·07	4·52	7·00	8·16
Galvanized	,,	3·26	3·99	5·88	9·10	10·62
Table H						
Black	,,	2·94	4·13	5·71	8·96	10·98
Galvanized	,,	3·82	5·37	6·99	11·65	14·27

Mechanical Installations – Market Prices of Materials

FLANGES TO B.S. 10

SOLID FORGED MILD STEEL BOSSED FLANGES DRILLED OR UNDRILLED – SCREWED

	Unit	Nominal size					
		15 mm £	20 mm £	25 mm £	32 mm £	40 mm £	50 mm £
Table D							
Black	No.	1·28	1·28	1·28	1·51	1·51	1·86
Galvanized	,,	1·66	1·66	1·66	1·96	1·96	2·41
Table E							
Black	,,	1·28	1·28	1·28	1·51	1·51	1·86
Galvanized	,,	1·62	1·62	1·62	1·96	1·96	2·41
Table F							
Black	,,	1·45	1·45	1·45	1·95	1·95	2·67
Galvanized	,,	1·89	1·89	1·89	2·53	2·53	3·47
Table H							
Black	,,	1·72	1·72	1·72	2·19	2·19	3·04
Galvanized	,,	2·24	2·24	2·24	2·84	2·84	3·95

		Nominal size				
		65 mm £	80 mm £	100 mm £	125 mm £	150 mm £
Table D						
Black	,,	1·97	2·41	3·47	6·89	6·89
Galvanized	,,	2·56	3·13	4·52	8·95	8·95
Table E						
Black	,,	2·11	2·54	3·78	7·35	7·35
Galvanized	,,	2·75	3·30	4·91	9·56	9·56
Table F						
Black	,,	3·47	3·99	5·27	11·06	11·06
Galvanized	,,	4·51	5·19	6·85	14·37	14·37
Table H						
Black	,,	3·99	4·90	6·46	12·57	12·57
Galvanized	,,	5·19	6·38	8·40	16·34	16·34

BOILERS

COMMERCIAL WATER BOILERS
PACKAGED BOILERS

		Rating kW			
		88	147	220	293
		£	£	£	£
Natural gas fired mild steel, stove enamelled casing, burner, gas train, thermostat drain cock, safety valve, including commissioning	Unit No.	1500·00	1700·00	1850·00	2150·00

		Rating kW			
		366	440	513	586
		£	£	£	£
	„	2600·00	3360·00	4200·00	4350·00

		Rating kW			
		88	147	220	293
		£	£	£	£
Boiler all as last but oil fired					
35 second oil	„	1060·00	1250·00	1500·00	1750·00
200/960 second oil	„				3250·00

		Rating kW			
		366	440	513	586
		£	£	£	£
35 second oil	„	1900·00	2900·00	3500·00	3650·00
200/960 second oil	„	3500·00	3700·00	3900·00	4050·00

		Rating kW			
		88	147	220	293
		£	£	£	£
Solid fuel fired mild steel boiler, stove enamelled casing, fan, thermostat, drain cock, safety valve, including commissioning but excluding stoker, grate and firebricks . . .	„	2350·00	2650·00	3300·00	4050·00

		Rating kW			
		366	440	513	586
		£	£	£	£
	„	4350·00	4800·00	5000·00	5800·00

		Rating kW			
		880	1172	1465	1759
		£	£	£	£
Natural gas fired mild steel boiler, stove enamelled casing, burner gas train, thermostat, drain cock, safety valve, including commissioning	„	5800·00	6600·00	7200·00	9000·00

		Rating kW			
		2052	2345	2638	2930
		£	£	£	£
	„	10000·00	11300·00	12750·00	13400·00

		Rating kW			
		880	1172	1465	1759
		£	£	£	£
Boiler all as last but oil fired					
35 second oil	„	4700·00	5300·00	5900·00	7700·00
200/960 second oil	„	4800·00	5300·00	6150·00	8300·00

		Rating kW			
		2052	2345	2638	2930
		£	£	£	£
35 second oil	„	9000·00	9700·00	10900·00	11800·00
200/960 second oil	„	9700·00	10300·00	11650·00	12500·00

		Rating kW			
		880	1172	1465	1759
		£	£	£	£
Solid fuel fired mild steel boiler, stove enamelled casing, fan, thermostat, drain cock, safety valve, including commissioning but excluding stoker, grate and firebricks . . .	„	7100·00	8050·00	9600·00	10800·00

FLANGES TO B.S. 10

SOLID FORGED MILD STEEL BOSSED FLANGES DRILLED OR UNDRILLED – SLIP ON FOR WELDING

	Unit	Nominal size					
		15 mm £	20 mm £	25 mm £	32 mm £	40 mm £	50 mm £
Table D Black	No.	1·16	1·16	1·16	1·39	1·39	1·71
Table E Black	,,	1·16	1·16	1·16	1·39	1·39	1·71
Table F Black	,,	1·31	1·31	1·31	1·77	1·77	2·45
Table H Black	,,	1·57	1·57	1·57	1·97	1·97	2·73

		Nominal size				
		65 mm £	80 mm £	100 mm £	125 mm £	150 mm £
Table D Black	,,	1·79	2·20	3·17	6·26	6·26
Table E Black	,,	1·92	2·32	3·42	6·68	6·68
Table F Black	,,	3·17	3·63	4·81	10·06	10·06
Table H Black	,,	3·63	4·46	5·89	11·42	11·42

SOLID FORGED MILD STEEL PLATE WELDING FLANGES

	Unit	Nominal size					
		15 mm £	20 mm £	25 mm £	32 mm £	40 mm £	50 mm £
Table D	No.	0·74	0·74	0·86	0·96	1·08	1·26
Table E	,,	0·74	0·74	0·86	0·96	1·08	1·26
Table F	,,	0·90	0·90	1·07	1·20	1·33	2·05
Table M	,,	1·06	1·06	1·20	1·46	1·64	2·38

		Nominal size				
		65 mm £	80 mm £	100 mm £	125 mm £	150 mm £
Table D	,,	1·51	1·93	2·31	3·49	4·15
Table E	,,	1·51	1·93	2·79	3·78	5·42
Table F	,,	2·46	2·93	4·42	6·91	8·02
Table H	,,	2·89	4·06	5·63	8·10	9·89

FLANGES TO B.S. 10

SOLID FORGED MILD STEEL WELDING NECK FLANGES

	Unit	*Nominal size*					
		15 mm £	20 mm £	25 mm £	32 mm £	40 mm £	50 mm £
Tables D and E	No.	1·88	1·88	1·88	2·63	2·63	3·26
Table F	,,	2·43	2·43	2·43	3·53	3·53	4·73
Table H	,,	2·93	2·93	2·93	4·10	4·10	5·19

		Nominal size				
		65 mm £	80 mm £	100 mm £	125 mm £	150 mm £
Tables D and E	,,	3·87	4·39	6·42	13·98	13·98
Table F	,,	5·78	6·86	9·34	19·83	19·83
Table H	,,	6·86	8·27	11·90	22·92	22·92

BRONZE SCREWED FLANGES DRILLED OR UNDRILLED

	Unit	*Nominal size*				
		15 mm £	22 mm £	28 mm £	35 mm £	42 mm £	54 mm £
Table D	No.	4·34	4·57	5·21	6·07	8·02	10·21
Table E	,,	4·34	4·57	5·21	6·07	8·02	10·21
Table F	,,	4·57	4·75	6·51	7·78	9·09	12·37
Table H	,,	6·07	6·30	7·61	8·91	9·96	14·28

		Nominal size				
		67 mm £	76 mm £	108 mm £	133 mm £	159 mm £
Table D	,,	13·23	16·70	25·35	32·29	42·45
Table E	,,	13·23	16·70	25·35	32·29	42·45
Table F	,,	17·14	21·24	29·89	47·45	61·74
Table H	,,	18·88	24·92	34·89	53·06	70·78

FLANGES TO B.S. 10

BRONZE WELDABLE BLANK FLANGES

	Unit	Nominal size					
		15 mm £	22 mm £	28 mm £	35 mm £	42 mm £	54 mm £
Table D	No.	4·34	5·08	6·39	8·02	11·91	12·11
Table E	,,	4·34	5·08	6·39	8·02	11·91	12·11
Table F	,,	4·57	5·53	8·97	11·94	14·00	17·14
Table H	,,	6·82	7·67	11·39	15·28	17·76	20·78

	Unit	Nominal size				
		67 mm £	76 mm £	108 mm £	133 mm £	159 mm £
Table D	,,	17·76	21·03	33·19	53·58	63·93
Table E	,,	17·76	21·03	33·19	53·58	63·93
Table F	,,	25·57	30·86	45·72	79·93	96·72
Table H	,,	27·42	34·23	50·26	92·38	117·83

BRONZE WELDABLE FLANGES DRILLED OR UNDRILLED

	Unit	Nominal size					
		15 mm £	22 mm £	28 mm £	35 mm £	42 mm £	54 mm £
Table D	No.	2·99	3·38	4·34	5·88	7·78	9·96
Table E	,,	2·99	3·38	4·34	5·88	7·78	9·96
Table F	,,	4·14	4·51	6·39	8·02	9·66	12·58
Table H	,,	6·28	6·51	8·48	10·08	12·37	15·47

	Unit	Nominal size				
		67 mm £	76 mm £	108 mm £	133 mm £	159 mm £
Table D	,,	12·81	16·26	24·92	31·32	39·10
Table E	,,	12·81	16·26	24·92	31·32	39·10
Table F	,,	17·41	21·43	30·86	49·08	62·91
Table H	,,	20·35	25·57	35·82	54·70	69·43

SUNDRIES
Black hex. round hex. bolts and nuts to B.S. 916

	Unit	Length							
		1¼″ £	1½″ £	1¾″ £	2″ £	2¼″ £	2½″ £	2¾″ £	3″ £	3¼″ £
½ in. diameter	100	10·98	12·09	12·63	13·20	13·79	15·03	15·62	16·24	17·15
⅝ in. diameter	,,		21·05	22·07	23·09	24·09	25·11	27·16	28·17	29·19	31·16
¾ in. diameter	,,			35·04	36·48	37·91	39·34	40·78	42·21	43·65	45·77

Mild steel round washers to B.S. 3410, table 7

	Unit	£
½ in. diameter	100	1·19
⅝ in. diameter	,,	1·96
¾ in. diameter	,,	3·01
⅞ in. diameter	,,	5·01
1 in. diameter	,,	6·70

FLANGES TO B.S. 10

SUNDRIES (*continued*)

		Nominal size				
		15 mm £	20 mm £	25 mm £	32 mm £	40 mm £	50 mm £
Corrugated brass joint rings full face, holed for bolts	Unit						
Tables D and E	100	14·55	16·30	20·37	22·13	26·20	32·02
Tables F and H	,,	23·28	23·28	26·20	27·95	29·11	37·84

		Nominal size				
		65 mm £	80 mm £	100 mm £	125 mm £	150 mm £
Tables D and E	,,	34·93	40·75	52·39	69·85	69·85
Tables F and H	,,	50·01	55·30	64·03	90·23	96·05

		Nominal size				
		15 mm £	20 mm £	25 mm £	32 mm £	40 mm £	50 mm £
Corrugated brass joint rings inside bolt circle							
Tables D and E.	,,	5·82	5·82	5·82	7·57	8·73	11·64
Tables F and H.	,,	5·82	5·82	5·82	7·57	8·73	11·64

		Nominal size				
		65 mm £	80 mm £	100 mm £	125 mm £	150 mm £
Tables D and E.	,,	14·55	17·46	23·28	29·11	34·93
Tables F and H.	,,	14·55	17·46	23·28	29·11	34·93

		Nominal size				
		15 mm £	20 mm £	25 mm £	32 mm £	40 mm £	50 mm £
Asbestos joint rings $\tfrac{1}{16}$ in. thick full face, holed for bolts							
Tables D and E	,,	11·26	11·99	18·04	18·98	20·96	24·18
Table F	,,	11·26	11·99	18·98	20·96	22·00	28·99
Table H	,,	18·04	18·04	18·98	20·96	22·00	28·99

		Nominal size				
		65 mm £	80 mm £	100 mm £	125 mm £	150 mm £
Tables D and E	,,	28·99	43·17	51·09	62·15	70·38
Table F	,,	43·17	47·56	54·53	70·38	79·45
Table H	,,	43·17	47·76	54·53	70·38	79·45

		Nominal size				
		15 mm £	20 mm £	25 mm £	32 mm £	40 mm £	50 mm £
Asbestos, joint rings $\tfrac{1}{16}$ in. thick, inside bolt circle							
Tables D and E	,,	7·30	7·72	8·56	8·97	10·02	11·26
Tables F.	,,	7·30	7·72	8·56	9·49	10·22	13·24
Table H.	,,	8·14	8·14	8·56	9·49	10·22	13·24

FLANGES TO B.S. 10

SUNDRIES (*continued*)

	Unit	Nominal size				
		65 mm £	80 mm £	100 mm £	125 mm £	150 mm £
Asbestos joint rings $\frac{1}{16}$ in. thick inside bolt circle (*continued*)						
Tables D and E	No.	13·24	19·92	27·74	46·62	50·26
Table F	,,	19·92	23·03	30·34	50·26	57·35
Table H	,,	19·92	23·03	30·34	50·26	57·35

	Unit	32 mm £	40 mm £	50 mm £	Nominal size			
					65 mm £	80 mm £	100 mm £	125 mm £	150 mm £
Rubber joint rings $\frac{1}{8}$ in. thick full face holed for bolts.									
Table D	100	13·50	16·20	18·90	24·30	24·30	29·70	54·00	63·45

EXPANSION JOINTS

STAINLESS STEEL SLEEVED BELLOWS TYPE EXPANSION JOINTS
(suitable for working pressures shown)

	Unit	15 mm 21 bar £	20 mm 21 bar £	*Nominal size* 25 mm 21 bar £	32 mm 21 bar £	40 mm 21 bar £	50 mm 21 bar £
Screwed ends for mild steel tube	No.	12·35	14·44	17·10	21·38	22·61	31·64
Screwed ends for copper tube	,,	15·58	18·43	20·52	25·75	26·51	37·43
Bevelled ends for welding	,,	11·50	13·40	18·91	19·26	20·71	28·98
Flanged ends to B.S. 10 table E	,,	23·56	24·99	26·70	28·03	29·26	38·67
Flanged ends to B.S. 10 table F	,,	24·99	25·18	27·22	30·31	31·73	45·60

		65 mm 17 bar £	*Nominal size* 80 mm 17 bar £	100 mm 17 bar £	125 mm 8 bar £	150 mm 8 bar £
Screwed ends for mild steel tube	,,	44·37	49·12			
Screwed ends for copper tube	,,	53·49	58·24			
Bevelled ends for welding	,,	33·54	39·24	48·74	68·31	82·75
Flanged ends to B.S. 10 table E	,,	45·98	51·68	67·17	89·59	108·02
Flanged ends to B.S. 10 table F	,,	54·82	62·23	71·73	112·58	131·96

WIRE REINFORCED RUBBER SLEEVED EXPANSION JOINTS
(suitable for working pressures up to 8 bar)

	Unit	32 mm £	40 mm £	50 mm £	*Nominal size* 65 mm £	80 mm £	100 mm £	125 mm £	150 mm £
Flanged ends to B.S. 10 table E	No.	46·36	46·61	48·34	49·95	57·55	63·14	76·97	85·14

SINGLE HINGED ANGULAR EXPANSION JOINTS

	Unit	15 mm £	20 mm £	*Nominal size* 25 mm £	32 mm £	40 mm £	50 mm £
Bevelled ends for welding	No.	46·45	47·01	47·57	51·81	53·20	59·12
Flanged ends to B.S. 10 table E	,,	50·40	51·37	52·39	56·12	58·93	63·54

		65 mm £	*Nominal size* 80 mm £	100 mm £	125 mm £	150 mm £
Bevelled ends for welding	,,	61·94	69·54	84·47	97·97	130·35
Flanged ends to B.S. 10 table E	,,	70·42	79·01	97·12	124·14	150·85

EXPANSION JOINTS

ARTICULATED TIED ANGULAR EXPANSION JOINTS

	Unit	15 mm £	20 mm £	*Nominal size* 25 mm £	32 mm £	40 mm £	50 mm £
Bevelled ends for welding.	No.	54·04	60·18	66·71	76·01	88·69	101·18
Flanged ends to B.S. 10 table E	,,	58·13	64·10	72·30	80·59	95·65	112·34

		65 mm £	*Nominal size* 80 mm £	100 mm £	125 mm £	150 mm £
Bevelled ends for welding.	,,	126·70	145·69	177·37	247·06	316·75
Flanged ends to B.S. 10 table E	,,	136·46	156·48	196·12	275·57	353·11

PIPE FIXINGS

FOR COPPER TUBE

Nominal size

Copper clips	Unit	15 mm £	22 mm £	28 mm £	35 mm £	42 mm £	54 mm £	67 mm £	76 mm £	108 mm £
Saddle band	100	1·93	2·31	3·47	4·62	8·86	12·71			
Single spacing clip	,,	3·08	3·47	6·16						
Two piece spacing clip	,,	4·24	5·01	7·32	10·40	17·71	23·87			
Brass brackets and rings										
Single pipe bracket screw on pattern	,,	30·80	34·65	42·35						
Single pipe bracket built in pattern	,,	46·20	53·90	57·75						
Single pipe ring	,,	53·90	57·75	69·30	73·15	80·85	96·25	226·80	285·60	442·40
Double pipe ring	,,	61·60	65·45	88·55	92·40	100·10	119·35	257·60	319·20	534·80
Wall bracket with screwed backplate	,,	77·00	92·40	111·65	142·45	188·65	238·70			
Wall bracket for building in	,,	61·60	73·15	80·85	84·70	134·75	157·85		330·40	604·80
Hospital brackets	,,	65·45	69·30	84·70	92·40	127·05	173·25			
Female backplate screw on pattern	,,	30·80	34·65	38·50	42·35	65·45	77·00	77·00	119·00	119·00
Male backplate screw on pattern	,,	26·95	30·80	34·65	38·50	57·75	73·15	73·15	119·00	119·00
Plastic										
Snap on	,,	2·45	3·08	5·25						
Hinged	,,	3·36	4·20	6·30						

FOR STEEL TUBE

Nominal size

	Unit	15 mm £	20 mm £	25 mm £	32 mm £	40 mm £	50 mm £
Single pipe bracket screw on pattern							
Black malleable	100	22·00	25·00	30·00	40·00	53·00	70·00
Galvanized	,,	30·00	34·00	40·00	53·00	71·00	94·00
Single pipe bracket build in pattern							
Black malleable	,,	22·00	25·00	30·00	40·00	53·00	63·00
Galvanized	,,	30·00	34·00	40·00	53·00	71·00	84·00
Pipe rings and backplates							
Single socket black malleable	,,	21·00	24·00	27·00	27·00	36·00	45·00
Single socket galvanized	,,	28·00	32·00	37·00	37·00	48·00	61·00
Double socket black malleable	,,	26·00	29·00	33·00	39·00	45·00	51·00
Double socket galvanized	,,	36·00	38·00	44·00	52·00	61·00	68·00
Screw on backplate black malleable	,,	27·00	27·00	27·00	27·00	27·00	27·00
Screw on backplate galvanized	,,	36·00	36·00	36·00	36·00	36·00	36·00

Nominal size

	Unit	65 mm £	80 mm £	100 mm £	125 mm £	150 mm £
Single pipe bracket screw on pattern						
Black malleable	,,	94·00	128·00	187·00		
Galvanized	,,	125·00	171·00	250·00		
Pipe rings and backplates						
Single socket black malleable	,,	66·00	80·00	122·00	245·00	273·00
Single socket galvanized	,,	89·00	107·00	163·00	327·00	365·00
Screw on backplate black malleable	,,	27·00	27·00	30·00	30·00	30·00
Screw on backplate galvanized	,,	36·00	36·00	41·00	41·00	41·00

PIPE FIXINGS

FLOOR OR CEILING COVER PLATES

	Unit	15 mm £	20 mm £	Nominal size 25 mm £	32 mm £	40 mm £	50 mm £
Plastic	No.	0·12	0·12	0·14	0·14	0·14	0·15
Chromium plated	,,	0·51	0·53	0·58	0·67	0·71	0·80

PIPE ROLLER AND CHAIR FOR DIAMETERS SHOWN

	Unit	Up to 50 mm £	80 mm £	100 mm £	150 mm £
Black malleable	100	29·00	37·00	93·00	169·00
Galvanized	,,	39·00	49·00	124·00	226·00

WALL BRACKET FOR PIPE ROLLER AND CHAIR

	Unit	Up to 50 mm £
Black malleable or galvanized	100	56·00

BOILERS

DOMESTIC WATER BOILERS

		Rating kW (Btu/hr)				
Natural gas fired cast iron boiler, floor standing, stove enamelled casing, electric controls, suitable for connection to conventional flue	Unit No.	8·79 (30000) £ 130·00	11·72 (40000) £ 145·00	13·19 (45000) £ 155·00	14·66 (50000) £ 165·00	16·12 (55000) £ 175·00
		Rating kw (Btu/hr)				
		17·59 (60000) £ 185·00	20·52 (70000) £ 205·00	23·45 (80000) £ 225·00	29·32 (100000) £ 265·00	43·98 (150000) £ 360·00
		Rating kW (Btu/hr)				
Boilers as last but with balanced flue .	,,	8·79 (30000) £ 175·00	11·72 (40000) £ 185·00	13·19 (45000) £ 190·00	14·66 (50000) £ 200·00	16·12 (55000) £ 210·00
		Rating kW (Btu/hr)				
		17·59 (60000) £ 225·00	20·52 (70000) £ 250·00	23·45 (80000) £ 285·00	29·32 (100000) £ 360·00	43·98 (150000) £ 600·00
		Rating kW (Btu/hr)				
Boiler as last but wall hung and for connection to conventional flue . .	,,	8·79 (30000) £ 120·00	11·72 (40000) £ 135·00	13·19 (45000) £ 140·00	14·66 (50000) £ 150·00	
Boiler as last but with balanced flue .	,,	145·00	150·00	170·00	185·00	
Fireplace mounted natural gas fire and back boiler, cast iron water boiler, electric controls, fire output 3·5 kW, with wood surround	,,	215·00	215·00	220·00	240·00	
			Rating kW (Btu/hr)			
Oil fired steel water boiler, 28 or 35 second oil, floor standing, stove enamelled casing, electric controls, fully automatic	,,	8·79 (30000) £ 282·00	11·72 (40000) £ 282·00	13·19 (45000) £ 282·00	14·66 (50000) £ 282·00	16·12 (55000) £ 312·00
		Rating kW (Btu/hr)				
		17·59 (60000) £ 312·00	20·52 (70000) £ 336·00	23·45 (80000) £ 366·00	29·32 (100000) £ 432·00	43·98 (150000) £ 498·00
		Rating kW (Btu/hr)				
Solid fuel fired cast iron water boiler, floor standing, stove enamelled casing, thermostat, draught stabilizer, electric controls, hand fired	,,	8·79 (30000) £ 175·00	11·72 (40000) £ 175·00	13·12 (45000) £ 220·00	14·66 (50000) £ 220·00	16·12 (55000) £ 225·00
		Rating kW (Btu/hr)				
		17·59 (60000) £ 240·00	20·52 (70000) £ 265·00	23·45 (80000) £ 295·00	29·32 (100000) £ 300·00	43·98 (150000) £ 410·00

Mechanical Installations – Market Prices of Materials 59

BOILERS

COMMERCIAL WATER BOILERS (*continued*)
SECTIONAL BOILERS

			Rating kW		
Natural gas fired cast iron boiler, insulated stove enamelled casing, burner, gas train, thermostat, drain cock, safety valve, including commissioning. . . .	Unit No.	88 £ 1600·00	117 £ 1900·00	147 £ 2050·00	220 £ 2300·00
			Rating kW		
	,,	293 £ 2900·00	440 £ 3500·00	586 £ 4100·00	880 £ 6000·00
			Rating kW		
Boiler all as last but oil fired 35 second oil	,,	88 £ 1150·00	117 £ 1450·00	147 £ 1700·00	220 £ 1900·00
			Rating kW		
	,,	293 £ 2350·00	440 £ 2950·00	586 £ 3600·00	880 £ 5200·00
			Rating kW		
Solid fuel fired cast iron boiler, insulated stove enamelled casing, fan, thermostat, drain cock, safety valve, including commissioning .	,,	88 £ 1700·00	117 £ 1700·00	147 £ 1900·00	2209 £ 2750·00
			Rating kW		
	,,	293 £ 3150·00	440 £ 4200·00		

BOILERS

COMMERCIAL STEAM BOILERS
PACKAGED BOILERS

			Rating kW (kgs/hr at 100 °C)			
Natural gas fired steel boiler, 10 bar maximum working pressure, insulated stove enamelled casing, burner, gas train, booster, controller and panel, feed pump and valved, all boiler mountings including commissioning	Unit No.	293 (455) £ 7000·00	586 (910) £ 9300·00	880 (1360) £ 10500·00	1172 (1815) £ 11300·00	1465 (2270) £ 13100·00
			Rating kW (kgs/hr. at 100 °C)			
		1760 (2720) £ 14500·00	2052 (3175) £ 15700·00	2345 (3630) £ 18300·00	2638 (4085) £ 18500·00	2930 (4535) £ 19500·00
			Rating kW (kgs/hr at 100 °C)			
		293 (455) £	586 (910) £	880 (1360) £	1172 (1815) £	1465 (2270) £
Boiler all as last but oil fired 35 second oil 200/960 second oil 3500 second oil	,, ,, ,,	6450·00 7000·00	8200·00 9600·00 9900·00	9400·00 9700·00 10200·00	10100·00 10800·00 10500·00	11500·00 11900·00 12100·00
			Rating kW (kgs/hr at 100 °C)			
		1760 (2720) £	2052 (3175) £	2345 (3630) £	2638 (4085) £	2930 (4535) £
35 second oil 200/960 second oil 3500 second oil	,, ,, ,,	12850·00 13400·00 13600·00	14200·00 14650·00 14900·00	15600·00 15600·00 15700·00	15600·00 17300·00	17700·00 18600·00

CISTERNS, TANKS AND CYLINDERS

GALVANIZED AFTER MADE MILD STEEL WELDED OPEN TOP COLD WATER CISTERN
(to B.S. 417)

	Capacity to water line litres	Unit	\multicolumn{6}{c}{Thickness of steel}
			1·2 mm £	1·6 mm £	2·0 mm £	2·5 mm £	3·2 mm £	4·8 mm £
Size No. SCM 45	18	No.	7·13	7·88	8·63	9·75	11·25	
SCM 70	36	,,	8·63	9·60	10·88	12·75	14·63	
SCM 90	54	,,	9·38	10·88	12·38	14·25	16·50	
SCM 110	68	,,	10·13	11·70	13·50	15·75	18·00	
SCM 135	86	,,	10·88	12·75	14·70	17·25	20·25	
SCM 180	114	,,	12·15	14·25	16·50	19·50	22·88	
SCM 230	159	,,	13·73	15·75	18·38	22·13	26·25	
SCM 270	191	,,	18·00	20·70	24·30	29·70	35·55	
SCM 320	227	,,	19·53	22·50	26·55	34·40	38·70	
SCM 360	264	,,	20·70	23·85	28·35	34·20	41·40	
SCM 450/1	327	,,		31·00	37·00	44·00	53·00	
SCM 680	491	,,		44·55	50·63	60·75	72·90	
SCM 910	709	,,		55·69	62·78	73·91	90·11	
SCM 1130	841	,,		64·80	72·90	85·05	101·25	
SCM 1600	1227	,,		81·00	91·13	105·30	126·56	
SCM 2270	1727	,,			117·45	134·66	162·00	227·81
SCM 2720	2137	,,			133·14	151·88	182·25	258·19
SCM 4540	3364	,,				216·63	253·13	354·38

	Unit	£
Plain hole cut in at works		
Up to and including 64 mm diameter	No.	0·50
Over 64 up to 150 mm diameter	,,	1·20
Screwed boss fitted at works		
15 mm diameter	,,	0·70
20 mm diameter	,,	0·80
25 mm diameter	,,	0·90
32 mm diameter	,,	1·00
40 mm diameter	,,	1·10
50 mm diameter	,,	1·20
75 mm diameter	,,	3·00
100 mm diameter	,,	6·00

CISTERNS, TANKS AND CYLINDERS

GALVANIZED MILD STEEL COVERS FOR OPEN TOP COLD WATER CISTERNS
(to B.S. 417)

		Thickness of steel				
		1·2 mm	1·6 mm	2·0 mm	2·5 mm	3·2 mm
Cistern size No.	Unit	£	£	£	£	£
SCM 45	No.	2·00	2·00	2·00	2·26	2·40
SCM 70	,,	2·00	2·00	2·00	2·60	3·00
SCM 90	,,	2·14	2·66	3·46	4·00	4·26
SCM 110	,,	2·26	2·84	3·68	4·26	4·54
SCM 135	,,	2·40	3·00	3·90	4·50	4·80
SCM 180	,,	3·00	3·76	4·88	5·62	6·00
SCM 230	,,	3·54	4·44	5·76	6·64	7·08
SCM 270	,,	4·80	6·00	7·80	9·00	9·60
SCM 320	,,	4·80	6·00	7·80	9·00	9·60
SCM 360	,,	5·20	6·50	8·46	9·76	10·40
SCM 450/1	,,	6·40	8·00	10·40	12·00	12·80
SCM 680	,,	8·22	10·29	13·37	15·41	16·44
SCM 910	,,	9·05	11·32	14·72	16·99	18·10
SCM 1130	,,	12·15	15·19	19·74	22·78	24·30
SCM 1600	,,		18·99	24·68	28·47	30·38
SCM 2270	,,			31·59	36·45	38·88
SCM 2720	,,			31·59	36·45	38·88
SCM 4540	,,			52·65	60·75	64·80

POLYPROPYLENE, RECTANGULAR TANKS
(complete with ballvalve fixing plate)
lots of 10 or more

Capacity to water line litres	Unit	Tank £	Cover £
18	No.	1·83	1·69
68	,,	6·47	3·80
91	,,	7·26	3·80
114	,,	8·69	4·54
182	,,	14·46	6·97
227	,,	16·60	6·97

FIBREGLASS, RECTANGULAR TANKS
(complete with ballvalve fixing plate)
lots of 10 or more

18	,,	8·11	2·49
68	,,	15·42	4·80
91	,,	17·09	4·90
114	,,	18·26	5·59
227	,,	32·54	7·62

POLYTHENE, CIRCULAR TANK
(complete with ballvalve fixing plate)
lots of 10 or more

114	,,	10·64	4·35
227	,,	16·32	5·56

GALVANIZED AFTER MADE MILD STEEL TANKS
(to B.S. 417)

		Capacity to water line litres	Unit	Grade A £	Grade B £
Size No. T25/1		95	No.	45·00	41·00
T30/A		114	,,	47·50	43·50
T40		155	,,	57·00	—
				Unit	£
Plain hole cut in at works				No.	0·50

CISTERNS, TANKS AND CYLINDERS

GALVANIZED AFTER MADE MILD STEEL WELDED HOT WATER STORAGE CYLINDERS
(to B.S. 417)

	Actual Capacity litres	Unit	Grade A £	Grade B £	Grade C £
Size No YM 91	73	No.	37·00	35·00	33·50
YM 114	100	,,	41·50	39·00	37·00
YM 127	114	,,	43·00	40·50	38·50
YM 141	123	,,	44·50	41·50	39·50
YM 150	136	,,	46·50	43·50	41·00
YM 117	159	,,	50·50	47·50	44·50
YM 218	195	,,	70·00	53·50	49·50
YM 264	241	,,	77·50	58·50	53·50
YM 355	332	,,	90·00	65·00	60·00
YM 455	441	,,	110·00	80·00	73·00

		Unit	£
Extra for manhole fitted at works			
150 mm diameter		No.	9·40
225 mm ,,		,,	10·70
300 mm ,,		,,	12·00
400 mm ,,		,,	13·40
450 mm ,,		,,	14·72
Extra for screwed immersion heater boss		,,	1·32
Extra for plug to immersion heater boss		,,	1·20

MILD STEEL FUEL STORAGE TANKS
(with all necessary screwed bosses and oil resisting joint ring)

Rectangular	Capacity litres	Unit	3 mm plate £	5 mm plate £	6 mm plate £
1219 × 610 × 610 mm	455	No.	63·25	80·15	—
1092 × 864 × 737 mm	682	,,	75·90	97·75	—
1219 × 915 × 813 mm	909	,,	87·40	115·00	—
1829 × 915 × 813 mm	1364	,,	117·30	155·25	—
1524 × 1219 × 1219 mm	2273	,,	149·50	201·25	—
1829 × 1219 × 1219 mm	2728	,,	169·05	227·70	—
2134 × 1219 × 1219 mm	3182	,,	188·03	255·30	—
2438 × 1219 × 1219 mm	3637	,,	207·00	281·75	333·50
2438 × 1524 × 1219 mm	4546	,,	235·75	322·00	379·50
2438 × 1829 × 1524 mm	6819	,,	—	414·00	494·50
3048 × 1829 × 1626 mm	9092	,,	—	—	615·25
Cylindrical (including cradles)					
1524 × 686 mm	455	,,	67·20	86·40	—
1676 × 762 mm	682	,,	78·00	102·00	—
1676 × 915 mm	909	,,	91·20	122·40	—
1676 × 1067 mm	1364	,,	—	146·40	—
1753 × 1372 mm	2273	,,	—	208·80	—
2057 × 1372 mm	2728	,,	—	232·80	—
2362 × 1372 mm	3182	,,	—	256·80	—
2667 × 1372 mm	3637	,,	—	282·00	—
3353 × 1372 mm	4546	,,	—	336·00	420·00
3962 × 1524 mm	6819	,,	—	444·00	540·00
3658 × 1829 mm	9092	,,	—	564·00	672·00

		Unit	£
Calibrated dipstick			
up to 1800 litres		No.	14·40
up to 6750 litres		,,	18·00
up to 9100 litres		,,	21·60
Extra for brass locking cap		,,	7·20

CISTERNS, TANKS AND CYLINDERS

GALVANIZED AFTER MADE INDIRECT CYLINDER WELDED THROUGHOUT
(with bolted end with up to 5 No screwed bosses)

Diameter mm	Length mm	Capacity litres	Heating surface m^2	Unit	2·5/2 mm £	3 mm plate £
381	762	77	0·6	No.	42·00	48·00
381	914	91	0·6	„	45·00	51·50
457	762	109	0·7	„	49·50	57·50
457	889	125	0·9	„	52·50	61·50
457	991	136	1·0	„	55·00	65·00
457	1092	159	1·1	„	62·00	73·50
457	1219	182	1·3	„	68·00	82·00

					3 mm plate £	5 mm plate £
508	1295	227	1·6	„	106·80	140·18
508	1473	273	1·9	„	119·04	159·09
610	1372	364	2·6	„	142·40	193·58
610	1549	409	2·9	„	157·98	213·60
610	1753	455	3·3	„	172·44	233·63
686	1626	568	4·0	„	196·91	267·00
762	1676	682	4·8	„	228·06	311·50
813	1676	796	5·7	„	255·88	356·00
813	1956	909	6·5	„	289·25	402·73
991	1956	1364	9·8	„	422·75	552·91

					5 mm plate £	6 mm plate £
1067	2286	1818	11·3	„	658·80	834·38
1219	2438	2728	15·8	„	885·55	1123·63
1372	2438	3182	18·6	„	1034·63	1312·75

Extra for boss for immersion heater				Unit No.	£ 1·32

CISTERNS, TANKS AND CYLINDERS

COPPER STORAGE CYLINDERS
(to B.S. 699)

B.S. Size No.	Capacity litres	Unit	Grade 1 Thickness mm	£	Grade 2 Thickness mm	£	Grade 3 Thickness mm	£
1	74	No.	1·2/1·6	44·86	0·9/1·4	36·00	0·7/1·2	30·44
2	98	„	1·2/1·8	52·20	0·9/1·6	41·58	0·7/1·2	34·16
4	86	„	1·6/2·0	61·77	1·0/1·6	43·40	0·7/1·2	31·63
5	98	„	1·6/2·0	66·20	1·0/1·6	45·60	0·7/1·2	33·82
6	109	„	1·6/2·0	68·40	1·0/1·6	50·00	0·7/1·2	36·03
7	120	„	1·6/2·0	72·06	1·0/1·6	52·93	0·7/1·2	38·25
8	144	„	1·6/2·0	80·90	1·0/1·6	58·83	0·7/1·2	42·68

Extra for additional boss fitted at works	Unit	£
15 mm diameter	No.	0·60
20 mm diameter	„	0·60
25 mm diameter	„	0·60
32 mm diameter	„	0·78
40 mm diameter	„	1·70
Extra for 57 mm diameter boss for immersion heater	„	0·90

BRAZED COPPER STORAGE CYLINDERS
(with screwed bosses)

Capacity litres	Unit	Tested 30 lb. Thickness mm	£	Tested 50 lb. Thickness mm	£
173	No.	1·2/1·6	73·91	1·6/2·0	91·80
273	„	1·2/1·6	94·99	1·6/2·0	120·49
341	„	1·6/2·3	153·00	2·0/2·6	195·08
386	„	1·6/2·3	165·75	2·0/2·6	211·65
414	„	1·6/2·3	178·50	2·0/2·9	227·59
455	„	1·6/2·3	184·88	2·0/2·9	235·88
568	„	1·6/2·3	212·93	2·0/2·9	272·85
682	„	2·0/3·3	308·55	2·6/4·1	376·13
796	„	2·0/4·1	363·38	2·6/4·5	433·50
909	„	2·0/4·1	387·60	2·6/4·5	463·46
1364	„	2·0/4·1	484·50	2·6/4·5	582·68
1818	„	2·6/4·47	705·08	3·3/5·4	841·50
2273	„	2·6/4·87	841·50	3·3/5·4	1000·88
2728	„	3·3/0·21	924·38	4·1/6·4	1099·69
3182	„	3·3/0·21	1224·00	4·1/6·4	1491·75

CISTERNS, TANKS AND CYLINDERS

COPPER INDIRECT CYLINDER
(with bolted top and up to 5 tappings for connections)

Capacity litres	Heating surface m^2	Unit	Tested 20 lb. £	Tested 30 lb. £	Tested 40 lb. £
77	0·6	No.	49·73	58·65	68·85
91	0·6	,,	52·28	61·20	73·95
109	0·7	,,	56·10	66·30	80·33
125	0·9	,,	61·20	73·95	89·25
136	1·0	,,	65·03	77·78	95·63
159	1·1	,,	69·49	82·88	102·00
182	1·3	,,	75·23	91·80	113·48

					Tested 50 lb. £
227	1·6	,,		156·83	187·43
273	1·9	,,		174·68	212·93
318	2·3	,,		253·73	314·93
364	2·6	,,		294·53	364·65
409	2·9	,,		320·03	396·53
455	3·3	,,		345·53	430·95
682	4·8	,,		573·75	711·45
796	5·7	,,		641·33	794·33
909	6·5	,,		721·01	896·33
1137	8·1	,,		977·29	1181·29
1364	9·8	,,		1129·65	1365·53
1591	9·8	,,		1303·05	1564·43
1818	11·8	,,		1459·88	1755·68
2046	12·1	,,		1565·70	1884·45
2273	13·0	,,		1854·49	2198·74

PRIMATIC SINGLE FEED INDIRECT COPPER CYLINDER
(to B.S. 1566: Part 2 1972 Grade-C including all standard connections)

B.S. Ref. No.	Capacity litres	Diameter mm	Height mm	Unit	£
3	104	1050	400	No.	48·89
5	86	750	450	,,	51·38
7	108	900	450	,,	54·84
8	130	1050	450	,,	64·28
9	152	1200	450	,,	68·91
10	180	1200	500	,,	102·76

Mechanical Installations – Market Prices of Materials 67

CISTERNS, TANKS AND CYLINDERS

DIRECT PATTERN COPPER COMBINATION HOT AND COLD WATER TANK
(to B.S. 3198 including all holes for connections, blanked off immersion boss drain and plug)

Capacity litres Hot	Cold	Diameter mm	Height mm	Unit	£
65	20	400	900	No.	34·23
85	25	450	900	,,	35·99
115	25	450	1075	,,	39·98
115	45	450	1200	,,	42·63
130	45	450	1300	,,	48·19
115	20	500	900	,,	51·50
150	45	500	1200	,,	55·60
115	115	500	1400	,,	64·65

INDIRECT PATTERN COPPER COMBINATION HOT AND COLD WATER TANK
(to B.S. 3198 coil heat exchanger including all holes for connections, drain boss and plug)

Capacity litres Hot	Cold	Diameter mm	Height mm	Unit	£
65	20	400	900	No.	44·55
85	25	450	900	,,	48·76
115	25	450	1075	,,	51·92
115	45	450	1200	,,	55·43
130	45	450	1300	,,	57·89
115	20	500	900	,,	57·48
150	45	500	1200	,,	66·82
115	115	500	1400	,,	75·24

Extra for aluminium protective anode . . . . ,, 1·29
Extra for immersion heater boss . . . . ,, 0·71

CALORIFIERS

COPPER STORAGE CALORIFIER
(generally to B.S. 853 with heating battery capable of raising contents from 10 °C to 65 °C in one hour, all connections provided, exclusive of fittings but including cradles or legs; static head not exceeding 1·35 bar)

		Capacity litres					
		400	1000	2000	3000	4000	4500
Horizontal	Unit	£	£	£	£	£	£
Primary L.P.H.W. at 82 °C on/71 °C off	. No.	637·00	1090·00	1670·00	2250·00	2650·00	3350·00
Primary steam at 3·2 bar	„	470·00	805·00	1390·00	1800·00	2120·00	3400·00
Vertical Primary L.P.H.W. at 82 °C on/71 °C off	„	605·00	1050·00	1620·00	2210·00	2615·00	3300·00
Primary steam at 3·2 bar	„	440·00	780·00	1340·00	1780·00	2100·00	2350·00

GALVANIZED STORAGE CALORIFIER
(specification as above)

		Capacity litres					
		400	1000	2000	3000	4000	4500
Horizontal	Unit	£	£	£	£	£	£
Primary L.P.H.W. at 82 °C on/71 °C off	. No.	545·00	830·00	1120·00	1450·00	1660·00	2100·00
Primary steam at 3·2 bar	„	405·00	590·00	880·00	1050·00	1180·00	1240·00
Vertical Primary L.P.H.W. at 82 °C on/71 °C off	„	525·00	805·00	1075·00	1430·00	1640·00	2080·00
Primary steam at 3·2 bar	„	380·00	570·00	840·00	1030·00	1160·00	1220·00

CAST IRON NON-STORAGE HEATING CALORIFIER
(generally to B.S. 853 heating secondary water with 82 °C flow and 71 °C return at 2 bar working pressure all connections provided, exclusive of fittings but including cradles or legs, static head not exceeding 1·35 bar)

		Capacity kW					
		88	176	293	586	879	1465
Horizontal	Unit	£	£	£	£	£	£
Steam at 3·2 bar	. No.	190·00	205·00	270·00	435·00	560·00	820·00
Steam at 4·8 bar	„	190·00	205·00	270·00	435·00	560·00	800·00
Vertical							
Steam at 3·2 bar	„	195·00	215·00	285·00	450·00	580·00	840·00
Steam at 4·8 bar	„	195·00	215·00	285·00	450·00	580·00	840·00
Horizontal Primary water at 116 °C on/93 °C off	„	280·00	470·00	640·00	920·00	1230·00	1770·00
Vertical Primary water at 116 °C on/93 °C off	„	300·00	490·00	660·00	945·00	1255·00	1800·00

Mechanical Installations – Market Prices of Materials 69

PUMPS, CIRCULATORS AND ACCELERATORS

SILENT RUNNING DIRECT DRIVE CENTRIFUGAL HEATING PUMP
(complete with bedplate, coupling guard and foundation bolts. Three phase electric motor to run at not exceeding 1450 r.p.m. Maximum working pressure 4 bar, maximum temperature 120 °C)

Pump size mm	Maximum delivery litres per second	Maximum head kN/m^2	Maximum motor rating kW	Unit	£
40	4·0	40·0	0·37	No.	385·00
40	4·0	65·0	1·50	,,	784·00
40	4·0	80·0	1·50	,,	784·00
40	4·0	120·0	2·20	,,	795·00
50	7·0	35·0	0·75	,,	436·00
50	7·0	80·0	1·10	,,	484·00
80	16·0	50·0	1·50	,,	657·00
80	16·0	80·0	2·20	,,	666·00
80	16·0	110·0	3·00	,,	682·00
80	16·0	150·0	4·00	,,	692·00
100	28·0	30·0	2·20	,,	765·00
100	28·0	45·0	2·20	,,	765·00
100	28·0	80·0	3·00	,,	773·00
100	28·0	110·0	4·00	,,	780·00
150	70·0	75·0	7·50	,,	1335·00
150	70·0	120·0	11·20	,,	1380·00
150	70·0	140·0	15·00	,,	1440·00

		Motor rating kW			
		0·37	0·55	0·75	1·10	1·50
	Unit	£	£	£	£	£
Extra for single phase motor	No.	6·00	9·00	42·00	50·00	58·00

Push button starters (for motors of the kW rating shown)

Single phase	,,	22·80	22·80	22·80	22·80	26·30
Three phase	,,	22·80	22·80	22·80	22·80	26·30

		Motor rating kW		
		2·20	3·00	4·00	5·50
		£	£	£	£
Single phase	,,	26·30	—	—	—
Three phase	,,	26·30	26·30	26·30	26·30

Star delta starters for motors of the following ratings

	Unit	£
7·50 kW	No.	79·00
11·20 kW	,,	117·30
15·00 kW	,,	135·30

PUMPS, CIRCULATORS AND ACCELERATORS

BELT-DRIVEN CENTRIFUGAL HEATING PUMP
(complete with bedplate, pulley guards and foundation bolts. Three phase electric motor to run at not exceeding 1450 r.p.m. Maximum working pressure 4 bar maximum temperature 110 °C)

Pump size mm	Maximum delivery litres per second	Maximum head kN/m^2	Maximum motor rating kW	Unit	£
40	4·0	40·0	0·37	No.	374·00
40	4·0	65·0	0·55	,,	376·00
40	4·0	80·0	0·75	,,	380·00
40	4·0	120·0	1·10	,,	390·00
50	7·0	35·0	0·75	,,	420·00
50	7·0	80·0	1·10	,,	435·00
80	16·0	50·0	1·50	,,	625·00
80	16·0	80·0	2·20	,,	634·00
80	16·0	110·0	3·00	,,	650·00
80	16·0	150·0	4·00	,,	660·00
100	25·0	30·0	1·50	,,	700·00
100	25·0	45·0	2·20	,,	715·00
100	25·0	80·0	3·00	,,	723·00
100	25·0	110·0	4·00	,,	730·00
150	70·0	75·0	7·50	,,	1255·00
150	70·0	120·0	11·20	,,	1300·00
150	70·0	140·0	15·00	,,	1360·00

SILENT RUNNING CLOSE COUPLED CENTRIFUGAL HEATING PUMP
(complete with bedplate and foundations bolts. Three phase electric motor to run at not exceeding 1450 r.p.m. Maximum working pressure 8 bar, maximum working temperature 110 °C)

Pump size mm	Maximum delivery litres per second	Maximum head kN/m^2	Maximum motor rating kW	Unit	£
40	4·0	40·0	0·55	No.	211·00
40	4·0	65·0	2·20	,,	479·00
40	4·0	80·0	2·20	,,	479·00
40	4·0	120·0	1·10	,,	479·00
50	7·0	35·0	0·75	,,	256·00
50	7·0	80·0	1·10	,,	344·00
80	16·0	50·0	2·20	,,	412·00
80	16·0	80·0	2·20	,,	412·00
80	16·0	110·0	3·00	,,	439·00
80	16·0	150·0	4·00	,,	646·00
100	28·0	30·0	2·20	,,	465·00
100	28·0	45·0	2·20	,,	465·00
100	28·0	80·0	3·00	,,	493·00
100	28·0	110·0	4·00	,,	493·00
150	70·0	75·0	7·50	,,	809·00
150	70·0	120·0	11·20	,,	915·00
150	70·0	140·0	15·00	,,	958·00

Two heads are usually better than one.

PULLEN PUMPS LTD
58 Beddington Lane, Croydon, Surrey CR9 4PT.
Tel: 01-684 9521

Particularly in situations where regular and speedy pump changeover is desirable without recourse to skilled maintenance staff. The Pullen 'Duopul' is a complete pumping system, ideal for use on central heating, HWS and chilled water applications in establishments such as old peoples' homes and schools. Pump changeover can be achieved electrically simply by pressing a button or throwing a switch.

The 'Duopul' has also been specifically designed to reduce capital cost, plant room space and installation time by reducing the number of valves and pipe fittings normally used on changeover systems. In the main, these savings are generated by the use of common suction and discharge branches, the latter incorporating a double action ball check valve which moves between two seats. An effective seal is thus produced preventing water recirculation through the standby pump. When both pumps are running together, the ball adopts a neutral position.

There are many pumps and pumping systems covered in the Pullen product range brochure — why not send for it?

The International Dictionary of Heating, Ventilating and Air Conditioning

Compiled by the **Documentation Committee of the Representatives of the European Heating and Ventilating Associations** under the **Chairmanship of Richard Eaves, A.L.A.**

The Dictionary comprises some 3,000 technical terms concerned with heating, ventilating and air conditioning. The terms are given in ten languages, the guide language being English and the others being French, German, Dutch, Italian, Hungarian, Polish, Russian, Swedish and Spanish. Containing more languages and terms than any other text, the translations are provided by practising engineers from each country concerned.

The main sequence of terms is series numbered and alphabetical indexes in each language refer the user to the appropriate term in the other languages by the serial numbers.

February 1981   410 pages
Hardback: 0 419 11650 8   £38.50

Third Edition
Building Techniques
Volume II Services

H King and **D Nield**
Revised by **J S Sansom**, Vauxhall College of Building

As the complexity of construction techniques increases there is a growing tendency for separate specialisms to develop as professional groups in their own right. This situation can lead to poor communications and inefficient building. Nowhere is this more true than in the field of services engineering.

Building Techniques was planned to provide an overall view of the construction process at an introductory level for those whose later specialisation will require them to concentrate on only a small part of that process. This long awaited new edition of Volume II covers all the services aspect of construction.

Volume 1: Structure (2nd edition 1976) is still available.

Third Edition December 1980   about 200 pages
Science Paperback: 0 412 21780 5
about £4.50

Chapman & Hall and E & F N Spon
The scientific, technical and medical divisions of Associated Book Publishers Ltd, 11 New Fetter Lane, London EC4P 4EE

Methuen Inc
733 Third Avenue
New York
NY 10017

PUMPS, CIRCULATORS AND ACCELERATORS

SILENT RUNNING PIPELINE MOUNTED CIRCULATOR
(for low and medium pressure heating and hot water services. Three phase electric motor to run at not exceeding 1450 r.p.m. Maximum working pressure 6 bar, maximum temperature 110 °C)

Pump size mm	Maximum delivery litres per second	Maximum head kN/m^2	Maximum motor rating kW	Unit	Standard cast iron construction £	Gunmetal construction £
32	2·0	17·0	0·25	No.	140·00	200·00
40	3·0	20·0	0·25	,,	151·00	205·00
50	5·0	30·0	0·37	,,	169·00	230·00
65	8·0	37·0	0·55	,,	198·00	270·00
80	12·0	42·0	1·10	,,	260·00	354·00
100	25·0	17·0	2·20	,,	429·00	580·00
100	25·0	37·0	4·00	,,	456·00	620·00

DUAL-MOUNTED SILENT RUNNING PIPELINE MOUNTED CIRCULATORS
(for low and medium pressure heating systems and hot water services complete with discharge and suction branches. Three phase electric motors to run at not exceeding 1450 r.p.m. Maximum working pressure 6 bar, maximum temperature 110 °C)

Pump sizes mm	Maximum delivery litres per second	Maximum head kN/m^2	Maximum motor rating kW	Unit	£
32	2·0	17·0	0·25	No.	392·00
40	3·0	20·0	0·25	,,	445·00
50	5·0	30·0	0·37	,,	480·00
65	8·0	37·0	0·55	,,	601·00
80	12·0	42·0	1·10	,,	746·00

SILENT RUNNING GLANDLESS HEATING ACCELERATOR
(for low and medium pressure heating and hot water services. Single or three phase electric motor to run at not exceeding 2800 r.p.m. Maximum working pressure 6 bar, maximum temperature 120 °C)

	Pump size mm	Maximum delivery litres per second	Maximum head kN/m^2	Maximum motor rating kW	Unit	Single phase £	Three phase £
With screwed connections	25	1·0	25·0	0·053	No.	81·66	74·88
	30	2·0	45·0	0·093	,,	99·36	92·40
With flanged connections including mating flanges	50	3·0	65·0	0·37	,,	—	207·30
	65	5·0	80·0	1·31	,,	—	241·00
	80	6·0	100·0	1·31	,,	—	250·80

PACKAGED PRESSURIZED COLD WATER SUPPLY SETS
(including pumps, membrane tank, valves, interconnecting pipework and control panel, mounted as mild steel frame and fully automatic, three phase electric motor)

Maximum delivery litres per second	Maximum head kN/m^2	Unit	£
3	600	No.	4000·00
7	600	,,	4200·00
16	600	,,	6000·00

PUMPS, CIRCULATORS AND ACCELERATORS

SELF-CONTAINED AUTOMATIC SUMP PUMP
(with single or three phase totally enclosed electric motor, float switch and gear to fit sump, totally submersible)

	Unit	Single phase £	Three phase £
	No.	250·00	250·00
As above but extended shaft or semi-submersible for sump depth of 2 m	,,	383·00	383·00

DOMESTIC HEATING GLANDLESS CIRCULATING PUMPS
(for domestic low pressure hot water heating systems, maximum working pressure 10 bar, maximum temperature 120 °C, 240 V 50 Hz electric motor)

	Unit	£
22 mm with copper tails	No.	21·20
22 mm with isolating valves	,,	23·14
28 mm with isolating valves	,,	23·85
25 mm with B.S.P. unions	,,	20·66
Extra for flanges to above pumps	Pair	4·54

AIR DISTRIBUTION EQUIPMENT

CEILING MOUNTING DIFFUSERS
Circular multi-cone aluminium diffuser, supply or extract

Nominal neck size	Unit	Without volume control £	With adjustable volume control £
150 mm	No.	14·50	20·85
275 mm	,,	18·21	24·50
300 mm	,,	24·76	31·83
375 mm	,,	31·84	39·30
450 mm	,,	38·12	45·85

Rectangular flush mounting aluminium diffuser, one-, two- or four-way flow.

Nominal neck size		Without volume control £	With opposed blade volume control damper £
150 × 150 mm	,,	12·12	16·77
300 × 150 mm	,,	14·35	19·72
300 × 300 mm	,,	19·26	26·27
450 × 150 mm	,,	16·18	22·80
450 × 300 mm	,,	22·27	31·31
450 × 450 mm	,,	29·28	40·55
600 × 150 mm	,,	18·14	26·14
600 × 300 mm	,,	25·35	36·03
600 × 450 mm	,,	33·34	47·30
600 × 600 mm	,,	40·81	63·08
900 × 300 mm	,,	31·25	48·41
900 × 600 mm	,,	51·42	86·79

SLOT DIFFUSERS
Continuous aluminium slot diffuser: including mounting brackets; average 3 m length

	Unit	1 slot £	2 slots £	3 slots £	4 slots £	6 slots £	8 slots £
Diffuser	m	10·80	14·96	19·65	23·85	32·40	41·57
Diffuser including equalizing deflectors	,,	21·68	26·36	31·90	37·34	47·66	58·14
Ends	Pair	2·95	3·08	3·28	3·47	3·75	4·06

PERFORATED DIFFUSERS

Aluminium perforated diffuser with hinged face plate

Nominal neck size		Without volume control £	With opposed blade volume control damper £
Rectangular neck			
150 × 150 mm	No.	21·95	27·00
200 × 200 mm	,,	36·75	42·80
250 × 250 mm	,,	37·60	44·60
300 × 300 mm	,,	38·32	46·20

Circular neck			With louvre volume control damper £
125 mm diameter	,,	28·75	41·72
150 mm ,,	,,	29·30	42·50
200 mm ,,	,,	45·06	58·82
250 mm ,,	,,	46·05	60·46
300 mm ,,	,,	47·00	62·00

AIR DISTRIBUTION EQUIPMENT

SUPPLY OR EXTRACT GRILLES
Aluminium grille with one set of vertical or horizontal individually adjustable blades

Size		Without volume control £	With opposed blade volume control damper £
150 × 150 mm.	No.	5·10	9·20
200 × 150 mm.	,,	5·40	9·60
200 × 200 mm.	,,	6·25	11·25
300 × 100 mm.	,,	4·80	9·20
300 × 150 mm.	,,	5·65	10·70
300 × 200 mm.	,,	6·62	12·25
300 × 300 mm.	,,	8·45	15·20
400 × 100 mm.	,,	5·75	10·80
400 × 150 mm.	,,	7·00	13·00
400 × 200 mm.	,,	8·20	14·80
400 × 300 mm.	,,	10·15	18·35
600 × 200 mm.	,,	10·75	19·25
600 × 300 mm.	,,	13·50	24·10
600 × 400 mm.	,,	16·65	29·30
600 × 500 mm.	,,	19·85	39·85
600 × 600 mm.	,,	22·95	44·95
800 × 300 mm.	,,	17·30	31·25
800 × 400 mm.	,,	21·40	37·70
800 × 600 mm.	,,	29·90	58·70
1000 × 300 mm.	,,	20·70	37·70
1000 × 400 mm.	,,	26·10	45·90
1000 × 600 mm.	,,	36·90	71·65
1000 × 800 mm.	,,	48·30	88·45
1200 × 600 mm.	,,	43·80	84·05
1200 × 800 mm.	,,	55·75	103·75
1200 × 1000 mm	,,	67·60	132·65

Mechanical Installations – Market Prices of Materials 75

AIR DISTRIBUTION EQUIPMENT

AUTOMATIC ROLL FILTER
(galvanized mild steel frame, single or three phase electric motor and pressure differential control, continuous acetate fibre filter media to retain 95% against B.S. 2831 Test Dust No 2 with an air velocity of 2·80 m^3/s)

Width mm	Height mm	Air volume m^3/hour	Unit	£
600	750	2500	No.	428·00
600	1500	5000	,,	448·00
900	1400	8500	,,	462·00
1150	1600	13500	,,	500·00

AUTOMATIC ROLL FILTER
(galvanized mild steel frame, vertically mounted including single phase electric motor and pressure differential control, continuous acetate fibre filter media to retain 95% against B.S. 2831 Test Dust No 2 with an air velocity of 2·80 m^3/s)

Width mm	Height mm	Air volume m^3/hour	Unit	£
914	1524	10000	No.	572·69
914	1829	12400	,,	599·13
914	2438	17000	,,	612·61
914	3048	21750	,,	627·80
1219	1524	14000	,,	621·06
1219	1829	17300	,,	626·11
1219	2438	24000	,,	642·97
1219	3048	30000	,,	656·46
1219	3658	37000	,,	669·96
1829	1829	27000	,,	696·93
1829	2438	37500	,,	712·11
1829	3048	48000	,,	728·93
1829	3658	58000	,,	745·83

PANEL FILTER
(galvanized mild steel side withdrawal frame, disposable two stage filter 50 mm thick to retain 92% against B.S. 2831 Test Dust No 2 with an air velocity of 2·80 m^3/hour)

Size mm	Air volume m^3/hour	Unit	Single panel £	Two panel £	Four panel £
500 × 500	1700	No.	12·38	15·98	24·60
600 × 600	1700	,,	15·10	19·86	31·07

HIGH EFFICIENCY FILTER
(sealed galvanized mild steel frame, disposable paper filter media with 0·01% penetration on a Sodium Flame Test to BS 3928)

	Size mm	Air volume m^3/hr.	Air velocity m^3/s	Unit	£
Front withdrawal including fixings for pre-filter	600 × 600 × 300	1700	2·80	No.	87·73
Side withdrawal	600 × 600 × 300	1700	2·80	,,	154·62

GREASE FILTER
(double sided vee bank unit, with washable filter media, suitable for hood and extract systems, with mounting frame and drip tray)

Size mm	Air volume m^3/hour	Unit	£
500 × 686 × 565	1360	No.	96·53
1000 × 686 × 565	2700	,,	148·60
1500 × 686 × 565	4100	,,	204·28

AIR DISTRIBUTION EQUIPMENT

BAG FILTER
(in 610 × 305 mm modules, in mild steel holding frames contained in mild steel installation frame; filter media to retain 99·2% against B.S. 2831 Test Dust No. 2 with an air velocity of 2·80 m^3/s)

	Width mm	Height mm	Length mm	Unit	£
Front withdrawal	610	315	380	No.	16·00
	610	315	635	,,	19·90
	610	315	900	,,	22·20
	610	610	380	,,	24·92
	610	610	635	,,	32·72
	610	610	900	,,	37·31
Side withdrawal					
Vertical	310	610	380	,,	17·39
	610	610	380	,,	29·29
	930	610	380	,,	40·72
	1240	610	380	,,	51·83
	1540	610	380	,,	63·63
	1860	610	380	,,	74·68
	2080	610	380	,,	85·97
	2376	610	380	,,	97·63
Horizontal	310	610	380	,,	21·28
	310	1240	380	,,	35·12
	310	1860	380	,,	49·42
	310	2400	380	,,	62·21

AIR DISTRIBUTION EQUIPMENT

FIRE DAMPERS
(steel shutter type fire damper with blades and frame outside the airstream complete with fusable link and actuating mechanism)

	Unit	Stainless steel £
Rectangular to fit duct size		
200 × 200 mm	No.	25·94
300 × 200 mm	,,	28·01
300 × 300 mm	,,	28·77
400 × 200 mm	,,	29·64
400 × 300 mm	,,	30·54
400 × 400 mm	,,	31·43
500 × 200 mm	,,	30·45
500 × 300 mm	,,	31·37
500 × 400 mm	,,	33·13
500 × 500 mm	,,	33·83
600 × 200 mm	,,	32·99
600 × 300 mm	,,	34·07
600 × 400 mm	,,	35·07
600 × 500 mm	,,	36·72
600 × 600 mm	,,	40·47
800 × 300 mm	,,	36·73
800 × 400 mm	,,	38·02
800 × 500 mm	,,	39·92
800 × 600 mm	,,	43·61
800 × 800 mm	,,	50·21
1000 × 300 mm	,,	39·68
1000 × 400 mm	,,	41·14
1000 × 500 mm	,,	43·23
1000 × 600 mm	,,	46·86
1000 × 800 mm	,,	54·28
1000 × 1000 mm	,,	60·30
Circular to fit duct size		
200 mm	,,	29·19
300 mm	,,	31·65
400 mm	,,	37·18
500 mm	,,	39·18
600 mm	,,	49·18
700 mm	,,	54·22
800 mm	,,	62·26
900 mm	,,	69·19
1000 mm	,,	76·39

AIR DISTRIBUTION EQUIPMENT

ATTENUATORS
(duct mounted galvanized sheet steel casing and frame, flanged ends for connection to duct work)

		Unit	900 mm £	Length 1500 mm £	2100 mm £
Rectangular					
300 × 150 mm	. . . . . .	No.	74·26		
300 × 300 mm	. . . . . .	,,	78·91	96·33	
600 × 300 mm	. . . . . .	,,	110·24	148·52	195·00
600 × 600 mm	. . . . . .	,,	168·35	216·97	276·25
900 × 300 mm	. . . . . .	,,	139·23	193·83	250·64
900 × 600 mm	. . . . . .	,,	219·31	286·59	357·37
900 × 900 mm	. . . . . .	,,	299·33	394·55	461·83
1200 × 300 mm	. . . . . .	,,	168·22	241·35	299·65
1200 × 600 mm	. . . . . .	,,	265·72	357·37	432·90
1200 × 900 mm	. . . . . .	,,	359·71	462·93	578·50
1200 × 1200 mm	. . . . . .	,,		585·00	741·40
1500 × 300 mm	. . . . . .	,,			349·31
1500 × 600 mm	. . . . . .	,,			512·20
1500 × 900 mm	. . . . . .	,,			697·45
1300 × 1200 mm	. . . . . .	,,			855·00

		Length Diameter × 1½ £	Diameter × 2 £
Circular, with pod			
150 mm	. . . . ,,	115·00	
300 mm	. . . . ,,	127·65	217·00
450 mm	. . . . ,,	162·15	263·00
600 mm	. . . . ,,	217·35	377·00
750 mm	. . . . ,,	299·00	474·00
900 mm	. . . . ,,	348·45	593·00
1200 mm	. . . . ,,	457·70	1236·25
1500 mm	. . . . ,,	535·90	1988·35

Mechanical Installations – Market Prices of Materials

FANS

SINGLE STAGE AXIAL FLOW FAN
(three phase 50 cycles, 380/440 volt electric motor, casing covering impeller and motor)

	Fan diameter	305 mm	483 mm	762 mm	1219 mm
	Motor speed	2800 r.p.m.	2900 r.p.m.	940 r.p.m.	725 r.p.m.
	Weight	29 kg	53 kg	98 kg	301 kg
Unit		£	£	£	£
No.		144·00	255·00	290·00	720·00

SINGLE STAGE AXIAL FLOW FAN
(as above but casing covering impeller only)

	Fan diameter	305 mm	483 mm	762 mm	1219 mm
	Motor speed	2800 r.p.m.	2900 r.p.m.	940 r.p.m.	725 r.p.m.
	Weight	23 kg	43 kg	77 kg	260 kg
Unit		£	£	£	£
No.		113·00	196·00	247·00	650·00

TWO STAGE AXIAL FLOW FAN
(as above with casing covering impeller and motor)

Unit No.		273·00	481·00	552·00	1366·00
Extra for special treatment					
Weatherproof treatment	,,	20·65	20·65	24·47	28·45
Chlorinated rubber paint	,,	11·24	20·65	45·51	71·11
Attachable feet, for fan of the diameter shown	Pair	5·70	9·40	18·77	32·71
Mounting plates, ditto	,,	11·24	14·11	48·70	64·01
Coned inlet, ditto	No.	13·73	15·00	40·19	65·46
Guard	,,	5·70	11·24	16·78	24·61

CENTRIFUGAL FAN
(Three phase electric motor, vee belt drive including drive guard and grillage)

	Fan duty	2·5 m³/s	5 m³/s	10 m³/s	15 m³/s
	static pressure	600 N/m²	650 N/m²	700 N/m²	750 N/m²
Unit		£	£	£	£
No.		759·00	1190·00	1760·00	3596·00
Extra for inspection door	,,	28·22	32·45	32·45	34·00

ROOF EXTRACT UNIT
(with propeller fan, three phase 50 cycles, 380/440 volt electric motor. Curved cowl including automatic shutters all welded construction, galvanized finish, for flat roof)

	Fan diameter	315 mm	450 mm	800 mm	1250 mm
	Motor speed	1360 r.p.m.	900 r.p.m.	560 r.p.m.	485 r.p.m.
	Weight	34 kg	69 kg	209 kg	458 kg
Unit		£	£	£	£
No.		136·80	195·00	427·65	1121·00

(as above for pitched roof including automatic shutters)

	Fan diameter	315 mm	450 mm	800 mm	1250 mm
	Motor speed	1360 r.p.m.	900 r.p.m.	560 r.p.m.	485 r.p.m.
	Weight	45 kg	73 kg	200 kg	508 kg
Unit		£	£	£	£
No.		149·60	203·40	454·50	1182·00

FANS

ANTI-VIBRATION MOUNTINGS
(per set of four)

	Unit	£
For fan 305–381 mm diameter		
Weight per mounting		
Up to 14 kg	Set	9·10
14–23 kg	,,	9·10
23–41 kg	,,	9·10
For fan 762 mm diameter		
Weight per mounting		
23–69 kg	,,	31·40
For fan 1219 mm diameter		
Weight per mounting		
54–69 kg	,,	31·40

BUTTERFLY TYPE DAMPER
(for fans of the diameter shown)

	Unit	Fan diameter 305 mm £	483 mm £	762 mm £	1219 mm £
Hand operated	No.	69·55	86·70	122·60	250·45
Air operated	,,	47·90	61·30	94·50	206·35

DIRECT ON LINE THREE PHASE STARTER
(380/440 volt)

	Unit	£
For single and two stage axial flow fan up to 3·72 kW		
With isolator	No.	54·50
Without isolator	,,	26·10
7·5 kW		
With isolator	,,	54·50
Without isolator	,,	26·10

WINDOW TYPE FAN
(of the diameter shown)

	Unit	152 mm £	229 mm £	305 mm £
Without shutter	No.	31·48	54·62	76·63
With shutter	,,	41·50	69·55	93·93

EXTERNAL LOUVRES

ACOUSTIC LOUVRES
(opening mounted, 300 mm deep galvanized steel louvres with blades packed with acoustic infill, suitable for screw fixing in opening)

Length mm	Height mm	Unit	Plain louvre £	With 12 mm galvanized mesh birdscreen £
900	600	No.	80·33	96·39
900	900	,,	120·49	144·59
900	1200	,,	110·16	137·70
900	1500	,,	137·70	172·13
900	1800	,,	165·24	206·55
900	2100	,,	192·78	240·98
900	2400	,,	220·32	275·40
900	2700	,,	248·20	309·83
900	3000	,,	275·40	344·25
1200	600	,,	107·10	128·52
1200	900	,,	110·16	137·70
1200	1200	,,	146·88	183·60
1200	1500	,,	183·66	229·50
1200	1800	,,	220·32	275·40
1200	2100	,,	257·04	321·30
1200	2400	,,	294·10	367·20
1500	600	,,	133·88	160·65
1500	900	,,	137·70	172·13
1500	1200	,,	183·60	229·50
1500	1500	,,	229·50	286·90
1500	1800	,,	275·40	344·25
1500	2000	,,	306·00	382·50
2000	600	,,	122·40	153·00
2000	900	,,	183·60	229·50
2000	1200	,,	244·80	306·00
2000	1500	,,	306·00	382·50

EXTERNAL LOUVRES

WEATHER LOUVRES
(opening mounted 100 mm deep galvanized steel louvres suitable for screw fixing into opening)

Length mm	Height mm	Unit	Plain louvre £	With 12 mm galvanized mesh birdscreen £
900	600	No.	36·72	50·49
900	900	,,	55·08	75·74
900	1200	,,	50·49	78·03
900	1500	,,	63·11	97·54
900	1800	,,	75·74	117·05
900	2100	,,	88·36	136·55
900	2400	,,	100·98	156·06
900	2700	,,	113·61	175·57
900	3000	,,	126·23	195·07
1200	600	,,	48·96	67·32
1200	900	,,	50·49	78·03
1200	1200	,,	67·32	104·04
1200	1500	,,	84·15	130·05
1200	1800	,,	100·98	156·06
1200	2100	,,	117·81	182·07
1200	2400	,,	134·64	208·08
1500	600	,,	61·20	84·15
1500	900	,,	63·11	97·54
1500	1200	,,	84·15	130·05
1500	1500	,,	105·20	162·56
1500	1800	,,	126·23	195·08
1500	2000	,,	140·25	216·75
2000	600	,,	56·10	86·70
2000	900	,,	84·15	130·05
2000	1200	,,	112·20	173·40
2000	1500	,,	140·25	216·75

RADIATORS

PRESSED STEEL PANEL TYPE RADIATORS
(including tappings and brackets)

Height mm	Length mm	Unit	Single Surface area m^2	£	Double Surface area m^2	£
305	800	No.	0·61	5·45	1·23	11·85
305	1120	,,	0·84	7·52	1·67	16·00
305	1440	,,	1·06	9·59	2·12	20·14
305	1760	,,	1·28	11·67	2·56	24·29
305	2080	,,	1·50	13·74	3·01	28·43
305	2400	,,	1·62	15·82	3·23	32·59
305	2720	,,	1·78	17·89	3·57	36·74
432	480	,,	0·40	4·23	0·80	9·41
432	800	,,	0·80	6·87	1·60	14·69
432	1280	,,	1·20	10·83	2·40	22·61
432	1600	,,	1·60	13·48	3·20	27·91
432	2080	,,	2·0	17·44	3·99	35·83
432	2560	,,	2·48	21·40	4·95	43·76
432	2880	,,	2·80	24·04	5·59	49·04
432	3200	,,	3·12	26·68	6·23	54·32
584	480	,,	0·53	5·52	1·06	11·99
584	800	,,	1·06	9·02	2·12	18·99
584	1280	,,	1·59	14·27	3·18	29·49
584	1600	,,	2·12	17·78	4·24	36·51
584	2080	,,	2·65	23·03	5·30	47·01
584	2560	,,	3·28	28·28	6·57	57·52
584	2880	,,	3·71	31·78	7·41	64·52
584	3200	,,	4·13	35·29	8·26	71·53
740	480	,,	0·66	6·81	1·32	14·57
740	800	,,	1·32	11·17	2·64	23·30
740	1280	,,	1·98	17·71	3·96	36·38
740	1600	,,	2·64	22·07	5·28	45·10
740	2080	,,	3·30	28·62	6·60	58·19
740	2560	,,	4·09	35·16		
740	2880	,,	4·62	39·53		
740	3200	,,	5·14	43·88		

SILL LINE NATURAL CONVECTORS

	Unit	406 mm high £	508 mm high £	610 mm high £	711 mm high £
Front panel with sloping front and damper	m	12·80	11·25	12·76	13·45
Inside corner	No.	7·47	7·63	9·05	8·63
Outside corner	,,	7·90	8·01	9·54	8·83
Valve box	,,	6·26	6·34	6·97	6·97
Stop end	,,	3·66	3·87	4·40	4·90
Front panel with flat front and top outlet	m	12·00	12·76	13·45	15·35
Inside corner	No.	8·54	8·83	9·05	9·23
Outside corner	,,	8·74	9·05	9·54	9·79
Valve box	,,	6·54	6·66	6·97	7·29
Stop end	,,	3·87	3·97	4·40	4·60

	Unit	Single element £	Double element £
32 mm black steel tube-finned	m	9·81	19·62
35 mm copper tube-finned	,,	11·29	22·58

RADIATORS

SKIRTING RADIATORS
(aluminium finned copper tube element; mild steel casing finished white stove enamelled; including tappings and brackets)

	Unit	600 mm £	900 mm £	1200 mm £	*Length* 1500 mm £	1800 mm £	2100 mm £
Unit assembly	No.		6·55	8·82	10·61	12·22	14·56
Casing only	,,	2·42	3·64	4·45	5·57	6·06	7·56
Element only	,,		3·50	4·69	6·06	7·00	8·26

	Unit	
End piece	No.	0·50
Splicer set 150 mm	,,	1·20
Jointing strip	,,	0·50
Outside corner	,,	0·90
Brackets	Pair	0·50

CONVECTOR HEATING UNITS

STEAM OR HOT WATER HEATING, FLOOR OR WALL TYPE MODEL PAINTED FINISH

	Unit	Height of unit 500 mm £	700 mm £
Unit 130 mm deep, single tube			
Length of unit 400 mm	No.	33·85	34·49
500 mm	,,	35·00	36·15
600 mm	,,	37·11	38·10
700 mm	,,	39·12	40·20
800 mm	,,	41·72	42·62
900 mm	,,	43·80	44·73
1000 mm	,,	45·72	46·75
1200 mm	,,	49·77	50·85
1400 mm	,,	53·96	54·85
Unit 130 mm deep, two tubes			
Length of unit 1200 mm	,,	58·90	60·90
1400 mm	,,	65·04	67·10

	Unit	£
Extra on convector heating unit for		
Air vent kit operating key	No.	3·25
Feet for conversion to floor mounting	Pair	2·19

UNIT HEATERS

The following is a brief selection of the many types of unit heaters and typical performance values available

RECIRCULATING, HORIZONTAL DISCHARGE TYPE HEATERS
(with individually adjustable louvres, elements of standard steel construction and totally enclosed normal speed (1400 r.p.m.) electric motor. Switchgear not included)

STEAM HEATING
(air entering at 15 °C)

Approximate size and weight of unit	Air m³/s	Steam at 1 bar Output kW	Temperature rise °C	Steam at 4 bar Output kW	Temperature rise °C	Unit No.	£
479 × 479 × 558 mm – 34 kg	0·40	11·45	22	14·95	29	,,	122·07
546 × 546 × 558 mm – 41 kg	0·80	19·00	19	24·75	25	,,	132·76
613 × 613 × 558 mm – 46 kg	1·20	26·50	19	34·65	24	,,	144·13
679 × 679 × 558 mm – 57 kg	2·00	37·50	15	48·75	20	,,	161·94
780 × 780 × 558 mm – 64 kg	2·05	50·30	20	65·60	26	,,	179·83

HOT WATER HEATING
(with a flow temperature of 80 °C, air entering at 15 °C)

Approximate size and weight of unit	Air m³/s	10 °C temperature drop Output kW	Temperature rise °C		£
479 × 479 × 558 mm – 37 kg	0·33	8·95	22	,,	125·87
546 × 546 × 558 mm – 43 kg	0·70	16·50	20	,,	138·07
613 × 613 × 558 mm – 50 kg	0·95	24·85	21	,,	154·75
679 × 679 × 558 mm – 62 kg	1·60	36·50	19	,,	174·74
780 × 780 × 558 mm – 71 kg	1·95	51·60	22	,,	200·47

HORIZONTAL DISCHARGE TYPE HEATERS
(as above but for connection to fresh air duct inlets)

STEAM HEATING
(air entering at −1 °C)

Approximate size and weight of unit	Air m³/s	Steam at 1 bar Output kW	Temperature rise °C	Steam at 4 bar Output kW	Temperature rise °C		£
479 × 479 × 558 mm – 34 kg	0·40	13·20	24	16·70	31	,,	124·63
546 × 546 × 558 mm – 41 kg	0·80	21·90	21	27·70	26	,,	135·90
613 × 613 × 558 mm – 46 kg	1·20	30·60	20	38·70	25	,,	147·68
679 × 679 × 558 mm – 57 kg	2·00	43·20	17	54·50	22	,,	165·78
780 × 780 × 558 mm – 64 kg	2·04	57·95	22	73·30	28	,,	183·97

UNIT HEATERS

HOT WATER HEATING
(with a flow temperature of 80 °C, air entering at −1 °C)

		10 °C temperature drop			
Approximate size and weight of unit	Air m³/s	Output kW	Temperature rise °C	Unit	£
479 × 479 × 558 mm − 37 kg	0·33	11·10	26	No.	128·43
546 × 546 × 558 mm − 43 kg	0·70	20·30	23	,,	141·21
613 × 613 × 558 mm − 50 kg	0·95	30·50	25	,,	158·30
679 × 679 × 558 mm − 62 kg	1·60	44·85	22	,,	178·58
780 × 780 × 558 mm − 71 kg	1·95	64·00	25	,,	204·61

RECIRCULATING DOWNWARD DISCHARGE TYPE HEATERS
(with adjustable outlet, elements of standard steel construction and protected normal speed (1400 r.p.m.) electric motor. Switchgear not included)

STEAM HEATING
(air entering at 15 °C)

		Steam at 1 bar		Steam at 4 bar			
Approximate size and weight of unit	Air m³/s	Output kW	Temperature rise °C	Output kW	Temperature rise °C	Unit	£
479 × 479 × 558 mm − 34 kg	0·40	10·90	22	14·20	29	No.	117·73
546 × 546 × 558 mm − 41 kg	0·75	18·00	19	23·50	25	,,	129·99
613 × 613 × 558 mm − 46 kg	1·10	25·25	19	32·90	24	,,	139·93
679 × 679 × 558 mm − 57 kg	1·80	35·50	16	46·33	21	,,	160·88
780 × 780 × 558 mm − 64 kg	1·90	47·80	20	62·33	26	,,	182·42

HOT WATER HEATING
(with a flow temperature of 80 °C, air entering at 15 °C)

		10 °C temperature drop			
Approximate size and weight of unit	Air m³/s	Output kW	Temperature rise °C		£
479 × 479 × 558 mm − 37 kg	0·33	8·95	22	,,	121·53
546 × 546 × 558 mm − 43 kg	0·70	16·50	20	,,	135·30
613 × 613 × 558 mm − 50 kg	0·95	24·85	21	,,	150·55
679 × 679 × 558 mm − 62 kg	1·60	36·50	19	,,	173·68
780 × 780 × 558 mm − 71 kg	1·95	51·60	22	,,	203·06

FIRE FIGHTING APPLIANCES

DRY RISING MAIN

Note: for tubing see earlier section
Bronze or gunmetal inlet breeching for pumping-in with 64 mm diameter male instantaneous coupling with cup and chain and 25 mm drain valve, screwed or flanged to B.S. table D

	Unit	£
Single inlet.	No.	71·20
Double inlet with back pressure valve	,,	102·91
Quadruple inlet with back pressure valve	,,	250·15

Steel inlet box for building in to external wall, with hinged wired glazed door suitably lettered

305 × 305 × 305 mm for single inlet	,,	54·20
610 × 406 × 305 mm for double inlet	,,	56·00
610 × 610 × 356 mm for quadruple inlet	,,	65·40

Bronze or gunmetal gate type outlet valve with 64 mm diameter female instantaneous coupling with cup and chain, inlet flanged or screwed. Wheelhead secured by padlock and leather strap

Screwed 64 mm male inlet	,,	81·51
Flanged B.S. table D inlet	,,	81·51

Bronze or gunmetal landing type outlet valve as above

Horizontal screwed 64 mm male inlet	,,	57·55
Horizontal flanged B.S. table D inlet	,,	64·35
Oblique screwed 64 mm male inlet	,,	51·60
Oblique flanged B.S. table D inlet	,,	58·15
Air valve screwed 20 mm or 25 mm.	,,	8·80

Sprinkler head pendant type with serrated deflector plate and 15 mm screwed connection

Standard finish	,,	2·66
Chromium plated	,,	2·90

FOAM INLET EQUIPMENT

Note: for tubing see earlier section

76 mm tapered inlet with screwed female outlet	,,	35·10
65 mm diameter gunmetal instantaneous foam inlet with 76 mm female threaded end.	,,	17·60
Cast iron foam spreaders screwed 76 mm female	,,	18·15

HOSE REELS

Automatic non-swing first aid pattern pedestal mounted reel fitted with 37 m of 20 mm rubber hose for working pressure up to 100 p.s.i.	,,	127·00
As above but swing pattern	,,	120·00
Automatic recess swing square cover pattern	,,	170·90
Fire brigade quality rubber lined canvas hose, 64 mm bore 12 ply, treated to prevent mildew and rot with and including two 64 mm diameter instantaneous gunmetal couplings	30 m length	112·04
Wrought iron cradle to take 30 m of 64 mm diameter hose	No.	68·00
Gunmetal branch pipe with 64 mm nozzle	,,	42·35

FIRE FIGHTING APPLIANCES

FIRE EXTINGUISHERS

	Unit	£
Water carbon dioxide type extinguisher to B.S. 1382		
4·6 litres size	No.	31·00
9 litres size	,,	31·00
Foam gas pressure type to B.S. 740, 9 litres size	,,	41·50
Carbon dioxide gas trigger type, 3 kg size	,,	59·00
Bromochlorodifluoromethane (B.C.F.), lever operated		
0·7 kg size	,,	20·40
1·5 kg size	,,	28·90
2·5 kg size	,,	43·30
5·0 kg size	,,	104.30
Dry powder type, 3 kg size	,,	29·80
Asbestos fire blanket in metal container		
914 × 914 mm	,,	14·60
1219 × 1219 mm	,,	21·88
1524 × 1524 mm	,,	31·80
1829 × 1829	,,	41·22

VALVES, TRAPS, REGULATORS AND GAUGES

COCKS

	Unit	15 mm £	20 mm £	25 mm £	32 mm £	40 mm £	50 mm £	65 mm £	80 mm £	100 mm £
Polished brass crutch head screw down bib cock to B.S. 1010	No.	2·37	3·56	6·98						
As above but with hose union	,,	3·37	5·56	10·25						
Bronze angle pattern draw-off cock ribbed for hose and male screw to B.S. 2879 type A	,,	2·50	2·72	6·12						
As above but with lockshield	,,	2·70	4·11	7·34						
Bronze plug pattern draw-off cock male screw and hose union with loose key	,,	5·72	7·67	9·66						
Bronze gland pattern ditto	,,	6·10	8·21	11·37	22·08	31·18	47·15			
Bronze cock, plug pattern, working pressure for cold services up to 9 bar with screwed ends	,,	6·37	9·02	13·72	21·88	30·18	42·98	119·52	165·38	
Bronze cock, gland pattern, three way, steam and cold services, working pressure up to 10 bar with screwed ends	,,	8·79	12·56	17·31	30·66	36·22	55·63			
Plug cock with screwed ends	,,	4·11	5·72	7·89	11·80	16·80	31·18	78·98	85·35	178·59

VALVES, TRAPS, REGULATORS AND GAUGES

COCKS (*continued*)

	Unit	15 mm £	22 mm £	28 mm £	35 mm £	42 mm £	54 mm £
Brass crutch-head stopcock with capillary joints for copper	No.	2·13	3·13	5·64			
Brass lockshield stopcock with capillary joints for copper	,,	2·61	3·73	6·64			
Gunmetal crutch-head stopcock with capillary joints for copper	,,	3·09	4·62	7·69	12·06	15·31	22·88
Gunmetal lockshield stopcock with capillary joints for copper	,,	3·51	5·22	8·69			
Brass double union stopcock with capillary joints for copper	,,	3·43	4·84	8·61			
Brass double union lockshield stopcock with capillary joints for copper	,,	3·84	5·40	9·63			
Gunmetal double union stopcock with capillary joints for copper	,,	5·60	6·90	12·76	21·28	27·88	43·86
Gunmetal double union lockshield stopcock with capillary joints for copper	,,	5·96	7·46	13·78			
Brass crutch-head easy-clean stopcock with capillary joints for copper	,,	2·73	3·88	6·83			
Brass combined stopcock and draincock with capillary joints for copper	,,	5·90	7·28				
Gunmetal ditto	,,	7·83					
Brass crutch head stopcock with compression joints for copper	,,	4·82	7·04	11·04	19·27	27·49	39·21
Brass lockshield stopcock with compression joints for copper	,,	2·95	4·10	7·56			
Gunmetal crutch-head stopcock with compression joints for copper	,,	3·52	5·43	8·11	14·93	21·34	29·10
Brass crutch-head easy-clean stopcock with compression joints for copper	,,	3·06	4·34	7·76			
Brass combined stopcock and draincock with compression joints for copper	,,	4·74	6·63				
Gunmetal ditto	,,	5·55					
Brass crutch-head stopcock with joints for polythene	,,	3·67	5·55	8·02			
Gunmetal ditto	,,	4·65	7·48	10·08	17·34	25·29	35·68
Brass crutch-head stopcock with compression joints for copper and polythene	,,	4·97	5·94				
Brass combined stopcock and draincock with joints for polythene	,,	6·63					
Brass draw-off coupling with compression joints for copper	,,	1·83					
Gunmetal ditto	,,	2·13	2·70				
Brass draincock	,,	1·50					
Brass lockshield draincock	,,	1·57					
Gunmetal lockshield draincock	,,	1·85					

VALVES, TRAPS, REGULATORS AND GAUGES

GAS VALVES

	Unit	15 mm £	20 mm £	25 mm £	32 mm £	40 mm £	50 mm £	65 mm £	80 mm £	100 mm £
Malleable iron ball type isolating valve, complete with lever, with screwed ends	No.	18·96	18·96	18·96	26·86	24·15	32·19	57·92	64·24	
As above but with ends flanged to B.S. table D	,,					33·76	42·89		73·13	135·65

	Unit	150 mm £
	,,	185·76

GATE VALVES

	Unit	15 mm £								
Bronze wedge disc non-rising stem gate valve, working pressure, saturated steam up to 9 bar, cold services up to 14 bar with screwed ends	,,	4·44	5·85	7·62	11·26	14·45	21·06	39·25	56·94	101·42
As above but working pressure, saturated steam up to 10 bar, cold services up to 21 bar and with ends flanged and drilled to B.S. table F	,,		15·44	20·60	26·29	32·84	44·27	67·38	84·24	
Cast iron wedge disc non-rising stem gate valve, working pressure, saturated steam up to 7 bar, cold services up to 12 bar with screwed ends	,,					27·46	30·02	34·60	46·21	

	Unit	40 mm £	50 mm £	65 mm £	80 mm £	100 mm £	125 mm £	150 mm £
As above but with ends flanged and drilled to B.S. table D or E	,,	27·66	27·66	30·08	34·74	46·21	65·84	76·30

	Unit			25 mm £	32 mm £	40 mm £	50 mm £			
Bronze parallel slide valve, working pressure, saturated steam up to 17 bar, cold services up to 28 bar with screwed ends	,,	22·15	23·65	35·50	54·65	63·70	80·90			
As above with ends flanged and drilled to B.S. table F	,,	30·15	32·80	44·70	62·90	74·85	98·20			

VALVES, TRAPS, REGULATORS AND GAUGES

GATE VALVES *continued*

	Unit	50 mm £	65 mm £	80 mm £	100 mm £	125 mm £	150 mm £
Cast iron parallel slide valve, working pressure, saturated steam up to 10 bar, cold services up to 17 bar with ends flanged and drilled to B.S. table F	No.	99·15	119·25	145·40	186·25	254·40	317·95

	Unit	15 mm £	22 mm £	28 mm £	35 mm £	42 mm £	54 mm £	67 mm £	76 mm £
Brass gate valve, working pressure cold services up to 14 bar with compression joints for copper	,,	7·06	9·89	13·75	19·86	26·23	41·05	76·44	107·19
Gunmetal fullway gate valve with capillary joints for copper	,,	3·32	3·93	5·36	11·97	14·30	20·73		
Ditto but lockshield pattern	,,	3·43	4·08	6·25	12·09	14·30	20·73		

GLOBE VALVES

	Unit	15 mm £	20 mm £	25 mm £	32 mm £	40 mm £	50 mm £	65 mm £	80 mm £
Bronze globe valve renewable disc, working pressure, saturated steam up to 9 bar, cold services up to 14 bar with screwed ends	,,	4·09	5·28	7·82	11·14	14·93	22·58	42·08	61·88
As above but with ends flanged and drilled to B.S. table D or E	,,	11·31	12·86	18·16	22·14	28·76	40·38		
As above but working pressure, saturated steam up to 14 bar, cold services up to 28 bar with ends flanged and drilled to B.S. table H	,,	22·71	24·74	31·75	42·98	51·75	77·29	117·93	153·96

VALVES, TRAPS, REGULATORS AND GAUGES

CHECK VALVES

	Unit	15 mm £	20 mm £	25 mm £	32 mm £	40 mm £	50 mm £	65 mm £	80 mm £
Bronze check valve swing pattern, working pressure for saturated steam up to 9 bar with screwed ends	No.	3·36	4·32	6·21	9·27	11·46	16·39	30·44	45·87
As above but renewable disc pattern, working pressure for saturated steam up to 7 bar, cold services up to 14 bar with ends flanged and drilled to B.S. table D or E	,,	14·13	17·72	22·45	27·87	37·60	54·01	82·25	119·34
As above but working pressure for saturated steam up to 14 bar, cold services up to 28 bar and ends flanged and drilled to B.S. table F or H	,,	21·51	23·26	29·07	39·45	47·00	68·20	104·75	142·61

		50 mm £	65 mm £	80 mm £	100 mm £	125 mm £	150 mm £
Cast iron swing pattern check valve, working pressure for cold services up to 14 bar with ends flanged and drilled to B.S. table D and E	,,	44·05	50·37	54·88	74·81	105·56	117·76
As above but with ends flanged and drilled to B.S. table F	,,	45·34	52·70	57·28	76·60	111·04	123·23

EQUILIBRIUM BALL VALVES

	Unit	32 mm £	40 mm £	50 mm £	65 mm £	80 mm £	100 mm £
Bronze equilibrium ball valve, hydraulic working pressure 10 bar with connection flanged to B.S. table D complete with copper float	No.	100·61	120·23	142·06	186·87	217·68	252·32	365·47

		40 mm £	50 mm £	65 mm £	80 mm £	100 mm £	125 mm £
Cast iron equilibrium ball valve, hydraulic working pressure 9 bar with connections flanged to B.S. table E complete with copper float	,,	96·96	105·00	131·64	182·60	228·50	329·88

VALVES, TRAPS, REGULATORS AND GAUGES

RADIATOR VALVES

		15 mm £	20 mm £	25 mm £	32 mm £
Bronze straight pattern radiator valve, screwed female and male union, wheelhead or lockshield matt finish	Unit No.	3·83	4·84	6·73	9·85
As above but angle pattern	„	3·06	3·90	5·49	
Chromium plated straight pattern radiator valve with compression joint for copper	„	7·56	9·79		
As above but angle pattern	„	5·82	7·56		
Brass wheelhead angle pattern radiator valve with capillary copper inlet and screwed union outlet	„	1·71			
Ditto but chromium plated	„	2·10			
Brass lockshield angle pattern radiator valve with capillary copper outlet and screwed union inlet	„	1·71			
Ditto but chromium plated	„	2·28			
Brass wheelhead angle pattern radiator valve with compression copper inlet and screwed union outlet	„	1·85			
Ditto but chromium plated	„	2·28			
Brass lockshield angle pattern radiator valve with compression copper outlet and screwed union inlet	„	1·85			
Ditto but chromium plated	„	2·10			

		8 mm £	10 mm £	12 mm £
Microbore twin entry brass radiator valve	„	3·59	3·92	
Ditto but chromium plated	„	4·34	4·77	
Microbore single entry brass radiator valve	„	1·64	1·64	1·64
Ditto but chromium plated	„	2·02	2·02	2·02

VALVES, TRAPS, REGULATORS AND GAUGES

SAFETY AND RELIEF VALVES

Minimum and maximum blow off pressure (p.s.i.)

	Unit	5 300 15 mm £	5 300 20 mm £	5 250 25 mm £	7 250 32 mm £	7 200 40 mm £	8 200 50 mm £	10 200 80 mm £
Bronze 'pop' type, top outlet, screwed	No.	15·28	18·95	22·90	29·13	37·86	51·59	96·38
As above but side outlet	,,	16·45	19·06	23·76	30·63	40·17	53·94	106·00
Bronze spring safety valve side outlet, screwed	,,	8·90	10·97	14·80	21·83	26·23	36·57	77·08
As above but open discharge type	,,	6·92	8·63	11·25	15·50	18·54	27·82	61·34

		15 mm £	20 mm £	25 mm £	32 mm £	40 mm £	50 mm £	65 mm £
'National' bronze spring safety valve, screwed end up to 80 mm flanged end up to 81 mm plus.	,,	16·05	20·56	27·70	33·22	39·78	53·28	80·92

	80 mm £	100 mm £
	121·44	296·76

AIR VALVES

	Unit	15 mm £
Automatic air vent for pressures up to 7 bar and 200 degrees Fahrenheit with screwed end	No.	32·39
As above with lock head isolating valve	,,	86·82
As above but for pressures up to 7 bar and 300 degrees Fahrenheit	,,	70·66
As above but for pressures up to 17 bar and 400 degrees Fahrenheit with end flanged to B.S. table H	,,	103·41
Air release valve with capillary joint to copper	,,	0·62

VALVES, TRAPS, REGULATORS AND GAUGES

CONTROL VALVES, THERMOSTATS AND REGULATORS

	Unit	15 mm £	20 mm £	25 mm £	32 mm £	40 mm £	50 mm £	65 mm £	80 mm £	100 mm £
Pressure reducing valve for steam for maximum range of 17 bar and 450 degrees Fahrenheit										
Screwed ends	No.	103·00	108·00	116·00						
Flanged ends to B.S. table H	„			139·00	160·00	189·00	220·00			
Cast iron body butterfly type control valve with two position electrically controlled motor for low pressure hot water with flanged ends complete with counter flanges	„			88·00	90·00	93·00	96·00	100·00	103·00	109·00
Cast iron body three-way motorized control valve for low pressure hot water with flanged ends and drilled to B.S. table F	„		103·00	104·00	106·00	110·00	115·00	148·00	174·00	231·00
Stem type air thermostat for control of motor incorporated in motorized valve	„	41·62								
Room thermostat for control of motor incorporated in motorized valve	„	6·40								
Thermostatic control valve for water or steam with 1·50 m of capillary tubing and phial, working pressure, steam 10 bar, water 17 bar and 400 degrees Fahrenheit										
Screwed ends	„	82·00	83·00	89·00						
Flanged ends to B.S. table H	„		104·00	105·00	119·00	125·00	129·00	191·50		
Extra for										
Mild steel pocket	„	9·45								
Stainless steel pocket	„	15·75								

	Unit	15 mm £	22 mm £	28 mm £	35 mm £	42 mm £	54 mm £	65 mm £
Horne temperature regulator for storage calorifier with screwed connections	No.	72·34	72·93	74·78	99·81	121·27	141·43	
As above but with flanged connections	„	106·15	106·98	114·34	147·99	174·84	202·88	
Temperature regulator for non-storage calorifiers with screwed connections including thermostat pocket and capillary tube	„	137·49	140·47	144·26	150·15	159·80	174·21	
As above but with flanged connections to B.S. table H	„	143·26	155·61	162·44	173·03	190·09	212·79	265·82

VALVES, TRAPS, REGULATORS AND GAUGES

STEAM TRAPS AND STRAINERS

	Unit	15 mm £	20 mm £	25 mm £	32 mm £	40 mm £	50 mm £	65 mm £	80 mm £	100 mm £
Cast iron inverted steam trap with pressure range up to 17 bar with screwed ends	No.	24·85	35·70	56·50		104·00	160·00			
As above but with ends flanged to B.S. table H	„	52·50	65·50	96·00		149·00	200·00			
Stainless steel thermodynamic trap with pressure range up to 24 bar and temperature range 550 degrees Fahrenheit with screwed ends	„	17·60	26·65							
As above but with ends flanged to B.S. table H	„	53·50	62·00	71·00						
Malleable iron pipe line strainer, maximum steam working pressure 14 bar with screwed ends	„	3·39	4·52	6·44	9·56	10·85				
Bronze pipe line strainer, maximum steam working pressure up to 17 bar with ends flanged to B.S. table H	„	30·15	35·45	44·70	68·50	81·50	125·00	161·00	203·00	363·00

SIGHT GLASSES

	Unit	15 mm £	20 mm £	25 mm £
Pressed brass, straight, single window sight glass with screwed ends	No.	7·08	7·90	9·88

THERMOMETERS

	Unit	15 mm £
200 mm D.o.E. pattern with 15 mm screwed tail, range 0 degrees to 120 degrees Centigrade (30/240. F)	No.	16·61
Vertical aluminium pressed case combined altitude gauge and thermometer graduated 10 to 120 degrees Centigrade and 15 mm head, screwed tail	„	19·27
Horizontal ditto	„	15·50

INSULATION

INSULATING JACKETS

50 mm nominal thickness glass-fibre flexible insulating sets for mild steel cisterns, finished externally with black polythene sheeting with bands and clips for fixing, for
Cistern, size mm

	Unit	£
460 × 310 × 310	No.	1·48
610 × 310 × 380	,,	1·97
610 × 610 × 310	,,	2·59
663 × 510 × 430	,,	2·68
663 × 510 × 480	,,	3·20
690 × 510 × 510	,,	3·20
740 × 560 × 560	,,	4·00
760 × 580 × 610	,,	4·20
970 × 690 × 690	,,	5·00
1071 × 690 × 652	,,	6·18
1220 × 610 × 610	,,	6·18

80 mm nominal thickness glass-fibre flexible insulating jackets for 'Fortic' cylinders, finished externally with flame retardant p.v.c. covering with bands and clips for fixing, for
Cylinders, size mm
Height × Diameter

900 × 440	,,	5·73
900 × 450	,,	6·38
1075 × 450	,,	7·36
1200 × 450	,,	9·65
900 × 500	,,	6·90
1200 × 500	,,	10·78
1300 × 500	,,	13·47
1400 × 500	,,	13·47

80 mm nominal thickness glass-fibre flexible insulating jackets, segmental type, for cylinders, finished externally with flame retardant p.v.c. covering with bands and clips for fixing, for
Cylinders, size mm
Height × Diameter

750 × 450	,,	4·61
900 × 400	,,	4·25
900 × 450	,,	4·25
975 × 400	,,	5·58
1050 × 400	,,	4·75
1050 × 450	,,	4·75
1200 × 400	,,	5·62
1200 × 450	,,	5·62
1200 × 500	,,	9·33
1200 × 600	,,	11·30
1350 × 450	,,	9·97
1350 × 500	,,	12·22
1500 × 500	,,	15·05
1500 × 600	,,	18·98
1800 × 600	,,	22·78

CAST IRON PIPES AND FITTINGS

CAST IRON SOIL, WASTE AND VENTILATING PIPES AND FITTINGS (to B.S. 416)

	Unit	50 mm £	75 mm £	100 mm £	150 mm £
Pipe 1·83 m length	No.	7·30	8·25	11·31	22·60
Bend	,,	3·28	4·30	6·32	10·86
Bend with access	,,	7·63	8·28	10·75	15·06
Single plain branch	,,	4·27	6·13	9·28	17·28
Ditto with access	,,	8·53	9·99	13·69	21·53
Double plain branch	,,	6·43	10·59	12·30	26·67
Ditto with access	,,	11·08	14·84	16·76	32·02
Parallel branch	,,		6·18	9·68	21·49
'Y' or breeches branch	,,	5·93	7·27	9·83	25·08
Ditto with access	,,	11·04	12·94	15·15	31·81
Single anti-syphon branch	,,	5·26	7·37	9·68	
Double anti-syphon branch	,,		9·36	10·52	24·74
Access pipe – oval door	,,	7·53	8·35	9·49	15·02
Shoes	,,	4·22	5·10	6·82	13·04
Shoes eared	,,	4·36	5·19	6·86	13·75
Socket reducer	,,		2·04	2·86	5·58
Socket plug	,,	1·21	1·35	1·61	2·69
Diminishing piece	,,		4·64	6·57	12·90
Offset					
75 mm projection	,,	2·62	3·57	5·68	11·96
150 mm projection	,,	3·71	4·78	6·80	13·28
225 mm projection	,,	4·20	5·78	7·76	15·97
300 mm projection	,,	5·36	6·57	8·75	20·53
450 mm projection	,,		11·47	16·06	
600 mm projection	,,	14·74	16·45	24·12	27·70
Boss pipe with 1 No. Boss	,,	8·10	8·70	10·58	16·75
Ditto with 2 No. bosses	,,	10·24	11·26	13·05	30·31
Plain P trap	,,	6·99	7·72	9·64	19·36
Access P trap	,,	11·14	11·77	13·79	25·39
Plain S or Q trap	,,	7·96	8·06	9·78	23·57
Access S or Q trap	,,	12·01	12·01	13·69	26·88
Vent pipe roof connector	,,	6·82	7·71	9·89	20·56
W.C. connectors					
150 mm long	,,			3·31	
225 mm long	,,			3·80	
300 mm long	,,			4·40	
450 mm long	,,			5·49	
600 mm long	,,			6·57	
W.C. connectors with anti-syphon arm					
150 mm long	,,			6·11	
225 mm long	,,			5·40	
300 mm long	,,			7·30	
450 mm long	,,			8·51	
600 mm long	,,			9·65	
Bent W.C. connectors with					
150 mm tail	,,			8·32	
450 mm tail	,,			11·00	
600 mm tail	,,			14·22	
Long tail bend	,,			16·82	
Long arm branch	,,			17·01	

U.P.V.C. PIPES AND FITTINGS

U.P.V.C. SOIL, WASTE AND VENTILATING PIPES AND FITTINGS
(to B.S. 4514: 1969 with solvent weld joints)

	Unit	Nominal bore 75 mm £	100 mm £	150 mm £
Pipe 4·00 m length	No.	7·68	9·12	19·28
Bend				
92°	,,	2·05	2·80	6·05
135°	,,	2·05	2·80	5·72
104°	,,		3·15	
112°	,,		3·15	
variable	,,		3·15	
Bend 92° with access	,,		6·29	
Single branch				
92°	,,	2·86	3·60	7·93
104°	,,	3·36	4·57	
135°	,,	3·18	4·57	13·89
Reducing branch				
92°	,,			8·87
104°	,,			10·26
135°	,,			8·87
Single branch 92° with access	,,		7·20	
Double branch				
92°	,,		8·67	13·54
104°	,,		9·07	15·11
135°	,,		9·07	
Straight expansion coupling	,,	0·84	1·03	3·01
W.C. connecting bend	,,	1·61	2·24	
W.C. connector	,,	2·28	2·00	
Access door	,,	2·48	2·48	4·45
P.V.C. to pitch fibre adaptor	,,	1·86	1·50	2·58
P.V.C. caulking bush	,,	1·75	1·75	
Metal caulking bush	,,	2·60	4·78	5·28
Bossed pipe with 35 mm boss	,,		0·99	
Ditto with 42 mm boss	,,		1·08	
Weathering apron	,,	0·57	0·68	2·04
Weathering slate	,,	9·48	9·48	
Vent cowl	,,	0·57	0·60	1·52

U.P.V.C. PIPES AND FITTINGS

U.P.V.C. SOIL, WASTE AND VENTILATING PIPES AND FITTINGS
(to B.S. 4514: 1969 with ring seal joints)

	Unit	75 mm £	100 mm £	150 mm £
Pipe 3·00 m length	No.	9·41	11·77	24·51
Bend				
92°	,,	1·86	2·61	6·38
135°	,,	1·24		
variable	,,	1·86	2·61	9·90
Bend 92° with access	,,		5·45	
Single branch				
92°	,,		3·25	8·08
104°	,,		3·25	8·08
135°	,,	2·32	2·78	6·23
adjustable	,,	2·48		
Single branch 92° with access	,,		6·98	
Single branch 104° with access	,,		4·68	
Double branch	,,		8·53	
Bossed pipe with 35 mm boss	,,		1·56	
Ditto with 42 mm boss	,,		1·04	
Access pipe	,,	3·76	4·04	
Access cap	,,	1·77	2·02	6·00
W.C. connector	,,	0·77	1·70	
Metal caulking bush	,,	2·33	3·09	5·75
Weathering apron	,,		0·68	
Vent cowl	,,	0·46	0·57	1·54

HIGH TEMPERATURE P.V.C. WASTE PIPE AND FITTINGS
(with solvent welded joints)

	Unit	32 mm £	38 mm £	50 mm £
Pipe 4·00 m length	No.	3·16	3·68	5·37
Bend	,,	0·59	0·62	0·96
Sweep tee				
90°	,,	0·80	0·99	1·86
135°	,,	0·89	1·08	1·96
Cross	,,		2·59	3·21
Union	,,	1·18	1·61	2·38
M I or F I to P.V.C. connector	,,	0·50	0·59	0·86
Access cap	,,	1·11	1·31	2·00
Vent cowl	,,			0·57
Level invert taper	,,	0·50	0·70	0·65

POLYPROPYLENE WASTE PIPE AND FITTINGS
(with ring seal joints)

	Unit			
Pipe 4·00 m length	No.	2·28	2·86	6·15
Bend	,,	0·67	0·73	1·02
Sweep tee				
90°	,,	0·82	1·01	1·92
112°	,,	0·89	1·08	1·96
135°	,,	0·89	1·08	1·96
M.I. to polypropylene connector	,,	0·50	0·59	0·86
Caulking bush	,,	0·46	0·62	0·70
Access cap	,,	0·29	0·29	0·99
Level invert taper	,,	0·50	0·70	0·65

U.P.V.C. PIPES AND FITTINGS

U.P.V.C. VENTILATING AND OVERFLOW PIPES AND FITTINGS

	Unit	Nominal bore 16 mm £	22 mm £
Pipe 4·00 m length.	No.	1·39	1·47
Bend	,,	0·22	0·36
Adjustable bend	,,	0·43	0·38
Branch	,,	0·23	0·27
Tank connector	,,		0·43
Bent tank connector	,,		0·48
Female iron to P.V.C. connector	,,		0·19

POLYPROPYLENE TRAPS

	Unit	32 mm £	Nominal size 38 mm £	50 mm £
Tubular P trap	No.	1·19	1·30	
Tubular S trap	,,	1·35	1·69	
Tubular running P trap	,,	1·92	1·97	4·62
Tubular running S trap	,,	2·20	2·38	
Bottle P trap	,,	1·19	1·38	
Bottle S trap	,,	1·57	1·84	
Combined bath trap and overflow	,,		4·37	

MECHANICAL AND PLUMBING INSTALLATIONS

Constants of Labour

'Constants' of labour are given for the major items of work for which prices are given in 'Prices for Measured Work' except where it has been assumed that the work would be carried out by sub-contractors.

MECHANICAL INSTALLATIONS

'Prices for Measured Work' have been based on the quantities of materials required at the rates given in 'Market Prices of Materials' plus 10% to cover overheads and 5% for profit, and the 'Constants of Labour' priced at a man hour rate of £5·22, this rate including an allowance of 40% to cover site and head office overheads and preliminary items and 5% for profit.

PLUMBING INSTALLATIONS

'Prices for Measured Work' have been based in a similar manner to those for MECHANICAL INSTALLATIONS except that the 'Constants of Labour' are priced at a man hour rate of £5·17.

Reference to the 'Constants' should assist the reader.

(1) To compare the prices given to those used in his own organization
(2) To calculate the effect of changes in wage rates
(3) To calculate analogous price for work similar to but differing in detail from the examples given.

STEEL TUBING AND FITTINGS

STEEL TUBING WITH SCREWED AND SOCKETED JOINTS
man hours per 100 m

	15 mm	20 mm	25 mm	32 mm	40 mm	50 mm
Tubing	52·50	54·00	59·00	66·00	72·00	78·00

	65 mm	80 mm	100 mm	125 mm	150 mm
Tubing	88·00	98·00	132·00	152·00	170·00

FITTINGS WITH SCREWED JOINTS
man hours per 100 fittings

	15 mm	20 mm	25 mm	32 mm	40 mm	50 mm
Elbows, bends, sockets and unions	66·00	88·00	100·00	116·00	134·00	160·00
Tees, pitcher tees and twin elbows	95·00	127·20	144·50	167·60	193·60	231·20
Single flange tables D, E, F and H	36·60	48·80	55·50	64·40	74·40	88·80

	65 mm	80 mm	100 mm	125 mm	150 mm
Elbows, bends, sockets and unions	196·00	232·00	320·00	460·00	600·00
Tees, pitcher tees and twin elbows	283·40	335·30	462·50	688·00	911·00
Single flange tables D, E, F and H	109·00	128·70	177·50	255·00	332·00

BOLTED CONNECTIONS BETWEEN FLANGES
man hours per 100 connections

Number of bolts	15 mm	20 mm	25 mm	32 mm	40 mm	50 mm
4	50·00	50·00	50·00	50·00	50·00	50·00

	65 mm	80 mm	100 mm	125 mm	150 mm
4	50·00	50·00	50·00		
8	88·00	88·00	88·00	88·00	88·00
12					152·00

STEEL TUBING WITH WELDED JOINTS
man hours per 100 m

	15 mm	20 mm	25 mm	32 mm	40 mm	50 mm
Tubing	54·60	57·00	62·80	72·00	80·00	92·00

	65 mm	80 mm	100 mm	125 mm	150 mm
Tubing	107·00	122·00	157·50	178·00	198·50

NOTE: *For pipe fixings see page 109*

Mechanical Installations – Constants of Labour

STEEL TUBING AND FITTINGS

FITTINGS WITH WELDED JOINTS
man hours per 100 fittings

	15 mm	20 mm	25 mm	32 mm	40 mm	50 mm
Elbow or return bend	51·60	68·80	85·00	106·60	133·40	186·00
Equal or reducing tees			122·50	154·00	193·40	286·00
Cap			47·50	59·10	73·40	108·00
Single flange tables D, E, F and H	29·70	38·80	47·50	59·10	73·40	108·00

	65 mm	80 mm	100 mm	125 mm	150 mm
Elbow or return bend	259·60	323·20	392·00	466·00	540·00
Equal or reducing tees	332·00	378·00	468·00	624·00	780·00
Cap	125·00	141·70	177·60	239·00	300·00
Single flange tables D, E, F and H	125·00	141·70	177·60	239·00	300·00

COPPER TUBING AND FITTINGS

COPPER TUBING TO B.S. 2871 WITH CAPILLARY JOINTS
man hours per 100 m

	12 mm	15 mm	22 mm	28 mm	35 mm	42 mm	54 mm	67 mm
Table X	41·22	41·22	42·66	46·74	52·14	57·20	62·32	71·82
Table Y		41·22	42·66	46·74	52·14	54·14	68·86	80·16
Table Z		43·94	45·58	49·88	56·50	61·76	67·22	80·16

COPPER TUBING TO B.S. 2871 WITH COMPRESSION JOINTS
man hours per 100 m

	15 mm	22 mm	28 mm	35 mm	42 mm	54 mm	67 mm	76 mm	108 mm
Table X	41·00	42·50	46·30	51·80	57·00	62·46	74·06	85·60	108·76
Table Y	41·08	42·52	46·42	51·94	57·14	67·56	81·74	95·86	121·26
Table Z	43·68	45·30	49·60	56·02	61·62	67·56	79·98	92·46	117·46

NOTE: *For pipe fixings see page 109*

COPPER TUBING AND FITTINGS

FITTINGS WITH COMPRESSION TYPE JOINTS
man hours per 100 fittings

	15 mm	22 mm	28 mm	35 mm	42 mm	54 mm
Reducing couplings, elbows, unions or return bends, copper to copper	18·30	21·70	25·00	30·00	35·00	40·00
Tank connectors	21·40	25·40	29·30	35·10	41·00	46·80
Adaptors or elbows, copper to iron	38·50	48·50	55·00	65·00	75·00	88·00
Stop ends	14·00	14·70	16·70	22·00	23·00	27·50
Equal reducing or swept tee	28·50	30·00	34·00	45·00	46·50	56·00

	67 mm	76 mm	108 mm	133 mm	159 mm
Reducing couplings, elbows, unions or return bends, copper to copper	58·40	77·83	87·20	102·74	118·30
Adaptors or elbows, copper to iron	114·00	139·50	181·50	248·50	316·00
Equal reducing or swept tee	86·00	115·50	127·00	140·00	165·50

FITTINGS WITH CAPILLARY TYPE JOINTS
man hours per 100 fittings

	12 mm	15 mm	22 mm	28 mm	35 mm	42 mm	54 mm	67 mm	76 mm	108 mm
Reducing couplings, elbows, unions or return bends, copper to copper	25·00	25·00	28·30	33·30	36·70	41·70	45·00	56·00	65·00	71·00
Tank connectors		31·70	35·40	41·60	45·80	50·80				
Adaptors or elbows, copper to iron		20·00	23·30	25·00	28·30	31·70	33·30	41·50	48·00	52·50
Tap connector or stop end	12·50	12·50	14·20	16·70	18·30	20·80	22·50	25·00	31·00	
Back plate elbow		50·00	53·30	59·00						
Equal, reducing or swept tee	36·70	36·70	40·00	44·50	58·70	61·70	66·70	83·30	108·00	115·00

NOTE: *For pipe fixings see page 109*

PIPE FIXINGS

PIPE FIXINGS FOR COPPER TUBE
man hours per 100

	15 mm	22 mm	28 mm	35 mm	42 mm	54 mm	67 mm	76 mm	108 mm
Saddle band screwed to wood	19·40	20·40	21·00	21·80	22·60	24·00	47·00	52·60	57·40
Single spacing pipe clip screwed to wood	19·40	20·40	21·00						
Two piece spacing clip screwed to wood	19·40	20·40	21·00	21·80	22·60	24·00			
Single pipe bracket screwed to wood	19·40	20·40	21·00						
Single pipe bracket plugged and screwed	36·60	37·60	38·40	39·20	41·00	41·40			

	15 mm	22 mm	28 mm	35 mm	42 mm	54 mm	67 mm	76 mm	108 mm
Backplate plugged and screwed	36·60	37·60	38·40	39·20	40·00	40·20	47·00	52·60	57·40

PIPE FIXINGS FOR STEEL TUBE
man hours per 100

	15 mm	20 mm	25 mm	32 mm	40 mm	50 mm
Single pipe bracket screwed to wood	19·40	20·40	21·00	21·80	22·60	24·00
Single pipe bracket plugged and screwed	36·60	37·60	38·40	39·20	40·00	41·40
Single pipe ring and fixing to threaded end	11·40	12·40	13·00	14·00	14·80	16·00

	65 mm	80 mm	100 mm	125 mm	150 mm
Single pipe bracket screwed to wood	29·60	35·20	40·00		
Single pipe bracket plugged and screwed	47·20	52·60	57·40	64·20	71·00
Single pipe ring and fixing to threaded end	29·60	35·20	40·00	50·00	60·00

	Up to 50 mm	Up to 80 mm	Up to 100 mm	Up to 150 mm
Pipe roller and chair bolted to support	48·00	70·00	80·00	108·00

	15 mm	22 mm	28 mm	35 mm	42 mm	54 mm
Single pipe ring and fixing to threaded end	19·40	20·40	21·00	21·80	22·60	24·00

	67 mm	76 mm	108 mm
	30·00	35·20	40·00

EXPANSION JOINTS

STAINLESS STEEL SLEEVED BELLOWS, HINGED ANGULAR OR ARTICULATED TIED TYPE EXPANSION JOINTS
man hours per unit

	15 mm	20 mm	25 mm	32 mm	40 mm	50 mm
Screwed joints	0·96	1·28	1·50	1·76	2·00	2·28
Welded joints	0·82	1·08	1·36	1·70	2·00	2·64
Flanged joints excluding bolted connections	0·36	0·48	0·60	0·72	0·78	0·84

	65 mm	80 mm	100 mm	125 mm	150 mm
Screwed joints	2·84	3·40			
Welded joints	3·50	4·32	5·52	7·00	8·40
Flanged joints excluding bolted connections	1·08	1·32	1·92	2·80	3·60

NOTE: *For bolted connections between flanges see page 106*

Mechanical Installations – Constants of Labour

BOILERS

BOILERS, PLACING IN POSITION ASSEMBLING AS NECESSARY AND COMMISSIONING
man hours per boiler

	Rating in kW (Btu/hr)						
	13·19 (45,000)	16·12 (55,000)	17·59 (60,000)	20·52 (70,000)	23·45 (80,000)	29·32 (100,000)	43·98 (150,000)
Domestic boilers							
Gas fired with conventional flue	8·00	9·00	10·00	10·00	12·00	14·00	17·00
Gas fired but with balanced flue	10·00	11·00	12·00	12·00	14·00	16·00	19·00
Oil fired	9·00	10·00	12·00	12·00	14·00	15·00	20·00
Solid fuel fired	8·00	9·00	10·00	10·00	12·00	14·00	17·00

		Rating in kW				
	147	293	440	586	880	1465
Commercial packaged water boilers						
Gas fired	20	24	32	40	40	50
Oil fired	20	24	32	40	40	50
Solid fuel excluding stoker		20	28	36	40	44

			Rating in kW				
	94	149	219	278	406	476	605	749
Commercial cast iron sectional water boilers								
Gas fired	46·00	70·00	108·00	130·00	194·00	222·00	322·00	422·00
Oil fired	50·00	74·00	112·00	134·00	198·00	226·00	326·00	426·00
Solid fuel	42·00	64·00	98·00	120·00	180·00	208·00	304·00	398·00

			Rating in kW			
	293	880	1172	1760	2345	2930
Packaged steam boilers gas or oil fired	80·00	110·00	120·00	160·00	200·00	200·00

CISTERNS, TANKS AND CYLINDERS

CISTERNS AND CYLINDERS AND HOISTING AND PLACING IN POSITION
man hours each

	SCM 70	SCM 230	SCM 270	B.S. type reference SCM 360	SCM 450	SCM 680	SCM 1600	SCM 2720	SCM 4540
Galvanized mild steel open top cistern	1·50	3·00	4·00	4·50	5·00	6·00	10·00	22·00	44·00

		227	273	364	*Capacity litres* 409 455		568	682	796	1364
Galvanized indirect cylinder		3·00	4·00	4·50	4·50	5·00	5·50	6·00	6·50	8·00

		1	2	*B.S. type reference* 4 5		6	7	8
Copper storage cylinder		1·50	1·50	1·50	1·50	1·50	2·00	2·80

	77	91	109	*Capacity litres* 125 137		159	164	227
Copper indirect cylinder	1·50	1·50	1·50	2·00	2·50	2·80	3·00	3·00

		273	318	*Capacity litres* 364 391		455	682	796
		4·00	4·50	4·50	4·50	5·00	6·00	6·50

		909	1137	*Capacity litres* 1364 1591		1818	2046	2273
		7·00	7·50	8·00	10·00	14·00	16·00	17·00

CALORIFIERS

CALORIFIERS AND FIXING IN POSITION
man hours each

	455	1362	*Capacity litres* 2723 3635		4546	
Storage calorifier copper or galvanized	7·00	9·00	19·00	32·00	48·00	

	88	176	*Capacity kW* 293 586		879	1465
Non-storage heating calorifier cast iron, horizontal or vertical						
steam at 3·2 bar	8·00	12·00	14·00	36·00	40·00	44·00
steam at 4·8 bar	5·00	8·00	12·00	22·00	28·00	40·00

PUMPS, CIRCULATORS AND ACCELERATORS

PUMPS, PLACING IN POSITION AND FIXING
man hours each

Pump size and weight (mm/kg)

	25/84	40/138	50/222	80/229	100/434	150/347
Direct drive centrifugal heating pump	9·00	11·00	14·00	15·00	21·80	38·50

	32/21	40/25	50/33	65/42	80/51	100/119
Pipeline mounted circulator	9·00	9·00	10·00	10·00	10·00	11·00

	25/7	30/11	50/29	65/43	89/48
Glandless heating accelerator pump	9·00	9·00	10·00	10·00	10·00

Self-contained automatic sump pump						20·00

FANS

FANS AND FIXING
man hour each

Diameter of fan

	305 mm	483 mm	762 mm	1219 mm
Single stage axial flow fan fixed to duct				
Long casing	16·00	17·00	19·00	28·00
Short casing	15·50	16·50	18·00	26·00
Two stage fan fixed to duct				
Long casing	17·00	18·50	22·00	39·00

	305 mm	457 mm	762 mm	1219 mm
Roof extract unit fixed to prepared kerb	11·00	13·00	19·00	30·00

	152 mm	229 mm	305 mm
Window type fan fixed in prepared opening	2·00	2·00	2·50

RADIATORS

PRESSED STEEL PANEL RADIATORS 914 MM LONG FIXED WITH BRACKETS
man hours per panel

Height of radiators

	305 mm	432 mm	584 mm	737 mm
Single panel	1·40	1·50	1·70	1·80
Double panel	1·60	1·70	1·90	2·00

CONVECTOR HEATING UNITS

CONVECTORS, WALL OR FLOOR TYPE, ASSEMBLING AND FIXING CASING AND CONNECTING TO MAINS
man hours per unit

	Wall type	Floor type	Height of unit 508 mm / 610 mm	660 mm / 762 mm
Unit 130 mm deep single tube				
Length of unit				
400 mm			2·10	2·20
500 mm			2·22	2·32
600 mm			2·30	2·40
700 mm			2·42	2·52
1100 mm			2·48	
1200 mm			2·52	2·62
1300 mm			2·56	2·66
Unit 130 mm deep two tubes				
Length of unit				
1200 mm			2·62	2·72
1400 mm			2·66	2·76

UNIT HEATERS

UNIT HEATERS FIXED TO SUSPENSION RODS AND CONNECTED TO HEATING SERVICES
man hours per unit

Approximate weight of unit

	34 kg	41 kg	45 kg	57 kg	64 kg
Horizontal discharge type	6·00	7·00	8·00	8·50	9·00

	34 kg	41 kg	48 kg	59 kg	68 kg
Downward discharge type	6·00	7·00	8·00	8·50	9·00

VALVES

VALVES AND JOINTS
man hours each

	15 mm	20 mm	25 mm	32 mm	40 mm	50 mm	80 mm
Cocks							
Draw-off cock with or without hose union	0·24	0·34	0·40				
Screwed stop cock crutch head	0·84	1·14	1·30	1·50	1·74	2·08	3·00
Stop cock with compression joints to copper	0·46	0·52	0·60	0·74	0·84	0·98	
As above but with capillary joints	0·46	0·54	0·64				
Plug cock with screwed joints	0·34	0·44	0·50	0·60	0·66	0·82	0·94
Plug cock with flanged ends	0·24	0·34	0·40	0·46	0·54	0·74	0·92
Three way cock with screwed joints	1·14	1·54	1·76	2·02	2·34	2·80	
Gate valves							
Gate valve with flanged joints	0·24	0·34	0·40	0·46	0·54	0·64	0·92
Gate valves with capillary joints	0·46	0·54	0·64	0·78	0·84	0·96	
Parallel slide valve with screwed joints	0·84	1·14	1·30	1·50	1·74	2·08	
As above but with flanged joints	0·24	0·34	0·40	0·46	0·54	0·64	
Safety valves							
Gunmetal safety valves with screwed joints	0·24	0·34	0·40	0·46	0·60	0·72	1·06
Air valves							
Automatic air vent with screwed joints	0·88						
Check valves							
Swing pattern check valve with screwed joints	0·84	1·14	1·30	1·50	1·74	2·08	3·00
As above but with flanged joints	0·24	0·34	0·40	0·46	0·54	0·64	0·92
Ball valves							
Equilibrium ball valve with screwed joint			0·40	0·46	0·54	0·64	0·92
Radiator valves							
Straight or angle pattern radiator valve with screwed joints	0·84	1·14	1·30	1·50			
Straight pattern valve with compression joints	0·46	0·54					

NOTE: *For bolted connections between flanges see page 106*

CAST IRON PIPE AND FITTINGS

CAST IRON PIPES WITH TYPE 'A' SOCKETS AND CAULKED LEAD JOINTS AND FIXED WITH HOLDERBATS
man hours per metre

	50 mm	75 mm	100 mm	150 mm
Pipe	0·33	0·60	0·75	0·83

CAST IRON FITTINGS WITH CAULKED LEAD JOINTS
man hours per fitting

	50 mm	75 mm	100 mm	150 mm
Bend	0·33	0·60	0·75	0·83
Bend with access	0·41	0·66	0·91	1·00
Bend long tail			1·25	
Single branch	0·50	0·75	1·00	1·25
Single branch with access	0·83	1·16	1·16	1·41
Double branch	0·90	1·15	1·30	1·40
Double branch with access	0·98	1·31	1·46	1·56
Parallel branch		1·00	1·00	1·25
Y or breeches branch		0·75	1·00	1·25
Y or breeches branch with access	0·58	0·81	1·16	1·41
Corner branch		1·00	1·50	
Single anti-syphon branch	0·75	1·00	1·00	
Double anti-syphon branch		0·90	1·50	1·65
Access pipe-oval door	0·41	0·90	1·00	1·25
Shoes	0·25	0·60	0·75	0·82
Diminishing piece	0·25	0·60	0·75	0·83
Offsets				
75 mm projection	0·25	0·60	0·75	0·83
150 mm projection	0·30	0·70	0·85	0·94
225 mm projection	0·30	0·70	0·85	0·94
300 mm projection	0·35	0·80	0·95	1·03
450 mm projection	0·35	0·80	1·00	1·03
600 mm projection	0·35	0·80	1·00	1·08
Boss pipe with 1 or 2 No bosses	0·33	0·60	0·75	0·83
Plain P trap	0·30	0·90	1·00	1·25
Access P trap	0·38	0·96	1·16	1·41
Plain S or Q trap	0·30	0·90	1·00	1·25
Access S or Q trap	0·38	0·96	1·16	1·41
W.C. connectors with or without anti-syphon arm				
150 mm long			0·75	
225 mm long			0·75	
300 mm long			0·75	
450 mm long			1·00	
600 mm long			1·00	

U.P.V.C. PIPE AND FITTINGS

U.P.V.C. PIPES WITH SOLVENT WELD JOINTS
man hours per metre

	32 mm	38 mm	50 mm	75 mm	100 mm	150 mm
Pipe	0·20	0·20	0·23	0·31	0·35	0·45

U.P.V.C. FITTINGS WITH SOLVENT WELD JOINTS
man hours per fitting

	32 mm	38 mm	50 mm	75 mm	100 mm	150 mm
Bend 92°	0·25	0·25	0·30	0·35	0·38	0·58
Bend 135°	0·25	0·25	0·30	0·35	0·38	0·58
Bend with access					0·40	
Single branch	0·30	0·30	0·33	0·41	0·50	0·66
Single branch with access					0·55	
Double branch 92°		0·40	0·41		0·55	0·72
Double branch 104°					0·55	0·72
Straight expansion coupling				0·25	0·25	0·30
W.C. connector				0·25	0·25	
Access door	0·16	0·20	0·20	0·25	0·30	0·30
P.V.C. caulking bush				0·45	0·45	
Metal caulking bush				0·40	0·40	0·45
Bossed pipe with 1 No boss					0·25	
Weathering apron				0·25	0·25	0·30
Weathering slate				1·00	1·00	
Vent cowl				0·30	0·30	0·30
Access cap	0·20	0·20	0·30	0·30	0·35	
Union coupling	0·25	0·25	0·30			
Level invert taper	0·25	0·25	0·25	0·30	0·30	

U.P.V.C. PIPES WITH RING SEAL JOINTS
man hours per metre

	32 mm	38 mm	50 mm	75 mm	100 mm	150 mm
Pipe	0·20	0·20	0·23	0·31	0·35	0·45

U.P.V.C. FITTINGS WITH RING SEAL JOINTS
man hours per fitting

	32 mm	38 mm	50 mm	75 mm	100 mm	150 mm
Bend 92° or 135°	0·25	0·25	0·30	0·35	0·38	0·58
Bend with access				0·34	0·40	0·60
Single branch	0·30	0·30	0·33	0·41	0·50	0·66
Single branch with access					0·45	
Double branch					0·55	
Bossed pipe with 1 No. boss					0·25	
Access pipe				0·35	0·40	
Access cap	0·20	0·20	0·20	0·25	0·30	0·40
W.C. connecting bend				0·25	0·25	
W.C. connector				0·25	0·25	
Metal caulking bush				0·40	0·45	
P.V.C. caulking bush	0·25	0·25	0·30			
Level invert taper	0·25	0·25	0·25			

U.P.V.C. PIPES AND FITTINGS

U.P.V.C. VENTILATING AND OVERFLOW PIPE WITH SOLVENT WELD JOINTS
man hours per metre

	16 mm	22 mm
Pipe	0·04	0·08

U.P.V.C. FITTINGS WITH SOLVENT WELD JOINTS
man hours per fitting

	16 mm	22 mm
Bend	0·04	0·08
Branch	0·06	0·10
Tank connector		0·10
Bent tank connector		0·10
Female iron to P.V.C. connector		0·10

P.V.C. TRAPS
man hours per fitting

	32 mm	38 mm
Tubular P trap	0·20	0·25
Tubular S trap	0·20	0·25
Tubular running P trap	0·20	0·25
Tubular running S trap	0·20	0·25
Bottle P trap	0·20	0·25
Bottle S trap	0·20	0·25
Combined bath trap and overflow		0·50

MECHANICAL AND PLUMBING INSTALLATIONS

Prices for Measured Work

These prices are intended to apply to new work in the London area and include allowances for all overhead charges, preliminary items and profit. The prices are for reasonable quantities of work and the user should make suitable adjustments if the quantities are especially small or especially large. Adjustments may also be required for locality (e.g. outside London) and for the market conditions (e.g. volume of work on hand or on offer) at the time of use.

Prices are given in metric terms. They are based on an inclusive labour rate of £5·22 per man hour for 'Mechanical' work and £5·17 per man hour for 'Plumbing' work and on the prices of materials given in 'Market Prices of Materials' or as indicated.

The labour element includes allowances of 40% to cover site and head office overheads and preliminary items and 5% for profit.

The material element includes allowances of 10% to cover overheads and 5% for profit.

The labour element, inclusive of overheads and profit, in any particular item can be ascertained by deducting the amount which appears in italics below the measured price from the measured price. The amount shown in italics represents the value of material content contained in the measured price including relevant allowances for waste, overheads and profit.

Where work is normally carried out by sub-contractors, such as ductwork and thermal insulation, prices have been based on sub-contractors' estimates with the addition of 15% to cover overheads, profit and any attendance necessary.

For details of the inclusive labour rate and assumptions made in calculating 'Prices for Measured Work' see 'Directions'.

Prices do not allow for any charges in respect of V.A.T.

STEEL TUBING AND FITTINGS

BLACK TUBING AND TUBULARS
(to B.S. 1387 with screwed and socketed joints – fixings given separately)

	Unit	15 mm £	20 mm £	25 mm £	32 mm £	40 mm £	50 mm £
Medium weight tubing	m	3·38 / 0·64	3·59 / 0·77	4·20 / 1·12	4·88 / 1·44	5·51 / 1·70	6·49 / 2·42
Extra over for Bend or spring	No.	2·32 / 0·37	3·04 / 0·49	3·59 / 0·70	4·50 / 1·14	5·19 / 1·30	7·03 / 2·19
Heavy weight tubing	m	3·48 / 0·74	3·71 / 0·89	4·38 / 1·30	5·12 / 1·67	5·79 / 1·98	6·89 / 2·81
Extra over for Bend or spring	No.	2·34 / 0·41	3·06 / 0·51	3·63 / 0·73	4·57 / 1·21	5·26 / 1·38	7·16 / 2·31
Extra over tubing for							
Labour bend	,,	2·09	2·09	2·51	3·34	4·18	4·59
Labour splay cut end	,,	0·73	0·84	0·94	1·31	1·40	1·61
Screwed joint to fitting and the like	,,	1·57	2·09	2·38	2·75	3·18	3·79

	Unit	65 mm £	80 mm £	100 mm £	125 mm £	150 mm £
Medium weight tubing	m	7·78 / 3·19	9·26 / 4·15	12·79 / 5·90	15·55 / 7·62	18·75 / 9·88
Extra over for Bend or spring	No.	10·09 / 4·19	13·18 / 6·46	20·66 / 11·40	53·09 / 42·40	74·23 / 60·45
Heavy weight tubing	m	8·31 / 3·72	9·95 / 4·83	13·77 / 6·88	16·81 / 8·88	20·39 / 11·51
Extra over for Bend or spring	No.	10·33 / 4·43	13·55 / 6·84	21·31 / 12·05	58·47 / 45·18	81·70 / 64·37
Extra over tubing for						
Labour bend	,,	5·22	5·85	16·70		
Labour splay cut end	,,	1·85	2·09	2·51		
Screwed joint to fitting and the like	,,	4·64	5·46	7·60	10·93	14·25

Designs & Estimates

Professionally Prepared

for Full Mechanical Services, Heating Installations, Plumbing Schemes

To Local Authority Works, Hospitals, Schools etc., etc.

- Commercial Properties
- Factories
- Churches
- Blocks of Flats
- Sports Centres
- Offices
- Garages
- Old People's Homes
- Hotels etc., etc.

Nation-Wide Service

Write, phone or call for
Free Brochure and
Briefing Form   DO IT NOW!

HAYWARD TECHNICAL SERVICES LTD.
HAY HOUSE
21 STROUD ROAD
GLOUCESTER GL1 5AD
TEL: 0452 27511

J. GARDNER & CO. LTD

Specialists in the manufacture and installation of ductwork and other sheet metal fabrications for the engineering industry for 100 years

Gardner Industrial Estate,
Kent House Lane,
Beckenham, Kent, BR3 1JL
Tel 01-778-6080   Telex 8951116

STEEL TUBING AND FITTINGS

BLACK TUBING AND TUBULARS
(to B.S. 1387 with screwed and socketed joints – fixings given separately) *continued*

	Unit	15 mm £	20 mm £	25 mm £	32 mm £	40 mm £	50 mm £
Extra over tubing for black wrought iron fittings to B.S.1740							
Elbow, male and female	No.	4·44	5·93	7·23	8·71	10·43	14·84
		1·00	*1·33*	*2·01*	*2·66*	*3·43*	*6·49*
Tee	,,	5·81	7·69	9·20	11·15	13·13	16·50
		0·85	*1·05*	*1·66*	*2·40*	*3·03*	*4·43*
Socket, plain	,,	3·58	4·75	5·42	6·35	7·35	8·92
		0·14	*0·16*	*0·20*	*0·30*	*0·36*	*0·57*
Cap	,,	2·26	2·97	3·56	4·29	5·01	6·20
		0·35	*0·42*	*0·66*	*0·92*	*1·13*	*1·56*
Backnut	,,	0·49	0·62	0·76	0·93	1·09	1·51
		0·15	*0·16*	*0·24*	*0·32*	*0·39*	*0·67*

		65 mm £	80 mm £	100 mm £	125 mm £	150 mm £
Elbow, male and female	,,	18·13	23·75	38·58	92·96	140·28
		7·90	*11·64*	*21·88*	*68·95*	*108·96*
Tee	,,	22·45	29·89	45·54	102·07	153·97
		7·66	*12·39*	*21·40*	*66·16*	*106·42*
Socket, plain	,,	11·27	13·56	19·47	29·54	39·59
		1·04	*1·45*	*2·76*	*5·52*	*8·27*
Cap	,,	8·67	10·77	16·39	35·57	50·72
		2·98	*4·05*	*7·12*	*22·26*	*33·39*
Backnut	,,	2·35	3·34	5·02	13·78	19·55
		1·02	*2·13*	*3·35*	*11·38*	*16·42*

STEEL TUBING AND FITTINGS

BLACK TUBING AND TUBULARS
(to B.S. 1387 with screwed and socketed joints – fixings given separately) *continued*

Extra over tubing for black malleable fittings to B.S. 143	Unit	15 mm £	20 mm £	25 mm £	32 mm £	40 mm £	50 mm £
Elbow	No.	3·71	4·95	5·75	6·91	8·18	9·98
		0·27	*0·35*	*0·53*	*0·86*	*1·18*	*1·62*
Tee	,,	5·31	7·11	8·19	9·78	11·43	14·14
		0·35	*0·47*	*0·65*	*1·03*	*1·33*	*2·07*
Socket	,,	3·64	4·83	5·54	6·53	7·64	9·53
		0·19	*0·24*	*0·32*	*0·47*	*0·65*	*1·18*
Reducing socket	,,	3·65	4·87	5·60	6·65	7·70	9·39
		0·27	*0·28*	*0·38*	*0·59*	*0·71*	*1·03*
Cap	,,	2·09	2·77	3·19	3·86	4·47	5·46
		0·18	*0·22*	*0·30*	*0·50*	*0·59*	*0·83*
Hexagonal bush	,,	2·06	2·75	3·13	3·80	4·42	5·49
		0·15	*0·21*	*0·24*	*0·44*	*0·53*	*0·86*
Hexagonal nipple	,,	0·55	0·71	0·85	1·08	1·29	1·75
		0·21	*0·25*	*0·32*	*0·47*	*0·59*	*0·92*
Hexagonal reducing nipple	,,	0·60	0·81	0·97	1·34	1·53	1·87
		0·25	*0·35*	*0·44*	*0·74*	*0·83*	*1·03*
Plug	,,	0·49	0·64	0·73	0·96	1·17	1·43
		0·15	*0·18*	*0·25*	*0·35*	*0·47*	*0·59*
Backnut	,,	0·55	0·68	0·77	0·90	1·11	1·43
		0·21	*0·22*	*0·25*	*0·30*	*0·41*	*0·59*
Twin elbow	,,	5·70	7·82	8·95	11·70	14·24	18·57
		0·74	*1·18*	*1·40*	*2·95*	*4·13*	*6·50*
Pitcher tee	,,	5·67	7·53	8·81	10·67	12·91	16·65
		0·71	*0·89*	*1·27*	*1·92*	*2·81*	*4·58*
Bend, female	,,	3·81	5·10	5·93	7·47	8·44	10·71
		0·37	*0·50*	*0·71*	*1·42*	*1·45*	*2·36*
Open return bend	,,	4·33	5·92	7·14	9·60	11·28	13·52
		0·89	*1·33*	*1·92*	*3·54*	*4·28*	*5·17*
Union, female	,,	4·70	6·01	7·07	8·61	10·10	12·27
		1·25	*1·42*	*1·85*	*2·56*	*3·10*	*3·92*
Union elbow male and female	,,	2·87	3·76	4·61	5·71	6·98	9·08
		0·96	*1·21*	*1·71*	*2·35*	*3·10*	*4·44*

STEEL TUBING AND FITTINGS

BLACK TUBING AND TUBULARS
(to B.S. 1387 with screwed and socketed joints – fixings given separately) *continued*

Extra over tubing for black malleable fittings to B.S. 143	Unit	65 mm £	80 mm £	100 mm £	125 mm £	150 mm £
Elbow	No.	13·48	16·54	24·97		
		3·25	4·43	8·27		
Tee	,,	18·93	22·23	24·14	56·29	80·04
		4·13	4·73	8·56	20·38	32·49
Socket	,,	12·00	14·62	20·84		
		1·77	2·51	4·13		
Reducing socket	,,	12·15	15·05	20·99		
		1·92	2·36	4·28		
Cap	,,	7·31	8·49	12·96		
		1·62	1·77	3·69		
Hexagonal bush	,,	7·26	9·08	13·40	21·88	27·97
		1·57	2·36	4·13	8·56	10·64
Hexagonal nipple	,,	2·65	3·72	5·66		
		1·62	2·51	3·99		
Hexagonal reducing nipple	,,	2·80	4·16			
		1·77	2·95			
Backnut	,,	2·26	2·63			
		1·24	1·42			
Twin elbow	,,	22·77	31·09			
		7·97	13·59			
Pitcher tee	,,	20·70	25·62	44·23		
		5·91	8·12	20·08		
Bend, female	,,	15·25	18·90	30·88	56·50	80·94
		5·02	6·79	14·18	32·49	49·62
Union, female	,,	18·23	25·18	42·31		
		8·00	13·07	25·61		

BLACK TUBING
(to B.S. 1387 with flanged joints – joints and fixings given separately)

	Unit	15 mm £	20 mm £	25 mm £	32 mm £	40 mm £	50 mm £
Medium weight tubing	m	3·21	3·37	3·91	4·54	5·09	6·04
		0·60	0·73	1·06	1·36	1·61	2·28
Heavy weight tubing	,,	3·31	3·48	4·08	4·76	5·35	6·42
		0·70	0·84	1·23	1·58	1·87	2·66

		65 mm £	80 mm £	100 mm £	125 mm £	150 mm £
Medium weight tubing	,,	7·22	8·58	11·74	14·09	16·96
		3·02	3·92	5·58	7·20	9·34
Heavy weight tubing	,,	7·72	9·22	12·66	15·28	18·50
		3·51	4·57	6·50	8·39	10·88

STEEL TUBING AND FITTINGS

BOLTED CONNECTION BETWEEN PAIR OF FLANGES
(including corrugated brass joint ring, bolts and nuts – flanges given separately)

	Unit	15 mm £	20 mm £	25 mm £	32 mm £	40 mm £	50 mm £
Table D	No.	3·47	3·48	3·52	3·54	3·58	4·29
		0·86	0·87	0·91	0·93	0·97	1·68
Table E	,,	3·47	3·48	3·57	3·59	3·63	4·34
		0·86	0·87	0·96	0·98	1·02	1·73
Table F	,,	3·60	3·60	4·28	4·34	4·35	4·49
		0·99	0·99	1·67	1·73	1·74	1·88
Table H	,,	4·29	4·29	4·32	4·39	4·40	4·59
		1·68	1·68	1·71	1·78	1·79	1·98

	Unit	65 mm £	80 mm £	100 mm £	125 mm £	150 mm £
Table D	,,	4·32	4·43	4·54	8·19	8·19
		1·71	1·82	1·93	3·60	3·60
Table E	,,	4·37	4·43	8·01	8·19	10·13
		1·76	1·82	3·42	3·60	5·54
Table F	,,	8·09	8·14	8·43	10·47	16·36
		3·50	3·55	3·84	5·88	8·43
Table H	,,	8·29	8·44	8·63	10·81	17·41
		3·70	3·85	4·04	6·22	9·48

MILD STEEL BLACK BOSSED FLANGE
(to B.S. 10 drilled for bolts and screwed and jointed to tubing – joint rings, nuts and bolts given separately)

	Unit	15 mm £	20 mm £	25 mm £	32 mm £	40 mm £	50 mm £
Table D	No.	3·50	4·14	4·49	5·24	5·76	6·94
		1·59	1·59	1·59	1·88	1·88	2·31
Table E	,,	3·50	4·14	4·49	5·24	5·76	6·94
		1·59	1·59	1·59	1·88	1·88	2·31
Table F	,,	3·72	4·35	4·70	5·78	6·30	7·96
		1·80	1·80	1·80	2·42	2·42	3·32
Table H	,,	4·05	4·69	5·04	6·08	6·60	8·41
		2·14	2·14	2·14	2·72	2·72	3·77

	Unit	65 mm £	80 mm £	100 mm £	125 mm £	150 mm £
Table D	,,	8·14	9·71	13·58	21·87	25·89
		2·45	2·99	4·32	8·56	8·56
Table E	,,	8·32	9·88	13·96	22·45	26·47
		2·63	3·16	4·70	9·14	9·14
Table F	,,	10·01	11·68	15·82	27·05	31·07
		4·32	4·96	6·55	13·74	13·74
Table H	,,	10·65	12·81	17·30	28·93	32·95
		4·96	6·10	8·03	15·62	15·62

Mechanical Installations – Prices for Measured Work

STEEL TUBING AND FITTINGS

MILD STEEL BLACK BLANK FLANGE
(to B.S. 10 drilled for bolts – joint ring, nuts and bolts given separately)

	Unit	15 mm £	20 mm £	25 mm £	32 mm £	40 mm £	50 mm £
Table D	No.	1·26	1·38	1·59	1·80	2·05	2·40
		0·92	*0·92*	*1·07*	*1·20*	*1·35*	*1·57*
Table E	,,	1·26	1·38	1·59	1·80	2·05	2·40
		0·92	*0·92*	*1·07*	*1·20*	*1·35*	*1·57*
Table F	,,	1·47	1·58	1·85	2·10	2·36	3·44
		1·12	*1·12*	*1·33*	*1·49*	*1·66*	*2·60*
Table H	,,	1·66	1·78	2·02	2·42	2·74	3·87
		1·32	*1·32*	*1·49*	*1·82*	*2·04*	*3·03*

	Unit	65 mm £	80 mm £	100 mm £	125 mm £	150 mm £
Table D	,,	2·92	3·61	4·54	7·12	9·05
		1·88	*2·40*	*2·87*	*4·72*	*5·92*
Table E	,,	2·92	3·61	5·13	7·50	10·50
		1·88	*2·40*	*3·46*	*5·10*	*7·34*
Table F	,,	4·17	5·03	7·29	11·10	13·28
		3·12	*3·82*	*5·62*	*8·70*	*10·15*
Table H	,,	4·70	6·34	8·77	13·54	16·78
		3·65	*5·13*	*7·10*	*11·14*	*13·64*

STEEL TUBING AND FITTINGS

GALVANIZED TUBING AND TUBULARS
(to B.S. 1387 with screwed and socketed joints – fixings given separately)

	Unit	15 mm £	20 mm £	25 mm £	32 mm £	40 mm £	50 mm £
Medium weight tubing	m	3·59 / 0·85	3·85 / 1·03	4·57 / 1·49	5·36 / 1·91	6·08 / 2·27	7·29 / 3·22
Extra over for							
Bend or spring	No.	2·42 / 0·49	3·17 / 0·62	3·79 / 0·89	4·83 / 1·47	5·55 / 1·67	7·66 / 2·81
Heavy weight tubing	m	3·73 / 0·99	4·02 / 1·20	4·81 / 1·73	5·67 / 2·23	6·45 / 2·64	7·82 / 3·75
Extra over for							
Bend or spring	No.	2·44 / 0·51	3·19 / 0·65	3·83 / 0·93	4·90 / 1·54	5·63 / 1·74	7·78 / 2·94
Extra over tubing for							
Labour bend	,,	2·09	2·09	2·51	3·34	4·18	4·59
Labour splay cut end	,,	0·73	0·84	0·94	1·31	1·40	1·61
Screwed joint to fitting and the like	,,	1·57	2·09	2·38	2·75	3·18	3·79

		65 mm £	80 mm £	100 mm £	125 mm £	150 mm £
Medium weight tubing	m	8·84 / 4·25	10·64 / 5·53	14·75 / 7·86	18·09 / 10·15	22·04 / 13·17
Extra over for						
Bend or spring	No.	11·26 / 5·37	15·01 / 8·29	23·86 / 14·60	68·42 / 55·12	95·85 / 78·52
Heavy weight tubing	m	9·55 / 4·95	11·55 / 6·44	16·05 / 9·16	19·76 / 11·83	24·21 / 15·34
Extra over for						
Bend or spring	No.	11·50 / 5·60	15·39 / 8·67	24·51 / 15·25	71·20 / 57·90	99·77 / 82·44
Extra over tubing for						
Labour bend	,,	5·22	5·85	16·70		
Labour splay cut end	,,	1·85	2·09	2·51	3·12	3·65
Screwed joint to fitting and the like	,,	4·64	5·46	7·60	10·93	14·25

STEEL TUBING AND FITTINGS

GALVANIZED TUBING AND TUBULARS
(to B.S. 1387 with screwed and socketed joints – fixings given separately) *continued*

Extra over tubing for galvanized wrought fittings to B.S. 1740	Unit	15 mm £	20 mm £	25 mm £	32 mm £	40 mm £	50 mm £
Elbow, male and female	No.	4·60	6·13	7·54	9·13	10·96	15·85
		1·15	*1·54*	*2·32*	*3·07*	*3·96*	*7·49*
Tee	,,	5·94	7·85	9·46	11·52	13·60	17·18
		0·98	*1·21*	*1·92*	*2·77*	*3·50*	*5·12*
Socket, plain	,,	3·60	4·78	5·45	6·40	7·44	9·01
		0·16	*0·18*	*0·23*	*0·34*	*0·44*	*0·66*
Cap	,,	2·31	3·03	3·66	4·43	5·19	6·44
		0·40	*0·49*	*0·77*	*1·06*	*1·30*	*1·80*
Backnut	,,	0·51	0·65	0·80	0·98	1·15	1·61
		0·17	*0·19*	*0·28*	*0·37*	*0·45*	*0·78*

		65 mm £	80 mm £	100 mm £	125 mm £	150 mm £
Elbow, male and female	,,	19·32	25·49	41·86	103·31	156·62
		9·09	*13·38*	*25·16*	*79·30*	*125·30*
Tee	,,	23·60	31·75	48·75	112·00	169·93
		8·81	*14·25*	*24·61*	*76·09*	*122·38*
Socket, plain	,,	11·43	13·78	19·88	30·36	40·83
		1·20	*1·67*	*3·18*	*6·35*	*9·51*
Cap	,,	9·12	11·37	17·46	38·91	55·73
		3·43	*4·65*	*8·19*	*25·60*	*38·40*
Backnut	,,	2·55	3·66	5·52	15·48	22·02
		1·53	*2·45*	*3·85*	*13·08*	*18·88*

STEEL TUBING AND FITTINGS

GALVANIZED TUBING AND TUBULARS
(to B.S. 1387 with screwed and socketed joints – fixings given separately) *continued*

Extra over tubing for galvanized malleable fittings to B.S. 143	Unit	15 mm £	20 mm £	25 mm £	32 mm £	40 mm £	50 mm £
Elbow	No.	3·76	5·03	5·89	7·12	8·37	10·32
		0·32	*0·43*	*0·67*	*1·06*	*1·38*	*1·97*
Tee	,,	5·43	7·27	8·41	10·13	11·88	14·83
		0·47	*0·63*	*0·87*	*1·38*	*1·77*	*2·76*
Socket	,,	3·70	4·91	5·65	6·69	7·86	9·93
		0·26	*0·32*	*0·43*	*0·63*	*0·87*	*1·58*
Reducing socket . . .	,,	3·72	4·97	5·73	6·84	7·94	9·73
		0·28	*0·37*	*0·51*	*0·79*	*0·95*	*1·38*
Cap	,,	2·15	2·84	3·29	4·03	4·67	5·74
		0·24	*0·30*	*0·39*	*0·67*	*0·79*	*1·10*
Hexagonal bush . . .	,,	2·11	2·82	3·21	3·95	4·59	5·78
		0·20	*0·28*	*0·32*	*0·59*	*0·71*	*1·14*
Hexagonal nipple . . .	,,	0·62	0·79	0·96	1·24	1·49	2·06
		0·28	*0·33*	*0·43*	*0·63*	*0·79*	*1·22*
Hexagonal reducing nipple .	,,	0·68	0·93	1·11	1·59	1·80	2·21
		0·33	*0·47*	*0·59*	*0·98*	*1·10*	*1·38*
Plug, solid	,,	0·56	0·72	0·90	1·14	1·39	1·70
		0·22	*0·26*	*0·37*	*0·53*	*0·69*	*0·87*
Backnut	,,	0·62	0·75	0·86	1·00	1·25	1·62
		0·28	*0·30*	*0·33*	*0·39*	*0·55*	*0·79*
Twin elbow	,,	5·94	8·21	9·41	12·69	15·62	20·73
		0·98	*1·58*	*1·87*	*3·94*	*5·51*	*8·66*
Pitcher tee	,,	5·90	7·82	9·24	11·31	13·85	18·17
		0·95	*1·18*	*1·69*	*2·56*	*3·74*	*6·10*
Bend, female	,,	3·94	5·26	6·17	7·95	8·92	11·50
		0·49	*0·67*	*0·95*	*1·89*	*1·93*	*3·15*
Open return bend . . .	,,	4·63	6·37	7·78	10·78	12·70	15·24
		1·18	*1·77*	*2·56*	*4·73*	*5·71*	*6·89*
Union, female . . .	,,	5·11	6·48	7·70	9·47	11·14	13·59
		1·67	*1·89*	*2·48*	*3·42*	*4·14*	*5·24*
Union elbow, male and female	,,	3·19	4·16	5·18	6·49	8·02	10·56
		1·28	*1·61*	*2·28*	*3·13*	*4·13*	*5·93*

STEEL TUBING AND FITTINGS

GALVANIZED TUBING AND TUBULARS
(to B.S. 1387 with screwed and socketed joints – fixings given separately) *continued*

Extra over tubing for galvanized malleable fittings to B.S. 143 continued	Unit	65 mm £	80 mm £	100 mm £	125 mm £	150 mm £
Elbow	No.	13·97	17·62	26·15	43·70	67·94
		3·74	*5·51*	*9·45*	*19·69*	*36·62*
Tee	,,	20·31	23·80	35·56	61·88	88·57
		5·51	*6·30*	*11·42*	*27·17*	*43·32*
Socket	,,	12·59	15·46	22·22		
		2·36	*3·35*	*5·51*		
Reducing socket	,,	12·79	15·26	22·41		
		2·56	*3·15*	*5·71*		
Cap	,,	7·86	9·08	14·19		
		2·17	*2·36*	*4·92*		
Hexagonal bush	,,	7·78	9·87	14·78	24·73	31·51
		2·09	*3·15*	*5·51*	*11·42*	*14·18*
Hexagonal nipple	,,	3·19	4·56	6·99		17·70
		2·17	*3·35*	*5·32*		*14·57*
Hexagonal reducing nipple	,,	3·39	5·15			
		2·36	*3·94*			
Backnut	,,	2·68	3·10			
		1·65	*1·89*			
Twin elbow	,,	25·43	35·62			
		10·63	*18·11*			
Pitcher tee	,,	22·67	28·33	50·92		
		7·88	*10·83*	*26·78*		
Bend, female	,,	16·93	21·17	35·61	67·33	97·48
		6·69	*9·06*	*18·90*	*43·32*	*66·16*
Union, female	,,	20·92	29·56	50·91		
		10·69	*17·45*	*34·20*		

GALVANIZED TUBING
(to B.S. 1387 with flanged joints – joints and fixings given separately)

	Unit	15 mm £	20 mm £	25 mm £	32 mm £	40 mm £	50 mm £
Medium weight tubing	m	3·41	3·61	4·26	4·99	5·62	6·80
		0·80	*0·97*	*1·41*	*1·81*	*2·14*	*3·04*
Heavy weight tubing	,,	3·55	3·77	4·49	5·29	5·98	7·30
		0·94	*1·13*	*1·64*	*2·11*	*2·50*	*3·54*

		65 mm £	80 mm £	100 mm £	125 mm £	150 mm £
Medium weight tubing	,,	8·23	9·88	13·60	16·49	20·07
		4·02	*5·23*	*7·43*	*9·60*	*12·45*
Heavy weight tubing	,,	8·89	10·74	14·82	18·07	22·12
		4·68	*6·09*	*8·66*	*11·18*	*14·50*

STEEL TUBING AND FITTINGS

BOLTED CONNECTION BETWEEN PAIR OF FLANGES
(including corrugated brass joint ring, bolts and nuts – flanges given separately)

	Unit	15 mm £	20 mm £	25 mm £	32 mm £	40 mm £	50 mm £
Table D	No.	3·47 / 0·86	3·48 / 0·87	3·52 / 0·91	3·54 / 0·93	3·58 / 0·97	4·29 / 1·68
Table E	,,	3·47 / 0·86	3·48 / 0·87	3·57 / 0·96	3·59 / 0·98	3·63 / 1·02	4·34 / 1·73
Table F	,,	3·60 / 0·99	3·60 / 0·99	4·28 / 1·67	4·34 / 1·73	4·35 / 1·74	4·49 / 1·88
Table H	,,	4·29 / 1·68	4·29 / 1·68	4·32 / 1·71	4·39 / 1·78	4·40 / 1·79	4·59 / 1·98

		65 mm £	80 mm £	100 mm £	125 mm £	150 mm £
Table D	,,	4·32 / 1·71	4·43 / 1·82	4·54 / 1·93	8·19 / 3·60	8·19 / 3·60
Table E	,,	4·37 / 1·76	4·43 / 1·82	8·01 / 3·42	8·19 / 3·60	10·13 / 5·54
Table F	,,	8·09 / 3·50	8·14 / 3·55	8·43 / 3·84	10·47 / 5·88	16·36 / 8·43
Table H	,,	8·29 / 3·70	8·44 / 3·85	8·63 / 4·04	10·81 / 6·22	17·41 / 9·48

MILD STEEL GALVANIZED BOSSED FLANGES
(to B.S. 10 drilled for bolts and screwed and jointed to tubing – joint ring, nuts and bolts given separately)

	Unit	15 mm £	20 mm £	25 mm £	32 mm £	40 mm £	50 mm £
Table D	No.	3·98 / 2·07	4·61 / 2·07	4·96 / 2·07	5·80 / 2·44	6·32 / 2·44	7·63 / 3·00
Table E	,,	3·98 / 2·07	4·61 / 2·07	4·96 / 2·07	5·80 / 2·44	6·32 / 2·44	7·63 / 3·00
Table F	,,	4·26 / 2·35	4·89 / 2·35	5·24 / 2·35	6·51 / 3·15	7·03 / 3·15	8·95 / 4·32
Table H	,,	4·70 / 2·79	5·33 / 2·79	5·68 / 2·79	6·89 / 3·53	7·42 / 3·53	9·54 / 4·91

		65 mm £	80 mm £	100 mm £	125 mm £	150 mm £
Table D	,,	8·88 / 3·19	10·61 / 3·89	14·88 / 5·61	24·44 / 11·13	28·46 / 11·13
Table E	,,	9·10 / 3·41	10·82 / 4·11	15·37 / 6·11	25·19 / 11·88	29·21 / 11·88
Table F	,,	11·30 / 5·61	13·17 / 6·45	17·78 / 8·52	31·18 / 17·87	35·20 / 17·87
Table H	,,	12·14 / 6·45	14·64 / 7·92	19·71 / 10·45	33·62 / 20·31	37·64 / 20·31

STEEL TUBING AND FITTINGS

MILD STEEL GALVANIZED BLANK FLANGES
(to B.S. 10 drilled for bolts – joint ring, nuts and bolts given separately)

	Unit	15 mm £	20 mm £	25 mm £	32 mm £	40 mm £	50 mm £
Table D	No.	1·54	1·65	1·91	2·16	2·45	2·87
		1·19	*1·19*	*1·38*	*1·56*	*1·75*	*2·04*
Table E	„	1·54	1·65	1·91	2·16	2·45	2·87
		1·19	*1·19*	*1·38*	*1·56*	*1·75*	*2·04*
Table F	„	1·81	1·92	2·25	2·55	2·85	4·22
		1·46	*1·46*	*1·73*	*1·94*	*2·15*	*3·38*
Table H	„	2·06	2·17	2·46	2·97	3·35	4·78
		1·71	*1·71*	*1·94*	*2·37*	*2·65*	*3·94*

	Unit	65 mm £	80 mm £	100 mm £	125 mm £	150 mm £
Table D	„	3·49	4·33	5·40	8·54	10·82
		2·44	*3·12*	*3·73*	*6·13*	*7·69*
Table E	„	3·48	4·33	6·17	9·04	12·71
		2·44	*3·12*	*4·50*	*6·64*	*9·58*
Table F	„	5·10	6·17	8·98	13·71	16·33
		4·06	*4·96*	*7·31*	*11·31*	*13·19*
Table H	„	5·79	7·88	10·36	16·88	20·87
		4·75	*6·67*	*8·69*	*14·48*	*17·74*

BLACK TUBING
(to B.S. 1387 with plain ends for welding with butt welded joints – fixings given separately)

	Unit	15 mm £	20 mm £	25 mm £	32 mm £	40 mm £	50 mm £
Medium weight tubing	m	3·45	3·70	4·33	5·12	5·78	7·09
		0·60	*0·73*	*1·06*	*1·36*	*1·61*	*2·28*
Heavy weight tubing	„	3·55	3·82	4·51	5·34	6·05	7·46
		0·70	*0·84*	*1·23*	*1·58*	*1·87*	*2·66*
Extra over tubing for							
Straight butt weld.	No.	1·57	2·09	2·61	3·45	4·18	6·26
Branch weld.	„	2·35	3·13	3·97	5·17	5·56	7·83
Welded reducing joint (priced on size of the larger tubing)	„	3·13	4·18	5·22	6·89	8·35	12·53

	Unit	65 mm £	80 mm £	100 mm £	125 mm £	150 mm £
Medium weight tubing	m	8·60	10·29	13·80	16·49	19·70
		3·02	*3·92*	*5·58*	*7·20*	*9·34*
Heavy weight tubing	„	9·10	10·94	14·72	17·68	21·24
		3·51	*4·57*	*6·50*	*8·39*	*10·88*
Extra over tubing for						
Straight butt weld.	No.	8·35	10·44	12·53	14·62	16·70
Branch weld.	„	10·44	13·05	15·66	18·27	20·88
Welded reducing joint (priced on size of the larger tubing)	„	16·70	20·88	22·97	25·06	33·41

STEEL TUBING AND FITTINGS

BLACK TUBING
(to B.S. 1387 with plain ends for welding with butt welded joints – fixings given separately) *continued*

Extra over medium weight tubing for medium weight weldable fittings to B.S. 1965 including welded joints

	Unit	15 mm £	20 mm £	25 mm £	32 mm £	40 mm £	50 mm £
Elbow, long radius 90 degree	,,	4·19 / 1·50	5·09 / 1·50	6·35 / 1·92	7·48 / 1·92	8·88 / 1·92	12·15 / 1·92
Elbow radius 45 degree	,,	4·19 / 1·50	5·09 / 1·50	6·18 / 1·74	7·31 / 1·74	8·71 / 1·74	11·98 / 1·74
Return bend, long radius 180 degree	No.	7·04 / 4·35	7·94 / 4·35	9·93 / 5·49	11·05 / 5·49	12·45 / 5·49	15·72 / 5·49
Branch bend	,,	7·34 / 3·65	8·57 / 3·65	10·79 / 4·60	12·55 / 4·60	13·65 / 4·60	17·55 / 4·60

	Unit	65 mm £	80 mm £	100 mm £	125 mm £	150 mm £
Elbow, long radius 90 degree	,,	16·30 / 2·74	19·86 / 2·99	25·72 / 5·26	33·70 / 9·38	41·75 / 13·57
Elbow radius 45 degree	,,	15·75 / 2·20	19·50 / 2·63	24·79 / 4·33	32·49 / 8·16	39·41 / 11·22
Return bend, long radius 180 degree	,,	20·53 / 6·98	25·43 / 8·56	35·69 / 15·22	51·18 / 26·86	64·67 / 36·48
Branch bend	,,	23·87 / 6·65	28·68 / 7·18	38·67 / 12·78	53·33 / 22·90	68·12 / 33·15

Extra over heavy weight tubing for heavy weight weldable fittings to B.S. 1965 including welded joints

	Unit	25 mm £	32 mm £	40 mm £	50 mm £	65 mm £	80 mm £	100 mm £	125 mm £	150 mm £
Elbow, short radius 90 degree	No.	9·77 / 5·33	10·90 / 5·33	12·30 / 5·33	15·56 / 5·33	18·88 / 5·33	23·62 / 6·75	31·60 / 11·14	42·45 / 18·14	54·63 / 26·44
Elbow, long radius 90 degree	,,	6·72 / 2·29	7·85 / 2·29	9·25 / 2·29	12·52 / 2·29	16·72 / 3·17	20·59 / 3·72	26·61 / 6·15	35·06 / 10·74	43·81 / 15·62
Elbow, long radius 45 degree	,,	6·41 / 1·97	7·54 / 1·97	8·94 / 1·97	12·20 / 1·97	16·14 / 2·59	20·09 / 3·22	25·68 / 5·22	33·52 / 9·19	41·91 / 13·72
Return bend, long radius 180 degree	,,	10·81 / 6·38	11·94 / 6·38	13·34 / 6·38	16·61 / 6·38	21·46 / 7·90	27·38 / 10·51	37·97 / 17·51	55·00 / 30·68	71·51 / 43·33
Branch bend	,,	11·67 / 5·49	13·44 / 5·49	14·54 / 5·49	13·21 / 5·49	24·83 / 7·60	30·41 / 8·91	40·36 / 14·47	55·88 / 25·44	72·00 / 37·02
Equal tee	,,	13·01 / 6·62	14·66 / 6·62	16·99 / 6·89	22·06 / 7·13	28·61 / 11·28	32·88 / 13·15	42·24 / 17·81	67·65 / 35·08	79·91 / 39·20
Reducing tee	,,	13·26 / 6·86	14·90 / 6·86	17·63 / 7·53	23·91 / 8·98	31·37 / 14·04	36·18 / 16·45	46·73 / 22·30	76·47 / 43·90	89·77 / 49·06
Cap	,,	6·37 / 3·89	6·97 / 3·89	7·72 / 3·89	9·87 / 4·23	11·53 / 5·00	12·57 / 5·17	15·95 / 6·68	20·72 / 8·25	25·89 / 10·23

STEEL TUBING AND FITTINGS

BLACK TUBING
(to B.S. 1387 with plain ends for welding with butt welded joints – fixings given separately) *continued*

Extra over medium or heavy weight tubing for weldable fittings to B.S. 1965 including welded joints	Unit	15 mm £	20 mm £	25 mm £	32 mm £	40 mm £	50 mm £
40 mm Concentric reducer	No.	9·42	9·26	8·79	9·42		
		4·42	3·86	3·00	3·00		
50 mm Concentric reducer	,,	11·12	11·78	10·86	11·49	11·87	
		5·53	5·10	3·37	3·37	3·20	
65 mm Concentric reducer	,,		13·43	14·03	12·93	14·22	15·88
			6·12	5·89	4·16	3·99	3·56
80 mm Concentric reducer	,,			16·21	17·91	16·68	17·85
				7·33	7·33	4·50	3·65
100 mm Concentric reducer	,,					23·24	25·06
						7·58	6·16
125 mm Concentric reducer	,,						29·30
							9·46
40 mm Eccentric reducer	,,	11·85	12·12	10·43	11·06		
		6·85	6·72	4·65	4·65		
50 mm Eccentric reducer	,,	14·10	14·01	13·09	12·76	13·31	
		8·52	7·33	5·60	4·65	4·65	
65 mm Eccentric reducer	,,		20·93	17·65	15·49	15·65	17·19
			13·62	9·51	6·72	5·42	4·87
80 mm Eccentric reducer	,,			19·07	20·77	19·99	19·87
				10·19	10·19	7·82	5·66
100 mm Eccentric reducer	,,					29·00	29·06
						13·34	10·16
125 mm Eccentric reducer	,,						39·82
							19·98

	Unit	65 mm £	80 mm £	100 mm £
80 mm Concentric reducer	No.	19·98		
		3·65		
100 mm Concentric reducer	,,	24·73	24·50	
		5·10	4·15	
125 mm Concentric reducer	,,	29·35	30·02	31·20
		8·99	7·58	7·19
150 mm Concentric reducer	,,		32·82	32·88
			9·33	7·82
80 mm Eccentric reducer	,,	22·00		
		5·66		
100 mm Eccentric reducer	,,	28·49	28·33	
		8·86	7·98	
125 mm Eccentric reducer	,,	38·67	38·00	36·11
		18·31	15·55	12·09
150 mm Eccentric reducer	,,		41·00	39·62
			17·51	14·62

STEEL TUBING AND FITTINGS

BOLTED CONNECTION BETWEEN PAIR OF FLANGES
(including corrugated brass joint ring, bolts and nuts – flanges given separately)

	Unit	15 mm £	20 mm £	25 mm £	32 mm £	40 mm £	50 mm £
Table D	No.	3·47 / 0·86	3·48 / 0·87	3·52 / 0·91	3·54 / 0·93	3·58 / 0·97	4·29 / 1·68
Table E	,,	3·47 / 0·86	3·48 / 0·87	3·57 / 0·96	3·59 / 0·98	3·63 / 1·02	4·34 / 1·73
Table F	,,	3·60 / 0·99	3·60 / 0·99	4·28 / 1·67	4·34 / 1·73	4·35 / 1·74	4·49 / 1·88
Table H	,,	4·29 / 1·68	4·29 / 1·68	4·32 / 1·71	4·39 / 1·78	4·40 / 1·79	4·59 / 1·98

	Unit	65 mm £	80 mm £	100 mm £	125 mm £	150 mm £
Table D	,,	4·32 / 1·71	4·43 / 1·82	4·54 / 1·93	8·19 / 3·60	8·19 / 3·60
Table E	,,	4·37 / 1·76	4·43 / 1·82	8·01 / 3·42	8·19 / 3·60	10·13 / 5·54
Table F	,,	8·09 / 3·50	8·14 / 3·55	8·43 / 3·84	10·47 / 5·88	16·36 / 8·43
Table H	,,	8·29 / 3·70	8·44 / 3·85	8·63 / 4·04	10·81 / 6·22	17·41 / 9·48

MILD STEEL BLACK WELDING NECK FLANGE
(to B.S. 10 drilled for bolts and welded to tubing – joint ring, nuts and bolts given separately)

	Unit	15 mm £	20 mm £	25 mm £	32 mm £	40 mm £	50 mm £
Table D	No.	3·89 / 2·34	4·36 / 2·34	4·81 / 2·34	6·35 / 3·26	7·10 / 3·26	9·69 / 4·06
Table E	,,	3·89 / 2·34	4·36 / 2·34	4·81 / 2·34	6·35 / 3·26	7·10 / 3·26	9·69 / 4·06
Table F	,,	4·56 / 3·01	5·04 / 3·01	5·49 / 3·01	7·49 / 4·38	8·22 / 4·38	11·51 / 5·88
Table H	,,	5·20 / 3·65	5·67 / 3·65	6·12 / 3·65	8·18 / 5·09	8·92 / 5·09	12·09 / 6·45

	Unit	65 mm £	80 mm £	100 mm £	125 mm £	150 mm £
Table D	,,	11·33 / 4·81	12·86 / 5·46	17·26 / 7·99	29·85 / 17·37	33·03 / 17·37
Table E	,,	11·33 / 4·81	12·86 / 5·46	17·26 / 7·99	29·85 / 17·37	33·03 / 17·37
Table F	,,	13·70 / 7·18	15·92 / 8·52	20·88 / 11·60	37·13 / 24·65	40·31 / 24·65
Table H	,,	15·05 / 8·52	17·68 / 10·28	24·07 / 14·80	40·97 / 28·49	44·15 / 28·49

Mechanical Installations – Prices for Measured Work

STEEL TUBING AND FITTINGS

MILD STEEL BLACK WELDING FLANGE
(to B.S. 10 drilled for bolts and welded to tubing – joint ring, nuts and bolts given separately)
continued

	Unit	15 mm £	20 mm £	25 mm £	32 mm £	40 mm £	50 mm £
Table D	No.	2·99	3·46	3·92	4·81	5·55	7·76
		1·44	*1·44*	*1·44*	*1·72*	*1·72*	*2·12*
Table E	,,	2·99	3·46	3·92	4·81	5·55	7·76
		1·44	*1·44*	*1·44*	*1·72*	*1·72*	*2·12*
Table F	,,	3·18	3·66	4·11	5·28	6·03	8·67
		1·63	*1·63*	*1·63*	*2·19*	*2·19*	*3·05*
Table H	,,	3·50	3·97	4·43	5·54	6·28	9·03
		1·95	*1·95*	*1·95*	*2·45*	*2·45*	*3·39*

	Unit	65 mm £	80 mm £	100 mm £	125 mm £	150 mm £
Table D	,,	8·75	10·12	13·21	20·26	23·45
		2·23	*2·73*	*3·94*	*7·79*	*7·79*
Table E	,,	8·91	10·28	13·53	20·78	23·96
		2·39	*2·88*	*4·26*	*8·30*	*8·30*
Table F	,,	10·46	11·91	15·25	24·99	28·17
		3·94	*4·51*	*5·98*	*12·51*	*12·51*
Table H	,,	11·04	12·95	16·59	26·68	29·86
		4·51	*5·55*	*7·32*	*14·20*	*14·20*

Mechanical Installations – Prices for Measured Work
COPPER TUBING AND FITTINGS

LIGHT GAUGE COPPER TUBING (LIGHT DRAWN)
(to B.S. 2871. Table X with capillary type joints – fixings given separately)

	Unit	12 mm £	15 mm £	22 mm £	28 mm £	35 mm £	42 mm £	54 mm £	67 mm £
Tubing	m	2·85 / 0·70	2·89 / 0·74	3·56 / 1·33	4·14 / 1·70	5·79 / 3·07	7·13 / 4·14	8·74 / 5·49	12·15 / 8·41

LIGHT GAUGE COPPER TUBING (LIGHT DRAWN)
(to B.S. 2871. Table X with compression type joints – fixings given separately)

	Unit	15 mm £	22 mm £	28 mm £	35 mm £	42 mm £	54 mm £	67 mm £	76 mm £	108 mm £
Tubing	m	3·00 / 0·85	3·71 / 1·49	4·37 / 1·95	6·24 / 3·53	7·69 / 4·70	10·00 / 6·74	14·23 / 10·40	19·76 / 15·29	28·09 / 22·41

LIGHT GAUGE COPPER TUBING (LIGHT DRAWN)
(to B.S. 2871. Table X with bronze welded butt joints – fixings given separately)

	Unit	15 mm £	22 mm £	28 mm £	35 mm £	42 mm £	54 mm £
Tubing	m	3·16 / 0·84	3·93 / 1·45	4·57 / 1·81	6·37 / 3·20	7·86 / 4·32	9·77 / 5·63

		67 mm £	76 mm £	108 mm £	133 mm £	159 mm £
Tubing	,,	13·04 / 8·11	16·06 / 10·35	21·89 / 14·62	26·07 / 17·67	37·60 / 28·08

COPPER TUBING
(to B.S. 2871. Table Y laid underground with capillary type joints)

	Unit	15 mm £	22 mm £	28 mm £	35 mm £	42 mm £	54 mm £	67 mm £
Tubing	m	3·28 / 1·13	4·19 / 1·97	4·96 / 2·52	6·76 / 4·04	7·78 / 4·96	12·23 / 8·63	15·59 / 11·40

COPPER TUBING
(to B.S. 2871. Table Y laid underground with compression type joints)

	Unit	15 mm £	22 mm £	28 mm £	35 mm £	42 mm £	54 mm £	67 mm £	76 mm £	108 mm £
Tubing	m	3·39 / 1·25	4·35 / 2·13	5·20 / 2·77	7·21 / 4·50	8·50 / 5·52	13·41 / 9·88	17·67 / 13·40	22·94 / 17·94	37·27 / 30·94

HARD-DRAWN THIN WALL COPPER TUBING
(to B.S. 2871. Table Z with capillary type joints – fixings given separately)

	Unit	15 mm £	22 mm £	28 mm £	35 mm £	42 mm £	54 mm £	67 mm £
Tubing	No.	2·93 / 0·64	3·45 / 1·08	3·96 / 1·35	4·99 / 2·04	6·31 / 3·08	8·66 / 5·15	11·51 / 7·32

Mechanical Installations – Prices for Measured Work

COPPER TUBING AND FITTINGS

HARD-DRAWN THIN WALL COPPER TUBING
(to B.S. 2871. Table Z with compression type joints – fixings given separately) *continued*

	Unit	15 mm £	22 mm £	28 mm £	35 mm £	42 mm £	54 mm £	67 mm £	76 mm £	108 mm £
Tubing	m	3·04	3·60	4·19	5·42	6·86	9·92	13·49	18·83	26·78
		0·76	*1·23*	*1·60*	*2·50*	*3·64*	*6·40*	*9·32*	*14·00*	*20·65*
Extra over tubing for										
Labour bend	No.	1·57	1·72	1·88	2·51	3·13	3·45	3·97	4·38	
Labour Splay cut end	„	0·60	0·63	0·70	0·98	1·05	1·21	1·41	1·57	

FITTINGS

Extra over tubing for capillary type fittings to B.S. 864 (where applicable) including joints

	Unit	15 mm £	22 mm £	28 mm £	35 mm £	42 mm £	54 mm £	67 mm £
Straight coupling, copper to copper	No.	1·47	1·73	2·11	2·72	3·33	4·69	8·41
		0·17	*0·26*	*0·37*	*0·81*	*1·15*	*2·34*	*5·80*
Concentric reducing coupling, copper to copper	„		1·71	2·00	2·93	3·88	5·04	
			0·40	*0·52*	*1·19*	*1·97*	*2·86*	
Straight connector, copper to iron	„	1·79	2·32	2·89	3·96	4·68	6·08	9·30
		0·49	*0·84*	*1·15*	*2·04*	*2·51*	*3·73*	*6·78*
Tank connector with long thread	„	3·11	3·86	4·99				
		1·45	*2·01*	*2·82*				
Adaptor, copper to iron	„	2·06	2·61	3·49	4·44	5·69	6·79	
		0·76	*1·14*	*1·75*	*2·52*	*3·51*	*4·44*	
Union coupling, copper to copper	„	3·08	4·03	5·17	6·69	8·76	12·98	19·55
		1·42	*2·18*	*3·00*	*4·30*	*6·11*	*9·94*	*16·06*
Stop end	„	1·06	1·34	1·89	2·53	3·14	4·22	
		0·41	*0·60*	*1·02*	*1·58*	*2·06*	*3·05*	
Elbow, copper to copper	„	1·59	1·96	2·52	3·68	4·85	8·45	16·19
		0·28	*0·49*	*0·79*	*1·76*	*2·68*	*6·10*	*13·58*
Backplate elbow, copper to copper	„	3·76	4·95					
		1·15	*2·17*					
Return bend, ditto	„	3·27	3·80	5·17				
		1·97	*2·32*	*3·44*				
Obtuse elbow, ditto	„	1·83	2·32	3·04	5·18	8·15		14·60
		0·52	*0·85*	*1·30*	*3·27*	*5·97*		*11·99*
Tee, all ends copper	„	2·44	2·89	3·80	5·80	7·31	10·86	21·03
		0·52	*0·80*	*1·49*	*2·75*	*4·09*	*7·38*	*16·68*
Reducing tee, ditto	„	2·87	3·75	5·64	7·99	11·02	11·02	17·49
			0·78	*1·45*	*2·60*	*4·77*	*7·54*	*13·14*
Sweep tee, ditto	„	3·06	3·75	5·45	8·12	10·35	12·33	18·63
		1·15	*1·66*	*3·14*	*5·08*	*7·13*	*8·85*	*14·28*
Straight tap connector	„	1·68	2·15					
		0·64	*0·94*					
Bent tap connector	„	1·81	2·41					
		0·77	*1·19*					

COPPER TUBING AND FITTINGS

FITTINGS *continued*

Extra over tubing for compression type fittings to B.S. 864 (where applicable) including joints

	Unit	15 mm £	22 mm £	28 mm £	35 mm £	42 mm £	54 mm £
Brass							
Straight coupling copper to copper	No.	1·67 / 0·72	2·09 / 0·96	2·88 / 1·58	4·16 / 2·59	5·31 / 3·48	7·55 / 5·46
Straight connector copper to imperial copper	„		2·14 / 1·00				
Male coupling	„	1·53 / 0·57	1·95 / 0·81	2·45 / 1·15			
Male coupling with long thread and backnut	„	2·36 / 1·05	2·91 / 1·44	3·65 / 1·91			
Female coupling	„	1·74 / 0·62	2·07 / 0·77	2·71 / 1·15	3·77 / 2·02	4·82 / 2·73	6·70 / 4·26
Lead coupling	„	3·50 / 0·72	3·88 / 1·00				
Elbow	„	1·77 / 0·81	2·38 / 1·24	3·12 / 1·82	4·73 / 3·17	6·72 / 4·89	9·99 / 7·90
Male elbow	„	1·72 / 0·77	2·19 / 1·05	2·98 / 1·67	4·79 / 3·22	6·91 / 5·08	
Female elbow	„	1·72 / 0·77	2·23 / 1·10	2·98 / 1·67	4·30 / 2·73	6·56 / 4·73	
Backplate elbow	„	3·92 / 1·10	4·86 / 1·96				
Tee equal or reducing	„	2·64 / 1·15	3·15 / 1·58	4·41 / 2·63	6·75 / 4·40	9·42 / 7·00	14·16 / 11·24
Tee with male iron branch	„	3·86 / 1·44	4·77 / 1·82				
Straight tap connector	„	2·82 / 0·81	3·58 / 1·10				
Backplate tee	„	4·79 / 1·58					
Reducing set	„	1·39 / 0·43	1·52 / 0·38	1·93 / 0·62	2·71 / 1·15	3·65 / 1·83	6·35 / 4·26
Tank coupling	„	2·92 / 0·91	3·63 / 1·10	4·55 / 1·67	6·40 / 3·01		
Tank coupling – long thread	„		4·02 / 1·48				

Mechanical Installations – Prices for Measured Work

COPPER TUBING AND FITTINGS

FITTINGS *continued*

Extra over tubing for compression type fittings to B.S. 864 (where applicable) including joints

Dezincifiable	Unit	15 mm £	22 mm £	28 mm £	35 mm £	42 mm £	54 mm £
Straight coupling copper to copper	No.	1·83 / 0·87	2·32 / 1·19	3·15 / 1·84	5·08 / 3·52	6·28 / 4·46	
Straight connector copper to imperial copper	,,		2·56 / 1·42				
Male coupling	,,	1·67 / 0·71	2·12 / 0·99	3·07 / 1·76	4·63 / 3·06	5·93 / 4·11	
Male coupling with long thread and backnut	,,		3·51 / 2·04	4·45 / 2·71			
Female coupling	,,	1·91 / 0·79	2·32 / 1·01	3·19 / 1·63	4·57 / 2·84	5·45 / 3·36	
Elbow	,,	1·94 / 0·99	2·51 / 1·37	3·72 / 2·41	5·80 / 4·24	7·50 / 5·67	
Male elbow	,,	2·14 / 1·19	2·57 / 1·44	3·68 / 2·37			
Female elbow	,,	1·97 / 1·01	2·62 / 1·48	3·59 / 2·29			
Backplate elbow	,,	4·28 / 1·46	5·44 / 2·54				
Tee equal	,,	2·97 / 1·48	3·69 / 2·12	5·28 / 3·51	7·78 / 5·43	9·75 / 7·33	
Tee reducing	,,	3·05 / 1·57	3·69 / 2·12	5·28 / 3·51			
Straight tap connector	,,	3·21 / 1·20	4·25 / 1·77				
Reducing set	,,	1·56 / 0·61	1·68 / 0·55	2·19 / 0·88	3·18 / 1·61	4·19 / 2·36	

COPPER TUBING AND FITTINGS

FITTINGS *continued*

Extra over tubing for high duty capillary type fittings to B.S. 864 (where applicable) including joints	Unit	15 mm £	22 mm £	28 mm £	35 mm £	42 mm £	54 mm £
Straight coupling, copper to copper	No.	2·15 / 0·84	2·83 / 1·36	3·67 / 1·93	5·29 / 3·37	5·88 / 3·70	7·77 / 5·42
Reducing coupling	,,	2·83 / 1·52	3·33 / 1·85	4·27 / 2·53			
Straight female connector	,,	3·29 / 1·98	3·76 / 2·28	5·11 / 3·37			
Straight male connector	,,	3·32 / 2·02	3·76 / 2·28	5·11 / 3·37		8·92 / 6·75	13·33 / 10·98
Adaptor male copper to female iron	,,	4·18 / 2·88	4·68 / 3·21				
Union coupling	,,	5·46 / 3·80	6·74 / 4·89	8·92 / 6·74	14·23 / 11·84	16·57 / 13·72	
Elbow	,,	3·75 / 2·45	4·09 / 2·61	5·64 / 3·90	8·00 / 6·08	9·78 / 7·60	15·28 / 12·93
Return bend	,,			9·77 / 8·03	11·21 / 9·29		
Equal tee	,,	4·78 / 2·86	5·63 / 3·54	6·97 / 4·66	11·08 / 8·03	13·37 / 10·15	19·52 / 16·04
Reducing tee	,,	5·96 / 3·87	6·87 / 4·56	9·56 / 6·51	13·60 / 10·38	16·67 / 13·19	25·28 / 20·93
Tee male iron branch	,,	7·81 / 5·72	9·05 / 6·75				
Stop end	,,	2·47 / 1·85					
Straight union adaptor	,,	2·73 / 1·69	3·50 / 2·28	4·35 / 3·04	6·97 / 5·49	8·60 / 6·94	
Bent union adaptor	,,	6·05 / 4·40	7·76 / 5·92	10·21 / 8·03			
Bent male union adaptor	,,	7·31 / 5·65	9·45 / 7·60	16·13 / 13·95			
Composite flange	,,		11·12 / 9·29	12·48 / 10·48	15·49 / 13·36	17·86 / 15·64	24·30 / 21·95

COPPER TUBING AND FITTINGS

FITTINGS *continued*

Extra over tubing for weldable type fittings to B.S. 864 (where applicable) including joints	Unit	15 mm £	22 mm £	28 mm £	35 mm £	42 mm £	54 mm £
Straight connector, copper to iron	No.	3·55 / 0·82	4·80 / 1·17	6·02 / 1·72	7·76 / 2·45	10·34 / 4·06	14·04 / 5·58
Straight union, copper to copper	,,	5·50 / 2·81	7·41 / 3·82	9·79 / 5·36	12·19 / 6·63	16·12 / 9·16	23·48 / 13·25
Elbow, copper to copper	,,	3·40 / 0·71	4·54 / 0·94	5·97 / 1·53	7·76 / 2·20	9·90 / 2·93	14·88 / 4·65
Elbow, copper to male iron	,,	4·25 / 1·53	5·61 / 1·98	7·41 / 3·11	9·82 / 4·51	12·63 / 6·35	19·02 / 10·57
Return bend, copper to copper	,,	4·36 / 1·67	6·25 / 2·66	8·04 / 3·60	12·07 / 6·50	15·68 / 8·71	23·12 / 12·89
Obtuse elbow, copper to copper	,,	3·44 / 0·75	4·54 / 0·94	5·97 / 1·53	7·76 / 2·20	9·90 / 2·93	14·78 / 4·55
Tee, all ends copper	,,	4·69 / 0·82	6·56 / 1·40	8·25 / 1·85	11·56 / 3·52	14·56 / 4·46	21·28 / 6·35
Sweep tee, all ends copper	,,	4·95 / 1·08	7·13 / 1·98	9·20 / 2·81	12·82 / 4·77	16·60 / 6·50	23·39 / 8·46
Extra over tubing for							
Bronze butt weld	,,	1·57	2·09	2·61	3·45	4·18	6·26
Bronze branch weld	,,	2·30	3·13	3·97	5·17	5·56	7·83
Bronze welded reducing joint	,,	3·13	4·18	5·22	6·89	8·35	12·53

	Unit	67 mm £	76 mm £	108 mm £	133 mm £	159 mm £
Straight connector, copper to iron	,,	19·60 / 8·70	23·96 / 10·66	32·36 / 15·45		
Straight union, copper to copper	,,	35·94 / 22·39	45·65 / 28·78	80·61 / 60·15		
Elbow, copper to copper	,,	19·90 / 6·35	30·12 / 13·25	46·91 / 26·45	73·86 / 49·53	94·43 / 66·24
Elbow, copper to male iron	,,	29·47 / 18·57	40·93 / 27·64	72·41 / 55·50		
Return bend, copper to copper	,,	35·20 / 21·65	50·54 / 33·67	92·52 / 72·06		
Obtuse elbow, copper to copper	,,	19·73 / 6·18	30·12 / 13·25	46·91 / 26·45	73·86 / 49·53	90·65 / 62·46
Tee, all ends copper	,,	26·49 / 9·15	34·40 / 14·64	53·27 / 28·80	90·30 / 57·68	112·77 / 72·06
Sweep tee, all ends copper	,,	30·59 / 13·25	43·78 / 24·03	62·89 / 38·42	111·39 / 78·77	141·56 / 100·84
Extra over tubing for						
Bronze butt weld	,,	8·35	10·44	12·53	14·62	16·70
Bronze branch weld	,,	10·44	13·05	15·66	18·27	20·88
Bronze welded reducing joint	,,	16·70	20·88	25·06	29·23	33·41

COPPER TUBING AND FITTINGS

BRONZE FLANGES
(to B.S. 10 drilled for bolts and welded to tubing – joint ring, nuts and bolts given separately)

	Unit No.	15 mm £	22 mm £	28 mm £	35 mm £	42 mm £	54 mm £
Table D	„	5·27	6·24	7·87	10·40	13·50	18·02
		3·72	*4·21*	*5·39*	*7·31*	*9·67*	*12·38*
Table E	„	5·27	6·24	7·87	10·40	13·50	18·02
		3·72	*4·21*	*5·39*	*7·31*	*9·67*	*12·38*
Table F	„	6·70	7·64	10·39	13·06	15·83	21·27
		5·15	*5·61*	*7·94*	*9·97*	*12·00*	*15·63*
Table H	„	9·36	10·12	13·02	15·62	19·20	24·87
		7·81	*8·09*	*10·54*	*12·53*	*15·37*	*19·23*

		67 mm £	76 mm £	108 mm £	133 mm £	159 mm £
Table D	„	22·46	27·61	40·25	51·41	64·26
		15·93	*20·21*	*30·98*	*38·93*	*48·60*
Table E	„	22·46	27·61	40·25	51·41	64·26
		15·93	*20·21*	*30·98*	*38·93*	*48·60*
Table F	„	28·17	34·04	47·63	73·49	93·85
		21·64	*26·64*	*38·36*	*61·01*	*78·19*
Table H	„	31·82	39·19	53·80	80·47	101·96
		25·29	*31·79*	*44·53*	*67·99*	*86·30*

BRONZE BLANK FLANGES
(to B.S. 10 drilled for bolts and welded to tubing – joint ring, nuts and bolts given separately)

	Unit No.	15 mm £	22 mm £	28 mm £	35 mm £	42 mm £	54 mm £
Table D	„	5·73	6·78	8·46	10·58	15·51	15·89
		5·39	*6·32*	*7·94*	*9·97*	*14·81*	*15·05*
Table E	„	5·73	6·78	8·46	10·58	15·51	15·89
		5·39	*6·32*	*7·94*	*9·97*	*14·81*	*15·05*
Table F	„	6·03	7·33	11·67	15·46	18·11	22·14
		5·69	*6·87*	*11·15*	*14·85*	*17·41*	*21·30*
Table H	„	8·82	9·99	14·67	19·60	22·78	26·67
		8·48	*9·53*	*14·15*	*18·99*	*22·08*	*25·83*

		67 mm £	76 mm £	108 mm £	133 mm £	159 mm £
Table D	„	23·12	27·35	42·93	69·00	82·60
		22·08	*26·14*	*41·26*	*66·60*	*79·47*
Table E	„	23·12	27·35	42·93	69·00	82·60
		22·08	*26·14*	*41·26*	*66·60*	*79·47*
Table F	„	32·83	39·57	58·49	107·76	123·35
		31·79	*38·36*	*56·83*	*99·36*	*120·22*
Table H	„	35·12	43·75	64·14	117·23	149·60
		34·08	*42·54*	*62·47*	*114·83*	*146·47*

COPPER TUBING AND FITTINGS

BOLTED CONNECTION BETWEEN PAIR OF FLANGES
(including corrugated brass joint ring, bolts and nuts – flanges given separately)

	Unit	15 mm £	22 mm £	28 mm £	35 mm £	42 mm £	54 mm £
Table D	No.	3·47	3·48	3·52	3·54	3·58	4·29
		0·86	0·87	0·91	0·93	0·97	1·68
Table E	,,	3·47	3·48	3·57	3·59	3·63	4·34
		0·86	0·87	0·96	0·98	1·02	1·73
Table F	,,	3·60	3·60	4·28	4·34	4·35	4·49
		0·99	0·99	1·67	1·73	1·74	1·88
Table H	,,	4·29	4·29	4·32	4·39	4·40	4·59
		1·68	1·68	1·71	1·78	1·79	1·98

	Unit	67 mm £	76 mm £	108 mm £	133 mm £	159 mm £
Table D	,,	4·32	4·43	4·54	8·19	8·19
		1·71	1·82	1·93	3·60	3·60
Table E	,,	4·37	4·43	8·01	8·19	10·13
		1·76	1·82	3·42	3·60	5·54
Table F	,,	8·09	8·14	8·43	10·47	16·36
		3·50	3·55	3·84	5·88	8·43
Table H	,,	8·29	8·44	8·63	10·81	17·41
		3·70	3·85	4·04	6·22	9·48

THERMOPLASTIC TUBING AND FITTINGS

LOW DENSITY POLYTHENE TUBE
(to B.S. 1972)

	Unit	½″ (13 mm) £	¾″ (19 mm) £	1″ (25 mm) £	1¼″ (32 mm) £	1½″ (38 mm) £	2″ (51 mm) £
Class B	m		1·91	2·29	2·88	3·43	4·43
			0·49	*0·75*	*1·17*	*1·54*	*2·37*
Class C	,,	1·79	2·07	2·55	3·32	3·97	5·32
		0·42	*0·65*	*1·01*	*1·60*	*2·09*	*3·26*
Class D	,,	1·89	2·19	2·77	3·69	4·47	
		0·52	*0·78*	*1·23*	*1·98*	*2·58*	
Extra over tubing for							
Labour bend . . .	No.	0·84	0·92	1·04	1·15	1·25	1·57
Labour splay cut end .	,,	0·34	0·42	0·52	0·57	0·65	0·76
Labour screwed end on Classs C tubing . . .	,,	1·04	1·20	1·36	1·57	1·74	2·09
Extra over Class B tubing for compression type fitting to B.S. 864							
Straight connector, polythene to polythene .	,,		5·25	7·20	9·78	14·11	17·58
			2·26	*3·79*	*5·71*	*9·45*	*12·09*
Elbow, polythene to polythene	,,		4·27	6·26	10·63	13·19	15·71
			2·88	*4·65*	*8·54*	*10·88*	*13·10*
Tee, all ends polythene .	,,		5·96	8·01	14·63	17·40	24·80
			4·11	*5·85*	*11·80*	*14·48*	*21·30*
Tank coupling . .	,,		4·27				
			2·71				
Straight tap connector .	,,		6·56				
			3·57				
Bent tap connector .	,,		6·99				
			4·00				
Extra over Class C tubing for compression type fitting to B.S. 864							
Straight connector, polythene to polythene .	,,	3·96	5·25	7·20	9·78	14·11	17·58
		1·64	*2·26*	*3·79*	*5·71*	*9·45*	*12·09*
Elbow, polythene to polythene	,,	3·28	4·27	6·26	10·63	13·19	15·71
		1·97	*2·88*	*4·65*	*8·54*	*10·87*	*13·10*
Tee all ends polythene .	,,	4·48	5·96	8·01	14·63	17·40	24·80
		2·69	*4·11*	*5·85*	*11·80*	*14·48*	*21·30*
Tank coupling . .	,,		4·27				
			2·71				
Straight tap connector .	,,	4·49	6·56				
		2·11	*3·57*				
Bent tap connector .	,,	4·92	6·99				
		2·54	*4·00*				

EXPANSION JOINTS

EXPANSION JOINTS AND FITTING IN PIPE LINES
(including screwed or welded joints or bolted connections)

STAINLESS STEEL SLEEVED BELLOWS TYPE EXPANSION JOINT
(suitable for working pressures shown)

	Unit	15 mm 21 bar £	20 mm 21 bar £	25 mm 21 bar £	32 mm 21 bar £	40 mm 21 bar £	50 mm 21 bar £
Screwed for mild steel	No.	20·36 *15·35*	24·63 *17·95*	29·09 *21·26*	35·76 *26·60*	38·54 *28·10*	51·22 *39·32*
Screwed for copper	,,	24·38 *19·37*	29·59 *22·91*	33·34 *25·51*	41·19 *32·00*	43·39 *32·96*	58·43 *46·53*
Welded	,,	18·57 *14·29*	22·29 *16·65*	30·60 *23·50*	33·43 *24·56*	36·18 *25·74*	49·80 *36·02*
Flanged to B.S. 10 table E (joints to mating flange given separately)	,,	31·17 *29·29*	33·57 *31·06*	36·31 *33·18*	38·60 *34·84*	40·44 *36·37*	52·44 *48·06*
Ditto but to B.S. 10 table F	,,	32·94 *31·06*	33·80 *31·29*	36·96 *33·83*	41·43 *37·67*	43·51 *39·44*	61·06 *56·68*

		65 mm 17 bar £	80 mm 17 bar £	100 mm 17 bar £	125 mm 8 bar £	150 mm 8 bar £
Screwed for mild steel	,,	69·97 *55·16*	78·80 *61·05*			
Screwed for copper	,,	81·30 *66·48*	90·14 *72·39*			
Welded	,,	59·95 *41·68*	71·32 *48·77*	89·39 *60·58*	121·44 *84·90*	146·70 *102·85*
Flanged to B.S. 10 table E joints to mating flange given separately).	,,	62·79 *57·15*	71·13 *64·24*	93·51 *83·49*	125·97 *111·35*	153·05 *134·26*
Ditto but to B.S. 10 table F.	,,	73·78 *68·14*	84·24 *77·35*	99·17 *89·15*	154·55 *139·93*	182·81 *164·02*

WIRE REINFORCED RUBBER SLEEVED EXPANSION JOINT
(with protective sleeves for working pressures up to 8 bar)

	Unit	32 mm £	40 mm £	50 mm £	65 mm £	80 mm £	100 mm £	125 mm £	150 mm £
Flanged to B.S. 10 table E with mild steel backing flange	No.	61·39 *57·63*	62·00 *57·93*	64·46 *60·08*	67·73 *62·09*	78·43 *71·54*	88·50 *78·48*	110·29 *95·67*	124·62 *105·83*

EXPANSION JOINTS

EXPANSION JOINTS AND FITTING IN PIPE LINES
(including screwed or welded joints or bolted connections) *continued*

SINGLE HINGED ANGULAR EXPANSION JOINTS

	Unit	15 mm £	20 mm £	25 mm £	32 mm £	40 mm £	50 mm £
Welded	No.	62·02	64·07	66·23	73·27	76·57	87·27
		57·74	*58·43*	*59·13*	*64·40*	*66·13*	*73·49*
Flanged to B.S. 10 table E	„	64·53	66·36	68·25	73·52	77·32	83·36
		62·65	*63·85*	*65·12*	*69·76*	*73·25*	*78·98*

		65 mm £	80 mm £	100 mm £	125 mm £	150 mm £
Welded	„	95·26	108·99	133·81	158·32	205·88
		76·99	*86·44*	*105·00*	*121·78*	*162·03*
Flanged to B.S. 10 table E	„	93·17	105·10	130·74	168·93	206·30
		87·53	*98·21*	*120·72*	*154·31*	*187·51*

ARTICULATED TIED ANGULAR EXPANSION JOINTS

	Unit	15 mm £	20 mm £	25 mm £	32 mm £	40 mm £	50 mm £
Welded	No.	71·45	80·44	90·02	103·35	120·68	139·55
		67·17	*74·80*	*82·92*	*94·48*	*110·24*	*125·77*
Flanged to B.S. 10 table E	„	74·14	82·19	93·00	103·93	122·96	144·02
		72·26	*79·68*	*89·87*	*100·17*	*118·89*	*139·64*

		65 mm £	80 mm £	100 mm £	125 mm £	150 mm £
Welded	„	175·76	203·64	249·28	343·64	437·57
		157·49	*181·09*	*220·47*	*307·10*	*393·72*
Flanged to B.S. 10 table E	„	175·26	201·39	253·80	357·15	457·71
		169·62	*194·50*	*243·78*	*342·53*	*438·92*

BOLTED CONNECTION BETWEEN PAIR OF FLANGES
(including corrugated brass joint ring, bolts and nuts – flanges given separately)

	Unit	15 mm £	20 mm £	25 mm £	32 mm £	40 mm £	50 mm £
Table E	No.	3·47	3·48	3·57	3·59	3·63	4·34
		0·86	*0·87*	*0·96*	*0·98*	*1·02*	*1·73*
Table F	„	3·60	3·60	4·28	4·34	4·35	4·49
		0·99	*0·99*	*1·67*	*1·73*	*1·74*	*1·88*

		65 mm £	80 mm £	100 mm £	125 mm £	150 mm £
Table E	„	4·37	4·43	8·01	8·19	10·13
		1·76	*1·82*	*3·42*	*3·60*	*5·54*
Table F	„	8·09	8·14	8·43	10·47	16·36
		3·50	*3·55*	*3·84*	*5·88*	*8·43*

Mechanical Installations – Prices for Measured Work

PIPE FIXINGS

FOR COPPER TUBE

	Unit	15 mm £	22 mm £	28 mm £	35 mm £	42 mm £	54 mm £	67 mm £	76 mm £	108 mm £
Copper clips										
Saddle band (screwed to wood)	No.	0·81 / 0·03	0·81 / 0·03	1·09 / 0·04	1·10 / 0·06	1·42 / 0·11	1·46 / 0·16			
Single spacing clip (screwed to wood)	„	0·82 / 0·04	0·83 / 0·04	1·12 / 0·08						
Two piece spacing clip (screwed to wood)	„	0·84 / 0·05	0·85 / 0·06	1·13 / 0·09	1·17 / 0·13	1·53 / 0·22	1·60 / 0·30			
Brass brackets and rings										
Single pipe bracket (screwed to wood)	„	1·17 / 0·38	1·21 / 0·43	1·57 / 0·53						
Single pipe bracket (hand to contractor for building in)	„	0·84 / 0·57	0·93 / 0·67	0·98 / 0·72						
Single pipe ring and fixing to threaded end	„	1·10 / 0·67	1·15 / 0·72	1·47 / 0·86	1·52 / 0·91	1·62 / 1·00	1·81 / 1·20	3·43 / 2·82	4·33 / 3·55	6·28 / 5·50
Double pipe ring ditto	„	1·20 / 0·77	1·25 / 0·81	1·71 / 1·10	1·76 / 1·15	1·85 / 1·24	2·09 / 1·48	3·81 / 3·20	4·75 / 3·97	7·43 / 6·65
Wall bracket (plugged and screwed)	„	2·00 / 0·96	2·19 / 1·15	2·43 / 1·39	3·08 / 1·77	3·65 / 2·34	4·27 / 2·97			
Hospital bracket ditto	„	1·95 / 0·81	1·99 / 0·86	2·45 / 1·05	2·54 / 1·15	2·97 / 1·58	3·55 / 2·15			
Wall bracket (hand to contractor for building in)	„	1·03 / 0·77	1·17 / 0·91	1·27 / 1·00	1·31 / 1·05	1·94 / 1·67	2·22 / 1·96			
Female backplate (plugged and screwed)	„	1·52 / 0·38	1·56 / 0·43	1·87 / 0·48	1·92 / 0·53	2·21 / 0·81	2·35 / 0·96	2·52 / 0·96	3·05 / 1·48	3·05 / 1·48
Male backplate (plugged and screwed)	„	1·47 / 0·33	1·52 / 0·38	1·82 / 0·43	1·87 / 0·48	2·11 / 0·72	2·30 / 0·91	2·48 / 0·91	3·05 / 1·48	3·05 / 1·48
Plastic										
Snap on (screwed to wood)	„	0·42 / 0·03	0·43 / 0·04	0·59 / 0·07						
Hinged (screwed to wood)	„	0·43 / 0·04	0·44 / 0·05	0·60 / 0·08						
Floor or ceiling cover plates (fixed over pipe)										
Plastic	„	1·15 / 0·14	1·21 / 0·15	1·27 / 0·17	1·31 / 0·17	1·35 / 0·17	1·43 / 0·18			
Chromium plated steel	„	1·64 / 0·63	1·72 / 0·66	1·82 / 0·72	1·98 / 0·84	2·06 / 0·88	2·24 / 0·99			

PIPE FIXINGS

FOR STEEL TUBE

		15 mm £	20 mm £	25 mm £	32 mm £	40 mm £	50 mm £	65 mm £	80 mm £	100 mm £
Single pipe brackets										
Black malleable (screwed to wood)	Unit No.	1·33 *0·30*	1·40 *0·34*	1·54 *0·44*	1·70 *0·58*	1·93 *0·75*	2·21 *0·96*	2·84 *1·29*	3·56 *1·72*	4·54 *2·45*
Galvanized (screwed to wood)	„	1·43 *0·40*	1·51 *0·45*	1·66 *0·56*	1·86 *0·72*	2·15 *0·97*	2·51 *1·26*	3·23 *1·68*	4·09 *2·25*	5·32 *3·23*
Black malleable (plugged and screwed)	„	2·27 *0·34*	2·33 *0·37*	2·46 *0·46*	2·63 *0·58*	2·90 *0·81*	3·18 *1·02*	3·80 *1·34*	4·52 *1·77*	5·50 *2·50*
Galvanized (plugged and screwed)	„	2·37 *0·44*	2·44 *0·48*	2·58 *0·58*	2·80 *0·75*	3·12 *1·03*	3·48 *1·32*	4·19 *1·73*	5·05 *2·30*	6·28 *3·28*
Black malleable (hand to contractor for building in)	„	0·87 *0·27*	0·96 *0·31*	1·05 *0·37*	1·23 *0·50*	1·43 *0·66*	1·62 *0·78*			
Galvanized (hand to contractor for building in)	„	0·97 *0·37*	1·07 *0·42*	1·18 *0·50*	1·39 *0·66*	1·66 *0·89*	1·88 *1·04*			

Pipe rings and backplates

		15 mm £	20 mm £	25 mm £	32 mm £	40 mm £	50 mm £
Single socket pipe ring, black malleable and fixing to threaded end	Unit No.	1·27 *0·26*	1·36 *0·30*	1·44 *0·34*	1·48 *0·34*	1·63 *0·45*	1·81 *0·56*

		65 mm £	80 mm £	100 mm £	125 mm £	150 mm £
	„	2·37 *0·82*	2·83 *0·99*	3·60 *1·51*	5·66 *3·05*	6·53 *3·40*

		15 mm £	20 mm £	25 mm £	32 mm £	40 mm £	50 mm £
As above but galvanized	„	1·36 *0·35*	1·46 *0·40*	1·56 *0·46*	1·60 *0·46*	1·78 *0·60*	2·00 *0·75*

		65 mm £	80 mm £	100 mm £	125 mm £	150 mm £
		2·65 *1·10*	3·17 *1·33*	4·11 *2·02*	6·68 *4·07*	7·67 *4·54*

		15 mm £	20 mm £	25 mm £	32 mm £	40 mm £	50 mm £
Double socket pipe ring: black malleable and fixing to threaded end	„	1·54 *0·32*	1·63 *0·36*	1·72 *0·41*	1·83 *0·48*	1·95 *0·56*	2·09 *0·63*
As above but galvanized	„	1·65 *0·43*	1·74 *0·47*	1·86 *0·55*	2·00 *0·65*	2·14 *0·75*	2·31 *0·85*
Backplate, black malleable (screwed to wood)	„	1·37 *0·36*	1·42 *0·36*	1·50 *0·40*	1·54 *0·40*	1·65 *0·47*	1·72 *0·47*

		65 mm £	80 mm £	100 mm £	125 mm £	150 mm £
	„	2·06 *0·51*	2·35 *0·51*	2·75 *0·66*	3·31 *0·70*	3·83 *0·70*

PIPE FIXINGS

FOR STEEL TUBE *continued*

	Unit No.	15 mm £	20 mm £	25 mm £	32 mm £	40 mm £	50 mm £
Backplate, galvanized (screwed to wood)	No.	1·48 *0·47*	1·53 *0·47*	1·61 *0·51*	1·65 *0·51*	1·79 *0·61*	1·86 *0·61*
		65 mm £	80 mm £	100 mm £	125 mm £	150 mm £	
	,,	2·22 *0·67*	2·51 *0·67*	2·95 *0·86*	3·51 *0·90*	4·03 *0·90*	
		15 mm £	20 mm £	25 mm £	32 mm £	40 mm £	50 mm £
Backplate, black malleable (plugged and screwed)	,,	2·31 *0·40*	2·36 *0·40*	2·46 *0·46*	2·51 *0·46*	2·66 *0·57*	2·73 *0·57*
		65 mm £	80 mm £	100 mm £	125 mm £	150 mm £	
	,,	3·07 *0·61*	3·36 *0·61*	3·84 *0·84*	4·19 *0·84*	4·55 *0·84*	
		15 mmm £	20 mm £	25 mm £	32 mm £	40 mm £	50 mm £
As above but galvanized . .	,,	2·42 *0·51*	2·47 *0·51*	2·57 *0·57*	2·62 *0·57*	2·80 *0·71*	2·87 *0·71*
		65 mm £	80 mm £	100 mm £	125 mm £	150 mm £	
	,,	3·23 *0·77*	3·52 *0·77*	4·02 *1·02*	4·37 *1·02*	4·73 *1·02*	

PIPE FIXINGS

FLOOR OR CEILING COVER PLATES
(fixed over pipe)

	Unit	15 mm £	20 mm £	25 mm £	32 mm £	40 mm £	50 mm £
Plastic	No.	1·15	1·21	1·27	1·31	1·18	1·43
		0·14	*0·15*	*0·17*	*0·17*	*0·17*	*0·18*
Chromium plated	,,	1·64	1·72	1·82	1·98	2·06	2·24
		0·63	*0·66*	*0·72*	*0·84*	*0·88*	*0·99*

PIPE ROLLER AND CHAIR BOLTED TO SUPPORT
(suitable for pipes of diameters shown)

	Unit	Up to 51 mm £	76 mm £	102 mm £	152 mm £
Black malleable	No.	3·00	4·26	5·73	8·45
		0·49	*0·61*	*1·55*	*2·81*

Mechanical Installations - Prices for Measured Work

PIPE FIXINGS

FABRICATED HANGERS AND BRACKETS
NOTE: *It has been assumed there would be sufficient quantities required to obtain the benefit of bulk purchase*

	Unit	£
Mild steel flats of various sizes including cutting to length, handling and fixing	kg	1·44
		0·74
Ditto angle	,,	1·62
		0·71
Ditto channel	,,	1·62
		0·71
Ditto rods up to 10 mm diameter	,,	1·62
		0·71

		3 mm	5 mm	6 mm	8 mm	10 mm	13 mm	16 mm
	Unit	£	£	£	£	£	£	£
Cut, other than initial cut, on metal of thickness shown	Per 25 mm			0·21	0·26	0·32	0·35	0·40
Ragged ends of Flats, of thickness shown	No.			0·61	0·66	0·71		
Angle, ditto	,,		0·93	0·93	1·00	1·06		
Bend on flat of thickness shown	,,			0·40	0·45	0·47		
Twist ditto	,,			0·40	0·45	0·47		
Drill hole not exceeding 6 mm diameter in steel not exceeding thickness shown	,,			0·21			0·26	
Ditto 13 mm diameter ditto	,,			0·26			0·32	
Ditto 25 mm diameter ditto	,,			0·40			0·47	
Thread end of rod of diameter shown	,,	0·35		0·40			0·47	
Bend in rod ditto	,,	0·32		0·32			0·32	
Form eye in rod ditto	,,	0·32		0·32			0·32	
Butt welding mild steel of the thickness shown	Per 25 mm		0·51	0·69			1·01	1·33
Welding ends of rods of the diameter shown to steel	No.	0·43		0·56			0·85	

MEDIUM QUALITY BLACK PIPE SLEEVES IN MILD STEEL
(to B.S. 1387 with plain ends and hand to contractor for building in)

		15 mm	20 mm	25 mm	32 mm	40 mm	50 mm
150 mm long to pass pipe of the diameter shown	Unit No.	£ 1·15	£ 1·55	£ 1·71	£ 2·00	£ 2·05	£ 2·45
		0·11	0·12	0·18	0·24	0·29	0·38
300 mm ditto	,,	1·24	1·67	1·89	2·24	2·32	2·85
		0·20	0·24	0·36	0·48	0·56	0·78
600 mm ditto	,,	1·36	1·79	2·07	2·48	2·62	3·22
		0·32	0·36	0·54	0·72	0·86	1·15

		65 mm	80 mm	100 mm	125 mm	150 mm
		£	£	£	£	£
150 mm ditto	,,	2·94	3·38	4·93	6·46	8·50
		0·53	0·67	0·96	1·24	1·61
300 mm ditto	,,	3·45	4·05	5·88	7·69	10·11
		1·04	1·34	1·91	2·47	3·22
600 mm ditto	,,	4·02	4·73	6·85	8·93	11·71
		1·61	2·02	2·88	3·71	4·82

DUCTWORK

Galvanized sheet metal rectangular section ductwork for low velocity systems constructed in accordance with Table 2 of Specification DW 141 for Sheet Metal Ductwork published by the H.V.C.A.

Length of side mm	Nominal sheet thickness mm	Maximum spacing between joints/stiffeners Without beading or cross breaking mm	With beading at 300 mm centres mm	With cross breaking mm	Minimum angle section for intermediate stiffeners mm
Up to 400	0·6	—	—	—	—
401 to 600	0·6	1500	—	—	25 × 25 × 3
601 to 800	0·8	1500	—	2000	25 × 25 × 3
801 to 1000	0·8	1200	1500	1500	25 × 25 × 3
1001 to 1500	1·0	800	1200	1200	40 × 40 × 4
1501 to 2250	1·0	800	800	800	40 × 40 × 4
2251 to 3000	1·2	600	600	600	50 × 50 × 5

For ducting galvanized after manufacture

Up to 300	1·2		As above
301 and over	1·6		As above

DUCTWORK

GALVANIZED SHEET METAL LOW VELOCITY RECTANGULAR SECTION DUCTWORK
(including all necessary stiffeners, joints, couplers in the running length and duct supports)

	Unit	100 × 100 mm £	150 × 100 mm £	150 × 150 mm £	200 × 100 mm £	200 × 150 mm £
Straight duct	m	17·86	17·86	17·86	17·86	19·39
Extra over for						
90° square bend with turning vanes	No.	24·41	24·41	24·41	24·41	29·13
90° radius bend without turning vanes	,,	23·25	23·25	23·25	23·25	27·74
45° ditto	,,	19·98	19·98	19·98	19·98	23·42
Stopped end	,,	3·59	3·59	3·59	3·59	3·59
Damper, butterfly pattern	,,	15·74	15·74	15·74	16·04	16·04
Fire damper, shutter type with access	,,	49·86	49·86	49·86	50·52	50·52
Diminishing piece, large end measured	,,	12·81	12·81	12·81	12·81	13·21
Square to round piece	,,	18·65	18·65	18·65	18·65	19·06
Branch, based on branch size	,,	9·73	9·73	9·73	9·73	10·67
Spigot	,,	6·42	6·42	6·42	6·42	7·05
Equal twin bend with turning vanes	,,	36·61	36·61	36·61	36·61	43·70
Flanged end	,,	3·59	3·59	3·59	3·59	3·59

		200 × 200 mm £	250 × 100 mm £	250 × 150 mm £	250 × 200 mm £	250 × 250 mm £
Straight duct	m	19·39	19·39	19·39	19·39	20·92
Extra over for						
90° square bend with turning vanes	No.	29·13	29·13	29·13	29·13	33·80
90° radius bend without turning vanes	,,	27·74	27·74	27·74	27·74	32·22
45° ditto	,,	23·42	23·42	23·42	23·42	26·87
Stopped end	,,	3·59	3·59	3·59	3·59	3·98
Damper, butterfly pattern	,,	16·04	16·04	16·04	16·04	17·11
Fire damper, shutter type with access	,,	50·52	50·52	50·52	50·52	54·13
Diminishing piece, large end measured	,,	13·21	13·21	13·21	13·21	13·61
Square to round piece	,,	19·06	19·06	19·06	19·06	19·45
Branch, based on branch size	,,	10·67	10·67	10·67	10·67	11·62
Spigot	,,	7·05	7·05	7·05	7·05	7·67
Equal twin bend with turning vanes	,,	43·70	43·70	43·70	43·70	50·70
Flanged end	,,	3·59	3·59	3·59	3·59	3·98

DUCTWORK

GALVANIZED SHEET METAL LOW VELOCITY RECTANGULAR SECTION DUCTWORK
(including all necessary stiffeners, joints, couplers in the running length and duct supports) *continued*

	Unit	300 × 100 mm £	300 × 150 mm £	300 × 200 mm £	300 × 250 mm £	300 × 300 mm £
Straight duct	m	19·39	19·39	20·92	20·92	20·92
Extra over for						
90° square bend with turning vanes	No.	29·13	29·13	33·80	33·80	33·80
90° radius bend without turning vanes	,,	27·74	27·74	32·22	32·22	32·22
45° ditto	,,	23·42	23·42	26·87	26·87	26·87
Stopped end	,,	3·59	3·59	3·98	3·98	3·98
Damper, butterfly pattern	,,	16·04	16·04	17·11	17·11	17·11
Fire damper, shutter type with access	,,	50·52	50·52	54·13	54·13	54·13
Diminishing piece, large end measured	,,	13·21	13·21	13·61	13·61	13·61
Square to round piece	,,	19·06	19·06	19·45	19·45	19·45
Branch, based on branch size	,,	10·67	10·67	11·62	11·62	11·62
Spigot	,,	7·05	7·05	7·67	7·67	7·67
Equal twin bend with turning vanes	,,	43·70	43·70	50·70	50·70	50·70
Flanged end	,,	3·59	3·59	3·98	3·98	3·98

		400 × 100 mm £	400 × 150 mm £	400 × 200 mm £	400 × 250 mm £	400 × 300 mm £
Straight duct	m	20·92	20·92	20·92	22·45	22·45
Extra over for						
90° square bend with turning vanes	No.	33·80	33·80	33·80	38·55	38·55
90° radius bend without turning vanes	,,	32·22	32·22	32·22	36·72	36·72
45° ditto	,,	26·87	26·87	26·87	30·31	30·31
Stopped end	,,	3·98	3·98	3·98	5·19	5·58
Damper, butterfly pattern	,,	17·11	17·11	17·11	17·72	17·72
Fire damper, shutter type with access	,,	54·13	54·13	54.13	55·47	55·47
Diminishing piece, large end measured	,,	13·61	13·61	13·61	14·02	14·02
Square to round piece	,,	19·45	19·45	19·45	19·85	19·85
Branch, based on branch size	,,	11·62	11·62	11·62	12·57	12·57
Spigot	,,	7·67	7·67	7·67	8·29	8·29
Equal twin bend with turning vanes	,,	50·70	50·70	50·70	57·82	57·82
Flanged end	,,	3·98	3·98	3·98	5·19	5·58

DUCTWORK

GALVANIZED SHEET METAL LOW VELOCITY RECTANGULAR SECTION DUCTWORK
(including all necessary stiffeners, joints, couplers in the running length and duct supports) *continued*

	Unit	400 × 400 mm £	500 × 150 mm £	500 × 200 mm £	500 × 250 mm £	500 × 300 mm £
Straight duct	m	23·99	22·45	22·45	22·45	23·99
Extra over for						
90° square bend with turning vanes	No.	43·26	38·55	38·55	38·55	43·26
90° radius bend without turning vanes	„	41·19	36·72	36·72	36·72	41·19
45° ditto	„	33·75	30·31	30·31	30·31	33·75
Stopped end	„	6·38	5·58	5·58	5·58	6·38
Damper, butterfly pattern	„	19·93	17·72	17·72	17·72	19·93
Fire damper, shutter type with access panel	„	60·32	55·47	55·47	55·47	60·32
Diminishing piece, large end measured	„	14·41	14·02	14·02	14·02	14·41
Square to round piece	„	20·26	19·85	19·85	19·85	20·26
Branch, based on branch size	„	13·50	12·57	12·57	12·57	13·50
Spigot	„	8·90	8·29	8·29	8·29	8·90
Equal twin bend with turning vanes	„	64·89	57·82	57·82	57·82	64·89
Flanged end	„	6·38	5·58	5·58	5·58	6·38

		500 × 400 mm £	500 × 500 mm £	600 × 150 mm £	600 × 200 mm £	600 × 250 mm £
Straight duct	m	23·99	25·51	22·45	23·99	23·99
Extra over for						
90° square bend with turning vanes	No.	43·26	47·98	38·55	43·26	43·26
90° radius bend without turning vanes	„	41·19	45·69	36·72	41·19	41·19
45° ditto	„	33·75	37·19	30·31	33·75	33·75
Stopped end	„	6·38	7·97	5·58	6·38	6·38
Damper, butterfly pattern	„	19·93	20·59	17·72	19·93	19·93
Fire damper, shutter type with access panel	„	60·32	61·80	55·47	60·32	60·32
Diminishing piece, large end measured	„	14·41	14·82	14·02	14·41	14·41
Square to round piece	„	20·26	20·65	19·85	20·26	20·26
Branch, based on branch size	„	13·50	14·45	12·57	13·50	13·50
Spigot	„	8·90	9·53	8·29	8·90	8·90
Equal twin bend with turning vanes	„	64·89	71·96	57·82	64·89	64·89
Flanged end	„	6·38	7·97	5·58	6·38	6·38

DUCTWORK

GALVANIZED SHEET METAL LOW VELOCITY RECTANGULAR SECTION DUCTWORK
(including all necessary stiffeners, joints, couplers in the running length and duct supports) *continued*

	Unit	600 × 300 mm £	600 × 400 mm £	600 × 500 mm £	600 × 600 mm £	800 × 200 mm £
Straight duct	m	23·99	25·51	25·51	35·23	31·22
Extra over for						
90° square bend with turning vanes	No.	43·26	43·26	43·26	68·31	58·28
90° radius bend without turning vanes	,,	41·19	41·19	41·19	65·06	55·74
45° ditto	,,	33·75	33·75	33·75	53·19	45·65
Stopped end	,,	6·38	6·38	6·38	9·56	7·97
Damper, butterfly pattern	,,	19·93	19·93	19·93	21·37	16·66
Fire damper, shutter type with access panel	,,	60·32	60·32	60·32	77·99	53·14
Diminishing piece, large end measured	,,	14·41	14·41	14·41	22·46	19·46
Square to round piece	,,	20·26	20·26	20·26	32·05	27·97
Branch, based on branch size	,,	13·50	13·50	13·50	18·06	16·34
Spigot	,,	8·90	8·90	8·90	11·92	10·78
Equal twin bend with turning vanes	,,	64·89	64·89	64·89	102·47	87·79
Flanged end	,,	6·38	6·38	6·38	9·56	7·97

	Unit	800 × 250 mm £	800 × 300 mm £	800 × 400 mm £	800 × 500 mm £	800 × 600 mm £
Straight duct	m	31·22	34·71	34·71	38·20	41·68
Extra over for						
90° square bend with turning vanes	No.	58·28	68·31	68·31	78·10	87·88
90° radius bend without turning vanes	,,	55·74	65·06	65·06	74·38	83·69
45° ditto	,,	45·65	53·19	53·19	60·73	68·26
Stopped end	,,	7·97	8·78	8·78	10·36	11·16
Damper						
butterfly pattern	,,	16·66	19·72			
multi-leaf pattern	,,			81·39	90·53	100·45
Fire damper, shutter type with access panel	,,	53·14	58·03	58·03	68·58	73·66
Diminishing piece, large end measured	,,	19·46	22·46	22·46	25·47	28·48
Square to round piece	,,	27·97	32·05	32·05	36·14	40·23
Branch, based on branch size	,,	16·34	18·06	18·06	19·80	21·52
Spigot	,,	10·78	11·92	11·92	13·06	14·19
Equal twin bend with turning vanes	,,	87·79	102·47	102·47	117·15	131·81
Flanged end	,,	7·97	8·78	8·78	10·36	11·16

DUCTWORK

GALVANIZED SHEET METAL LOW VELOCITY RECTANGULAR SECTION DUCTWORK
(including all necessary stiffeners, joints, couplers in the running length and duct supports) *continued*

	Unit	800 × 800 mm £	1000 × 250 mm £	1000 × 300 mm £	1000 × 400 mm £	1000 × 500 mm £
Straight duct	m	45·17	38·20	38·20	41·68	41·68
Extra over for						
90° square bend with turning vanes	No.	97·66	78·10	78·10	87·88	87·88
90° radius bend without turning vanes	,,	93·01	74·38	74·38	83·69	83·69
45° ditto	,,	75·80	60·73	60·73	68·26	68·26
Stopped end	,,	12·75	9·97	10·36	11·16	11·95
Damper, multi-leaf pattern	,,	121·66	77·55	80·05	90·33	100·05
Fire damper, shutter type with access panel	,,	167·34	112·63	115·76	116·87	127·81
Diminishing piece, large end measured	,,	31·49	25·46	25·46	28·48	28·48
Square to round piece	,,	44·32	36·14	36·14	40·23	40·23
Branch, based on branch size	,,	23·24	19·80	19·80	21·52	21·52
Spigot	,,	15·33	13·06	13·06	14·19	14·19
Equal twin bend with turning vanes	,,	146·49	117·15	117·15	131·81	131·81
Flanged end	,,	12·75	9·97	10·36	11·16	11·95

		1000 × 600 mm £	1000 × 800 mm £	1000 × 1000 mm £	1200 × 300 mm £	1200 × 400 mm £
Straight duct	m	45·17	48·66	52·15	41·68	45·17
Extra over for						
90° square bend with turning vanes	No.	97·66	107·43	117·22	87·88	97·66
90° radius bend without turning vanes	,,	93·01	102·32	111·64	83·69	93·01
45° ditto	,,	75·80	83·34	90·89	68·26	75·80
Stopped end	,,	12·75	14·35	15·94	11·95	12·75
Damper, multi-leaf pattern	,,	110·50	133·17	156·50	127·96	144·68
Fire damper, shutter type with access panel	,,	153·46	181·88	210·58	175·15	179·37
Diminishing piece, large end measured	,,	31·49	34·49	37·49	28·48	31·49
Square to round piece	,,	44·32	48·39	52·48	40·23	44·32
Branch, based on branch size	,,	23·24	24·96	26·68	21·52	23·24
Spigot	,,	15·33	16·47	17·60	14·19	15·33
Equal twin bend with turning vanes	,,	146·49	161·16	175·83	131·81	146·49
Flanged end	,,	12·75	14·35	15·94	11·95	12·75

DUCTWORK

GALVANIZED SHEET METAL LOW VELOCITY RECTANGULAR SECTION DUCTWORK
(including all necessary stiffeners, joints, couplers in the running length and duct supports) *continued*

	Unit	1200 × 500 mm £	1200 × 600 mm £	1200 × 800 mm £	1200 × 1000 mm £	1200 × 1200 mm £
Straight duct	m	58·78	58·78	71·25	77·49	83·73
Extra over for						
90° square bend with turning vanes	No.	110·76	110·76	135·62	148·04	160·46
90° radius bend without turning vanes	,,	105·49	105·49	129·16	140·99	152·82
45° ditto	,,	62·90	79·31	95·74	103·95	112·16
Stopped end	,,	13·55	14·35	17·53	19·13	19·13
Damper, multi-leaf pattern	,,	162·19	180·65	221·88	243·67	361·31
Fire damper, shutter type with access panel	,,	217·59	240·46	249·05	273·86	464·44
Diminishing piece, large end measured	,,	27·06	31·38	35·69	37·83	40·00
Square to round piece	,,	39·34	46·35	53·36	56·86	60·37
Branch, based on branch size	,,	21·77	24·29	26·81	28·06	29·33
Spigot	,,	14·37	16·03	17·69	18·53	19·36
Equal twin bend with turning vanes	,,	166·15	166·15	203·43	222·06	240·69
Flanged end	,,	13·55	14·35	17·53	19·13	19·13

		1400 × 400 mm £	1400 × 500 mm £	1400 × 600 mm £	1400 × 800 mm £	1400 × 1000 mm £
Straight duct	m	58·78	65·02	71·25	77·49	83·73
Extra over for						
90° square bend with turning vanes	No.	110·76	123·19	135·62	148·04	160·46
90° radius bend without turning vanes	,,	105·49	117·32	129·16	140·99	152·82
45° ditto	,,	62·90	87·53	95·74	103·95	112·61
Stopped end	,,	14·35	15·14	15·94	17·54	19·13
Damper, multi-leaf pattern	,,	153·64	171·71	190·76	221·88	243·67
Fire damper, shutter type with access panel	,,	206·97	229·40	253·00	291·57	289·74
Diminishing piece, large end measured	,,	27·06	33·53	35·69	37·83	40·00
Square to round piece	,,	39·34	49·85	53·36	56·86	60·37
Branch, based on branch size	,,	21·77	25·55	28·81	28·06	29·33
Spigot	,,	14·37	16·86	17·69	18·53	19·36
Equal twin bend with turning vanes	,,	166·15	184·79	203·43	222·06	240·69
Flanged end	,,	14·35	15·14	15·94	17·54	19·13

DUCTWORK

GALVANIZED SHEET METAL LOW VELOCITY RECTANGULAR SECTION DUCTWORK
(including all necessary stiffeners, joints, couplers in the running length and duct supports) *continued*

	Unit	1400 × 1200 mm £	1600 × 400 mm £	1600 × 500 mm £	1600 × 600 mm £	1600 × 800 mm £
Straight duct	m	96·20	71·25	71·25	77·49	77·49
Extra over for						
90° square bend with turning vanes	No.	185·30	135·62	135·62	148·04	148·04
90° radius bend without turning vanes	„	176·47	129·16	129·16	140·99	140·99
45° ditto	„	128·57	95·74	95·74	103·95	103·95
Stopped end	„	20·72	15·94	16·73	17·54	17·54
Damper, multi-leaf pattern	„	381·54	162·77	180·27	200·73	243·27
Fire damper, shutter type with access panel	„	446·71	250·67	277·62	309·12	374·60
Diminishing piece, large end measured	„	44·30	35·69	35·69	37·83	37·83
Square to round piece	„	67·38	53·36	53·36	56·86	56·86
Branch, based on branch size	„	31·84	26·81	26·81	28·06	28·06
Spigot	„	21·02	17·69	17·69	18·53	18·53
Equal twin bend with turning vanes	„	277·94	203·43	203·43	222·06	222·06
Flanged end	„	20·72	15·94	16·73	17·54	17·54

	Unit	1600 × 1000 mm £	1600 × 1200 mm £	1800 × 500 mm £	1800 × 600 mm £	1800 × 800 mm £
Straight duct	m	96·20	102·43	83·72	83·72	96·20
Extra over for						
90° square bend with turning vanes	No.	185·30	197·71	160·46	160·46	185·30
90° radius bend without turning vanes	„	176·47	188·30	152·82	152·82	176·47
45° dittos	„	128·57	136·79	112·61	112·61	128·57
Stopped end	„	20·72	22·31	18·33	19·13	20·72
Damper, multi-leaf pattern	„	266·80	401·14	191·74	210·99	255·14
Fire damper, shutter type with access panel	„	347·51	531·08	252·79	261·56	316·30
Diminishing piece, large end measured	„	44·30	46·47	40·00	40·00	44·30
Square to round piece	„	67·38	70·88	60·37	60·37	67·38
Branch, based on branch size	„	31·84	33·11	29·33	29·33	31·84
Spigot	„	21·02	21·85	19·36	19·36	21·02
Equal twin bend with turning vanes	„	277·94	296·57	240·69	240·69	277·94
Flanged end	„	20·72	22·31	18·33	19·13	20·72

DUCTWORK

GALVANIZED SHEET METAL LOW VELOCITY RECTANGULAR SECTION DUCTWORK
(including all necessary stiffeners, joints, couplers in the running length and duct supports) *continued*

	Unit	1800 × 1000 mm £	1800 × 1200 mm £	2000 × 500 mm £	2000 × 600 mm £	2000 × 800 mm £
Straight duct	m	102·44	108·66	89·96	96·20	102·44
Extra over for						
90° square bend with turning vanes	No.	197·71	210·13	172·87	185·30	197·71
90° radius bend without turning vanes	,,	188·30	200·12	164·65	176·47	188·30
45° ditto	,,	136·79	144·99	120·36	128·57	136·79
Stopped end	,,	22·31	23·90	19·92	20·72	22·31
Damper, multi-leaf pattern	,,	289·94	349·25	200·11	223·23	266·80
Fire damper, shutter type with access panel	,,	375·92	539·75	264·16	289·56	346·71
Diminishing piece, large end measured	,,	46·47	48·62	42·16	44·30	46·47
Square to round piece	,,	70·88	74·38	61·56	67·38	70·88
Branch, based on branch size	,,	33·11	34·37	30·58	31·84	33·11
Spigot	,,	21·85	22·67	20·19	21·02	21·85
Equal twin bend with turning vanes	,,	296·57	315·20	259·31	277·94	296·57
Flanged end	,,	22·31	23·90	19·92	20·72	22·31

		2000 × 1000 mm £	2000 × 1200 mm £
Straight duct	m	108·66	121·14
Extra over for			
90° square bend with turning vanes	No.	210·13	234·97
90° radius bend without turning vanes	,,	200·12	223·78
45° ditto	,,	144·99	161·42
Stopped end	,,	21·25	25·48
Damper, multi-leaf pattern	,,	313·08	441·78
Fire damper, shutter type with access panel	,,	403·86	563·88
Diminishing piece, large end measured	,,	48·62	52·93
Square to round piece	,,	74·38	81·39
Branch, based on branch size	,,	34·37	36·89
Spigot	,,	22·67	24·34
Equal twin bend without turning vanes	,,	315·20	352·46
Flanged end	,,	21·25	25·48

Mechanical Installations – Prices for Measured Work

DUCTWORK

Galvanized sheet metal rectangular section ductwork for high velocity systems constructed in accordance with Table 3 of specification DW 141 for Sheet Metal Ductwork published by the H.V.C.A.

Length of side mm	Nominal sheet thickness mm	Maximum spacing between joints/stiffeners mm	Minimum angle section for joints/stiffeners mm
Up to 300	0·8	—	—
301 to 450	0·8	1200	25 × 25 × 3
451 to 600	0·8	1200	25 × 25 × 3
601 to 800	0·8	1000	40 × 40 × 4
801 to 1000	0·8	800	40 × 40 × 4
1001 to 1500	1·0	600	50 × 50 × 5
1501 to 2250	1·2	600	50 × 50 × 5 (plus 8 mm dia. central tie rod)

DUCTWORK

GALVANIZED SHEET METAL HIGH VELOCITY RECTANGULAR SECTION DUCTWORK
(including all necessary stiffeners, joints, couplers in the running length and duct supports)

	Unit	150 × 100 mm £	150 × 150 mm £	200 × 100 mm £	200 × 150 mm £	200 × 200 mm £
Straight duct	m	29·20	29·20	29·20	35·14	35·14
Extra over for						
90° square bend with turning vanes	No.	39·39	39·39	39·39	49·14	49·14
90° radius bend without turning vanes	„	37·51	37·51	37·51	46·80	46·80
45° ditto	„	31·65	31·65	31·65	39·19	39·19
Stopped end	„	3·59	3·59	3·59	3·59	3·59
Damper, butterfly pattern	„	18·10	18·10	18·10	18·45	18·45
Fire damper, shutter type with access panel	„	54·94	54·94	54·94	54·94	54·94
Diminishing piece large end measured	„	16·81	16·81	16·81	21·00	21·00
Square to round piece	„	23·50	23·50	23·50	29·26	29·26
Branch, based on branch size	„	11·60	11·60	11·60	13·76	13·76
Spigot	„	5·80	5·80	5·80	6·87	6·87
Equal twin bend with turning vanes	„	59·08	59·08	59·08	73·71	73·71
Flanged end	„	3·59	3·59	3·59	3·59	3·59

		250 × 100 mm £	250 × 150 mm £	250 × 200 mm £	250 × 250 mm £	300 × 100 mm £
Straight duct	m	35·14	35·14	35·14	41·07	35·14
Extra over for						
90° square bend with turning vanes	No.	49·14	49·14	49·14	58·88	49·14
90° radius bend without turning vanes	„	46·80	46·80	46·80	56·07	46·80
45° ditto	„	39·19	39·19	39·19	46·73	39·19
Stopped end	„	3·59	3·59	3·59	3·98	3·59
Damper, butterfly pattern	„	18·45	18·45	18·45	19·76	18·45
Fire damper, shutter type with access panel	„	54·94	60·40	61·16	63·30	59·94
Diminishing piece large end measured	„	21·00	21·00	21·00	25·19	21·00
Square to round piece	„	29·26	29·26	29·26	35·00	29·26
Branch, based on branch size	„	13·76	13·76	13·76	15·93	13·76
Spigot	„	6·87	6·87	6·87	7·96	6·87
Equal twin bend with turning vanes	„	73·71	73·71	73·71	88·32	73·71
Flanged end	„	3·59	3·59	3·59	3·98	3·59

DUCTWORK

GALVANIZED SHEET METAL HIGH VELOCITY RECTANGULAR SECTION DUCTWORK
(inlcuding all necessary stiffeners, joints, couplers in the running length and duct supports) *continued*

	Unit	300 × 150 mm £	300 × 200 mm £	300 × 250 mm £	300 × 300 mm £	400 × 100 mm £
Straight duct	m	35·14	41·07	41·07	41·07	41·07
Extra over for						
90° square bend with turning vanes	No.	49·14	58·88	58·88	58·88	58·88
90° radius bend without turning vanes	,,	46·80	56·07	56·07	56·07	56·07
45° ditto	,,	39·19	46·73	46·73	46·73	46·73
Stopped end	,,	3·59	3·98	4·39	4·78	3·98
Damper, butterfly pattern	,,	18·45	18·45	19·76	19·76	19·76
Fire damper, shutter type with access panel	,,	54·94	61·16	63·30	63·30	63·30
Diminishing piece, large end measured	,,	21·00	25·19	25·19	25·19	25·19
Square to round piece	,,	29·26	35·00	35·00	35·00	35·00
Branch, based on branch size	,,	13·76	15·93	15·93	15·93	15·93
Spigot	,,	6·87	7·96	7·96	7·96	7·96
Equal twin bend with turning vanes	,,	73·71	88·32	88·32	88·32	88·32
Flanged end	,,	3·59	3·98	4·39	4·78	3·98

		400 × 150 mm £	400 × 200 mm £	400 × 250 mm £	400 × 300 mm £	400 × 400 mm £
Straight duct	m	41·07	41·07	47·00	47·00	52·95
Extra over for						
90° square bend with turning vanes	No.	58·88	58·88	68·62	68·62	78·35
90° radius bend without turning vanes	,,	56·07	56·07	65·35	65·35	74·62
45° ditto	,,	46·73	46·73	54·26	54·26	61·80
Stopped end	,,	4·38	4·78	5·19	5·19	6·38
Damper, butterfly pattern	,,	19·76	19·76	20·47	20·47	23·01
Fire damper, shutter type with access panel	,,	63·30	63·30	63·86	63·86	69·44
Diminishing piece, large end measured	,,	25·19	25·19	29·37	29·37	33·56
Square to round piece	,,	35·00	35·00	40·76	40·76	46·51
Branch, based on branch size	,,	15·93	15·93	18·10	18·10	20·26
Spigot	,,	7·96	7·96	9·04	9·04	10·13
Equal twin bend with turning vanes	,,	88·32	88·32	102·92	102·92	117·52
Flanged end	,,	4·38	4·78	5·19	5·58	6·38

DUCTWORK

GALVANIZED SHEET METAL HIGH VELOCITY RECTANGULAR SECTION DUCTWORK
(including all necessary stiffeners, joints, couplers in the running length and duct supports) *continued*

	Unit	500 × 150 mm £	500 × 200 mm £	500 × 250 mm £	500 × 300 mm £	500 × 400 mm £
Straight duct	m	47·00	47·00	47·00	52·95	52·95
Extra over for						
90° square bend with turning vanes	No.	68·62	68·62	68·62	78·35	78·35
90° radius bend without turning vanes	,,	65·35	65·35	65·35	74·62	74·62
45° ditto	,,	54·26	54·26	54·26	61·80	61·80
Stopped end	,,	5·18	5·58	5·98	6·38	7·17
Damper, butterfly pattern	,,	20·47	20·47	20·47	23·01	23·01
Fire damper, shutter type with access panel	,,	63·09	63·86	64·64	64·64	70·01
Diminishing piece, large end measured	,,	29·37	29·37	29·37	33·56	33·56
Square to round piece	,,	40·76	40·76	40·76	46·51	46·51
Branch, based on branch size	,,	18·10	18·10	18·10	20·26	20·26
Spigot	,,	9·04	9·04	9·04	10·13	10·13
Equal twin bend with turning vanes	,,	102·92	102·92	102·92	117·52	117·52
Flanged end	,,	5·18	5·58	5·98	6·38	7·17

	Unit	500 × 500 mm £	600 × 150 mm £	600 × 200 mm £	600 × 250 mm £	600 × 300 mm £
Straight duct	m	58·87	47·00	52·95	52·95	52·95
Extra over for						
90° square bend with turning vanes	No.	88·11	68·62	78·35	78·35	78·35
90° radius bend without turning vanes	,,	83·91	65·35	74·62	74·62	74·62
45° ditto	,,	69·33	54·26	61·80	61·80	61·80
Stopped end	,,	7·97	5·98	6·36	6·78	7·17
Damper, butterfly pattern	,,	23·78	20·47	23·09	23·09	23·09
Fire damper, shutter type with access panel	,,	72·13	63·86	64·64	65·81	65·81
Diminishing piece, large end measured	,,	37·75	29·37	33·56	33·56	33·56
Square to round piece	,,	52·25	40·76	46·51	46·51	46·51
Branch, based on branch size	,,	22·43	18·10	20·26	20·26	20·26
Spigot	,,	11·20	9·04	10·13	10·13	10·13
Equal twin bend with turning vanes	,,	132·16	102·92	117·52	117·52	117·52
Flanged end	,,	7·97	5·98	6·36	6·78	7·17

DUCTWORK

GALVANIZED SHEET METAL HIGH VELOCITY RECTANGULAR SECTION DUCTWORK
(including all necessary stiffeners, joints, couplers in the running length and duct supports) *continued*

	Unit	600 × 400 mm £	600 × 500 mm £	600 × 600 mm £	800 × 200 mm £	800 × 250 mm £
Straight duct	m	58·87	64·80	64·80	58·87	58·87
Extra over for						
90° square bend with turning vanes	No.	88·11	98·06	98·06	88·11	88·11
90° radius bend without turning vanes	„	83·91	93·39	93·39	83·91	83·91
45° ditto	„	69·33	76·88	76·88	69·33	69·33
Stopped end	„	7·97	8·77	9·56	7·97	8·37
Damper, butterfly pattern	„	23·01	23·01	23·01	19·24	19·24
Fire damper, shutter type with access panel	„	71·12	72·54	74·47	67·51	75·26
Diminishing piece, large end measured	„	37·75	41·93	41·93	37·75	37·75
Square to round piece	„	52·25	58·00	58·00	52·25	52·25
Branch, based on branch size	„	22·43	24·59	24·59	22·43	22·43
Spigot	„	11·20	12·29	12·29	11·20	11·20
Equal twin bend with turning vane	„	132·16	147·09	147·09	132·16	132·16
Flanged end	„	7·97	8·77	9·56	7·97	8·37

		800 × 300 mm £	800 × 400 mm £	800 × 500 mm £	800 × 600 mm £	800 × 800 mm £
Straight duct	m	64·80	64·80	84·41	94·87	105·32
Extra over for						
90° square bend with turning vanes	No.	98·06	98·06	103·13	111·73	120·33
90° radius bend without turning vanes	„	93·39	93·39	98·22	106·41	114·60
45° ditto	„	76·88	76·88	81·05	87·42	93·79
Stopped end	„	8·77	9·56	10·36	11·15	12·75
Damper						
butterfly pattern	„	22·77				
multi-leaf pattern	„		94·00	104·56	116·02	141·00
Fire damper, shutter type with access panel	„	91·44	116·04	140·97	158·75	189·23
Diminishing piece, large end measured	„	41·93	41·93	47·52	51·49	55·46
Square to round piece	„	58·00	58·00	67·32	72·88	78·45
Branch, based on branch size	„	24·59	24·59	24·59	26·25	27·90
Spigot	„	12·29	12·29	12·29	13·12	13·94
Equal twin bend with turning vanes	„	147·09	147·09	154·70	167·60	180·49
Flanged end	„	8·77	9·56	10·36	11·15	12·75

Mechanical Installations – Prices for Measured Work

DUCTWORK

GALVANIZED SHEET METAL HIGH VELOCITY RECTANGULAR SECTION DUCTWORK
(including all necessary stiffeners, joints, couplers in the running length and duct supports) *continued*

	Unit	1000 × 250 mm £	1000 × 300 mm £	1000 × 400 mm £	1000 × 500 mm £	1000 × 600 mm £
Straight duct	m	84·42	84·42	94·87	94·87	105·32
Extra over for						
90° square bend with turning vanes	No.	103·13	103·13	111·73	111·73	120·33
90° radius bend without turning vanes	,,	98·22	98·22	106·41	106·41	114·60
45° ditto	,,	81·04	81·04	87·42	87·42	93·79
Stopped end	,,	9·97	10·36	11·15	11·95	12·75
Damper, multi-leaf pattern	,,	89·57	92·45	104·33	115·56	127·63
Fire damper, shutter type with access panel	,,	127·06	129·54	144·78	159·16	158·24
Diminishing piece, large end measured	,,	47·52	47·52	51·49	51·49	55·46
Square to round piece	,,	67·32	67·32	72·88	72·88	78·45
Branch, based on branch size	,,	24·59	24·59	26·25	26·25	27·90
Spigot	,,	12·29	12·29	13·12	13·12	13·94
Equal twin bend with turning vanes	,,	154·70	154·70	167·60	167·60	180·49
Flanged end	,,	9·97	10·36	11·15	11·95	12·75

		1000 × 800 mm £	1000 × 1000 mm £	1200 × 300 mm £	1200 × 400 mm £	1200 × 500 mm £
Straight duct	m	115·78	136·69	105·32	105·32	115·78
Extra over for						
90° square bend with turning vanes	No.	128·92	146·11	120·33	120·33	128·92
90° radius bend without turning vanes	,,	122·78	139·15	114·60	114·60	122·78
45° ditto	,,	100·16	112·91	93·97	93·97	100·16
Stopped end	,,	14·35	15·94	11·95	12·75	13·55
Damper, multi-leaf pattern	,,	153·81	180·76	146·68	166·32	187·33
Fire damper, shutter type with access panel	,,	205·74	238·76	198·88	222·25	247·65
Diminishing piece, large end measured	,,	59·42	67·37	55·46	55·46	59·42
Square to round piece	,,	84·01	95·13	78·45	78·45	84·01
Branch, based on branch size	,,	29·57	32·88	27·90	27·90	29·57
Spigot	,,	14·77	16·44	13·94	13·94	14·77
Equal twin bend with turning vanes	,,	193·37	219·17	180·49	180·49	193·37
Flanged end	,,	14·35	15·94	11·95	12·75	13·55

DUCTWORK

GALVANIZED SHEET METAL HIGH VELOCITY RECTANGULAR SECTION DUCTWORK
(including all necessary stiffeners, joints, couplers in the running length and duct supports) *continued*

	Unit	1200 × 600 mm £	1200 × 800 mm £	1200 × 1000 mm £	1200 × 1200 mm £	1400 × 400 mm £
Straight duct	m	115·78	136·69	147·15	157·60	123·98
Extra over for						
90° square bend with turning vanes	No.	128·92	146·11	154·71	163·30	150·05
90° radius bend without turning vanes	„	122·78	139·15	147·34	155·52	142·91
45° ditto	„	100·16	112·91	119·29	125·66	118·26
Stopped end	„	14·35	15·94	17·53	19·13	14·35
Damper, multi-leaf pattern	„	207·90	255·26	280·66	416·96	176·72
Fire damper, shutter type with access panel	„	273·05	302·26	331·47	514·35	234·95
Diminishing piece, large end measured	„	59·42	67·37	71·34	75·31	65·22
Square to round piece	„	84·01	95·11	100·70	106·27	89·77
Branch, based on branch size	„	29·57	32·88	34·53	36·20	32·50
Spigot	„	14·77	16·44	17·27	18·10	16·24
Equal twin bend with turning vanes	„	193·37	219·17	232·07	244·95	225·08
Flanged end	„	14·35	15·94	17·53	19·13	14·35

	Unit	1400 × 500 mm £	1400 × 600 mm £	1400 × 800 mm £	1400 × 1000 mm £	1400 × 1200 mm £
Straight duct	m	136·09	148·22	160·35	172·46	196·71
Extra over for						
90° square bend with turning vanes	No.	157·62	165·20	172·78	180·35	195·51
90° radius bend without turning vanes	„	150·12	157·33	164·55	171·76	186·20
45° ditto	„	123·81	129·38	134·94	140·49	151·62
Stopped end	„	15·14	15·94	17·53	19·13	20·72
Damper, multi-leaf pattern	„	197·50	219·45	255·26	280·66	438·90
Fire damper, shutter type with access panel	„	260·35	288·29	316·23	363·22	560·07
Diminishing piece, large end measured	„	70·29	75·35	80·41	85·48	95·60
Square to round piece	„	97·24	104·71	112·19	119·66	134·59
Branch, based on branch size	„	34·33	36·15	37·98	39·80	43·45
Spigot	„	17·15	18·06	18·98	19·89	21·71
Equal twin bend with turning vanes	„	236·43	247·80	259·17	270·52	293·26
Flanged end	„	15·14	15·94	17·53	19·13	20·72

DUCTWORK

GALVANIZED SHEET METAL HIGH VELOCITY RECTANGULAR SECTION DUCTWORK
(including all necessary stiffeners, joints, couplers in the running length and duct supports) *continued*

	Unit	1600 × 400 mm £	1600 × 500 mm £	1600 × 600 mm £	1600 × 800 mm £	1600 × 1000 mm £
Straight duct	m	148·22	148·22	160·35	172·46	196·71
Extra over for						
90° square bend with turning vanes	No.	165·20	165·20	172·78	180·35	195·51
90° radius bend without turning vanes	„	157·33	157·33	164·55	171·76	186·20
45° ditto	„	129·38	129·38	134·94	140·49	151·62
Stopped end	„	15·94	15·94	17·53	19·13	20·72
Damper, multi-leaf pattern	„	187·11	207·90	254·00	280·66	307·23
Fire damper, shutter type with access panel	„	247·65	249·48	302·26	363·22	394·97
Diminishing piece, large end measured	„	75·35	75·35	80·41	85·48	95·60
Square to round piece	„	104·71	104·71	112·91	119·66	134·59
Branch, based on branch size	„	36·15	36·15	37·98	39·80	43·45
Spigot	„	18·06	18·06	18·98	19·89	21·71
Equal twin bend with turning vanes	„	247·80	247·80	259·17	270·52	293·26
Flanged end	„	15·94	15·94	17·53	19·13	20·72

	Unit	1600 × 1200 mm £	1800 × 500 mm £	1800 × 600 mm £	1800 × 800 mm £	1800 × 1000 mm £
Straight duct	m	208·85	172·46	172·46	208·85	208·85
Extra over for						
90° square bend with turning vanes	No.	203·09	180·35	180·35	203·09	203·09
90° radius bend without turning vanes	„	193·41	171·76	171·76	193·41	193·41
45° dittos	„	157·17	140·49	140·49	157·17	157·17
Stopped end	„	22·31	18·39	19·13	22·32	22·32
Damper, multi-leaf pattern	„	364·90	220·61	242·55	294·53	294·53
Fire damper, shutter type with access panel	„	496·57	261·03	317·50	345·35	390·39
Diminishing piece, large end measured	„	100·66	85·48	85·48	100·66	100·66
Square to round piece	„	142·06	119·66	119·66	142·06	142·06
Branch, based on branch size	„	45·27	39·80	39·80	45·27	45·27
Spigot	„	22·63	19·89	19·89	22·63	22·63
Equal twin bend with turning vanes	„	304·63	270·52	270·52	304·63	304·63
Flanged end	„	22·31	18·39	19·13	22·32	22·32

Mechanical Installations – Prices for Measured Work 169

DUCTWORK

GALVANIZED SHEET METAL HIGH VELOCITY RECTANGULAR SECTION DUCTWORK
(including all necessary stiffeners, joints, couplers in the running length and duct supports) *continued*

	Unit	1800 × 1200 mm £	2000 × 500 mm £	2000 × 600 mm £	2000 × 800 mm £	2000 × 1000 mm £
Straight duct	m	220·96	184·59	196·70	208·84	220·96
Extra over for						
90° square bend with turning vanes	No.	210·66	187·93	195·51	203·09	210·66
90° radius bend without turning vanes	,,	200·62	178·98	186·20	193·41	200·62
45° ditto	,,	162·73	146·04	151·62	157·17	162·73
Stopped end	,,	23·91	19·92	20·72	22·32	21·25
Damper, multi-leaf pattern	,,	352·22	231·13	257·57	307·23	361·52
Fire damper, shutter type with access panel	,,	527·05	300·98	301·46	397·51	462·28
Diminishing piece, large end measured	,,	105·73	90·54	95·60	100·66	105·73
Square to round piece	,,	149·54	127·13	134·59	142·06	149·54
Branch, based on branch size	,,	47·10	41·63	43·45	45·27	47·10
Spigot	,,	23·54	20·80	21·71	22·63	23·54
Equal twin bend with turning vanes	,,	315·98	281·89	293·26	304·62	315·98
Flanged end	,,	23·91	19·92	20·72	22·32	21·25

	Unit	2000 × 1200 mm £
Straight duct	m	245·22
Extra over for		
90° square bend with turning vanes	No.	225·81
90° radius bend without turning vanes	,,	215·06
45° ditto	,,	173·85
Stopped end	,,	25·50
Damper, multi-leaf pattern	,,	364·50
Fire damper, shutter type with access panel	,,	552·45
Diminishing piece, large end measured	,,	115·85
Square to round piece	,,	164·48
Branch, based on branch size	,,	50·75
Spigot	,,	23·36
Equal twin bend with turning vanes	,,	338·72
Flanged end	,,	25·50

DUCTWORK

Galvanized sheet metal spirally-wound circular section ductwork for low velocity systems constructed in accordance with Table 12 of Specification DW 141 for Sheet Metal Ductwork published by the H.V.C.A.

Nominal diameter mm	Nominal sheet thickness mm	Maximum spacing between joints/stiffeners mm	Angle sections for joints/stiffeners mm
Up to 205	0·6	Unlimited	25 × 25 × 3
205 to 760	0·8	Unlimited	25 × 25 × 3
761 to 1020	1·0	Unlimited	30 × 30 × 3
1021 to 1525	1·2	1800	40 × 40 × 4

GALVANIZED SHEET METAL LOW VELOCITY SPIRALLY-WOUND CIRCULAR SECTION DUCTWORK
(including all necessary stiffeners, joints, couplers in the running length and duct supports)

	Unit	75 mm £	100 mm £	125 mm £	150 mm £	200 mm £	250 mm £
Straight duct	m	14·51	15·35	16·08	16·57	18·21	20·67
Extra over for							
90° radius bend	No.	7·52	8·51	8·27	9·97	11·21	13·53
45° ditto	,,	6·29	7·29	7·74	8·18	9·28	11·42
90° equal twin bend	,,	7·79	12·77	13·79	14·96	16·82	15·62
Stopped end	,,	1·20	1·74	2·29	2·82	3·37	3·91
Diminishing piece large end measured	,,	8·57	9·32	10·04	11·21	13·25	14·74
Transformation piece round to square	,,	12·85	13·97	15·05	16·82	19·88	22·11
Conical tee	,,	5·28	8·24	11·34	14·45	17·54	20·66
Conical cross	,,	7·46	11·76	16·25	20·75	25·24	29·77
Fire damper, shutter type with access panel	,,	56·03	56·03	56·03	56·03	64·59	64·59
Angle ring	,,	8·44	8·44	8·44	8·44	9·69	10·13
45° branch	,,	4·19	7·18	10·31	13·42	16·25	19·66

		300 mm £	350 mm £	400 mm £	450 mm £	500 mm £	600 mm £
Straight duct	m	22·25	23·69	24·97	26·67	28·49	45·18
Extra over for							
90° radius bend	No.	16·26	17·70	21·03	23·74	27·11	35·68
45° ditto	,,	13·82	15·03	16·25	17·16	20·05	25·57
90° equal twin bend	,,	24·40	26·55	31·53	37·99	40·66	57·08
Stopped end	,,	4·46	5·00	5·54	6·08	6·63	7·17
Diminishing piece large end measured	,,	17·60	19·59	22·16	23·90	25·62	30·03
Transformation piece round to square	,,	26·41	29·39	33·25	35·80	38·44	45·06
Conical tee	,,	24·40	27·59	30·78	33·97	37·16	40·35
Conical cross	,,	35·20	39·81	44·45	49·07	53·30	58·33
Fire damper, shutter type with access panel	,,	76·35	82·21	88·08	99·86	111·60	123·32
Angle ring	,,	10·43	10·72	11·02	11·32	11·62	12·21
45° branch	,,	23·43	26·64	29·86	33·07	36·31	39·47

DUCTWORK

GALVANIZED SHEET METAL LOW VELOCITY SPIRALLY-WOUND CIRCULAR SECTION DUCTWORK
(including all necessary stiffeners, joints, couplers in the running length and duct supports) *continued*

	Unit	700 mm £	800 mm £	900 mm £	1000 mm £	1100 mm £	1200 mm £
Straight duct	m	52·56	60·14	67·61	75·10	82·58	90·08
Extra over for							
90° radius bend	No.	61·26	86·88	112·47	138·07	163·66	189·27
45° ditto	„	42·38	59·19	76·00	92·80	109·63	126·43
90° equal twin bend	„	91·90	130·32	168·71	207·10	245·48	283·91
Stopped end	„	7·72	8·26	7·17	9·34	9·90	10·44
Diminishing piece large end measured	„	36·92	43·84	50·74	57·66	64·57	71·48
Transformation piece round to square	„	55·04	65·75	76·12	86·49	96·84	107·22
Conical tee	„	43·51	46·72	49·91	53·10	56·29	59·46
Conical cross	„	62·96	63·05	72·21	76·82	81·42	86·08
Fire damper, shutter type with access panel	„	146·70	158·40	188·10	199·80	223·20	241·20
Angle ring	„	11·33	11·86	12·38	12·92	13·44	13·96
45° branch	„	42·69	45·90	52·53	55·53	58·74	58·73

Galvanized sheet metal spirally-wound circular section ductwork for high velocity systems constructed in accordance with Table 12 of Specification DW 141 for Sheet Metal Ductwork published by the H.V.C.A.

Nominal diameter mm	Nominal sheet thickness mm	Maximum spacing between joints/stiffeners mm	Angle sections for joints/stiffeners mm
Up to 205	0·6	Unlimited	25 × 25 × 3
206 to 760	0·8	Unlimited	25 × 25 × 3
761 to 1020	1·0	Unlimited	30 × 30 × 3
1021 to 1525	1·2	1800	40 × 40 × 4

GALVANIZED SHEET METAL HIGH VELOCITY SPIRALLY-WOUND CIRCULAR SECTION DUCTWORK
(including all necessary stiffeners, joints, couplers in the running length and duct supports)

	Unit	75 mm £	100 mm £	125 mm £	150 mm £	200 mm £	250 mm £
Straight duct	m	16·13	17·05	17·87	18·41	20·33	22·97
Extra over for							
90° radius bend	No.	8·36	9·46	10·21	11·08	12·46	15·03
45° ditto	„	6·99	8·10	8·60	9·09	10·31	12·69
90° equal twin bend	„	12·54	14·19	15·32	16·62	18·69	17·36
Stopped end	„	1·33	1·93	2·54	3·13	3·74	4·34
Diminishing piece large end measured	„	9·52	10·35	11·15	12·46	14·72	16·38
Transformation piece round to square	„	14·28	15·52	16·72	18·69	22·09	24·57
Conical tee	„	5·87	9·16	12·60	16·05	19·49	22·95
Conical cross	„	8·29	13·07	18·05	23·06	28·05	33·08
Fire damper, shutter type with access panel	„	62·25	62·25	62·25	62·25	71·77	71·77
Angle ring	„	9·38	9·38	9·38	9·38	10·77	11·26
45° branch	„	4·66	7·98	11·46	14·91	16·39	21·84

DUCTWORK

GALVANIZED SHEET METAL HIGH VELOCITY SPIRALLY-WOUND CIRCULAR SECTION DUCTWORK
(including all necessary stiffeners, joints, couplers in the running length and duct supports) *continued*

	Unit	300 mm £	350 mm £	400 mm £	450 mm £	500 mm £	600 mm £
Straight duct	m	24·72	26·32	27·74	29·63	31·66	50·20
Extra over for							
90° radius bend	No.	18·07	19·67	23·37	26·38	30·12	39·64
45° ditto	,,	15·36	16·70	18·06	19·07	22·28	28·41
90° equal twin bend	,,	27·11	29·50	35·05	42·21	45·18	63·42
Stopped end	,,	4·95	5·56	6·15	6·76	7·37	7·97
Diminishing piece large end measured	,,	19·56	21·77	24·62	26·52	28·47	33·37
Transformation piece round to square	,,	29·34	32·66	36·94	39·78	42·71	50·05
Conical tee	,,	27·11	30·66	34·20	37·74	41·29	44·83
Conical cross	,,	39·11	44·24	49·39	54·52	59·66	64·81
Fire damper, shutter type with access panel	,,	84·83	91·35	97·87	110·95	124·00	137·03
Angle ring	,,	11·59	11·91	12·24	12·58	12·91	13·57
45° branch	,,	26·03	29·60	33·17	36·74	40·34	43·86

	Unit	700 mm £	800 mm £	900 mm £	1000 mm £	1100 mm £	1200 mm £
Straight duct	m	58·40	66·82	75·12	83·45	91·76	100·09
Extra over for							
90° radius bend	No.	68·07	96·53	124·97	153·41	181·84	210·30
45° ditto	,,	47·09	65·77	84·44	103·11	121·81	140·48
90° equal twin bend	,,	102·11	144·80	187·46	230·11	272·76	315·45
Stopped end	,,	8·58	9·18	9·79	10·38	11·00	11·60
Diminishing piece large end measured	,,	41·03	48·71	56·38	64·07	71·74	79·42
Transformation piece round to square	,,	61·56	73·06	84·58	96·10	107·60	119·13
Conical tee	,,	48·36	51·91	55·45	59·00	62·54	66·07
Conical cross	,,	69·95	70·06	80·23	85·36	90·49	95·64
Fire damper, shutter type with access panel	,,	163·00	176·00	209·00	222·00	248·00	268·00
Angle ring	,,	12·59	13·18	13·76	14·35	14·93	15·51
45° branch	,,	47·43	51·00	58·14	61·70	65·27	65·26

Mechanical Installations – Prices for Measured Work

DUCTWORK

Galvanized sheet metal straight seamed circular section ductwork for low velocity systems constructed in accordance with Table 10 of Specification DW 141 for Sheet Metal Ductwork published by the H.V.C.A.

Nominal diameter mm	Nominal sheet thickness mm	Maximum spacing between joints/stiffeners mm	Angle sections for joints/stiffeners mm
Up to 510	0·6	Unlimited	25 × 25 × 3
510 to 760	0·8	Unlimited	25 × 25 × 3
761 to 1020	1·0	1250	30 × 30 × 3
1021 to 1525	1·2	1250	40 × 40 × 4

GALVANIZED SHEET METAL LOW VELOCITY STRAIGHT-SEAMED CIRCULAR SECTION DUCTWORK
(including all necessary stiffeners, joints, couplers in the running length and duct supports)

	Unit	1300 mm £	1400 mm £	1600 mm £	1800 mm £
Straight duct	m	97·56	110·13	122·69	135·27
Extra over for					
90° radius bend	No.	215·06	240·45	291·65	342·85
45° ditto	„	143·24	160·06	193·68	227·30
90° equal twin bend	„	322·61	360·68	437·49	514·27
Stopped end	„	10·98	11·52	12·62	13·70
Diminishing piece large end measured	„	78·37	85·28	99·11	112·91
Transformation piece round to square	„	117·57	127·92	148·66	169·34
Conical tee	„	56·40	59·26	65·00	70·74
Conical cross	„	81·62	85·80	94·11	102·44
Fire damper, shutter type with access panel	„	234·90	276·00	293·85	328·50
Angle ring	„	14·49	13·52	14·45	15·41
45° branch	„	55·57	58·61	64·43	70·20

Galvanized sheet metal straight-seamed circular section ductwork high velocity systems constructed in accordance with Table 11 of Specification DW 141 for Sheet Metal Ductwork published by the H.V.C.A.

Nominal diameter mm	Nominal sheet thickness mm	Maximum spacing between joints/stiffeners mm	Angle sections for joints/stiffeners mm
Up to 510	0·8	Unlimited	25 × 25 × 3
511 to 760	1·0	Unlimited	25 × 25 × 3
761 to 1020	1·2	1250	30 × 30 × 3
1021 to 1525	1·2	1250	40 × 40 × 4

DUCTWORK

GALVANIZED SHEET METAL HIGH VELOCITY STRAIGHT-SEAMED CIRCULAR SECTION DUCTWORK
(including all necessary stiffeners, joints, couplers in the running length and duct supports)

	Unit	1300 mm £	1400 mm £	1600 mm £	1800 mm £
Straight duct	m	108·40	122·37	136·32	150·30
Extra over for					
90° radius bend	No.	238·96	267·17	324·06	380·94
45° ditto	,,	159·16	177·84	215·20	252·55
90° equal twin bend	,,	358·45	400·76	486·10	571·41
Stopped end	,,	12·20	12·81	14·02	15·22
Diminishing piece large end measured	,,	87·08	94·76	110·12	125·46
Transformation piece round to square	,,	130·63	142·17	165·18	188·18
Conical tee	,,	62·67	65·84	72·22	78·60
Conical cross	,,	90·69	95·33	104·57	113·82
Fire damper, shutter type with access panel	,,	261·00	306·67	326·25	365·40
Angle ring	,,	16·10	15·02	16·05	17·12
45° branch	,,	61·74	65·16	71·59	78·00

Galvanized sheet metal spirally wound flat oval section ductwork for high velocity systems constructed in accordance with Table 16 and 18 of Specification DW 141 for Sheet Metal Ductwork published by the H.V.C.A.

GALVANIZED SHEET METAL SPIRALLY WOUND HIGH VELOCITY FLAT OVAL SECTION DUCTWORK
(including all necessary stiffeners, joints, couplers in the running length and duct supports)

	Unit	345 × 102 mm £	427 × 102 mm £	508 × 102 mm £	531 × 203 mm £	559 × 152 mm £
Straight duct	m	22·05	23·55	24·56	36·25	36·25
Extra over for						
90° square bend with turning vanes	No.	20·98	21·79	22·63	29·21	29·21
90° bend without turning vanes	,,	34·62	37·78	40·60	48·10	47·47
45° ditto	,,	20·65	22·35	24·00	30·75	30·75
Stopped end	,,	5·14	5·92	7·00	7·92	7·92
Diminishing piece, large end measured	,,	18·90	19·32	19·75	22·12	22·12
Oval to round piece	,,	28·35	28·96	29·63	33·19	33·19
45° tee	,,	12·47	12·70	12·90	15·43	15·43
45° regular 'Y' tee	,,	13·88	14·13	14·44	17·28	17·28
90° tee	,,	21·66	22·35	23·01	25·84	25·84

DUCTWORK

GALVANIZED SHEET METAL SPIRALLY WOUND HIGH VELOCITY FLAT OVAL SECTION DUCTWORK
(including all necessary stiffeners, joints, couplers in the running length and duct supports) *continued*

	Unit	582 × 254 mm £	632 × 305 mm £	678 × 508 mm £	709 × 457 mm £	737 × 406 mm £
Straight duct	m	39·38	42·58	48·96	54·76	48·96
Extra over for						
90° square bend with turning vanes	No.	30·72	31·67	34·31	34·79	35·05
90° bend without turning vanes	,,	51·00	58·09	61·53	62·05	62·64
45° ditto	,,	32·42	61·44	37·27	37·62	37·78
Stopped end	,,	9·03	9·51	11·64	11·64	11·64
Diminishing piece, large end measured	,,	23·49	21·35	26·03	26·03	26·03
Oval to round piece	,,	35·27	35·93	39·04	39·04	39·04
45° tee	,,	21·06	23·33	35·59	33·90	29·88
45° regular 'Y' tee	,,	23·99	27·26	46·75	43·80	41·29
90° tee	,,	28·33	29·23	40·01	39·50	38·52

		765 × 356 mm £	818 × 406 mm £	851 × 203 mm £	978 × 406 mm £	1247 × 356 mm £
Straight duct	m	48·96	50·28	53·30	68·47	93·71
Extra over for						
90° square bend with turning vanes	No.	35·27	36·37	41·19	46·54	84·43
90° bend without turning vanes	,,	63·17	69·14	80·24	89·18	137·85
45° ditto	,,	38·07	39·71	45·58	51·33	82·32
Stopped end	,,	11·64	13·05	11·64	15·58	22·54
Diminishing piece, large end measured	,,	26·03	26·52	28·16	29·11	46·08
Oval to round piece	,,	39·04	39·75	42·26	43·69	69·11
45° tee	,,	26·27	30·07	22·97	33·16	44·10
45° regular 'Y' tee	,,	36·76	41·98	38·16	45·39	61·53
90° tee	,,	36·96	39·02	37·40	41·63	67·00

	Unit	1275 × 305 mm £	1303 × 254 mm £	1387 × 406 mm £	1671 × 457 mm £	1699 × 406 mm £	1727 × 356 mm £
Straight duct	m	93·71	108·21	98·33	153·05	153·05	153·05
Extra over for							
90° square bend with turning vanes	No.	84·43	84·43	149·60	109·62	109·62	109·62
90° bend without turning vanes	,,	138·73	138·73	90·96	184·61	184·05	185·98
45° ditto	,,	83·67	83·67	91·35	114·33	114·86	115·69
Stopped end	,,	22·54	22·54	24·61	32·81	32·81	32·81
Diminishing piece, large end measured	,,	46·07	46·07	48·12	54·13	54·13	53·35
Oval to round piece	,,	69·11	69·11	72·18	81·22	81·22	80·00
45° tee	,,	40·66	37·60	50·24	62·90	56·74	51·61
45° regular 'Y' tee	,,	48·52	43·76	69·37	82·87	77·96	71·78
90° tee	,,	67·26	66·17	73·30	82·32	80·91	77·96

DUCTWORK

FLEXIBLE CIRCULAR SECTION DUCTWORK
(of galvanized spring steel covered with neoprene coated fibreglass)

	Unit	50 mm £	100 mm £	150 mm £	200 mm £	250 mm £
Duct	m	8·08	9·49	10·84	12·24	13·61
Extra over for Connection to equipment	No.	5·38	6·37	7·32	8·31	9·28

	Unit	300 mm £	350 mm £	400 mm £	450 mm £
Duct	m	15·00	16·39	17·77	19·14
Extra over for Connection to equipment	No.	10·27	11·22	12·21	13·18

ACOUSTIC LINING

	Unit	£
25 mm thick foamed barafoam fixed to ductwork with adhesive	m^2	21·01

GRILLES AND DIFFUSERS

	Unit	150 × 100 mm £	150 × 150 mm £	200 × 200 mm £	250 × 250 mm £	300 × 200 mm £
Fix only extract or supply grille to nozzle of ductwork	No.	5·20	6·25	8·31	10·41	10·41
	„	300 × 300 mm £ 12·47	450 × 300 mm £ 15·56	500 × 350 mm £ 17·66	550 × 400 mm £ 19·72	700 × 450 mm £ 23·88
	„		800 × 500 mm £ 27·00	900 × 550 mm £ 30·10	1100 × 600 mm £ 33·23	1200 × 600 mm £ 37·38

		200 mm £	300 mm £	400 mm £	500 mm £	600 mm £
Fix only circular pattern ceiling diffuser with volume control sub-frame and adjustable deflector frame	„	14·47	18·35	24·48	30·60	36·73

ACCESS DOORS
(including frame fitted to ducts)

	Unit	450 × 450 mm £	600 × 600 mm £	900 × 600 mm £
To straight	„	74·07	89·26	98·79
To curled	„	112·16	134·87	148·14
Test holes	„	2·87		

Mechanical Installations – Prices for Measured Work

BOILERS

DOMESTIC WATER BOILERS
(stove enamelled casing, electric controls, placing in position, assembling and connecting – electrical work measured separately)

		Rating kW (Btu/hr)				
		8·79 (30000)	11·72 (40000)	13·19 (45000)	14·66 (50000)	16·12 (55000)
Gas fired, floor standing connected to conventional flue	Unit No.	£ 191·91 *150·15*	£ 209·24 *167·48*	£ 220·79 *179·03*	£ 237·56 *190·58*	£ 249·11 *202·13*

		Rating kW (Btu/hr)				
		17·59 (60000)	20·52 (70000)	23·45 (80000)	29·32 (100000)	43·98 (150000)
	,,	£ 265·88 *213·68*	£ 288·98 *236·78*	£ 322·52 *259·88*	£ 379·16 *306·08*	£ 504·54 *415·80*

		Rating kW (Btu/hr)				
		8·79 (30000)	11·72 (40000)	13·19 (45000)	14·66 (50000)	16·12 (55000)
Gas fired, floor standing including balanced flue	,,	£ 254·33 *202·13*	£ 265·88 *213·68*	£ 271·65 *219·45*	£ 288·42 *231·00*	£ 299·97 *242·55*

		Rating kW (Btu/hr)				
		17·59 (60000)	20·52 (70000)	23·45 (80000)	29·32 (100000)	43·98 (150000)
	,,	£ 322·52 *259·88*	£ 351·39 *288·75*	£ 402·26 *329·18*	£ 499·32 *415·80*	£ 792·18 *693·00*

		Rating kW (Btu/hr)				
		8·79 (30000)	11·72 (40000)	13·19 (45000)	14·66 (50000)	16·12 (55000)
Oil fired, floor standing connected to conventional flue	,,	£ 372·69 *325·71*	£ 372·69 *325·71*	£ 372·69 *325·71*	£ 377·91 *325·71*	£ 412·56 *360·36*

		Rating kW (Btu/hr)				
		17·59 (60000)	20·52 (70000)	23·45 (80000)	29·32 (100000)	43·98 (150000)
	,,	£ 423·00 *360·36*	£ 450·72 *388·08*	£ 495·81 *422·73*	£ 577·26 *498·96*	£ 679·59 *575·19*

		Rating kW (Btu/hr)				
		8·79 (30000)	11·72 (40000)	13·19 (45000)	14·66 (50000)	16·12 (55000)
Solid fuel fired, floor standing connected to conventional flue	,,	243·89 *202·13*	243·89 *202·13*	295·86 *254·10*	301·08 *254·10*	306·86 *259·88*

		Rating kW (Btu/hr)				
		17·59 (60000)	20·52 (70000)	23·45 (80000)	29·32 (100000)	43·98 (150000)
	,,	£ 329·40 *277·20*	£ 358·28 *306·08*	£ 403·37 *340·73*	£ 419·58 *346·50*	£ 562·29 *473·55*

Mechanical Installations – Prices for Measured Work
BOILERS

PACKAGED WATER BOILERS
(with boiler mountings, controls, enamelled casing, burner, insulation, all connections and commissioning – electrical work measured separately)

		Unit	Rating kW 88 £	147 £	220 £	293 £
Gas fired	. . . . . .	No.	1835·00	2070·00	2260·00	2610·00
			1730·00	*1965·00*	*2140·00*	*2485·00*
Oil fired.	. . . . . .	,,	1330·00	1550·00	1860·00	2145·00
			1225·00	*1445·00*	*1730·00*	*2020·00*
Solid fuel	. . . . . .	,,	2820·00	3165·00	3960·00	4825·00
			2715·00	*3060·00*	*3810·00*	*4680·00*

			Rating kW 366 £	440 £	513 £	586 £
Gas fired	. . . . . .	,,	3170·00	4050·00	5060·00	5235·00
			3005·00	*3880·00*	*4850·00*	*5025·00*
Oil fired.	. . . . . .	,,	2360·00	3515·00	4250·00	4425·00
			2195·00	*3350·00*	*4040·00*	*4215·00*
Solid fuel	. . . . . .	,,	5210·00	5730·00	5985·00	6910·00
			5025·00	*5545·00*	*5775·00*	*6700·00*

			Rating kW 880 £	1172 £	1465 £
Gas fired	. . . . . .	,,	6910·00	7885·00	8575·00
			6700·00	*7625·00*	*8315·00*
Oil fired.	. . . . . .	,,	4715·00	5345·00	5920·00
			4505·00	*5080·00*	*5660·00*
Solid fuel	. . . . . .	,,	7045·00	7970·00	9510·00
			6815·00	*7740·00*	*9240·00*

CAST IRON SECTIONAL PACKAGED BOILERS
(with controls, enamelled jacket, insulation, combined thermometer and altitude gauge – assembled on prepared base and commissioned by supplier – electrical work measured separately)

		Unit	Rating kW 88 £	117 £	147 £	220 £
Gas fired	. . . . . .	No.	2090·00	2510·00	2735·00	3220·00
			1850·00	*2195·00*	*2370·00*	*2655·00*
Oil fired.	. . . . . .	,,	1590·00	2000·00	2350·00	2780·00
			1330·00	*1675·00*	*1965·00*	*2195·00*
Solid fuel	. . . . . .	,,	2185·00	2235·00	2530·00	3690·00
			1965·00	*1965·00*	*2195·00*	*3175·00*

			Rating kW 293 £	440 £	586 £	880 £
Gas fired	. . . . . .	,,	4080·00	5140·00	6405·00	9540·00
			3350·00	*4040·00*	*4735·00*	*6930·00*
Oil fired.	. . . . . .	,,	3475·00	4555·00	5870·00	8670·00
			2715·00	*3405·00*	*4160·00*	*6005·00*
Solid fuel	. . . . . .	,,	4315·00	5945·00		
			3640·00	*4850·00*		

Mechanical Installations – Prices for Measured Work 179

BOILERS

PACKAGED STEAM BOILERS
(with boiler mountings, centrifugal water feed pump, insulation and sheet steel wrap round casing, plastic coated – electrical work measured separately)

	Unit	293 £	586 £	*Rating kW* 880 £	1172 £	1465 £
Gas fired	,,	8505·00 *8085·00*	11265·00 *10740·00*	12700·00 *12125·00*	13680·00 *13050·00*	15860·00 *15130·00*
Oil fired 35 second	,,	7865·00 *7450·00*	9995·00 *9470·00*	11430·00 *10855·00*	12290·00 *11665·00*	14015·00 *13280·00*

	Unit	1760 £	2052 £	*Rating kW* 2345 £	2638 £	2930 £
Gas fired	,,	17585·00 *16745·00*	19075·00 *18135·00*	22180·00 *21135·00*	22410·00 *21365·00*	23565·00 *22520·00*
Oil fired 35 second	,,	15675·00 *14840·00*	17340·00 *16400·00*	19060·00 *18020·00*	20215·00 *19175·00*	21485·00 *20445·00*

CISTERNS, TANKS AND CYLINDERS

GALVANIZED AFTER MADE MILD STEEL OPEN TOP CISTERNS
(with loose covers to B.S. 417 Grade 'A', and hoisting and placing in position)

B.S. size number and capacity to water line litres

	Unit	SCM 45 18 £	SCM 70 36 £	SCM 90 54 £	SCM 110 68 £	SCM 135 86 £	SCM 180 114 £
Cistern	No.	22·77 *14·93*	27·42 *19·58*	30·52 *22·68*	35·31 *24·87*	37·48 *27·04*	44·28 *31·22*
		SCM 230 159 £	SCM 270 191 £	SCM 320 227 £	SCM 360 264 £	SCM 450/1 327 £	SCM 680 491 £
	,,	51·42 *35·76*	68·98 *48·10*	74·96 *51·46*	78·14 *54·64*	95·71 *69·61*	125·99 *94·67*
		SCM 910 709 £	SCM 1130 841 £	SCM 1600 1227 £	SCM 2270 1727 £	SCM 2720 2137 £	SCM 4540 3364 £
	,,	154·75 *112·99*	181·01 *134·03*	218·48 *166·28*	296·21 *212·69*	348·93 *234·09*	569·49 *339·81*

		Unit	£
Extra for plain holes cut in at works			
Up to and including 64 mm diameter		No.	0·50
Over 64 mm up to 150 mm diameter		,,	1·20

			15 mm £	20 mm £	25 mm £	32 mm £	40 mm £	50 mm £	80 mm £	100 mm £
Extra for screwed boss fitted at works	Unit No.		0·70	0·80	0·90	1·00	1·10	1·20	3·00	6·00

GALVANIZED AFTER MADE MILD STEEL TANKS
(to B.S. 417, and hoisting and placing in position)

B.S. size number and capacity litres

	Unit	T 25/1 95 £	T 30/1 114 £	T 40 155 £
Tank Grade 'A'	No.	63·78 *55·94*	72·10 *59·04*	86·51 *70·85*
Tank Grade 'B'	,,	58·80 *50·96*	67·13 *54·07*	
Extra for plain hole cut in at works	,,	0·50		

Mechanical Installations – Prices for Measured Work

CISTERNS, TANKS AND CYLINDERS

GALVANIZED AFTER MADE MILD STEEL CYLINDERS
(to B.S. 417 and hoisting and placing in position)

B.S. size number and capacity litres

	Unit	YM 91 73 £	YM 114 100 £	YM 127 114 £	YM 141 123 £	YM 150 136 £
Cylinder Grade A	No.	53·83	59·42	63·89	65·75	70·86
		45·99	*51·58*	*53·45*	*55·31*	*57·80*
Cylinder Grade B	,,	51·35	56·32	60·78	62·02	67·13
		43·51	*48·48*	*50·34*	*51·58*	*54·07*
Cylinder Grade C	,,	49·48	53·83	58·30	59·54	64·02
		41·64	*45·99*	*47·86*	*49·10*	*50·96*

		YM 177 159 £	YM 218 195 £	YM 264 241 £	YM 355 332 £	YM 455 441 £
Cylinder Grade A	,,	75·83	102·67	117·21	135·37	162·83
		62·77	*87·01*	*96·33*	*111·87*	*136·73*
Cylinder Grade B	,,	72·10	82·16	93·60	104·30	125·54
		59·04	*66·50*	*72·72*	*80·80*	*99·44*
Cylinder Grade C	,,	68·37	77·19	87·38	98·08	116·84
		55·31	*61·53*	*66·50*	*74·58*	*90·74*

Extra for handhole fitted at works	Unit	£
150 mm diameter	No.	9·40
225 mm diameter	,,	10·70
300 mm diameter	,,	12·00
400 mm diameter	,,	13·40
450 mm diameter	,,	14·72

GALVANIZED AFTER MADE INDIRECT CYLINDERS WELDED THROUGHOUT
(with bolted heads and placing in position)

Capacity litres

	Unit	77 £	91 £	109 £	125 £	136 £	159 £	182 £
Cylinder 2/2·6 mm	No.	60·04	63·77	69·36	75·70	81·42	91·69	100·18
		52·21	*55·94*	*61·53*	*65·26*	*68·37*	*77·07*	*84·52*
Cylinder 3 mm plate	,,	67·49	71·84	79·30	86·88	93·85	105·98	117·59
		59·66	*64·01*	*71·47*	*76·44*	*80·80*	*91·36*	*101·93*

		227 £	273 £	364 £	409 £	455 £	568 £	682 £	796 £	909 £
Cylinder 3 mm plate	Unit No.	148·41	168·84	200·49	219·85	240·44	273·47	314·80	351·98	396·08
		132·75	*147·96*	*177·00*	*196·36*	*214·34*	*244·76*	*283·48*	*318·05*	*359·54*
Cylinder 5 mm plate	,,	189·90	218·63	264·10	288·99	316·50	360·59	418·51	476·44	537·13
		174·24	*197·75*	*240·61*	*265·50*	*290·40*	*331·88*	*387·19*	*442·51*	*500·59*

	Unit	1364 £	1818 £	2728 £	3182 £
Cylinder 5 mm plate	No.	729·03	891·66	1215·58	1432·20
		687·27	*818·64*	*1100·74*	*1286·04*
Cylinder 6 mm plate	,,		1110·15	1511·51	1777·91
			1037·13	*1396·67*	*1631·75*

	Unit	£
Extra for boss for immersion heater	No.	1·32

CISTERNS, TANKS AND CYLINDERS

COPPER STORAGE CYLINDERS
(to B.S. 699 and placing in position)

		B.S. size and capacity litres					
		1	2	4	5	6	7	8
		74	98	86	98	109	120	144
	Unit	£	£	£	£	£	£	£
Cylinder Grade 1 . . .	No.	63·60	75·21	84·61	90·10	92·84	100·00	115·17
		55·77	*67·38*	*76·78*	*82·27*	*85·01*	*89·56*	*100·55*
Cylinder Grade 2 . . .	„	52·62	59·51	61·76	64·50	69·98	76·24	87·75
		44·79	*51·68*	*53·93*	*56·67*	*62·15*	*65·80*	*73·13*
Cylinder Grade 3 . . .	„	45·67	50·29	47·15	49·87	52·62	57·97	67·67
		37·84	*42·46*	*39·32*	*42·04*	*44·79*	*47·53*	*53·05*

		15 mm	20 mm	25 mm	32 mm	40 mm
	Unit	£	£	£	£	£
Extra for additional boss fitted at works	No.	0·60	0·60	0·60	0·80	1·70
Extra for boss for immersion heater	„	0·90				

COPPER INDIRECT CYLINDERS
(with bolted top and up to 5 tappings for connections and placing in position)

		Capacity litres					
		77	91	109	125	136	159	182
	Unit	£	£	£	£	£	£	£
Cylinder tested 20 lb.	No.	69·64	72·81	77·56	86·51	93·88	100·99	109·17
		61·81	*64·98*	*69·73*	*76·07*	*80·83*	*86·37*	*93·51*
Cylinder tested 30 lb.	„	80·73	83·90	90·24	102·36	109·73	117·63	129·77
		72·90	*76·07*	*82·41*	*91·92*	*96·68*	*103·01*	*114·11*
Cylinder tested 40 lb.	„	93·41	99·75	107·67	121·38	131·91	141·41	156·71
		85·58	*91·92*	*99·84*	*110·94*	*118·86*	*126·79*	*141·05*

		227	273	318	364	409	455	682	796
		£	£	£	£	£	£	£	£
Cylinder tested 30 lb.	„	210·59	238·00	338·87	389·58	421·28	455·59	744·49	831·10
		194·93	*217·12*	*315·38*	*366·09*	*397·79*	*429·49*	*713·17*	*797·17*
Cylinder tested 50 lb.	„	248·63	285·55	414·94	476·75	514·79	561·77	915·65	1021·28
		232·97	*264·67*	*391·45*	*453·26*	*491·30*	*535·67*	*884·33*	*987·35*

		909	1137	1364	1591	1818	2046	2273
		£	£	£	£	£	£	£
Cylinder tested 30 lb.	„	932·76	1253·92	1445·92	1671·89	1887·70	2029·69	2393·87
		896·22	*1214·77*	*1404·16*	*1619·69*	*1814·62*	*1946·17*	*2305·13*
Cylinder tested 50 lb.	„	1150·67	1507·49	1739·11	1996·78	2255·38	2425·89	2821·77
		1114·13	*1468·34*	*1697·35*	*1944·58*	*2182·30*	*2342·37*	*2733·03*

CISTERNS, TANKS AND CYLINDERS

SUPPLY AND ERECT

Pressed steel sectional tank of standard construction with external flanges, consisting of 5 mm plates 1220 × 1220 mm, stays, cleats and jointing materials. Weatherproof cover 3 mm thick fitted with one 460 mm diameter manhole with hinged lid and one 150 mm cowl ventilator. An allowance has been made for pipe connections. Finished one coat bituminous primer

	Unit	£
6100 × 3660 × 1220 mm deep, 25,460 litres capacity		
On prepared base at ground level or on prepared independent structure 15 m up	No.	4600·00
4880 × 4880 × 3660 mm deep, 83,400 litres capacity		
On prepared base at ground level or on prepared independent structure 15 m up	,,	9050·00
All as before but two 150 mm cowl ventilators		
12,200 × 10,980 × 3660 mm deep, 477,715 litres capacity		
On prepared base at ground level or on prepared independent structure 15 m up	,,	31750·00
All as before but cover 6 mm thick and two 150 mm cowl ventilators		
14,640 × 14,640 × 4880 mm deep, 1,026,760 litres capacity		
On prepared base at ground level or on prepared independent structure 15 m up	,,	57120·00

All as before but with internal flanges and flat non-weather proof covers, erected under cover

5180 × 3960 × 1370 mm deep, 28,090 litres capacity		
On prepared base at ground level or on prepared independent structure 15 m up	,,	5475·00
8840 × 5180 × 1370 mm deep, 62,170 litres capacity		
On prepared base at ground level or on prepared independent structure 15 m up	,,	10160·00
6400 × 5180 × 2590 mm deep, 85,660 litres capacity		
On prepared base at ground level or on prepared independent structure 15 m up	,,	10950·00
7620 × 7620 × 2590 mm deep, 150,460 litres capacity		
On prepared base at ground level or on prepared independent structure 15 m up	,,	16050·00

Glass fibre sectional tank of standard construction with external flanges, consisting of 12 mm plates 1220 × 1220 mm and 1220 × 610 mm, stays, cleats and jointing materials. Galvanized steel external bracing and supports. Weatherproof sealed cover fitted with one 610 × 610 mm manhole with hinged lid and one 100 mm cowl ventilator

2440 × 2440 × 1220 mm deep, 7300 litres capacity		
On prepared base at ground level or on prepared independent structure 15 m up	,,	1740·00
3660 × 2440 × 1830 mm deep, 16,300 litres capacity		
On prepared base at ground level or on prepared independent structure 15 m up	,,	3000·00
4880 × 3660 × 1830 mm deep, 32,700 litres capacity		
On prepared base at ground level or on prepared independent structure 15 m up	,,	4270·00
7320 × 3660 × 2440 mm deep, 65,500 litres		
On prepared base at ground level or on prepared independent structure 15 m up	,,	7455·00
7320 × 6100 × 2440 mm deep, 109,100 litres capacity		
On prepared base at ground level or on prepared independent structure 15 m up	,,	9934·00
9750 × 6100 × 3050 mm deep, 180,000 litres capacity		
On prepared base at ground level	,,	14920·00
Extra; on prepared independent structure 15 m up	,,	420·00
8530 × 7320 × 3660 mm deep, 229,000 litres capacity		
On prepared base at ground level	,,	18470·00
Extra; on prepared independent structure 15 m up	,,	436·00
12,190 × 10,970 × 3660 mm deep, 486,000 litres capacity		
On prepared base at ground level	,,	29720·00
Extra; on prepared independent structure 15 m up	,,	970·00

CALORIFIERS

COPPER STORAGE CALORIFIER
(generally to B.S. 853 with heating battery capable of raising contents from 10 °C to 65 °C in one hour, all connections provided, fittings excluded, static head not exceeding 1·35 bar and fixing in position on and including cradles or legs)

		\multicolumn{6}{c}{Capacity litres}
Horizontal	Unit	400 £	1000 £	2000 £	3000 £	4000 £	5000 £
Primary L.P.H.W. at 82 °C on/71 °C off	No.	828·33 *791·79*	1396·63 *1354·87*	2148·89 *2075·81*	2927·25 *2796·75*	3502·75 *3293·95*	4414·61 *4164·05*
Primary steam at 3·2 bar	,,	620·75 *584·21*	1042·38 *1000·62*	1800·85 *1727·77*	2367·90 *2237·40*	2834·96 *2635·16*	4476·76 *4226·20*
Vertical Primary L.P.H.W. at 82 °C on/71 °C off	,,	788·56 *752·02*	1346·91 *1305·15*	2086·74 *2013·66*	2877·53 *2747·03*	3459·25 *3250·45*	4352·46 *4101·90*
Primary steam at 3·2 bar	,,	583·46 *546·92*	1011·30 *969·54*	1738·70 *1665·62*	2343·04 *2212·54*	2819·10 *2610·30*	3171·61 *2921·05*

GALVANIZED STORAGE CALORIFIER
(all as above)

		\multicolumn{6}{c}{Capacity litres}
Horizontal	Unit	400 £	1000 £	2000 £	3000 £	4000 £	5000 £
Primary L.P.H.W. at 82 °C on/71 °C off	No.	713·98 *677·44*	1073·45 *1031·69*	1465·24 *1392·16*	1932·85 *1802·35*	2272·18 *2063·38*	2860·86 *2610·30*
Primary steam at 3·2 bar	,,	539·96 *503·42*	775·13 *733·37*	1166·92 *1093·84*	1435·65 *1305·15*	1675·54 *1466·74*	1791·88 *1541·32*
Vertical Primary L.P.H.W. at 82 °C on/71 °C off	,,	689·12 *652·58*	1042·38 *1000·62*	1409·31 *1336·23*	1907·99 *1777·49*	2247·32 *2038·52*	2836·00 *2585·44*
Primary steam at 3·2 bar	,,	508·88 *472·34*	750·27 *708·5!*	1117·20 *1044·12*	1410·79 *1280·29*	1650·68 *1441·88*	1767·02 *1516·46*

CALORIFIERS

CAST IRON NON-STORAGE HEATING CALORIFIER
(all as above heating secondary water with 82 °C flow and 71 °C return at 2 bar working pressure)

		Capacity kW					
			88	176	293	586	879	1465
Horizontal	Unit	£	£	£	£	£	£
Steam at 3·2 bar.	. No.	262·27	291·36	382·59	655·55	842·24	1269·82
		236·17	*254·82*	*335·61*	*540·71*	*696·08*	*1019·26*
Steam at 4·8 bar.	,,	262·27	291·36	382·59	655·55	842·24	1244·96
		236·17	*254·82*	*335·61*	*540·71*	*696·08*	*994·40*
Vertical							
Steam at 3·2 bar.	,,	268·49	303·79	401·24	674·19	867·10	1294·68
		242·39	*267·25*	*354·26*	*559·35*	*720·94*	*1044·12*
Steam at 4·8 bar.	,,	268·49	303·79	401·24	674·19	867·10	1294·68
		242·39	*267·25*	*354·26*	*559·35*	*720·94*	*1044·12*
Horizontal Primary water at 116 °C on/93 °C off	,,	374·14	620·75	842·50	1258·40	1675·05	2450·67
		348·04	*584·21*	*795·52*	*1143·56*	*1528·89*	*2200·11*
Vertical Primary water at 116 °C on/93 °C off	,,	399·00	645·61	867·36	1289·48	1706·13	2487·96
		372·90	*609·07*	*820·38*	*1174·64*	*1559·97*	*2237·40*

PUMPS, CIRCULATORS AND ACCELERATORS

SILENT RUNNING DIRECT DRIVE CENTRIFUGAL HEATING PUMP
(with three phase electric motor to run at not exceeding 1450 r.p.m. complete with bedplate, coupling guard and foundation bolts and fixing on prepared base, including bolted connections and supply only of mating flanges to work at maximum pressure of 400 kN/m^2 and maximum temperature of 120°C – electrical work measured separately)

Pump size mm	Maximum delivery litres/second	Maximum head kN/m^2	Maximum motor rating kW	Unit	£
40	4·0	80·0	1·50	No.	1021·49
					974·51
50	7·0	80·0	1·10	,,	674·69
					601·61
80	16·0	80·0	2·20	,,	906·14
					827·84
100	28·0	80·0	3·00	,,	1074·43
					960·84
150	70·0	120·0	11·20	,,	1916·41
					1715·34

BELT-DRIVEN CENTRIFUGAL HEATING PUMP
(with three phase electric motor to run at not exceeding 1450 r.p.m. complete with bedplate, pulley guard and foundation bolts and fixing on prepared base including bolted connections and supply only of mating flanges to work at maximum pressure of 400 kN/m^2 and a maximum temperature of 100°C – electrical work measured separately)

Pump size mm	Maximum delivery litres/second	Maximum head kN/m^2	Maximum motor rating kW	Unit	£
40	4·0	80·0	0·75	No.	519·32
					472·34
50	7·0	80·0	1·10	,,	613·79
					540·71
80	16·0	80·0	2·20	,,	866·36
					788·06
100	25·0	80·0	3·00	,,	1012·28
					898·69
150	70·0	120·0	11·20	,,	1816·97
					1615·90

Push button starter for three phase motor of the kW shown and handing to electricians for fixing and connections	Unit No.	0·37 kW £ 30·95 *28·34*	0·55 kW £ 30·95 *28·34*	0·75 kW £ 30·95 *28·34*	1·10 kW £ 30·95 *28·34*	1·50 kW £ 35·30 *32·69*	2·20 kW £ 35·30 *32·69*	3·00 kW £ 35·30 *32·69*	4·00 kW £ 35·30 *32·69*	5·50 kW £ 35·30 *32·69*

Star Delta starter for motor of the kW shown and handing to electricians for fixing and connections	Unit No.	7·5 kW £ 101·65 *98·20*	11·20 kW £ 149·25 *145·80*	15·00 kW £ 171·63 *168·18*

Mechanical Installations – Prices for Measured Work

PUMPS, CIRCULATORS AND ACCELERATORS

SILENT RUNNING PIPELINE MOUNTED CIRCULATOR
(for low and medium pressure heating and hot water services with three phase electric motor to run at not exceeding 1450 r.p.m. and fixing, including bolted connections and supply only of mating flanges, to work at maximum pressure of 600 kN/m² and maximum temperature of 100 °C – electrical work measured separately)

Pump size mm	Maximum delivery litres/second	Maximum head kN/m²	Maximum motor rating kW	Weight kg	Unit	Standard cast iron construction £	Gunmetal construction £
32	2·0	17·0	0·25	21	No.	221·00	295·58
						174·02	*248·60*
40	3·0	20·0	0·25	25	,,	234·67	301·80
						187·69	*254·82*
50	5·0	30·0	0·37	33	,,	262·27	338·09
						210·07	*285·89*
65	8·0	37·0	0·55	42	,,	298·31	387·81
						246·11	*335·61*
80	12·0	42·0	1·10	51	,,	375·38	492·22
						323·18	*440·02*
100	25·0	37·0	4·00	119	,,	619·01	828·08
						566·81	*770·66*

SILENT RUNNING GLANDLESS HEATING ACCELERATOR PUMP
(for low and medium pressure heating and hot water services, with electric motor to run at 2800 r.p.m. and fixing, including connections and supply only of mating flanges where applicable to work at maximum pressure of 600 kN/m² and maximum temperature of 120 °C – electrical work measured separately)

	Pump size mm	Maximum delivery litres/second	Maximum head kN/m²	Maximum motor rating kW	Weight kg	Unit	Single phase £	Three phase £
Screwed connections	25	1·0	25·0	0·053	7	No.	148·48	140·06
							101·50	*93·08*
	32	2·0	45·0	0·093	11	,,	178·48	161·83
							123·50	*114·85*
Flanged connections	50	3·0	65·0	0·37	29	,,		309·87
								257·67
	65	5·0	80·0	1·31	43	,,		351·76
								299·56
	80	6·0	100·0	1·31	48	,,		363·94
								311·74

Push button starter for single phase motor of the kW shown and handing to electrician for fixing and connections . . .	Unit No.	0·37 kW £ 30·95 *28·34*	0·55 kW £ 30·95 *28·34*	0·75 kW £ 30·95 *28·34*	1·10 kW £ 30·95 *28·34*	1·50 kW £ 35·30 *32·69*	2·20 kW £ 35·30 *32·69*

SELF-CONTAINED AUTOMATIC SUMP PUMP
(with totally enclosed electric motor, and fixing, float switch and gear – electrical work measured separately)

	Unit	£
Totally submersible	No.	415·15
		310·75
Extended shaft or semi-submersible for sump depth of 2 m	,,	580·47
		476·07

FANS

SINGLE STAGE AXIAL FLOW FAN
(three phase 50 cycle 380/440 volt electric motor, long casing and fixing to duct – electrical work measured separately)

	Unit	\multicolumn{4}{c}{Diameter of fan}
		305 mm	483 mm	762 mm	1219 mm
		\multicolumn{4}{c}{Speed of motor (r.p.m.)}
		2800	2900	940	725
		\multicolumn{4}{c}{Weight kg}
		29	53	98	301
		£	£	£	£
Fan	No.	262·51	405·71	459·65	1041·12
		178·99	316·97	360·47	894·96

SINGLE STAGE FAN
(as above but short casing)

Fan	,,	221·37	329·76	400·98	943·67
		140·46	243·63	307·02	807·95

TWO STAGE FAN
(as above but long casing)

Fan	,,	428·08	694·45	800·98	1901·52
		339·34	597·88	686·14	1697·94

ROOF EXTRACT UNIT
(for flat roof with propeller fan, three phase 50 cycle 380/440 volt electric motor with curved cowl and including automatic shutters with fusible link fixed to prepared kerb – electrical work measured separately)

	Unit	\multicolumn{4}{c}{Diameter of fan}
		315 mm	450 mm	800 mm	1250 mm
		\multicolumn{4}{c}{Speed of motor (r.p.m.)}
		1360	900	560	485
		\multicolumn{4}{c}{Weight kg}
		34	69	209	458
		£	£	£	£
	No.	227·46	310·25	630·75	1550·00
		170·04	242·39	531·57	1393·40

WINDOW TYPE FAN
(and fixing in prepared opening – electrical work measured separately)

	Unit	\multicolumn{3}{c}{Diameter of fan}
		152 mm	229 mm	305 mm
		£	£	£
Without shutter	No.	49·57	78·33	108·30
		39·13	67·89	95·25
With shutter	,,	62·02	96·89	129·80
		51·58	86·45	116·75

PUSH BUTTON STARTER
(for single or two stage axial flow fan 305–762 mm diameter and hand to electrician for fixing and connection)

	Unit	£
Up to 3·72 kW motors		
Without isolator.	No.	30·02
With isolator	,,	62·68
7·5 kW motors		
Without isolator.	,,	30·02
With isolator	,,	62·68

Mechanical Installations – Prices for Measured Work

RADIATORS

PRESSED STEEL PANEL TYPE RADIATORS
(fixed with and including brackets)

Height mm	Length mm	Unit	Single Surface area m²	£	Double Surface area m²	£
305	800	No.	0·56	13·60 *6·29*	1·11	22·04 *13·69*
305	1120	,,	0·78	16·00 *8·69*	1·56	26·83 *18·48*
305	1440	,,	1·00	18·39 *11·08*	2·01	31·61 *23·26*
305	1760	,,	1·23	23·92 *13·48*	2·45	39·54 *28·06*
305	2080	,,	1·45	31·53 *15·87*	2·90	49·54 *32·84*
305	2400	,,	1·67	34·97 *18·27*	3·34	55·39 *37·64*
305	2720	,,	1·90	37·37 *20·67*	3·79	60·18 *42·43*
432	480	,,	0·48	12·72 *4·89*	0·96	19·74 *10·87*
432	800	,,	0·80	15·76 *7·93*	1·60	25·84 *16·97*
432	1280	,,	1·28	20·34 *12·51*	2·56	34·99 *26·12*
432	1600	,,	1·60	28·09 *15·56*	3·20	45·80 *32·23*
432	2080	,,	2·08	35·80 *20·14*	4·15	58·08 *41·38*
432	2560	,,	2·56	45·60 *24·72*	5·11	72·46 *50·54*
432	2880	,,	2·88	48·65 *27·77*	5·75	78·56 *56·64*
432	3200	,,	3·20	53·79 *30·82*	6·39	86·75 *62·74*

RADIATORS

PRESSED STEEL PANEL TYPE RADIATORS
(fixed with and including brackets) *continued*

Height mm	Length mm	Unit	Single Surface area m²	£	Double Surface area m²	£
584	480	No.	0·64	15·25 *6·38*	1·27	23·77 *13·85*
584	800	,,	1·06	19·82 *10·42*	2·12	32·37 *21·93*
584	1280	,,	1·69	30·05 *16·48*	3·39	48·68 *34·06*
584	1600	,,	2·12	39·32 *20·53*	4·24	62·01 *42·17*
584	2080	,,	2·75	50·61 *26·60*	5·51	79·36 *54·30*
584	2560	,,	3·39	61·90 *32·67*	6·78	96·72 *66·44*
584	2880	,,	3·81	68·03 *36·71*	7·63	105·84 *74·52*
584	3200	,,	4·24	77·30 *40·76*	8·47	120·19 *82·61*
740	480	,,	0·79	17·27 *7·87*	1·58	27·27 *16·83*
740	800	,,	1·32	25·43 *12·90*	2·64	40·48 *26·91*
740	1280	,,	2·11	39·25 *20·46*	4·22	61·86 *42·02*
740	1600	,,	2·64	50·56 *25·50*	5·28	78·19 *52·09*
740	2080	,,	3·43	64·38 *33·06*	6·86	99·57 *67·21*
740	2560	,,	4·22	71·93 *40·61*		
740	2880	,,	4·75	76·97 *45·65*		
740	3200	,,	5·28	88·27 *50·69*		

CONVECTOR HEATING UNITS

SILL LINE NATURAL CONVECTORS
(including assembling and fixing panels and connecting to mains)

		Height of unit			
		406 mm	508 mm	610 mm	711 mm
	Unit	£	£	£	£
Front panel with sloping front and damper	m	18·52	16·90	19·20	20·37
		15·91	*13·98*	*15·86*	*16·72*
Inside corner	No.	12·74	13·34	15·53	15·43
		9·29	*9·48*	*11·25*	*10·73*
Outside corner	,,	13·27	13·82	16·14	15·68
		9·82	*9·96*	*11·86*	*10·98*
Valve box	,,	10·39	10·80	12·00	12·31
		7·78	*7·88*	*8·66*	*8·66*
Stop end	,,	7·16	7·73	8·81	9·74
		4·55	*4·81*	*5·47*	*6·09*
Front panel with flat front and top outlet	m	17·53	18·78	20·06	22·73
		14·92	*15·86*	*16·72*	*19·08*
Inside corner	No.	14·07	14·81	15·53	16·17
		10·62	*10·98*	*11·25*	*11·47*
Outside corner	,,	14·31	15·11	16·14	16·87
		10·86	*11·25*	*11·86*	*12·17*
Valve box	,,	10·74	11·20	12·00	12·71
		8·13	*8·28*	*8·66*	*9·06*
Stop end	,,	7·42	7·85	8·81	9·37
		4·81	*4·93*	*5·47*	*5·72*

		Single element	Double element
	Unit	£	£
32 mm black steel tube – finned	m	15·95	30·64
		12·19	*24·38*
35 mm copper tube – finned	,,	17·79	34·32
		14·03	*28·06*

CONVECTOR HEATING UNITS

CONVECTORS FOR STEAM OR HOT WATER HEATING FLOOR OR WALL TYPE
(including assembling and fixing casings and connecting to mains)

		Unit	Height of Unit 500 mm £	700 mm £
Unit 130 mm deep single tube				
Length of unit:				
400 mm	. . . .	No.	53·04	54·35
			42·08	*42·87*
500 mm	. . . .	,,	55·10	57·04
			43·51	*44·93*
600 mm	. . . .	,,	57·72	59·78
			46·13	*47·36*
700 mm	. . . .	,,	60·53	62·39
			48·63	*49·97*
800 mm	. . . .	,,	63·76	65·40
			51·86	*52·98*
900 mm	. . . .	,,	66·76	68·44
			54·44	*55·60*
1000 mm	. . . .	,,	69·36	71·16
			56·83	*58·11*
1200 mm	. . . .	,,	75·01	76·89
			61·86	*63·21*
1400 mm	. . . .	,,	79·08	80·71
			67·07	*68·18*
Unit 130 mm deep, two tubes				
Length of unit:				
1200 mm	. . . .	,,	86·89	89·90
			73·21	*75·70*
1400 mm	. . . .	,,	94·73	97·82
			80·84	*83·41*
Extra for				
Feet for conversion to floor mounting	. . . . .	,,	5·23	
			2·72	

UNIT HEATERS

HORIZONTAL DISCHARGE UNIT HEATER
(recirculating type fixed to suspension rods (measured separately) and connecting to steam or hot water services – electrical connections and switchgear measured separately)

	Unit No.	Approx. weight				
		34 kg	41 kg	45 kg	57 kg	64 kg
				Output		
		15 kW	25 kW	35 kW	49 kW	65 kW
		£	£	£	£	£
Steam at 4 bar	. . . . No.	179·37	197·55	216·56	240·77	265·08
		148·05	161·01	174·80	196·40	218·10

		Approx. weight				
		37 kg	43 kg	50 kg	62 kg	71 kg
				Output		
		9 kW	16 kW	25 kW	37 kW	52 kW
		£	£	£	£	£
L.P.H.W.	. . . . . ,,	183·98	209·21	229·44	256·29	290·11
		152·66	167·45	187·68	211·92	243·13

UNIT HEATER
(as above, but including connecting to fresh air duct)

	Unit No.	Approx. weight				
		34 kg	41 kg	45 kg	57 kg	64 kg
				Output		
		17 kW	28 kW	39 kW	54 kW	73 kW
		£	£	£	£	£
Steam at 4 bar	. . . . No.	213·79	237·90	262·63	289·80	322·30
		151·15	164·82	179·11	201·06	223·12

		Approx. weight				
		37 kg	43 kg	50 kg	62 kg	71 kg
				Output		
		11 kW	20 kW	30 kW	45 kW	64 kW
		£	£	£	£	£
L.P.H.W.	. . . . . ,,	218·40	244·34	275·51	305·32	347·33
		155·76	171·26	191·99	216·58	248·15

DOWNWARD DISCHARGE UNIT HEATER
(recirculating type fixed to suspension rods (measured separately) and connecting to steam or hot water services – electrical connections and switchgear measured separately)

	Unit No.	Approx. weight				
		34 kg	41 kg	46 kg	57 kg	64 kg
				Output		
		14 kW	24 kW	33 kW	46 kW	62 kW
		£	£	£	£	£
Steam at 4 bar	. . . . No.	174·10	194·19	211·47	239·49	268·22
		142·78	157·65	169·71	195·12	221·24

		Approx. weight				
		37 kg	43 kg	50 kg	62 kg	71 kg
				Output		
		9 kW	16 kW	25 kW	36 kW	52 kW
		£	£	£	£	£
L.P.H.W.	. . . . . ,,	178·11	205·85	224·35	255·00	293·25
		147·39	164·09	182·59	210·64	246·27

FIRE FIGHTING APPLIANCES

DRY RISING MAIN

Note
For tubing and flanged connections see earlier sections

	Unit	£
Bronze or gunmetal inlet breeching for pumping in with 64 mm diameter instantaneous male coupling with cap and chain and 25 mm drain valve		
Single inlet with 76 mm screwed joint to steel	No.	96·79
		86·35
As above, but 76 mm flanged to B.S. table D (bolted connection to counter flanges measured separately)	,,	91·57
		86·35
Double inlet with back pressure valve and with screwed joint to steel	,,	139·01
		124·81
As above but flanged as before	,,	131·49
		124·81
Quadruple inlet with back pressure valve and with screwed joint to steel	,,	320·92
		303·38
As above but flanged as before	,,	315·91
		303·38
Steel dry riser inlet box with hinged, wire glazed door suitably lettered and hand to main contractor for building in		
305 × 305 × 305 mm for single inlet	,,	68·34
		65·73
610 × 406 × 305 mm for double inlet	,,	70·53
		67·92
610 × 610 × 356 mm for quadruple inlet	,,	81·93
		79·32
Bronze or gunmetal gate type outlet valve with 64 mm diameter female instantaneous coupling with cap and chain, inlet flanged or screwed, wheelhead secured by padlock and leather strap		
Screwed 64 mm male inlet and joint to steel	,,	103·04
		98·86
Flanged B.S. table D inlet (bolted connection to counter flanges measured separately)	,,	103·04
		98·86
Bronze or gunmetal landing type valve as above		
Horizontal, screwed 64 mm male inlet and joint to steel	,,	73·98
		69·80
As above but flanged to B.S. table D (bolted connections to counter flanges measured separately)	,,	82·22
		78·04
Oblique, screwed as before	,,	66·76
		62·58
Oblique, flanged as before	,,	74·70
		70·52
Air valve and screwed joint to steel		
25 mm diameter	,,	15·16
		10·67

FIRE FIGHTING APPLIANCES

HOSE REELS

	Unit	£
Automatic non-swing first aid pedestal mounted reel and hand to main contractor for fixing. Connect reel to supply with 25 mm screwed joint. Reel fitted with 37 metres, 20 mm rubber hose suitable for working pressure up to 7 bar	No.	164·47 *154·03*
As above but swinging pattern	„	155·98 *145·54*
As above but automatic recessed swing type	„	217·71 *207·27*
30 metres length of fire brigade quality rubber-lined canvas hose 64 mm bore 12 ply, treated to prevent mildew and rot and fitted with two 64 mm diameter instantaneous gun metal couplings and place on hose cradle (measured separately)	„	141·10 *135·88*
Wrought iron hose cradle to hold 30 metres of 64 mm bore hose (measured separately) and hand to main contractor for fixing	„	85·08 *82·47*
Gun metal branch pipe with 64 mm nozzle and place in position on hose	„	53·97 *51·36*

FIRE EXTINGUISHERS

	Unit	£
Water carbon dioxide type to B.S. 1382 and place in position		
4·6 litres size	„	40·21 *37·60*
9 litres size	„	40·21 *37·60*
Foam gas pressure type to B.S. 740, 9 litres size and place in position	„	52·94 *50·33*
Carbon dioxide gas trigger type, 3 kg size and place in position	„	74·17 *71·56*
Bromochlorodifluoromethane (B.C.F.) lever operated and place in position		
0·75 kg size	„	27·35 *24·74*
1·5 kg size	„	37·66 *35·05*
2·5 kg size	„	55·12 *52·51*
5·0 kg size	„	129·11 *126·50*
Dry powder type, 3 kg size and place in position	„	38·75 *36·14*
Asbestos fire blanket in metal container and place in position		
Size 914 × 914 mm	„	20·32 *17·71*
Size 1219 × 1219 mm	„	29·15 *26·54*
Size 1524 × 1524 mm	„	41·18 *38·57*
Size 1829 × 1829 mm	„	52·60 *49·99*

Mechanical Installations – Prices for Measured Work
VALVES, REGULATORS, TRAPS AND GAUGES

COCKS

		15 mm £	20 mm £	25 mm £	32 mm £	40 mm £	50 mm £	65 mm £	80 mm £
Polished brass crutch head screw down bibcock to B.S. 1010 and screwed joint to steel	Unit No.	4·09 *2·84*	6·09 *4·32*	10·56 *8·47*					
Bibcock as above but with hose union	,,	5·34 *4·09*	8·62 *6·85*	14·52 *12·43*					
Bronze angle pattern draw-off cock, ribbed for hose and screwed joint to steel	,,	4·29 *3·04*	5·07 *3·30*	9·51 *7·42*					
Draw-off cock as above but with lockshield	,,	4·53 *3·28*	6·76 *4·99*	10·99 *8·90*					
Bronze plug pattern draw-off cock with hose union and loose key and screwed joint to steel	,,	8·19 *6·94*	11·07 *9·30*	13·81 *11·72*					
Draw-off cock as above but gland pattern	,,	10·22 *7·40*	13·82 *9·96*	18·28 *13·79*	31·89 *26·77*	43·76 *37·81*	64·29 *57·19*		
Bronze cock, plug pattern working pressure for cold services up to 9 bar and screwed joints to steel	,,	12·11 *7·73*	16·89 *10·94*	23·43 *16·64*	34·37 *26·54*	45·68 *36·60*	62·99 *52·13*	158·52 *144·95*	216·23 *200·57*

		15 mm £	22 mm £	28 mm £	35 mm £	42 mm £	54 mm £	67 mm £	76 mm £	108 mm £
Plug cock with screwed joints to steel	Unit No.	6·76 *4·99*	9·24 *6·94*	12·18 *9·57*	17·44 *14·31*	23·82 *20·37*	42·09 *37·81*	100·38 *95·79*	108·42 *103·51*	222·34 *216·60*

		15 mm £	20 mm £	25 mm £	32 mm £	40 mm £	50 mm £
Bronze cock, three way gland pattern working pressure for steam and cold services up to 10 bar and screwed joints to steel	Unit No.	16·61 *10·66*	23·28 *15·24*	30·18 *20·99*	47·72 *37·18*	56·14 *43·93*	82·09 *67·47*

VALVES, REGULATORS, TRAPS AND GAUGES

COCKS *continued*

	Unit	15 mm £	22 mm £	28 mm £	35 mm £	42 mm £	54 mm £
Brass crutch-head stopcock with capillary joints for copper	No.	5·26 *2·65*	6·50 *3·89*	9·77 *7·01*			
Brass lockshield stopcock with capillary joints for copper	,,	5·85 *3·24*	7·24 *4·63*	11·01 *8·25*			
Gunmetal crutch-head stopcock with capillary joints for copper	,,	6·45 *3·84*	8·36 *5·75*	12·32 *9·56*	18·43 *14·99*	22·47 *19·03*	32·34 *28·43*
Gunmetal lockshield stopcock with capillary joints for copper.	,,	6·97 *4·36*	9·10 *6·49*	13·57 *10·81*			
Brass double union stopcock with capillary joints for copper	,,	7·70 *4·26*	9·46 *6·02*	14·36 *10·71*			
Brass double union lockshield stopcock with capillary joints for copper.	,,	8·22 *4·78*	10·16 *6·72*	15·62 *11·97*			
Gunmetal double union stopcock with capillary joints for copper	,,	10·40 *6·96*	12·02 *8·58*	19·51 *15·86*	30·10 *26·45*	38·57 *34·66*	58·85 *54·52*
Gunmetal double union lockshield stopcock with capillary joints for copper	,,	10·85 *7·41*	12·71 *9·27*	20·77 *17·12*			
Brass crutch-head easy-clean stopcock with capillary joints for copper	,,	6·00 *3·39*	7·44 *4·83*	11·25 *8·49*			
Brass combined stopcock and draincock with capillary joints for copper.	,,	9·94 *7·33*	11·65 *9·04*				
Gunmetal ditto	,,	12·35 *9·74*					
Brass crutch-head stopcock with compression joints for copper	,,	8·25 *5·85*	11·25 *8·54*	16·52 *13·39*	27·23 *23·37*	37·72 *33·34*	52·67 *47·55*
Brass lockshield stopcock with compression joints for copper	,,	5·80 *3·66*	7·24 *5·10*	11·75 *9·40*			
Gunmetal crutch-head stopcock with compression joints for copper	,,	6·51 *4·37*	8·89 *6·75*	12·43 *10·08*	21·16 *18·55*	29·55 *26·53*	39·62 *36·18*
Brass crutch-head easy-clean stopcock with compression joints for copper	,,	5·95 *3·81*	7·53 *5·39*	11·99 *9·64*			
Brass combined stopcock and draincock with compression joints to copper	,,	8·03 *5·89*	10·38 *8·24*				
Gunmetal ditto	,,	9·03 *6·89*					
Brass crutch-head stopcock with joints for polythene.	,,	6·91 *4·57*	9·23 *6·89*	12·47 *9·97*			
Gunmetal ditto	,,	8·13 *5·78*	11·65 *9·30*	15·03 *12·53*	24·31 *21·55*	34·62 *31·44*	47·95 *44·35*

Mechanical Installations – Prices for Measured Work
VALVES, REGULATORS, TRAPS AND GAUGES

COCKS continued

	Unit	15 mm £	22 mm £
Brass crutch-head stopcock with compression joints for copper and polythene	No.	8·41 / 6·17	9·62 / 7·38
Brass combined stopcock and draincock with joints for polythene	„	10·58 / 8·24	
Brass draw-off coupling with compression joints for copper	„	3·57 / 2·27	
Gunmetal ditto	„	3·94 / 2·64	4·66 / 3·36
Brass draincock	„	2·54 / 1·87	
Brass lockshield draincock	„	2·62 / 1·95	
Gunmetal lockshield draincock	„	2·96 / 2·29	

GAS VALVES

	Unit	15 mm £	20 mm £	25 mm £	32 mm £	40 mm £	50 mm £	65 mm £	80 mm £
Malleable iron ball type isolating valve with screwed connections	No.	27·37 / 22·99	28·94 / 22·99	29·78 / 22·99	40·41 / 32·58	38·37 / 29·29	49·90 / 39·04	83·82 / 70·25	93·57 / 77·91

	Unit	40 mm £	50 mm £	80 mm £	100 mm £	150 mm £
Malleable iron ball type isolating valve with flanged ends to B.S. table D including bolted connection	No.	50·92 / 40·94	63·94 / 52·02	102·35 / 88·69	180·27 / 164·51	254·20 / 225·29

GATE VALVES

	Unit	15 mm £	20 mm £	25 mm £	32 mm £	40 mm £	50 mm £	65 mm £	80 mm £	100 mm £
Bronze wedge disc non-rising stem gate valve, working pressure saturated steam up to 14 bar, cold services up to 28 bar and screwed joints to steel	No.	9·76 / 5·38	13·05 / 7·10	16·03 / 9·24	21·49 / 13·66	26·60 / 17·52	36·41 / 25·55	61·17 / 47·60	84·72 / 69·06	144·72 / 123·00
Gate valve as above but working pressure saturated steam up to 10 bar, cold services up to 21 bar but with flanged ends to B.S. table F including bolted connections	„			27·69 / 18·72	35·63 / 24·98	42·97 / 31·89	51·35 / 39·83	66·00 / 53·69	100·08 / 81·72	123·24 / 102·16

VALVES, REGULATORS, TRAPS AND GAUGES

GATE VALVES continued

		15 mm £	22 mm £	28 mm £	35 mm £	42 mm £	54 mm £	67 mm £	80 mm £
Brass gate valve, working pressure cold services up to 14 bar, and with compression joints to copper	Unit No.	10·96 8·56	14·70 11·99	19·81 16·68	27·95 24·09	36·19 31·81	54·91 49·79	98·45 92·71	136·26 130·00
Gunmetal fullway gate valve with capillary joints for copper .	,,	6·74 4·13	7·50 4·89	9·43 6·67	18·32 14·88	21·21 17·77	29·68 25·77		
Ditto but lockshield pattern .	,,	6·87 4·62	7·68 5·07	10·53 7·77	18·46 15·02	21·23 17·79	29·67 25·76		

		50 mm £	65 mm £	80 mm £	100 mm £	125 mm £	150 mm £
Cast iron wedge disc non-rising stem gate valve, working pressure, saturated steam up to 7 bar, cold services up to 12 bar and screwed joints to steel . .	Unit No.	44·17 33·31	49·97 36·40	57·62 41·96	77·76 56·04		
Gate valve as above but with flanged ends to B.S. table D including bolted connections	,,	36·89 33·55	40·66 36·48	46·94 42·14	62·72 56·04	89·77 79·85	105·06 92·53
Bronze parallel slide valve, working pressure, saturated steam up to 17 bar, cold services up to 28 bar and screwed joints to steel		15 mm £	20 mm £	25 mm £	32 mm £	40 mm £	50 mm £
	,,	31·24 26·86	34·63 28·68	49·84 43·05	74·11 66·28	86·34 77·26	108·98 98·12
Slide valve as above but with flanged ends to B.S. table F including bolted connections	,,	45·02 36·57	48·75 39·78	64·86 54·21	87·37 76·29	102·30 90·78	131·42 119·10
Slide valve as above but cast iron, working pressure, saturated steam up to 10 bar, cold services up to 17 bar with flanged ends to B.S. table F including bolted connections		50 mm £	65 mm £	80 mm £	100 mm £	125 mm £	150 mm £
	,,	132·57 120·25	164·99 144·63	197·42 176·34	249·42 225·88	339·40 308·54	430·86 385·61

Mechanical Installations – Prices for Measured Work
VALVES, REGULATORS, TRAPS AND GAUGES

GLOBE VALVES

	Unit	15 mm £	20 mm £	25 mm £	32 mm £	40 mm £	50 mm £	65 mm £	80 mm £
Bronze globe valve, renewable disc, working pressure, saturated steam up to 9 bar, cold services up to 14 bar and screwed joints to steel . . .	No.	9·34 *4·96*	12·35 *6·40*	16·27 *9·48*	21·34 *13·51*	27·18 *18·10*	38·25 *27·39*	64·61 *51·04*	90·71 *75·05*
Globe valve as above but with flanged ends to B.S. table D including bolted connections . . .	,,	21·90 *13·71*	24·33 *15·60*	31·26 *22·03*	36·43 *26·85*	44·96 *34·88*	61·00 *48·98*		
Globe valve as above but working pressure, saturated steam up to 14 bar, cold services up to 28 bar but with flanged ends to B.S. table H including bolted connections	,,	37·38 *27·55*	40·35 *30·00*	49·18 *38·45*	63·31 *52·13*	74·38 *62·76*	106·26 *93·74*	163·78 *143·02*	203·60 *186·72*

CHECK VALVES

	Unit	15 mm £	20 mm £	25 mm £	32 mm £	40 mm £	50 mm £	65 mm £	80 mm £
Bronze check valve swing pattern, working pressure, saturated steam up to 9 bar and screwed joints to steel . .	No.	8·45 *4·07*	11·19 *5·24*	14·33 *7·54*	19·07 *11·24*	22·98 *13·90*	30·72 *19·86*	50·49 *36·92*	71·30 *55·64*
Check valve as above but renewable disc pattern, working pressure, saturated steam up to 7 bar, cold services up to 14 bar with flanged ends to B.S. table D including bolted connections .	,,	25·32 *17·13*	30·22 *21·49*	36·46 *27·23*	43·38 *33·80*	55·68 *45·60*	77·52 *65·50*	112·68 *99·76*	158·40 *144·74*
Check valve as above but working pressure, saturated steam up to 14 bar, cold services up to 28 bar with flanged ends to B.S. table H including bolted connections	,,	35·92 *26·09*	38·57 *28·22*	45·99 *35·26*	59·02 *47·84*	68·62 *57·00*	95·24 *82·72*	147·80 *127·04*	194·64 *172·96*

	Unit	50 mm £	65 mm £	80 mm £	100 mm £	125 mm £	150 mm £
Cast iron swing pattern check valve, working pressure for cold services up to 14 bar with flanged ends to B.S. table D including bolted connections	No.	65·34 *53·42*	73·89 *61·07*	80·22 *66·56*	106·47 *90·73*	154·32 *128·02*	171·73 *142·82*
Check valve as above but B.S. table F .	,,	67·30 *54·98*	84·28 *63·92*	90·55 *69·47*	116·44 *92·90*	165·53 *134·67*	194·71 *149·46*

Mechanical Installations – Prices for Measured Work
VALVES, REGULATORS, TRAPS AND GAUGES

EQUILIBRIUM BALL VALVES

		25 mm £	32 mm £	40 mm £	50 mm £	65 mm £	80 mm £	100 mm £
Bronze equilibrium ball valve, hydraulic working pressure 10 bar with flanged end to B.S. table D including bolted connection	Unit No.	131·15 *122·02*	155·30 *145·82*	182·27 *172·29*	238·55 *226·63*	276·83 *264·01*	319·68 *306·02*	459·00 *443·24*

RADIATOR VALVES

		15 mm £	20 mm £	25 mm £	32 mm £
Bronze straight pattern, wheelhead or lockshield, matt finish and screwed joints to steel	Unit No.	8·80 *4·42*	11·54 *5·59*	14·56 *7·77*	19·00 *11·38*
Radiator valve as above but angle pattern	„	7·91 *3·53*	10·45 *4·50*	13·13 *6·34*	
Straight pattern valve with compression joints to copper	„	11·13 *8·73*	14·13 *11·31*		
Angle pattern valve with capillary copper inlet and screwed union outlet	„	9·12 *6·72*	11·55 *8·73*		
Brass wheelhead angle pattern radiator valve with capillary copper inlet and screwed union outlet	„	4·73 *2·12*			
Ditto but chromium plated	„	5·22 *2·61*			
Brass lockshield angle pattern radiator valve with capillary copper outlet and screwed union inlet	„	4·73 *2·12*			
Ditto but chromium plated	„	5·45 *2·84*			
Brass wheelhead angle pattern radiator valve with compression copper inlet and screwed union outlet	„	4·92 *2·31*			
Ditto but chromium plated	„	5·45 *2·84*			
Brass lockshield angle pattern radiator valve with compression copper outlet and screwed union inlet	„	4·92 *2·31*			
Ditto but chromium plated	„	5·22 *2·61*			

		8 mm £	10 mm £	12 mm £
Microbore twin entry brass radiator valve	„	5·76 *4·45*	6·18 *4·87*	
As above but chromium plated	„	6·70 *5·39*	7·24 *5·93*	
Microbore single entry brass radiator valve	„	3·34 *2·03*	3·34 *2·03*	3·34 *2·03*
As above but chromium plated	„	3·82 *2·51*	3·82 *2·51*	3·82 *2·51*

VALVES, REGULATORS, TRAPS AND GAUGES

SAFETY VALVES

	Unit	15 mm £	20 mm £	25 mm £	32 mm £	40 mm £	50 mm £	65 mm £	80 mm £	100 mm £
Bronze 'pop' type valve top outlet and screwed joint to steel	No.	19·78 *18·53*	24·75 *22·98*	29·86 *27·77*	37·73 *35·33*	49·05 *45·92*	66·33 *62·57*	95·47 *90·77*	122·43 *116·90*	
Valve as above but side outlet	,,	21·20 *19·95*	24·88 *23·11*	30·90 *28·81*	39·55 *37·15*	51·85 *48·72*	69·18 *65·42*	127·50 *122·80*	134·09 *128·56*	
Valve as above but spring type side outlet	,,	12·04 *10·79*	15·07 *13·30*	20·04 *17·95*	28·88 *26·48*	34·94 *31·81*	48·11 *44·35*	74·80 *70·10*	97·80 *92·27*	
'National' bronze valve and screwed joint to steel	,,	22·29 *19·47*	28·80 *24·94*	38·08 *33·59*	45·41 *40·29*	54·20 *48·25*	71·72 *64·62*	107·01 *98·14*	157·51 *147·28*	371·92 *359·91*

AIR VALVES

		15 mm
Automatic air vent for pressure up to 7 bar and 200 degrees Fahrenheit (93° Celsius) and screw joint to steel including regulating, adjusting and testing	Unit No.	£ 43·87 *39·28*
Air vent as above with lockhead isolating valve	,,	50·82 *44·66*
Air vent as above but for pressure up to 7 bar and 300 degrees Fahrenheit (149° Celsius)	,,	91·86 *85·70*
Automatic air vent for pressure up to 17 bar and 400 degrees Fahrenheit (204° Celsius) with flanged end to B.S. Table H including bolted connection to counter flange (measured separately) and regulating, adjusting and testing	,,	126·67 *125·42*

VALVES, REGULATORS, TRAPS AND GAUGES

CONTROL VALVES

	Unit	20 mm £	25 mm £	32 mm £	40 mm £	50 mm £	65 mm £	80 mm £	100 mm £
Pressure reducing valve for steam for maximum range of 17 bar and 450 degrees Fahrenheit (232° Celsius) with flanged ends to B.S. table H and bolted connections to counter flanges (measured separately) . . .	No.		185·78 *168·58*	212·23 *194·05*	248·67 *229·22*	288·95 *266·82*			
Cast iron body butterfly type control valve with two position electrically controlled motor for low pressure hot water and flanged complete with counter flanges (electrical connection measured separately) . . .	,,		117·63 *106·73*	120·20 *109·15*	124·33 *112·79*	130·44 *116·43*	136·18 *121·28*	140·57 *124·92*	149·63 *132·20*
Stem type air thermostat for control of motorized valve and fixing by grub screws (electrical connection measured separately) . .	,,	53·09 *50·48*							
Room thermostat for control of motorized valve fixing to any surface (electrical connection measured separately) . . .	,,	10·37 *7·76*							

	Unit	15 mm £	20 mm £	25 mm £
Thermostatic control valve for water or steam with screwed ends complete with 1·5 metres of capillary tube and phial . . .	,,	104·67 *99·45*	107·45 *100·66*	115·46 *107·94*

203

VALVES, REGULATORS, TRAPS AND GAUGES

STEAM TRAPS AND STRAINERS

		15 mm	20 mm	25 mm	32 mm	40 mm	50 mm	65 mm	80 mm	100 mm
Cast iron inverted bucket type with pressure gauge up to 17 bar with screwed joints to steel	Unit No.	£ 34·52 30·14	£ 49·25 43·30	£ 75·31 68·52	£	£ 135·21 126·13	£ 204·91 194·05	£	£	£
Trap as above but with ends flanged to B.S. table H including bolted connections	„	73·50 63·67	89·79 79·44	127·16 116·43		192·64 180·71	255·50 242·56			
Stainless steel thermodynamic trap with pressure range up to 24 bar and temperature range 550 degrees Fahrenheit (228° Celsius) and screwed joints to steel	„	25·73 21·35	38·27 32·32							
Trap as above but with ends flanged to B.S. table H including bolted connections	„	74·71 64·88	85·54 75·19	96·84 86·11						
Malleable iron pipe line strainer, maximum steam working pressure 14 bar with screwed joints to steel	„	17·07 4·11	20·01 5·48	23·24 7·81		28·20 11·59	31·04 13·16			
Bronze pipe line strainer maximum steam working pressure 17 bar and ends flanged to B.S. table H including bolted connections	„	46·40 36·57	53·34 42·99	64·94 54·21	94·26 83·08	110·46 98·84	164·12 151·60	216·02 195·26	267·88 246·20	464·19 440·25

THERMOMETERS

		£
200 mm D.o.E. pattern with 15 mm screwed tail and pocket, range 0–120 degrees Centigrade (30–240 °F) and screwed joint to steel	Unit No.	22·96 20·14
Vertical combined altitude gauge and thermometer, graduated 15 m head and 10–120 degrees Centigrade (50–250 °F) and screwed joint to steel	„	26·19 23·37
Horizontal ditto	„	21·62 18·80

THERMAL INSULATION

THERMAL INSULATION APPLIED TO GENERAL SURFACES

	Unit	48 kg/m³ £	80 kg/m³ £
Mineral fibre glass fibre rigid slab plain finish fixed to ductwork with adhesive and 25 mm mesh galvanized wire netting			
25 mm thick	m²	4·70	4·94
38 mm thick	,,	5·19	5·69
51 mm thick	,,	5·69	6·19
Extra for			£
Layer of 0·9 mm thick aluminium sheet secured with rivets or self-tapping screws	,,		13·21

			80 kg/m³ £
Mineral fibre glass fibre rigid slab fixed to ductwork with adhesive, secured with 25 mm mesh galvanized wire netting, edges protected with expanded metal beading, supercoated with 12 mm thick self-setting cement, trowelled smooth, scrim embedded in surface			
25 mm thick	,,		23·38
38 mm thick	,,		24·12
51 mm thick	,,		24·61
Extra for			£
One coat of emulsion paint on cement surface	,,		1·26

		32 kg/m³ £	48 kg/m³ £
Aluminium foil faced mineral fibre glass fibre rigid slab, fixed to ductwork with adhesive and 25 mm mesh galvanized wire netting, joints and ends sealed with self-adhesive tape			
25 mm thick	,,	6·26	6·51
38 mm thick	,,	6·81	7·07
51 mm thick	,,	7·30	7·54
Extra for			£
0·9 mm thick aluminium sheet secured with rivets or self-tapping screws	,,		13·21

		16 kg/m³ £	32 kg/m³ £
Aluminium foil faced mineral fibre glass fibre mattress, fixed to ductwork with adhesive and secured with 50 mm mesh galvanized wire netting, joints and ends sealed with self-adhesive tape			
25 mm thick	,,	4·90	5·45
38 mm thick	,,	5·22	6·13
51 mm thick	,,	5·53	6·84

		48 kg/m³ £	128 kg/m³ £
Mineral fibre glass fibre mattress, plain finish fixed to ductwork with adhesive and secured with 50 mm mesh galvanized wire netting, joints and ends taped			
25 mm thick	,,	5·16	6·13
38 mm thick	,,	5·68	6·63
51 mm thick	,,	6·22	7·85

THERMAL INSULATION

THERMAL INSULATION APPLIED TO PIPELINES

	Unit	Thickness 19 mm £	25 mm £	38 mm £
Plain-faced pre-formed mineral fibre glass fibre rigid sections, fixed with aluminium bands at 450 mm intervals Tube, including bends, to the following diameters				
15 mm	m	1·64	1·74	2·36
20 mm	,,	1·72	1·83	2·45
25 mm	,,	1·80	1·92	2·56
32 mm	,,	1·89	2·03	2·69
40 mm	,,	1·96	2·11	2·78
50 mm	,,	2·16	2·32	3·06
65 mm	,,	2·36	2·55	3·34
80 mm	,,	2·52	2·73	3·55
100 mm	,,		3·23	4·11
125 mm	,,		3·59	4·54
150 mm	,,		4·08	5·15
Extra for Fitting around pipe fittings with insulation of the thickness indicated				
Pair of flanges				
15 mm	No.		1·74	2·36
20 mm	,,		1·82	2·45
25 mm	,,		1·92	2·56
32 mm	,,		2·03	2·69
40 mm	,,		2·11	2·78
50 mm	,,		2·32	3·06
65 mm	,,		2·55	3·34
80 mm	,,		2·73	3·55
100 mm	,,		3·23	4·11
125 mm	,,		3·59	4·54
150 mm	,,		4·08	5·15
Screwed valves				
15 mm	,,		1·74	2·36
20 mm	,,		1·82	2·45
25 mm	,,		1·92	2·56
32 mm	,,		2·03	2·69
40 mm	,,		2·11	2·78
50 mm	,,		2·32	3·06
Flanged valves				
15 mm	,,		2·62	3·53
20 mm	,,		2·74	3·67
25 mm	,,		2·88	3·85
32 mm	,,		3·05	4·04
40 mm	,,		3·18	4·18
50 mm	,,		3·49	4·60
65 mm	,,		3·84	5·01
80 mm	,,		4·09	5·32
100 mm	,,		4·97	6·17
125 mm	,,		5·39	6·81
150 mm	,,		6·12	7·73

THE INSULATION SPECIALISTS.

We provide a unique Insulation Contracting Service to the Heating, Ventilating and Air-Conditioning sector of the Construction Industry.

Contact CAPE......................
for advice on design, pricing and application, plus the benefit of our long experience as Major Insulation Contractors.

Cape Building Services Limited
MECHANICAL SERVICES DIVISION
Freshwater Road, Dagenham, Essex, RM8 1SR
Telephone 01-597 2221   Telex 897436
Offices at: Manchester, London, Dundee, Reading.

Heat Pumps

Heat Recovery Systems

A directory of equipment and techniques

R D Heap, Formerly Research Scientist, Electricity Council Research Centre, Chester

This book explains clearly the principles behind various types of heat pumps, and describes their main components and design features. The history of heat pump development and application is outlined, and methods of evaluating heat pump economics are discussed.

No specialised technical knowledge is assumed so that any potential heat pump user will be able to assess the feasibility of any particular application. Extensive references and a bibliography cater for the reader seeking further information on specific points.

1979   168 pages   Illustrated
Hardback: 0 419 11330 4   £8.75

D A Reay, Business Development Manager of International Research and Development Co. Ltd.

The recovery of waste heat has been practised for many decades particularly in power generation and the energy intensive industries. During the past few years the interest in energy conservation has attracted an increasing number of products on to the market, and the potential user of such equipment has been set an enormous task in identifying the correct type of system to meet his requirements. **Heat Recovery Systems** firstly describes the various types of waste heat recovery equipment available for use on industrial processes and in heating ventilating and air consideration. Secondly, and in this respect it is unique, it details heat recovery systems made by over 200 companies worldwide, this data being obtained with the assistance of questionnaires sent to each manufacturer.

1979   608 pages   illustrated
Hardback: 0 419 11400 9   £22.50

E & F N Spon Ltd
The technical division
of Associated Book Publishers
Ltd, 11 New Fetter Lane,
London EC4P 4EE

Methuen Inc
733 Third Avenue
New York
NY 10017

Mechanical Installations – Prices for Measured Work

THERMAL INSULATION

THERMAL INSULATION APPLIED TO PIPELINES *continued*

	Unit	Thickness 19 mm £	25 mm £	38 mm £
Scrim-faced pre-formed mineral fibre glass fibre rigid sections, fixed with aluminium bands at 450 mm intervals Tube, including bends, to the following diameters				
15 mm	m	1·77	1·91	2·56
20 mm	,,	1·86	2·00	2·67
25 mm	,,	1·95	2·07	2·80
32 mm	,,	2·07	2·23	2·95
40 mm	,,	2·16	2·32	3·05
50 mm	,,	2·37	2·56	3·35
65 mm	,,	2·60	2·82	3·65
80 mm	,,	2·80	3·03	3·89
100 mm	,,		3·66	4·61
125 mm	,,		4·09	5·11
150 mm	,,		4·64	5·80
Extra for Fitting around pipe fittings with insulation of the thickness indicated				
Pair of flanges				
15 mm	No.		1·91	2·56
20 mm	,,		2·00	2·67
25 mm	,,		2·07	2·80
32 mm	,,		2·23	2·95
40 mm	,,		2·32	3·05
50 mm	,,		2·56	3·35
65 mm	,,		2·82	3·65
80 mm	,,		3·03	3·89
100 mm	,,		3·66	4·61
125 mm	,,		4·09	5·11
150 mm	,,		4·64	5·80
Screwed valves				
15 mm	,,		1·91	2·56
20 mm	,,		2·00	2·67
25 mm	,,		2·07	2·80
32 mm	,,		2·23	2·95
40 mm	,,		2·32	3·05
50 mm	,,		2·56	3·35
Flanged valves				
15 mm	,,		2·86	3·85
20 mm	,,		3·00	4·00
25 mm	,,		3·11	4·19
32 mm	,,		3·35	4·42
40 mm	,,		3·49	4·57
50 mm	,,		3·85	5·02
65 mm	,,		4·23	5·48
80 mm	,,		4·54	5·84
100 mm	,,		5·50	6·92
125 mm	,,		6·13	7·66
150 mm	,,		6·96	8·70

THERMAL INSULATION

THERMAL INSULATION APPLIED TO PIPELINES continued

Foil-faced pre-formed mineral fibre glass fibre rigid sections, fixed with adhesive and 25 mm mesh galvanized wire netting and aluminium bands at 450 mm intervals all joints and ends vapour sealed
Tube including bends to the following diameters

	Unit	Thickness 19 mm £	25 mm £	38 mm £
15 mm	m	2·34	2·50	3·20
20 mm	,,	2·44	2·61	3·33
25 mm	,,	2·55	2·73	3·47
32 mm	,,	2·68	2·86	3·63
40 mm	,,	2·77	2·97	3·74
50 mm	,,	3·01	3·23	4·05
65 mm	,,	3·28	3·51	4·39
80 mm	,,	3·49	3·74	4·65
100 mm	,,		4·56	5·57
125 mm	,,		5·05	6·12
150 mm	,,		5·66	6·84

Extra for
Fitting around pipe fittings with insulation of the thickness indicated

Pair of flanges

	Unit			
15 mm	No.		2·50	3·20
20 mm	,,		2·61	3·33
25 mm	,,		2·73	3·47
32 mm	,,		2·86	3·63
40 mm	,,		2·97	3·74
50 mm	,,		3·23	4·05
65 mm	,,		3·51	4·39
80 mm	,,		3·74	4·65
100 mm	,,		4·56	5·57
125 mm	,,		5·05	6·12
150 mm	,,		5·66	6·84

Screwed valves

15 mm	,,		2·50	3·20
20 mm	,,		2·61	3·33
25 mm	,,		2·73	3·47
32 mm	,,		2·86	3·63
40 mm	,,		2·97	3·74
50 mm	,,		3·23	4·05

Flanged valves

15 mm	,,		3·74	4·81
20 mm	,,		3·92	4·99
25 mm	,,		4·09	5·20
32 mm	,,		4·30	5·44
40 mm	,,		4·46	5·61
50 mm	,,		4·85	6·09
65 mm	,,		5·27	6·58
80 mm	,,		5·61	6·99
100 mm	,,		6·85	8·35
125 mm	,,		7·58	9·18
150 mm	,,		8·49	10·26

THERMAL INSULATION

THERMAL INSULATION APPLIED TO PIPELINES
(of the nominal bores shown)

	Unit	Thickness 19 mm £	Thickness 25 mm £	Thickness 38 mm £
Canvas-faced pre-formed mineral fibre glass fibre rigid sections, fixed with aluminium bands at 450 mm intervals Tube, including bends to the following diameters				
15 mm	m	2·33	2·50	2·75
20 mm	,,	2·43	2·60	3·31
25 mm	,,	2·55	2·71	3·45
32 mm	,,	2·68	2·85	3·62
40 mm	,,	2·77	2·96	3·73
50 mm	,,	3·00	3·22	4·05
65 mm	,,	3·27	3·50	4·38
80 mm	,,	3·48	3·73	4·64
100 mm	,,		4·56	5·56
125 mm	,,		5·02	6·09
150 mm	,,		5·62	6·84

Extra for
Fitting around pipe fittings with insulation of the thickness indicated

Pair of flanges

	Unit			
15 mm	No.		2·50	2·75
20 mm	,,		2·60	3·31
25 mm	,,		2·71	3·45
32 mm	,,		2·85	3·62
40 mm	,,		2·96	3·73
50 mm	,,		3·22	4·05
65 mm	,,		3·50	4·38
80 mm	,,		3·73	4·64
100 mm	,,		4·56	5·56
125 mm	,,		5·02	6·09
150 mm	,,		5·62	6·84

Screwed valves

15 mm	,,		2·50	2·75
20 mm	,,		2·60	3·31
25 mm	,,		2·71	3·45
32 mm	,,		2·85	3·62
40 mm	,,		2·96	3·73
50 mm	,,		3·22	4·05

Flanged valves

15 mm	,,		3·74	4·12
20 mm	,,		3·90	4·98
25 mm	,,		4·08	5·19
32 mm	,,		4·29	5·43
40 mm	,,		4·44	5·60
50 mm	,,		4·84	6·09
65 mm	,,		5·26	6·57
80 mm	,,		5·60	6·96
100 mm	,,		6·85	8·34
125 mm	,,		7·54	9·15
150 mm	,,		8·44	10·26

THERMAL INSULATION

THERMAL INSULATION APPLIED TO PIPELINES
(of the nominal bores shown) *continued*

	Unit	Thickness 25 mm £	Thickness 38 mm £
Asbestos-free calcium silicate rigid sections, laps fixed with aluminium bands at 450 mm intervals			
Tube, including bends, to the following diameters			
15 mm	m	2·91	3·81
20 mm	,,	3·04	3·92
25 mm	,,	3·16	4·09
32 mm	,,	3·30	4·30
40 mm	,,	3·42	4·41
50 mm	,,	3·66	4·69
65 mm	,,	3·93	5·05
80 mm	,,	4·18	5·35
100 mm	,,	4·91	6·15
125 mm	,,	5·37	6·75
150 mm	,,	5·91	7·39
Extra for			
Fitting around pipe fittings with insulation of the thickness indicated			
Pair of flanges			
15 mm	No.	2·91	3·81
20 mm	,,	3·04	3·92
25 mm	,,	3·16	4·09
32 mm	,,	3·30	4·30
40 mm	,,	3·42	4·41
50 mm	,,	3·66	4·69
65 mm	,,	3·93	5·05
80 mm	,,	4·18	5·35
100 mm	,,	4·91	6·15
125 mm	,,	5·37	6·75
150 mm	,,	5·91	7·39
Screwed valves			
15 mm	,,	2·91	3·81
20 mm	,,	3·04	3·92
25 mm	,,	3·16	4·09
32 mm	,,	3·30	4·30
40 mm	,,	3·42	4·41
50 mm	,,	3·66	4·69
Flanged valves			
15 mm	,,	4·37	5·72
20 mm	,,	4·56	5·88
25 mm	,,	4·75	6·13
32 mm	,,	4·95	6·45
40 mm	,,	5·13	6·62
50 mm	,,	5·50	7·03
65 mm	,,	5·89	7·58
80 mm	,,	6·27	8·03
100 mm	,,	7·37	9·22
125 mm	,,	8·06	10·12
150 mm	,,	8·87	11·09

THERMAL INSULATION

THERMAL INSULATION APPLIED TO PIPELINES
(of the nominal bores shown) *continued*

			Thickness	
Extra for		19	25	38
0·5 mm thick aluminium sheet secured with rivets or self tapping screws to any of the above sectional insulations of the thickness indicated	Unit	mm £	mm £	mm £
15 mm	m	3·69	3·78	3·88
20 mm	,,	3·72	3·82	3·94
25 mm	,,	3·78	3·79	3·98
32 mm	,,	3·82	3·84	4·04
40 mm	,,	3·79	3·88	4·09
50 mm	,,	3·88	3·98	4·20
65 mm	,,	3·98	4·09	4·31
80 mm	,,	4·09	4·20	4·40
100 mm	,,	4·94	5·04	5·23
125 mm	,,	5·14	5·23	5·50
150 mm	,,	5·34	5·50	5·69

CAST IRON PIPE AND FITTINGS

CAST IRON SOIL, WASTE AND VENTILATING PIPE AND FITTINGS
(type 'A' sockets with caulked lead joints and fixed with holderbats)

	Unit	50 mm £	75 mm £	100 mm £	150 mm £
Pipe	m	7·77	9·76	12·55	20·66
		6·07	*6·66*	*8·68*	*16·37*
Extra over pipe for the following fittings					
Bend	No.	5·67	8·31	11·53	17·46
		3·97	*5·21*	*7·66*	*13·17*
Bend with access	,,	11·59	13·70	18·06	23·88
		9·48	*10·29*	*13·36*	*18·71*
Single plain branch	,,	9·17	12·79	16·70	27·94
		5·30	*7·62*	*11·53*	*21·48*
Ditto with access	,,	14·89	18·40	23·00	34·04
		10·60	*12·41*	*17·01*	*26·76*
Double plain branch	,,	12·64	19·10	22·01	40·38
		7·99	*13·16*	*15·29*	*33·15*
Ditto with access	,,	18·71	25·21	28·37	47·86
		13·64	*18·44*	*20·83*	*39·80*
Parallel branch	,,		12·85	17·20	33·17
			7·68	*12·03*	*26·71*
'Y' or breeches branch	,,	10·55	13·79	18·59	39·10
		7·37	*9·04*	*12·22*	*31·17*
Ditto with access	,,	17·37	21·20	26·19	48·47
		13·72	*16·08*	*18·83*	*39·54*
Single anti-syphon branch	,,	10·40	14·33	17·20	
		6·53	*9·16*	*12·03*	
Double anti-syphon branch	,,		17·37	23·42	41·24
			11·63	*13·08*	*30·75*
Access pipe-oval door	,,	11·46	15·00	16·96	25·12
		9·35	*10·37*	*11·79*	*18·66*
Shoes	,,	6·53	9·43	12·34	20·43
		5·24	*6·33*	*8·47*	*16·20*
Socket reducer	,,		5·63	7·42	11·22
			2·53	*3·55*	*6·93*
Socket plug	,,	5·31	5·49	7·27	8·61
		1·50	*1·68*	*2·00*	*3·34*
Diminishing piece	,,		8·87	12·04	20·32
			5·77	*8·17*	*16·03*
Offsets					
75 mm projection	,,	4·55	7·54	10·93	19·16
		3·26	*4·44*	*7·06*	*14·87*
150 mm projection	,,	6·16	9·55	12·84	21·37
		4·61	*5·94*	*8·45*	*16·51*
225 mm projection	,,	6·77	10·79	14·03	24·71
		5·22	*7·18*	*9·65*	*19·85*
300 mm projection	,,	8·46	12·30	15·79	30·84
		6·66	*8·17*	*10·88*	*25·52*
450 mm projection	,,		18·39	25·13	
			14·26	*19·96*	
600 mm projection	,,	20·51	25·51	36·35	41·27
		18·32	*20·45*	*29·98*	*34·43*

CAST IRON PIPE AND FITTINGS

CAST IRON SOIL, WASTE AND VENTILATING PIPE AND FITTINGS
(type 'A' sockets with caulked lead joints and fixed with holderbats) *continued*

Extra over pipe for the following fittings *continued*	Unit	50 mm £	75 mm £	100 mm £	150 mm £
Boss pipe with 1 No. boss.	No.	11·84	13·91	17·02	25·11
		10·14	*10·81*	*13·15*	*20·82*
Ditto with 2 No. bosses	,,	14·30	17·81	20·97	42·95
		12·73	*14·00*	*16·22*	*37·68*
Plain 'P' trap	,,	10·23	14·24	17·07	30·52
		8·68	*9·59*	*11·98*	*24·06*
Access 'P' trap	,,	15·81	19·59	23·13	38·85
		13·85	*14·63*	*17·14*	*31·56*
Plain 'S' or 'Q' trap	,,	11·44	14·67	17·33	35·76
		9·89	*10·02*	*12·16*	*29·30*
Access 'S' or 'Q' trap	,,	16·89	19·89	23·01	40·70
		14·93	*14·93*	*17·02*	*33·41*
W.C. connectors					
150 mm long	,,			7·98	
				4·11	
225 mm long	,,			8·59	
				4·72	
300 mm long	,,			9·34	
				5·47	
450 mm long	,,			11·99	
				6·82	
600 mm long	,,			14·63	
				8·17	
W.C. connectors with anti-syphon arm					
150 mm long	,,			12·34	
				7·59	
225 mm long	,,			11·46	
				6·71	
300 mm long	,,			13·82	
				9·07	
450 mm long	,,			16·95	
				10·58	
600 mm long	,,			18·36	
				11·99	
Bent W.C. connectors with 300 mm tail	,,			15·09	
				10·34	
Ditto with 450 mm tail	,,			20·04	
				13·67	
Ditto with 600 mm tail	,,			24·05	
				17·68	

U.P.V.C. PIPE AND FITTINGS

U.P.V.C. SOIL, WASTE AND VENTILATING PIPE
(with ring seal joints, fixed with P.V.C. clips, plugged and screwed)

	Unit	32 mm £	38 mm £	50 mm £	75 mm £	100 mm £	150 mm £
Pipe	m	1·74	1·91	3·09	4·52	5·45	9·93
		0·71	0·88	1·91	2·92	3·65	7·61
Extra over pipe for the following fittings							
Bend 92°	No.	2·12	2·19	2·81	4·11	5·20	10·92
		0·83	0·90	1·26	2·31	3·24	7·93
Bend 92° with access	,,					9·31	
						6·77	
Single branch 92°	,,	2·71	2·97	4·32	5·30	7·20	14·23
		1·16	1·42	2·62	3·19	4·04	10·04
Single branch with access	,,					11·00	
						8·68	
Double branch	,,					13·45	
						10·61	
Bossed pipe with 1 No. 35 mm boss	,,					3·23	
						1·94	
Access pipe	,,				6·42	7·08	
					4·67	5·02	
Access cap	,,	1·39	1·39	2·26	3·49	4·06	9·51
		0·36	0·36	1·23	2·20	2·51	7·45
W.C. connector	,,				2·55	3·70	
					0·96	2·11	
Metal caulking bush	,,				5·44	6·70	10·01
					2·90	3·84	7·15
P.V.C. caulking bush	,,	2·16	2·36	2·78			
		0·57	0·77	0·87			
Union	,,	2·26	2·36	2·91			
		0·67	0·77	1·32			
Level invert taper	,,	2·21	2·46	2·40			
		0·62	0·87	0·81			
Weathering apron	,,					2·76	
						0·85	

Mechanical Installations – Prices for Measured Work

U.P.V.C. PIPE AND FITTINGS

U.P.V.C. SOIL, WASTE AND VENTILATING PIPE AND FITTINGS
(with solvent welded joints, fixed with P.V.C. pipe clips plugged and screwed)

	Unit	32 mm £	38 mm £	50 mm £	75 mm £	100 mm £	150 mm £
Pipe	m	2·01	2·17	2·84	3·98	4·63	8·31
		0·98	1·14	1·66	2·38	2·83	5·99
Extra over pipe for the following fittings							
Bend 92 degree	No.	2·02	2·06	2·74	4·34	5·44	10·51
		0·73	0·77	1·19	2·54	3·48	7·52
Bend 135 degree	,,	2·02	2·06	2·74	4·34	5·44	10·09
		0·73	0·77	1·26	2·54	3·48	7·10
Bend 92 degree with access	,,					9·88	
						7·82	
Single branch 92 degree	,,	2·55	2·78	4·01	5·66	7·05	13·27
		1·00	1·23	2·31	3·55	4·47	9·86
Double branch 92 degree	,,		5·19	5·72		13·61	20·55
			3·13	3·61		10·77	16·83
Double branch 104°	,,					14·11	22·50
						11·27	18·78
Straight expansion coupling	,,				2·33	2·57	5·29
					1·04	1·28	3·74
W.C. connector	,,				4·12	3·78	
					2·83	2·49	
Access door	,,	2·15	2·62	2·44	4·37	4·63	7·05
		1·33	1·59	1·49	3·08	3·08	5·50
P.V.C. caulking bush	,,				4·49	4·49	
					2·17	2·17	
Metal caulking bush	,,				5·29	8·00	8·89
					3·23	5·94	6·57
Bossed pipe with 35 mm boss	,,					2·52	
						1·23	
Access cap	,,	2·62	2·88	2·96	4·63	5·42	
		1·33	1·59	1·41	3·08	3·62	
Union	,,	2·70	3·22	4·43			
		1·41	1·93	2·88			
Weathering apron	,,				1·97	2·13	4·04
					0·70	0·84	2·49
Weathering slate	,,				16·95	16·95	
					11·78	11·78	
Vent cowl	,,				2·25	2·29	3·43
					0·70	0·74	1·88
Level invert taper	,,	1·89	2·09	2·96	3·52	5·94	
		0·60	0·80	1·67	1·97	4·39	

U.P.V.C. PIPE AND FITTINGS

U.P.V.C. VENTILATING AND OVERFLOW PIPE AND FITTINGS
(with solvent weld joints and fixed with P.V.C. clips plugged and screwed)

	Unit	16 mm £	22 mm £
Pipe	m	0·64	0·87
		0·43	*0·46*
Extra over pipe for the following fittings			
Bend	No.	0·48	0·86
		0·27	*0·45*
Branch	,,	0·60	0·86
		0·29	*0·34*
Tank connector	,,		1·05
			0·53
Bent tank connector	,,		1·12
			0·60
Female iron to P.V.C. connector	,,		0·76
			0·24

TRAPS

POLYPROPYLENE TRAPS

	Unit	Nominal bore 32 mm £	38 mm £
Tubular 'P' trap	No.	2·51	2·91
		1·48	1·62
Tubular 'S' trap	,,	2·71	3·39
		1·68	2·10
Tubular running 'P' trap	,,	3·42	3·74
		2·39	2·45
Tubular running 'S' trap	,,	3·76	4·25
		2·73	2·96
Bottle 'P' trap	,,	2·51	3·00
		1·48	1·72
Bottle 'S' trap	,,	2·98	3·58
		1·95	2·29
Combined bath trap and overflow	,,		8·00
			5·43

PART TWO

Electrical Installations

Direction, *page 221*
Rates of Wages and Working Rules, *page 225*
Market Prices of Materials, *page 235*
Labour Constants, *page 259*
Prices for Measured Work, *page 271*

ELECTRICAL INSTALLATIONS

Directions

RATES OF WAGES AND WORKING RULES

This section gives rates of wages current at January 1981 and extracts from the working rules of the Electrical Contracting Industry as amended up to February 1980.

MARKET PRICES OF MATERIALS

The prices given, unless otherwise stated, include for delivery to sites in the London area at March 1980 and represent the prices paid by contractors after the deduction of all trade discounts but exclude any charges in respect of V.A.T.

LABOUR CONSTANTS

Labour constants are given for the major items of work for which prices are given in Prices for Measured Work.

PRICES FOR MEASURED WORK

These prices are intended to apply to new work in the London area and include allowances for all overhead charges, preliminary items and profit. The prices are for reasonable quantities of work and the user should make suitable adjustments if the quantities are especially small or especially large. Adjustments may also be required for locality (e.g. outside London) and for the market conditions (e.g. volume of work on hand or on offer) at the time of use.

The labour rate on which these prices have been based is £4·71 per man hour which is the London Standard Wage rate at January 1981 plus allowances for all other emoluments and expenses. To this rate has been added 40% to cover site and head office overheads and preliminary items together with a further 5% for profit, resulting in an inclusive rate of £6·92 per man hour. The rate of £4·71 per man hour has been calculated on a working year of 2043 hours; a detailed build-up of the rate is given at the end of these Directions.

In calculating 'Prices for Measured Work' the following assumptions have been made:
   (*a*) That the work is carried out as a sub-contract under the Standard Form of Building Contract and that such facilities as are usual would be afforded by the main contractor.
   (*b*) That, unless otherwise stated, the work is being carried out in open areas at a height which would not require more than simple scaffolding.
   (*c*) That the building in which the work is being carried out is no more than six storeys high.

Where these assumptions are not valid, as for example where work is carried out in ducts and similar confined spaces or in multi-storey structures when additional time is needed to get to and from upper floors, then an appropriate adjustment must be made to the prices. Such adjustment will normally be to the labour element only.

No allowance has been made in the prices for any cash discount to the main contractor.

DIRECTIONS

The labour element, inclusive of overheads and profit, in any particular item can be ascertained by deducting the amount which appears in italics below the measured price from the measured price. The amount shown in italics represents the value of material content contained in the measured price including relevant allowances for waste, overheads and profit.

The prices of materials upon which 'Prices for Measured Work' are based are as shown in 'Market Prices of Materials', or as indicated, with the addition of 10% to cover overheads and a further 5% for profit. Allowance has been made for waste where necessary.

LABOUR RATE

The following detail shows how the labour rate of £4·71 per man hour has been calculated.

Total annual cost of notional eleven man gang

London hourly rate

Comprising: £
1 Technician (including London weighting). . . . . 3·33
4 Approved Electricians (including London weighting). . . . 2·91
6 Electricians (including London weighting) . . . . . 2·66

Hours actually worked £ £
1021½ hours Technician . . . . @ 3·33   3401·60
8127 hours Approved Electricians . . . . . . @ 2·91   23780·52
12258 hours Electricians . . . . @ 2·66   32606·28

Non-productive overtime
113½ hours Technician . . . . @ 3·33   377·96
454 hours Approved Electricians . . . . @ 2·91   1321·14
681 hours Electricians . . . . @ 2·66   1811·46

Daily travelling allowance
75 minutes per day per Operative
  Technician 227 days × 75 mins = 283¾ hours . . . . @ 3·33   944·89
  Approved Electricians 908 days × 75 mins = 1135 hours . . @ 2·91   3302·85
  Electricians 1362 days × 75 mins = 1702½ hours . . . @ 2·66   4528·65

Daily fares
2497 Days, Technician, Approved Electricians and Electricians . . @ 1·90   4744·30

Trade supervision
1021½ hours Technician . . . . @ 3·33   3401·60

National Insurance Not contracted out Contributions based on estimated weekly gross pay shown in brackets
Technician (£180·00). . . . . 48 weeks @ 22·60   1084·80
Approved Electricians (£156·00). . . . . . 192 weeks @ 21·34   4097·28
Electricians (£143·00) . . . . 288 weeks @ 19·63   5653·44

C/f   91056·77

Electrical Installations

DIRECTIONS

	B/f	£ 91056·77
J.I.B. Benefits Scheme		
49 weeks Technician . . . . @	10·34	506·66
196 weeks Approved Electricians . . . . @	9·20	1803·20
294 weeks Electricians . . . . @	8·63	2537·22
Recognized holidays with pay		
64 hours Technician . . . . @	3·33	213·12
256 hours Approved Electricians . . . . @	2·91	744·96
384 hours Electricians . . . . @	2·66	1021·44

Training
C.I.T.B. Levy 11 @ £43·20 . . . . 475·20

98358·57

Severance pay and sundry costs
Add 1% . . . . 983·59

99342·16

Employer's liability and third party insurance
Add say £1·60% . . . . 1589·47

Annual cost of notional 11 man gang (10½ men actual working) . . . . 100931·63

Average Annual Cost per man . . . . 9612·54

All-in hour cost. . . . . 4·71

ELECTRICAL INSTALLATIONS
Rates of Wages and Working Rules

ELECTRICAL CONTRACTING INDUSTRY

Extracts from National Working Rules determined by:
Joint Industry Board for the Electrical Contracting Industry
Kingswood House
47/51 Sidcup Hill, Sidcup, Kent, DA14 6HP
Telephone 01-302 0031

RATES OF WAGES

EFFECTIVE FROM FIRST PAYWEEK FOLLOWING 1st JANUARY 1981

	Standard wage rates £	Hourly rates Large engineering construction sites wage rates £	London weighting from November 1979 £
Technician	3·19	4·26	0·14
Approved Electrician	2·74	3·81	0·14
Electrician	2·52	3·59	0·14
Labourer	1·96	3·03	0·14

Apprentice rates
Age

16	1·01	—	0·05
17	1·13	—	0·06
18	1·26	2·31	0·07
19	1·64	2·68	0·09
20	2·02	3·05	0·11

Responsibility money
Approved Electricians in charge of work, who undertake supervision of other operatives, shall be paid 'Responsibility Money' of not less than 2½p and not more than 25p per hour.

Combined J.I.B. Benefits Stamp Value (*Contribution period September 1979 to September 1980*)

	Weekly stamp value £	Annual Holiday Payment* £
Technician	10·34	385·92
Approved Electrician	9·20	331·20
Electrician	8·63	303·84
Labourer	7·21	235·68

* Actual amount of Annual Holidays' Payment is dependent upon continuity of employment.

WORKING RULES

SECTION I: INTRODUCTION

The J.I.B. National Working Rules are made under Rule 76 of the Rules of the Joint Industry Board (contained in booklet JIB/1), as the National Joint Industrial Council for the Electrical Contracting Industry.

The principal objects of the Joint Industry Board are to regulate the relations between employers and employees engaged in the industry and to provide all kinds of benefits for persons concerned with the Industry in such ways as the Joint Industry Board may think fit, for the purpose of stimulating and furthering the improvement and progress of the Industry for the mutual advantage of the employers and employees engaged therein, and in particular, for the purpose aforesaid, and in the public interest, to regulate and control employment and productive capacity within the Industry and the levels of skill and proficiency, wages, and welfare benefits of persons concerned in the Industry. 'The Industry' means the Electrical Contracting Industry in all its branches in England, Wales, Northern Ireland, the Isle of Man and the Channel Islands and such other places as may from time to time be determined by the Joint Industry Board, including the design, manufacture, sale, distribution, installation, maintenance, repair and renewal of all kinds of electrical installations, equipment and appliances and ancillary activities.

SECTION II: NATIONAL WORKING RULES

1: General

These J.I.B. National Working Rules and Industrial Determinations supersede previous Rules and Agreements made between the constituent parties of the National Joint Industrial Council for the Electrical Contracting Industry and shall govern and control the conditions for electrical installation work and ancillary activities covered by the Joint Industry Board for the Electrical Contracting Industry ('J.I.B.') and come into effect in respect of work performed on and after Monday, 2 March 1970. These Rules apply nationally.

2: Grading

The Definitions of the grading of operatives are set out in Section V of these Rules and the appropriate National Standard J.I.B. Graded Rates of Wages shall be paid. Grading shall only be valid by the possession of a Grade Card issued by the Joint Industry Board.

Nothing in these rules shall prevent the maximum flexibility in the employment of skilled operatives.

3: Working Hours

The working week shall be $37\frac{1}{2}$ hours per week (from first payweek following 1st January 1981) worked on five days, Monday to Friday inclusive.

The normal day shall be eight hours worked during any consecutive nine hours between 7.30 a.m. and 6.30 p.m. Where shifts are required which fall outside these limits, payments and conditions of work shall be determined by the Joint Industry Board.

Meal breaks, including washing time, shall be unpaid of one hour duration or lesser period at the Employer's discretion and shall not be exceeded. The Employer shall declare the working days and hours (including breaks) on each job.

WORKING RULES

4: Utilization of Working Hours

There shall be full utilization of working hours which shall not be subject to unauthorized 'breaks'. Time permitted for tea breaks shall not be exceeded.

Bad timekeeping and/or unauthorized absence from the place of work, during working hours, shall be construed as Industrial Misconduct.

Meetings of operatives shall not be held during working hours except by arrangement with the Job/Shop Representative and with the prior permission of the Employer or the Employer's site management or the Employer's representative.

6: Wages

The National Standard J.I.B. Graded Rates of Wages shall be those from time to time determined by the Joint Industry Board.

The J.I.B. Rates of Wages appropriate to operatives shall be such rates for their grades in the Zone where their shop is situated as the Joint Industry Board may from time to time determine to be appropriate for their grade in the place where they are working and they shall be paid no more and no less wages.

This rule does not permit of the introduction of any scheme of payment by result or production bonuses except as may be determined from time to time by the J.I.B. National Board.

7: Payment of Wages

Wages shall normally be paid by credit transfer. Alternatively, another method of payment may be adopted by mutual arrangement between Employer and Operative.

Wages shall be calculated for weekly periods and paid to the Operative within 5 normal working days of week termination unless alternative arrangements are agreed.

9: Overtime

(a) Hours

Overtime is deprecated by the Joint Industry Board; systematic overtime is to be particularly avoided.

A Regional Joint Industry Board may, from time to time, declare permissible hours of overtime in the Region which shall not be exceeded without permission of the Regional Joint Industry Board.

Overtime will not be restricted in the case of Breakdowns or Urgent Maintenance and Repairs.

(b) Payment

(*i*) 40 hours (from first payweek following 1st January 1981) shall be worked at normal rates in any one week (Monday to Friday) before any overtime premium is calculated. Premium time shall be paid at time-and-a-half. All hours worked between 1 p.m. on Saturday and normal starting time on Monday shall be paid at double time.

Exceptions: For the purpose of premium payment, an Operative shall be deemed to have worked normal hours on days where, although no payment is made by the Employer, the Operative:

(*a*) has lost time through certified sickness.

(*b*) was on a rest period for the day following continuous working all the previous night.

(*c*) was absent with the Employer's permission.

WORKING RULES

(*ii*) Any Operative who has not worked five days (as determined in Rule 3) from Monday to Friday, taking into account the exceptions detailed above, is precluded from working the following Saturday or Sunday.

(*c*) Call out

Notwithstanding the previous Clause, when an operative is called upon to return to work after his normal finishing time and before his next normal starting time he shall be paid at time-and-a-half for all the hours involved (home-to-home), subject to the guaranteed minimum payment. Between 1 p.m. on Saturday, and the normal starting time on Monday the appropriate premium rate shall be double time.

The guaranteed minimum payment for a single call out under this clause shall be the equivalent of 4 hours at the Operative's National Standard J.I.B. Graded Rate of Wages.

11: Definition of 'Shop'

Employers shall declare the Branches of their business as the Shop from which entitlement to travelling time and fares shall be calculated, subject to the Branches fulfilling the following conditions:

The premises are owned or rented by the Employer.

The premises are used for the purpose of general trading or personnel management as distinct from the management of one contract or one site.

There shall be a full-time staff available during normal working hours capable of dealing with, and resolving, enquiries relating to recruitment, payment of wages and other matters affecting employment.

The place of employment is the Shop and by custom and practice all Shop Recruited Operatives are transferable from job to job.

Employers are required to notify the Joint Industry Board when they establish a new Branch, Office or Shop in any region.

12: Travelling Time and Fares, Country and Lodging Allowances

(*a*) National Rule

Operatives who are required to book on and off at the Employers' Shop shall be entitled to time from booking on until booking off with overtime if the time so booked exceeds the normal working day. They shall also be entitled to actual fares from the Shop to Job and from the Job to the Shop.

(*b*) Operatives who are required to start and finish at the normal starting and finishing time on jobs up to and including 25 miles from the Shop shall receive payment for travelling time and fares on the following basis:

Distance (point to point)	Total daily travelling time (Ordinary Rates) per day plus actual fares
Return journey of up to	
1 mile each way	Nil
2 miles each way	20 minutes
3 miles each way	30 minutes
4 miles each way	40 minutes
5 miles each way	50 minutes

WORKING RULES

Return journeys over 5 miles and up to 10 miles each way	75 minutes
Return journeys over 10 miles and up to and including 15 miles each way	100 minutes
Return journeys over 15 miles and up to 20 miles each way	110 minutes
Return journeys over 20 miles and up to and including 25 miles each way	120 minutes

(*c*) Country Allowance

(*i*) Operatives sent from the Employer's Shop who are required to start and finish at the normal starting and finishing time on jobs over 25 miles from the Shop and who elect to travel daily to the job instead of taking lodgings will be paid Country Allowance of £7·00 per day worked in lieu of travelling time and all fares. In the case of London jobs this is over and above the J.I.B. Rates of Wages for the London Zone.

(*d*) Lodging Allowance

(*i*) Operatives sent from the Employer's Shop who are required to start and finish at the normal starting and finishing time on jobs over 25 miles from the Shop, who elect to lodge away from home and provide proof of lodging to the Employer's satisfaction will be paid a Lodging Allowance of £7·00 per night.

(*ii*) The Employee shall be reimbursed for any Value Added Tax charged on the cost of Lodging, subject to provision by the Operative of a Valid Tax invoice on which the Employer can claim Input Credit.

(*iii*) Travelling time and fares between the lodgings and the job shall not be paid except where it is proved to the Employer's satisfaction that suitable lodging accommodation is not available near the job when travelling time between the lodgings and the job in excess of half an hour each way and actual fares in excess of £0·10 per day shall be paid.

(*iv*) On being sent to the job the Operative shall receive his actual fare and travelling time at ordinary rates from the Employer's Shop and when he returns to the Employer's Shop except that when, of his own free will, he leaves the job within one calendar month from the date of his arrival and in cases where he is dismissed by the Employer for proved bad timekeeping, improper work or similar misconduct no return travelling time or fares shall be paid.

(*v*) The payment of Lodging Allowance shall not be made when suitable board and lodging is arranged by the Employer at no cost to the Operative.

(*vi*) The payment of Lodging allowance shall not be made during absence from employment unless a Medical Certificate is produced for the whole of the period claimed. When an operative is sent home by the firm at their cost the payment of Lodging Allowance shall cease.

(*vii*) No payment for the retention of lodgings during Annual Paid Holiday shall be made by the Employer except in cases where the Operative is required to pay a retention fee during Annual Paid Holiday when reimbursement shall be of the amount actually paid to a maximum of £1·00 per week upon production of proof of payment to the Employer's satisfaction.

(*viii*) Where an Operative is away from his lodgings at a weekend under Rule 12(*e*) but has to pay a retention fee for his lodgings, reimbursement shall be the amount actually paid, to a maximum of £6·00 per night upon production of proof of payment to the Employer's satisfaction.

WORKING RULES

(*e*) Period return fares for Operatives who lodge

(*i*) On jobs up to and including 100 miles from the Employer's Shop, return railway fares from the Job to the Employer's Shop, without travelling time, shall be paid for every two weeks.

(*ii*) On jobs over 100 miles and up to and including 250 miles from the Employer's Shop, return railway fares from the Job to the Employer's Shop, without travelling time, shall be paid every four weeks.

(*iii*) On jobs over 250 miles from the Employer's Shop, return railway fares from the Job to the Employer's Shop, with 8 hours travelling time at ordinary rates, shall be paid every four weeks.

(*iv*) In cases under sub-clauses (*ii*) and (*iii*) above, where the Employer, through necessity or expediency, requires his Operatives to work during the specified weekend leave period, he shall arrange that they shall have another period in substitution but this provision shall not apply under sub-clause (*i*) above.

N.B. All distances shall be calculated in a straight line (point to point).

When Annual Holiday with pay are taken the period returns may be moved forward or backward from the date upon which they become due, to enable the period returns to coincide with the date of the Annual Paid Holiday.

Special consideration shall be given to Operatives where it is necessary for them to return home on compassionate grounds, e.g. domestic illness.

(*f*) Locally engaged labour

Where an Employer does not have a Shop within 25 miles of the Job, he can engage labour domiciled within a 25 miles radius of that Job. Operatives shall receive the J.I.B. Rate of Wages applicable to the Zone of the Job and travelling time and fares in accordance with Clause (*b*), but with the exception of 'home' being substituted for 'shop' in Clause (*b*).

Locally engaged labour, domiciled within a 25 miles radius of the Job, can be transferred to other Jobs within that radius without affecting their entitlements under this Rule. Operatives transferred to a Job outside that radius and within the Zone Rate will be entitled to Country Allowance or Lodging Allowance in accordance with the Rules.

13: Statutory Holidays

(*a*) Qualification

Eight hours pay at the appropriate National Standard J.I.B. Graded Rate of Wages shall be paid for a maximum of eight Statutory Holidays per annum. In general, the following shall constitute such paid holidays:

Good Friday; Easter Monday; Spring Time Bank Holiday; May Day; Late Summer Bank Holiday; Christmas Day; Boxing Day; New Year's Day.

In areas where any of these days are not normally observed as holidays in the Electrical Contracting Industry, traditional local holidays may be substituted by mutual agreement and subject to the determination of the appropriate Regional Joint Industry Board.

When Christmas Day and/or Boxing Day or New Year's Day falls on a Saturday or Sunday, the following provisions apply:

Christmas Day

When Christmas Day falls on a Saturday or Sunday, the Tuesday next following shall be deemed to be a paid holiday.

WORKING RULES

Boxing Day or New Year's Day
When Boxing Day or New Year's Day falls on a Saturday or Sunday, the Monday next following shall be deemed to be a paid holiday.
In order to qualify for payment, operatives must work full time for the normal day on the working days preceding and following the holiday.
For the purpose of this Rule, an operative shall be deemed to have worked on one or both of the qualifying days when the Operative
(*i*) has lost time through certified sickness.
(*ii*) was on a rest period for the day following continuous working all the previous night.
(*iii*) was absent with the Employer's permission.

(*b*) Payment for working Statutory Holidays
When operatives are required to work on a Paid Holiday within the scope of this Agreement, they shall receive wages at the following for all hours worked –
Christmas Day – Double time and a day or shift off in lieu for which they shall be paid wages at bare time rates for the hours constituting a normal working day. The alternative day hereunder shall be mutually agreed between the Employer and the Operatives concerned.
In respect of all other days: either
(*a*) Time-and-a-half plus a day or shift off in lieu for which they shall be paid wages at bare time rates for the hours constituting a normal working day. The alternative day hereunder shall be mutually agreed between the Employer and the Operatives concerned; or
(*b*) at the discretion of the Employer 2½ times the bare time rate in which event no alternative day is to be given.
In the case of night shift workers required to work on a Statutory Holiday, the premiums mentioned above shall be calculated upon the night shift rate of time-and-a-third. Time off in lieu of Statutory Holidays shall be paid at bare time day rates.

14: Annual Holidays

Operatives shall be entitled to payment for Annual Holidays as determined from time to time under the J.I.B. Annual Holiday with Pay Scheme, depending upon their continuity of service in the Industry.
The J.I.B. Annual Holiday with Pay Scheme is carried out by the Joint Industry Board by the sale of J.I.B. Holiday Credit Stamps which are fixed to the Operative's J.I.B. Annual Holiday Card. Details of the Scheme are given in a separate J.I.B. combined welfare/insurance scheme booklet (the 'gold' book ref. JIB/74BS).

15: Sickness with Pay and Group Life Insurance Scheme

Operatives shall be entitled to payment at the time of sickness as determined from time to time under the J.I.B. Sickness with Pay Scheme, depending upon their continuity of service in the Industry. Similarly, operatives shall be entitled to Death, Accidental Death, Dismemberment and Permanent and Total Disability Benefits as determined from time to time under the J.I.B. Group Life Insurance Schemes depending upon their continuity of service in the Industry.
The J.I.B. Sickness with Pay, Group Life Insurance, Accidental Death, Dismemberment and Permanent and Total Disability Schemes are carried out by the Joint Industry Board by the sale of J.I.B. Combined Benefit Stamps which are fixed to the Operative's J.I.B.

WORKING RULES

Benefits Card. Details of the Schemes are given in a separate J.I.B. combined welfare/insurance scheme booklet (the 'gold' book ref. JIB/74BS).

SECTION V: GRADING DEFINITIONS

1: Technician

Must have been a Registered Apprentice for at least four years and have had practical training in electrical installation work. Must have obtained the City and Guilds of London Institute Course 51 C Certificate in Electrical Installation Work (or approved equivalent), *and* London Institute Course 51 C Certificate in Electrical Installation Work (or approved equivalent), *and* either

(*a*) Must be at least 27 years of age.

Must have had at least five years' subsequent experience as a Foreman, Chargehand and/or Approved Electrician with 'responsibility money', with a minimum of three years in a supervisory capacity in charge of electrical engineering installations of such a complexity and dimension as to require wide technical experience and organizational ability. Technicians must have knowledge of the most economical and effective layout of such installations together with the ability to achieve a high level of productivity in the work which they control. They must also have a thorough working knowledge of the National Working Rules for the Electrical Contracting Industry, of the current I.E.E. Regulations for the Electrical Equipment of Buildings, of the Electricity (Factories Act) Special Regulations, 1908 and 1944, the Electricity Supply Regulations, 1937, issued by the Electricity Commissioners so far as they deal with consumers' installations (i.e. Regulations 22–29 inclusive and 31), of any Regulations dealing with consumers' installations which may be issued, relevant British Standards and Codes of Practice, and of the Construction Industry Safety Regulations; or

(*b*) Must have exceptional technical skill, ability and experience beyond that expected of an Approved Electrician, and whose value to the Technician under (*a*) above and, with the support of his present Employer, may be granted this grade by the Joint Industry Board.

2: Approved Electrician

Must have been a Registered Apprentice and have had practical training in electrical installation work. Must have obtained at least the City and Guilds Electrical Installation Work Course 'B' Certificate (or approved equivalent) and have been granted the Electrician's Certificate (or approved equivalent).

Must have had two years' experience working as Electrician subsequent to the satisfactory completion of the Apprenticeship immediately prior to application for this grade, or be 22 years of age, whichever is the sooner.

Approved Electricians must possess particular practical, productive and electrical engineering skills with adequate technical and supervisory knowledge so as to be able to work on their own proficiently and carry out electrical installation work without detailed supervision in the most efficient and economical manner; be able to set out jobs from drawings and specifications and requisition the necessary installation materials. They must also have a thorough working knowledge of the National Working Rules for the Electrical Contracting Industry, of the Electricity Supply Regulations, 1937, issued by the Electricity Commissioners so far as they deal with consumers' installations (i.e. Regulations 22–29 inclusive and 31), of any Regulations dealing with consumers' installations which may be

Electrical Installations – Rates of Wages and Working Rules

WORKING RULES

issued, relevant British Standards and Codes of Practice, and of the Construction Industry Safety Regulations.

3: Electrician

One of the following qualifications; either

(*a*) Must have been a registered Apprentice and have had adequate training in electrical installation work.

Must have obtained at least the City and Guilds Electrical Installation Work Course 'B' Certificate (or approved equivalent).

Must have obtained the Employer's recommendation for grading, which shall not be unreasonably withheld.

Must be able, with the application for grading and any other relevant supporting evidence which may be required, to satisfy the Grading Committee of experience and ability.

Must be able to carry out electrical installation work efficiently in accordance with the National Working Rules for the Electrical Contracting Industry, the current I.E.E. Regulations for the Electrical Equipment of Buildings, and the Construction Industry Safety Regulations; or

(*b*) Must be at least 21 years of age.

Must have been a registered Apprentice and have had adequate practical training in electrical installation work.

Must have obtained the City and Guilds Electrical Installation Work Course 'A' Certificate (or approved equivalent).

Must be able with the application for grading, to satisfy the Grading Committee of experience and ability.

Must be able to carry out electrical installation work efficiently in accordance with the National Working Rules for the Electrical Contracting Industry, the current I.E.E. Regulations for the Electrical Equipment of Buildings, and the Construction Industry Safety Regulations.

4: Labourers

Labourers may be employed to assist in the installation of cables in accordance with Section IV of these Rules, and to do other unskilled work under supervision provided that they should not be used to re-introduce pair working. Nothing in these rules should be taken to imply that labourers must be employed where there is not sufficient unskilled work to justify their employment nor to prevent skilled men from doing a complete electrical installation job including the unskilled elements in these circumstances. On any Site at any time there shall be employed in total no more than one Labourer to four skilled J.I.B. Graded Operatives. This particular requirement may be reviewed in the light of the particular circumstances in respect of a particular site upon application, by either Party to the appropriate Regional Joint Industry Board.

ELECTRICAL INSTALLATIONS
Market Prices of Materials

In the section which follows the prices given, unless otherwise stated, include the cost of delivery to sites in the London area at March/April 1980 and represent the prices paid by contractors after the deduction of all trade discounts. Prices do not allow for any charges in respect of V.A.T.

SWITCHGEAR AND DISTRIBUTION

BUS BAR CHAMBERS

Sheet steel case with copper bars (length 1524 mm)	Unit	100 amp £	200 amp £	400 amp £	600 amp £	800 amp £	1200 amp £
	No.	84·26	120·33	153·42	241·80	359·56	492·52

DISTRIBUTION BOARDS

Sheet steel 415/440 volt grade

20 amp H.R.C. fuse	Unit	4 way £	6 way £	8 way £	10 way £	12 way £	14 way £	16 way £
S.P. & N.	No.	40·99	50·06	59·17	69·67	79·87	103·62	114·15
T.P. & N.	,,	71·93	86·10	100·28	121·16	137·25	196·74	261·62
30 amp H.R.C. fuse								
S.P. & N.	,,	43·72	56·67	71·60	84·17	100·10		
T.P. & N.	,,	93·00	119·75	145·87	173·92	200·87		

		2 way £	3 way £	4 way £	6 way £	8 way £	10 way £
60 amp H.R.C. fuse	Unit						
T.P. & N.	No.	111·56	118·33	128·24	171·75	211·85	297·83
100 amp H.R.C. fuse							
T.P. & N.	,,	212·44	226·00	247·54	313·28	358·85	

Sheet steel non fully shrouded 400/415 volt grade rewirable

20 amp ways		3 way £	4 way £	6 way £	8 way £	10 way £	12 way £
S.P. & N.	,,		19·22	25·13	29·03	34·43	37·74
T.P. & N.	,,	29·47	35·21	44·60	56·87	65·82	75·84
30 amps ways							
S.P. & N.	,,		23·69	27·77	31·60	42·81	48·53
T.P. & N.	,,	38·09	45·23	60·57	73·90	88·45	102·80

Sheet steel non fully shrouded 400/415 volt grade rewirable

20 amp ways		14 way £	16 way £	18 way £	20 way £	24 way £
S.P. & N.	,,	47·93	58·63	66·80		
T.P. & N.	,,	129·60	159·05	171·60	189·12	208·85
30 amp ways						
S.P. & N.	,,					
T.P. & N.	,,	164·03	183·06	195·24	213·53	

SWITCHGEAR AND DISTRIBUTION

DISTRIBUTION BOARDS continued

60 amp ways	Unit	2 way £	3 way £	4 way £	6 way £	8 way £	10 way £	12 way £
S.P. & N.	No.	35·93	42·54	47·28	67·13	82·45		
T.P. & N.	„	60·27	80·79	91·92	133·56	172·03	200·55	239·00
100 amp ways								
T.P. & N.	„	94·31	124·68	164·28	216·97	298·97	358·59	417·51
Cartridge (H.R.C.)								
200 amp ways								
T.P. & N.	„	336·98	405·37	552·55	712·00			
300 amp ways								
T.P. & N.	„	423·46	534·92	725·48	970·64			

SWITCHES, ISOLATORS

		20 amp £	30 amp £	60 amp £	100 amp £
Cast iron enclosure 415/440 volt	Unit				
Double pole	No.	10·90	12·32	20·66	33·81
Triple pole	„	12·79	15·69	26·24	39·92
Triple pole and neutral	„	14·26	17·29	28·75	42·46
Sheet steel case 415/440 volt					
Double pole	„	11·37	13·83	20·34	30·18
Triple pole	„	12·28	14·94	24·18	35·77
Triple pole and neutral	„	12·96	15·72	26·13	38·60
Switch fuses, sheet steel case, rewirable 415/440 volt					
S.P. & N.	„	12·98	15·71	25·26	38·51
D.P.	„	13·77	17·01	27·75	44·52
T.P.	„	15·51	20·04	34·76	57·67
T.P. & N.	„	16·42	21·36	36·79	60·32
Switch fuse, cast iron case, rewirable 415/440 volt					
S.P. & N.	„	13·50	16·30	27·03	46·79
D.P.	„	14·55	17·57	30·07	50·71
T.P.	„	16·60	21·79	41·89	67·45
T.P. & N.	„	18·07	23·50	43·75	70·26

Fuse switches sheet steel case 415/440 volt H.R.C. cartridges	Unit	60 amp £	100 amp £	200 amp £	300 amp £	400 amp £	600 amp £	800 amp £
S.P. & N.	No.	46·77	64·88	90·92				
D.P.	„	46·77	64·88	90·92				
T.P.	„	49·41	67·45	94·27	167·52	240·85	403·88	447·44
T.P. & N.	„	54·65	73·51	99·95	173·92	246·38	412·69	456·17
T.P.S. & N.	„			120·07		277·29	494·12	545·20

CONDUIT

ALUMINIUM

	Unit	16 mm £	20 mm £	25 mm £
Aluminium pliable conduit in 10 m lengths	100 metres	86·67	95·58	122·31
Zinc alloy coupler with screwed male end Metric thread	100	36·54	41·04	45·00

STEEL

Heavy gauge screwed and socketed	Unit	16 mm £	20 mm £	25 mm £	32 mm £	1¼" mm £	2" mm £
Welded	100 metres						
Black		55·63	64·83	86·68	119·99	153·19	249·56
Galvanized		80·29	93·22	124·29	156·68	200·56	296·75

		⅝" £	¾" £	1" £	1¼" £	1½" £	2" £
Solid drawn							
Black	,,	214·76	246·60	279·98	344·23	457·69	681·38
Galvanized	,,	217·98	251·49	290·98	349·39	461·62	688·99

PVC

High impact, unscrewed, no couplings		16 mm £	20 mm £	25 mm £	32 mm £
Heavy gauge	,,	39·02	46·46	63·00	99·23
Light gauge	,,	25·40	31·64	53·15	70·88

CONDUIT FITTINGS

STEEL, SCREWED

	Unit	16 mm £	20 mm £	25 mm £	32 mm £	1¼″ £	2″ £
Coupling							
Black	100	12·00	12·00	13·80	29·10	36·50	60·00
Galvanized	,,	15·20	15·20	18·20	38·30	49·90	78·70
Flanged coupling							
Black	,,		76·40	113·20	172·30		
Galvanized	,,		92·80	129·20	185·30		
Earthing coupling							
Black	,,		34·40	41·80			
Galvanized	,,		44·70	54·10			
Inspection coupling							
Black	,,		76·00	113·10	208·00	416·10	
Galvanized	,,		95·50	142·90	262·10	526·20	
Elbow							
Black	,,	33·40	33·40	52·50	99·50	158·00	253·00
Galvanized	,,	44·90	44·90	71·10	141·60	214·80	341·70
Inspection elbow, channel type							
Black	,,	54·50	54·50	68·90	110·90	158·30	310·90
Galvanized	,,	72·60	72·60	93·50	151·30	213·50	417·20
Elbow, top inspection type							
Black	,,		83·20				
Galvanized	,,		112·50				
Bend, standard							
Black	,,		41·70	65·60			
Galvanized	,,		56·30	89·40			
Inspection bend							
Black	,,		61·10	83·00	164·70	242·20	434·20
Galvanized	,,		80·30	108·90	216·70	317·50	568·80
Inspection tee							
Black	,,	68·60	68·60	84·80	159·20	244·50	420·20
Galvanized	,,	87·60	87·60	113·20	217·00	330·60	566·60

PVC

Coupling							
Heavy gauge	,,	6·50	7·50	10·00	20·10	38·70	67·50
Inspection	,,	14·40	15·40	20·10	44·90	75·10	134·70
Bend							
Heavy gauge	,,	20·10	23·10	34·70	53·00	131·90	236·00
Inspection	,,	31·80	31·80	60·20			

CONDUIT BOXES

SCREWED CIRCULAR STEEL BOXES

Small pattern without covers	Unit	16 mm £	20 mm £	25 mm £	32 mm £
Back outlet					
Black	100	75·80	75·80	112·60	
Galvanized	,,	101·50	101·50	149·40	
Terminal					
Black	,,	63·90	63·90	80·80	193·60
Galvanized	,,	84·50	84·50	106·40	241·60
Terminal and back outlet					
Black	,,		94·90	130·50	
Galvanized	,,		122·00	161·80	
Angle					
Black	,,	69·90	69·90	88·40	221·60
Galvanized	,,	92·20	92·20	117·00	276·20
Through way					
Black	,,	69·80	69·80	88·40	221·60
Galvanized	,,	92·20	92·20	117·00	276·20
Through way and back outlet					
Black	,,		101·00	158·40	
Galvanized	,,		131·30	206·80	
Light steel cover					
Black	,,	12·00	12·00	12·00	12·00
Galvanized	,,	15·60	15·60	15·60	15·60
Deep pattern without covers					
Terminal					
Black	,,		140·80	160·90	
Galvanized	,,		176·00	210·40	
Angle					
Black	,,		153·00	171·20	
Galvanized	,,		190·80	223·90	
Through way					
Black	,,		150·20	171·20	
Galvanized	,,		187·00	223·90	
Three way					
Black	,,		157·30	179·10	
Galvanized	,,		195·30	233·10	
Four way					
Black	,,		164·90	187·60	
Galvanized	,,		205·10	245·60	
Heavy steel cover					
Black	,,		105·30	105·30	
Galvanized	,,		131·50	131·50	
Pendant plate					
Black	,,		44·70	47·10	
Galvanized	,,		58·30	60·80	
Ball and socket					
Black	,,		92·30		
Galvanized	,,		111·70		

CONDUIT BOXES

PVC CIRCULAR BOXES

		16 mm £	20 mm £	25 mm £
Back outlet	100	36·40	36·40	43·10
Terminal	,,	26·60	26·60	40·30
Through way	,,	30·30	30·30	44·20
Angle	,,	30·30	30·30	44·20
Tee	,,	32·90	32·90	48·80
Terminal and back outlet	,,	42·70	42·70	61·50
Four way	,,	38·80	38·80	55·10

LOOPING-BOXES

Standard pattern 16 mm or 20 mm

		1 Hole £	2 Hole £	3 Hole £	4 Hole £
Black	100	80·50	80·50	80·50	80·50
Galvanized	,,	105·50	105·50	105·50	105·50

GREY CAST IRON ADAPTABLE BOXES
(with heavy covers)

Square pattern

		150 × 150 × 75 mm £	150 × 150 × 100 mm £	225 × 225 × 75 mm £	300 × 300 × 100 mm £
Black	No.	5·21	7·08	11·28	27·71
Galvanized	,,	7.60	10.10	15·95	37·06

Rectangular pattern

		150 × 75 × 50 mm £	150 × 100 × 50 mm £	150 × 100 × 75 mm £	225 × 150 × 75 mm £
Black	,,	3·02	4·68	5·08	8·84
Galvanized	,,	4·23	6·66	7·94	12·64

		225 × 150 × 100 mm	300 × 150 × 75 mm
Black	,,	12·62	16·43
Galvanized	,,	17·59	24·63

Electrical Installations – Market Prices of Materials

CONDUIT BOXES

SHEET STEEL ADAPTABLE BOXES

Square pattern	Unit	50 × 50 × 37·5 mm £	75 × 75 × 37·5 mm £	75 × 75 × 50 mm £	75 × 75 × 75 mm £
Black	No.	0·52	0·62	0·64	0·83
Galvanized	,,	0·75	0·90	0·93	1·19

		100 × 100 × 50 mm £	150 × 150 × 50 mm £	150 × 150 × 75 mm £	150 × 150 × 100 mm £
Black	,,	0·74	1·06	1·28	1·91
Galvanized	,,	1·10	1·56	1·90	1·56

		225 × 225 × 50 mm £	225 × 225 × 100 mm £	300 × 300 × 100 mm £	
Black	,,	2.22	2.85	4.94	
Galvanized	,,	3·26	4·19	7·30	

Rectangular pattern		100 × 75 × 37·5 mm £	100 × 75 × 50 mm £	150 × 75 × 50 mm £	150 × 75 × 75 mm £
Black	,,	0·64	0·60	0·75	0·92
Galvanized	,,	0·93	0·97	1·11	1·37

		150 × 100 × 75 mm £	225 × 75 × 50 mm £	225 × 150 × 75 mm £	225 × 150 × 100 mm £
Black	,,	1·11	1·15	1·99	2·29
Galvanized	,,	1·61	1·70	2·94	3·38

		300 × 150 × 50 mm £	300 × 150 × 75 mm £	300 × 225 × 100 mm £	
Black	,,	2·46	2·52	3·60	
Galvanized	,,	3·62	3·74	5·30	

WEATHERPROOF JUNCTION BOXES

Sheet steel enclosure, galvanized finish, rail mounted plastic terminal blocks, gland plates and gaskets, with side hung door and padlock

	Unit	10 way £	20 way £	40 way £	60 way £
	No.	26·99	29·32	39·65	44·32

		100 way £	150 way £	200 way £	250 way £
	,,	63·84	75·48	103·47	116·54

		300 way £	350 way £	400 way £	450 way £
	,,	128·17	139·81	151·44	163·08

CONDUIT BOXES
CONDUIT CLIPS AND SADDLES

Clip, B.S. 31	Unit	16 mm £	20 mm £	25 mm £	32 mm £	1½" £	2" £
Black	100	2·02	2·02	2·77	7·17	10·43	13·87
Galvanized	,,	2·58	2·58	3·51	9·58	13·06	17·33
Saddle							
Black	,,	11·41	11·41	14·23	26·46	28·25	35·74
Galvanized	,,	14·68	14·68	18·32	31·92	33·21	46·03
Splayed distance saddle ½ in. spacing							
One way							
Black	,,	34·00	34·00	44·00	57·80	75·60	82·60
Galvanized	,,	46·30	46·30	60·40	78·80	101·40	113·30

Electrical Installations – Market Prices of Materials

CABLE TRUNKING

GALVANIZED OR STOVE ENAMELLED MILD STEEL TRUNKING

	Unit	50 × 50 mm	75 × 50 mm	75 × 75 mm	100 × 50 mm	100 × 75 mm	100 × 100 mm	150 × 50 mm	150 × 75 mm	150 × 100 mm	150 × 150 mm	225 × 75 mm
Single compartment	Metre	2·04	2·25	2·46	2·58	2·70	3·05	3·90	4·40	4·70	5·50	6·31
Flanged connector	No.	0·26	0·27	0·27	0·33	0·33	0·33	0·37	0·37	0·37	0·57	0·61
Sealing end	,,	0·26	0·27	0·27	0·33	0·33	0·33	0·37	0·37	0·37	0·57	0·61

		225 × 100 mm	225 × 150 mm	225 × 225 mm	300 × 75 mm	300 × 100 mm	300 × 150 mm	300 × 225 mm	300 × 300 mm
Single compartment	Metre	6·51	7·49	9·44	7·49	8·93	9·44	10·45	11·42
Flanged connector	No.	0·73	0·87	1·04	0·87	1·04	1·04	1·04	1·17
Sealing end	,,	0·73	0·87	1·04	0·87	1·04	1·04	1·04	1·17

		50 × 50 mm	75 × 50 mm	75 × 75 mm	100 × 50 mm	100 × 75 mm	100 × 100 mm	150 × 50 mm	150 × 75 mm	150 × 100 mm	150 × 150 mm	225 × 75 mm
Double compartment	Metre	2·67	2·87	3·20	3·23	3·47	3·99	4·67	5·22	5·80	7·15	
Flanged connector	No.	0·26	0·27	0·27	0·33	0·33	0·33	0·37	0·37	0·37	0·57	0·61
Sealing end	,,	0·26	0·27	0·27	0·33	0·33	0·33	0·37	0·37	0·37	0·57	0·61
Triple compartment	Metre		3·50	3·99	3·91	4·22	4·93	5·30	5·99	6·72	8·35	
Flanged connector	No.		0·27	0·27	0·33	0·33	0·33	0·37	0·37	0·37	0·57	
Sealing end	,,		0·27	0·27	0·33	0·33	0·33	0·37	0·37	0·37	0·57	
Four compartment	Metre				4·58	4·98	5·88	5·92	6·74	7·63	9·53	
Flanged connector	No.				0·33	0·33	0·33	0·37	0·37	0·37	0·57	
Sealing end	,,				0·33	0·33	0·33	0·37	0·37	0·37	0·57	
Thickness of trunking		1 mm	1·2 mm	1·2 mm	1·2 mm	16 swg	16 swg	16 swg	16 swg	16 swg	16 swg	16 swg

CABLE TRUNKING FITTINGS

GALVANIZED OR STOVE ENAMELLED MILD STEEL FITTINGS

90 DEGREE BEND

	Unit	50 × 50 mm £	75 × 50 mm £	75 × 75 mm £	100 × 50 mm £	100 × 75 mm £	100 × 100 mm £	150 × 50 mm £	150 × 75 mm £	150 × 100 mm £	150 × 150 mm £	225 × 75 mm £
Single compartment	No.	1·66	1·90	2·08	2·13	2·17	2·28	2·43	2·99	3·30	3·67	5·10
Double compartment	,,	2·01	2·27	2·29	2·45	2·47	2·59	2·79	3·32	3·55	3·92	
Triple compartment	,,		2·59	2·64	2·74	2·77	2·89	3·07	3·63	3·86	4·24	
Four compartment	,,				3·07	3·16	3·23	3·37	3·92	4·18	4·55	

		225 × 100 mm £	225 × 150 mm £	225 × 225 mm £	300 × 75 mm £	300 × 100 mm £	300 × 150 mm £	300 × 225 mm £	300 × 300 mm £
Single compartment	,,	6·04	6·69	7·79	7·20	7·34	7·79	9·05	11·36

TEE PIECE

		50 × 50 mm £	75 × 50 mm £	75 × 75 mm £	100 × 50 mm £	100 × 75 mm £	100 × 100 mm £	150 × 50 mm £	150 × 75 mm £	150 × 100 mm £	150 × 150 mm £	225 × 75 mm £
Single compartment	,,	1·91	2·22	2·30	2·52	2·59	2·67	2·83	3·49	3·96	4·32	6·42
Double compartment	,,	2·59	2·93	2·95	3·17	3·23	3·30	3·49	4·14	4·55	4·89	
Triple compartment	,,		3·55	3·56	3·83	3·88	3·92	4·12	4·75	5·20	5·54	
Four compartment	,,				4·49	4·53	4·58	4·77	5·40	5·84	6·20	

		225 × 100 mm £	225 × 150 mm £	225 × 225 mm £	300 × 75 mm £	300 × 100 mm £	300 × 150 mm £	300 × 225 mm £	300 × 300 mm £
Single compartment	,,	8·40	9·43	10·99	10·12	10·61	11·06	12·74	15·26

CROSS HORIZONTAL

	Unit	50 × 50 mm £	75 × 50 mm £	75 × 75 mm £	100 × 50 mm £	100 × 75 mm £	100 × 100 mm £	150 × 50 mm £	150 × 75 mm £	150 × 100 mm £	150 × 150 mm £	225 × 75 mm £
Single compartment	No.	2·21	2·89	3·35	3·68	3·81	4·04	4·22	4·55	4·93	5·22	8·69
Double compartment	,,	2·85	3·55	3·93	4·34	4·42	4·67	4·89	5·20	5·47	5·78	
Triple compartment	,,		4·20	4·59	5·01	5·06	5·31	5·51	5·83	6·08	6·42	
Four compartment	,,				5·64	5·71	5·98	6·15	6·35	6·72	7·06	

		225 × 100 mm £	225 × 150 mm £	225 × 225 mm £	300 × 75 mm £	300 × 100 mm £	300 × 150 mm £	300 × 225 mm £	300 × 300 mm £
Single compartment	,,	10·50	11·55	13·25	12·45	12·84	13·25	15·09	17·98

CABLE TRUNKING

PVC TRUNKING, FINISH GREY, CLIP ON LID

	Unit	50 × 50 mm £	75 × 50 mm £	75 × 75 mm £	100 × 50 mm £	100 × 75 mm £
Single compartment	Metre	1·91	2·42	3·14	4·03	4·48
Crossovers	No.	5·64	6·18	6·71	10·42	11·61
Stop ends	,,	0·24	0·34	0·47	0·66	1·05
Flanged couplings	,,	1·43	1·62	2·06	2·34	2·91
Internal couplings	,,	0·52	0·58	0·58	0·86	1·06
External couplings	,,	0·65	0·80	0·98	1·62	2·03
Angle, top cover	,,	1·02	1·55	1·91	3·90	5·58
Angle, external cover	,,	1·02	2·84	3·77	4·25	7·62
Angle, internal cover	,,	1·02	2·84	3·77	4·25	7·62
Tee, top cover	,,	1·66	2·52	2·84	4·61	6·60
Tee, external cover	,,	3·83	4·18	4·99	5·94	7·94
Tee, internal cover	,,	3·83	4·18	4·99	5·94	7·94

		100 × 100 mm £	150 × 75 mm £	150 × 100 mm £	150 × 150 mm £
Single compartment	Metre	5·18	8·74	10·95	11·42
Crossovers	No.	11·61	13·54	16·98	21·43
Stop ends	,,	1·05	2·76	3·39	4·71
Flanged couplings	,,	2·91	3·39	4·22	5·34
Internal couplings	,,	1·06			
External couplings	,,	2·03	2·99	3·70	4·58
Angle, top cover	,,	5·58	9·21	10·92	16·03
Angle, external cover	,,	7·62	8·83	10·50	16·03
Angle, internal cover	,,	7·62	8·83	10·50	16·03
Tee, top cover	,,	6·60	11·39	14·62	18·37
Tee, external cover	,,	7·94	11·39	14·62	18·37
Tee, internal cover	,,	7·94	11·39	14·62	18·37

		50 mm £	75 mm £	100 mm £
Dividing strips (1·8 m long)	,,	1·55	2·06	2·62

CABLE TRAYS

CABLE TRAYS

	Unit	50 mm £	75 mm £	100 mm £	150 mm £	200 mm £	225 mm £
Standard galvanized mild steel light duty cable tray	Metre	0·99	1·08	1·26	1·77	1·93	2·47
Ditto with return flange	,,			2·58	3·17		3·73
Bend	,,	1·32	1·36	1·46	1·90	2·31	2·78
Ditto with return flange	,,			9·45	12·10		14·26
Tee	,,	1·99	2·04	2·19	2·85	3·47	4·18
Ditto with return flange	,,			13·86	15·39		18·60

	Unit	300 mm £	375 mm £	450 mm £	600 mm £	750 mm £	900 mm £
Standard galvanized mild steel light duty cable tray	,,	3·25	3·95	4·55	7·72	10·22	12·28
Ditto with return flange	,,	4·41		6·97	8·40	9·67	
Bend	,,	4·11	6·03	7·17	11·19	17·14	24·86
Ditto with return flange	,,	16·43		26·98	37·62	46·78	
Tee	,,	6·17	9·05	10·76	16·78	25·70	37·29
Ditto with return flange	,,	20·52		34·06	47·97	60·10	
Coupling pieces any size	Pair	1·05					

LADDER RACK

LADDER RACK, GALVANIZED FINISH

Width

	Unit	150 mm £	200 mm £	300 mm £	400 mm £	600 mm £	1000 mm £
Straight	Metre	4·57	4·79	4·90	5·00	5·55	9·23
Bend (radius)	No.	11·83	12·16	12·93	13·17	15·05	29·94
Bend (outside radius)	,,		19·23	22·86	25·24	27·83	46·15
T-Junction	,,		20·43	20·72	21·00	21·39	33·36
X-Junction	,,		27·09	27·76	28·03	29·02	
Riser	,,	12·28	13·09	13·41	13·83	14·52	16·76
Cantilever bracket	,,	1·20	1·30	1·93	2·57	3·29	10·33
Wall bracket	,,	1·83	1·89	1·96	2·06	2·25	
End connector	,,	0·77					
Universal coupling	,,	1·52					

CABLE AND CORDS

SINGLE CORE PVC INSULATED COPPER CABLE

450/750 volt grade	Unit	1·0 mm² £	1·5 mm² £	2·5 mm² £	4 mm² £	6 mm² £	10 mm² £	16 mm² £
	100 metres	3·19	4·34	6·66	12·12	17·08	28·96	44·08
		25 mm² £	35 mm² £	50 mm² £	70 mm² £	95 mm² £	120 mm² £	150 mm² £
	„	73·52	101·84	138·00	196·16	277·76	384·48	475·68
		185 mm² £	240 mm² £	300 mm² £	400 mm² £	500 mm² £	630 mm² £	
	„	586·16	769·04	958·80	1466·40	1790·16	2330·40	

SINGLE CORE BUTYL RUBBER INSULATED BRAIDED AND COMPOUNDED

450/750 volt grade	Unit 100 metres	1·0 mm² £	1·5 mm² £	2·5 mm² £	4 mm² £	6 mm² £	10 mm² £	16 mm² £	35 mm² £	40 mm² £
		14·60	16·88	22·80	34·96	47·28	81·60	119·76	186·80	249·28
		50 mm² £	70 mm² £							
	„	365·70	523·52							

PVC INSULATED AND SHEATHED COPPER

450/750 volt grade	Unit	1·0 mm² £	1·5 mm² £	2·5 mm² £	4 mm² £	6 mm² £	10 mm² £	16 mm² £
Single core . . .	100 metres	5·47	7·12	10·68	17·60	23·68	38·96	55·12
Flat twin core . .	„	10·20	13·40	19·52	35·36	50·24	81·36	123·52
Twin core, and E.C.C.	„	11·08	13·92	19·88	38·00	54·32	88·64	134·72
Flat three core . .	„	15·12	20·24	32·40	53·92	72·88	123·52	194·32
Three core, and E.C.C.	„	18·16	22·72	35·52	57·76	88·48	144·96	228·64

ARMOURED PVC INSULATED AND SHEATHED

600/1000 volt grade	Unit	1·5 mm² £	2·5 mm² £	4 mm² £	6 mm² £	10 mm² £	16 mm² £
Two core . . .	100 metres	39·68	50·48	70·32	88·08	136·88	162·88
Cable gland . .	100	70·00	70·00	92·00	111·00	161·00	161·00
Brass locknut . .	„	12·00	12·00	12·00	12·00	13·00	13·00
PVC shroud . .	„ 100	11·00	11·00	14·00	11·00	20·00	20·00
Three core . . .	metres	46·80	59·84	85·92	116·32	171·92	225·84
Cable gland . .	100	70·00	92·00	92·00	111·00	161·00	161·00
Brass locknut . .	„	12·00	12·00	12·00	12·00	13·00	13·00
PVC shroud . .	„ 100	11·00	11·00	14·00	14·00	20·00	20·00
Four core . . .	metres	52·88	69·44	112·96	141·84	211·84	303·28
Cable gland . .	100	70·00	92·00	111·00	111·00	161·00	161·00
Brass locknut . .	„	12·00	12·00	12·00	12·00	13·00	13·00
PVC shroud . .	„	11·00	14·00	11·00	11·00	20·00	20·00

FP200
Fire resistant wiring Cables

- Fire resistant
- Surge resistant
- Moisture resistant
- Non-ageing
- Fully screened
- Low installed cost
- Easily terminated
- Mechanically strong
- ANTS certified

FP 200 cables are now accepted by many Authorities as the 'alternative cables' for wiring Fire Alarms to the new British Standard 5839 : Part 1 : 1980.

For cable enquiries and further information contact John Vaughan, Product Sales Manager, Special Cables at Aberdare, or any Pirelli General regional office.

PIRELLI GENERAL

PIRELLI GENERAL
CABLE WORKS LTD
P.O. BOX 1, ABERDARE,
MID GLAMORGAN, SOUTH WALES
Telephone: Aberdare 872416 Telex: 498567

P5609

4 will go into the 80's

The Graham Group will enter the 1980s restructured into 4 holding companies.

These 4 companies will account for 140 branches throughout the British Isles, with a product range of over 100,000 items, ensuring unrivalled service to their customers.

The new companies comprise:

Scottish Region
GRAHAM BOYD LTD.,
23, Silvergrove Street, Glasgow G40 1EP.
Tel: 041-554 1194.

North Region
GRAHAM GRATRIX LTD.,
Regional Head Office, P.O. Box 36,
96 Leeds Road, Huddersfield HD1 4RH.
Tel: 0484 37366

South East Region
GRAHAM FORD LTD.,
Regional Head Office,
13/17 Station Parade, Eastbourne
BN21 1BB.
Tel: 0323 21311

South West Region
GRAHAM REEVES LTD.,
Manor House, Totnes.
TQ9 5DQ
Tel: 0803 863900

Europe's Largest Suppliers of Building and Construction Materials.

Graham Group

TRY GRAHAMS FIRST! in 1980

Electrical Installations – Market Prices of Materials

CABLE AND CORDS

ARMOURED PVC INSULATED AND SHEATHED continued

	Unit	1·5 mm² £	2·5 mm² £	4 mm² £	6 mm² £
600/1000 volt grade	100				
Five core	metres	72·32	95·68	162·64	218·08
Cable gland	100	92·00	92·00	111·00	161·10
Brass locknut	,,	12·00	12·00	12·00	13·00
PVC shroud	,,	14·00	14·00	11·00	20·00
Seven core	100 metres	87·60	116·56	195·28	278·48
Cable gland	100	92·00	111·00	161·00	161·00
Brass locknut	,,	12·00	12·00	13·00	13·00
PVC shroud	,,	14·00	11·00	20·00	20·00
Eight core	100 metres	110·40	154·80	240·24	
Cable gland	100	92·00	111·00	161·00	
Brass locknut	,,	12·00	12·00	13·00	
PVC shroud	,,	14·00	11·00	20·00	
Ten core	100 metres	130·96	176·32	306·88	
Cable gland	100	111·00	161·00	161·00	
Brass locknut	,,	12·00	13·00	13·00	
PVC shroud	,,	11·00	20·00	20·00	
Twelve core	100 metres	142·40	196·00	341·04	
Cable gland	100	111·00	161·00	161·00	
Brass locknut	,,	12·00	13·00	13·00	
PVC shroud	,,	11·00	20·00	20·00	
Fourteen core	100 metres	190·56	255·76	398·16	
Cable gland	100	111·00	161·00	236·00	
Brass locknut	,,	12·00	13·00	22·00	
PVC shroud	,,	11·00	20·00	31·00	
Twenty-two core	100 metres	272·00	286·16	610·40	
Cable gland	100	161·00	236·00	416·00	
Brass locknut	,,	13·00	22·00	43·00	
PVC shroud	,,	20·00	31·00	34·00	
Thirty-seven core	100 metres	397·60	545·84		
Cable gland	100	236·00	236·00		
Brass locknut	,,	22·00	22·00		
PVC shroud	,,	31·00	31·00		
Forty-eight core	100 metres	478·24	705·76		
Cable gland	100	236·00	416·00		
Brass locknut	,,	22·00	43·00		
PVC shroud	,,	31·00	34·00		

HEAT-RESISTING FLEXIBLE CORD

	Unit	0·5 mm² £	0·75 mm² £	1·0 mm² £	1·5 mm² £	2·5 mm² £
300–500 volt grade	100					
Two core	metres	15·04	22·16	23·20	32·64	48·72
Three core	,,	20·24	24·00	30·16	44·72	60·56

CABLE AND CORDS

VULCANIZED RUBBER INSULATED TOUGH RUBBER SHEATHED FLEXIBLE CORD

	Unit	0·5 mm²	0·75 mm²	1·0 mm²	1·5 mm²	2·5 mm²
300–500 volt grade	100					
Two core	metres	13·60	16·36	19·12	26·96	39·52
Three core	,,		22·40	26·16	34·96	54·00
Four core	,,		28·08	32·88	45·36	70·16
Five core	,,		55·52	61·28	83·52	131·52

VULCANIZED RUBBER INSULATED HEAVY RAYON BRAIDED FLEXIBLE CORD

	Unit	0·5 mm² £	0·75 mm² £	1 mm² £	1·5 mm² £
300–500 volt grade	100				
Two core circular	metres	23·84	30·08	32·88	
Three core circular	,,	36·56	43·92	51·52	61·68

MINERAL INSULATED COPPER SHEATHED CABLE WITH COPPER CONDUCTORS

	Unit	2L 1·0 £	2L 1·5 £	2L 2·5 £	2L 4 £	3L 1·0 £
Light duty 600 volt grade	100					
Bare	metres	37·32	48·44	63·00	86·64	46·44
PVC sheathed	,,	41·92	53·64	68·80	94·08	51·72
Standard seal	100	15·93	15·93	15·93	15·93	15·93
Standard gland	,,	41·31	41·31	41·31	41·31	41·31
Gland shroud	,,	14·13	14·13	14·13	14·13	14·13
PVC extension	3 metres	0·22	0·22	0·22	0·22	0·22
Two way copper saddles						
Bare	100	1·45	1·45	1·45	1·45	1·45
Plastic covered	,,	4·32	4·82	5·30	5·36	4·82
One hole clips						
Bare	,,	1·35	1·35	1·62	1·95	1·35
Plastic covered	,,	5·03	5·56	5·68	5·87	5·56

		3L 1·5 £	4L 1·0 £	4L 1·5 £	7L 1·0 £	7L 1·5 £
Light duty 600 volt grade	100					
Bare	metres	58·72	55·24	71·52	80·76	100·20
PVC sheathed	,,	64·44	60·52	77·76	88·04	108·20
Standard seal	100	15·93	15·93	15·93	41·85	41·85
Standard gland	,,	41·31	41·31	41·31	67·05	67·05
Gland shroud	,,	14·13	14·13	14·13	17·64	17·64
PVC extension	3 metres	0·22	0·22	0·22	0·22	0·22
Two way copper saddle						
Bare	100	1·45	1·45	1·45	1·45	1·99
Plastic covered	,,	5·30	5·30	5·30	5·36	5·58
One hole clips						
Bare	,,	1·62	1·62	1·85	1·95	2·10
Plastic covered	,,	5·68	5·68	5·70	5·87	7·72

Electrical Installations – Market Prices of Materials

CABLE AND CORDS

MINERAL INSULATED COPPER SHEATHED CABLE WITH COPPER CONDUCTORS *continued*

	Unit	1H 6 £	1H 10 £	1H 16 £	1H 25 £	1H 35 £	1H 50 £	1H 70 £
Heavy duty 1000 volt grade	100							
Bare . . .	metres	61·64	79·60	109·60	154·56	204·24	260·20	339·56
PVC sheathed	,,	67·36	86·64	117·96	163·88	214·44	271·40	352·08
Standard seal .	100	15·93	15·93	15·93	15·93	15·93	41·85	41·85
Standard gland .	,,	41·31	41·31	41·31	41·31	41·31	67·05	67·05
Gland shroud .	,,	14·13	14·13	14·13	14·13	14·13	17·64	17·64
PVC extension .	3 metres	0·28	0·30	0·35	0·37	0·45	0·53	0·61
Two way copper saddle Bare . . .	100	1·45	1·45	1·99	2·16	2·37	5·47	5·58
Plastic covered .	,,	5·30	5·36	5·58	9·18	9·57	9·64	11·06
One hole clips Bare . . .	,,	1·62	1·85	2·10	2·25	2·39	4·47	5·22
Plastic covered .	,,	5·68	5·70	7·72	8·60	8·82	9·34	10·14

		1H 95 £	1H 120 £	1H 150 £	2H 1·5 £	2H 2·5 £	2H 4 £	2H 6 £
Heavy duty 1000 volt grade	100							
Bare . . .	metres	445·96	540·84	669·88	69·08	83·92	108·44	139·48
PVC sheathed	,,	464·80	561·48	692·20	77·16	92·64	117·92	145·44
Standard seal .	100	41·85	90·00	90·00	15·93	15·93	15·93	15·93
Standard gland .	,,	67·05	129·60	129·60	41·31	41·31	41·31	41·31
Gland shroud .	,,	17·64	21·87	21·87	14·13	14·13	14·13	14·13
PVC extension .	3 metres	0·68	0·89	1·05	0·22	0·22	0·23	0·28

		1H 95 £	1H 120 £	1H 150 £	2H 1·5 £	2H 2·5 £	2H 4 £	2H 6 £
Two way copper saddle Bare . . .	100	6·05	9·99	10·30	1·99	1·99	2·37	5·08
Plastic covered .	,,	15·02	15·31	15·58	5·36	5·58	9·18	9·57
One hole clips Bare . . .	,,	5·28	5·65	9·77	1·95	2·16	2·25	4·19
Plastic covered .	,,	14·39	15·43	16·16	5·87	7·72	8·60	9·34

		2H 10 £	2H 16 £	2H 25 £	3H 1·5 £	3H 2·5 £	3H 4 £	3H 6 £
Heavy duty 1000 volt grade	100							
Bare . . .	metres	182·56	262·36	366·84	76·56	96·96	123·44	154·12
PVC sheathed	,,	194·24	276·44	387·72	84·84	106·24	133·44	164·96
Standard seal .	100	41·85	41·85	90·00	15·93	15·93	15·93	41·85
Standard gland .	,,	67·05	67·05	129·60	41·31	41·31	41·31	67·05
Gland shroud .	,,	17·64	17·64	21·87	14·13	14·13	14·13	17·64
PVC extension .	3 metres	0·30	0·35	0·37	0·22	0·22	0·23	0·28
Two way copper saddle Bare . . .	100	5·47	5·74	9·99	1·99	2·16	2·37	5·08
Plastic covered .	,,	9·83	15·02	15·58	5·58	9·18	9·57	9·64
One hole clips Bare . . .	,,	4·47	5·22	9·47	2·10	2·25	2·39	4·19
Plastic covered .	,,	9·42	14·37	15·43	7·72	8·60	8·82	9·34

CABLE AND CORDS

MINERAL INSULATED COPPER SHEATHED CABLE WITH COPPER CONDUCTORS *continued*

	Unit	3H 10	3H 16	3H 25	4H 1·5	4H 2·5	4H 4	4H 6
Heavy duty 1000 volt grade	100	£	£	£	£	£	£	£
Bare	metres	224·84	317·24	482·92	95·00	120·44	147·32	195·48
PVC sheathed	,,	237·32	336·44	505·00	103·92	130·36	158·60	207·56
Standard seal	100	41·85	41·85	166·05	15·93	15·93	41·85	41·85
Standard gland	,,	67·05	67·05	288·45	40·86	41·31	67·05	67·05
Gland shroud	,, 3	17·64	17·64	33·48	14·13	14·13	17·64	17·64
PVC extension	metres	0·30	0·35	0·37	0·22	0·22	0·26	0·28
Two way copper saddle								
Bare	100	5·58	6·05	10·30	2·16	2·37	5·09	5·47
Plastic covered	,,	11·06	15·31	15·58	9·18	9·18	9·64	9·83
One hole clips								
Bare	,,	5·22	5·28	9·72	2·16	2·39	4·19	4·47
Plastic covered	,,	10·13	14·39	16·15	8·61	8·82	9·34	9·42

		4H 10	4H 16	4H 25	7H 1·5	7H 2·5	12H 2·5	19H 1·5
Heavy duty 1000 volt grade	100	£	£	£	£	£	£	£
Bare	metres	279·96	404·64	591·24	130·40	177·84	311·32	477·72
PVC sheathed	,,	294·04	427·44	622·60	140·76	189·12	330·40	497·96
Standard seal	100	41·85	90·00	166·05	41·85	41·85	90·00	166·05
Standard gland	,,	67·05	129·60	288·45	67·05	67·05	129·60	288·45
Gland shroud	,, 3	17·64	21·87	33·48	17·64	17·64	21·87	33·48
PVC extension	metres	0·30	0·35	0·37	0·22	0·22	0·22	0·22
Two way copper saddle								
Bare	100	5·74	9·99	10·62	5·09	5·47	6·05	9·99
Plastic covered	,,	15·02	15·58	19·34	9·57	9·64	15·31	15·31
One hole clips								
Bare	,,	5·22	9·47	10·66	4·19	4·47	5·28	5·65
Plastic covered	,,	14·37	15·43	16·98	8·82	9·34	14·39	15·21

Electrical Installations – Market Prices of Materials

FITTINGS AND ACCESSORIES

LIGHTING SWITCHES

Metal clad, surface mounted 5 amp protected switch to BS 3676:1963 20 mm tapped entry

	Unit	1 gang £	2 gang £	3 gang £	4 gang £	6 gang £
One way	10	20·30	35·00	52·50	70·00	105·10
Two way	„	23·10	40·60	60·90	81·30	121·90

		8 gang £	9 gang £	10 gang £	12 gang £
One way	„	140·10	157·60	175·10	210·10
Two way	„	162·50	182·80	203·20	243·80

As above but weatherproof

		1 gang £	2 gang £	3 gang £	4 gang £
One way	„	43·70	84·80	126·00	164·60
Two way	„	46·50	90·40	134·40	175·80

Modular type switches comprising galvanized steel box and switch mounting harness with interchangeable switch units and bronze or satin chrome coverplate, for flush mounting

		1 gang £	2 gang £	3 gang £	4 gang £	6 gang £
One way	„	21·40	25·60	40·70	44·90	69·40
Two way	„	23·00	28·80	45·60	51·30	79·10

		8 gang £	9 gang £	12 gang £	18 gang £	24 gang £
One way	„	77·80	94·40	106·90	180·00	216·60
Two way	„	90·70	109·00	126·30	209·10	255·30

As above but surface mounting and grey finish

		1 gang £	2 gang £	3 gang £	4 gang £	6 gang £
One way	„	13·10	17·40	27·70	31·90	51·90
Two way	„	14·70	20·70	32·60	38·30	61·60

		8 gang £	9 gang £	12 gang £	18 gang £	24 gang £
One way	„	60·30	76·20	88·70	147·40	183·90
Two way	„	73·20	90·70	108·00	176·50	222·70

Ceiling switch, white moulded plastic mounting block and cover, pull cord for surface mounting

	Unit	5 amp 1 way single pole £	5 amp 2 way single pole £	15 amp 1 way double pole £	15 amp 2 way double pole £
Standard unit	10	14·60	16·70	24·40	
With neon indicator	„	25·90	28·10	29·60	48·00

Electrical Installations – Market Prices of Materials
FITTINGS AND ACCESSORIES

SOCKET OUTLETS

13 amp metal clad socket outlet to B.S. 1363: 1967; galvanized steel box and coverplate with white plastic inserts; suitable for surface mounting

	Unit	1 gang £	2 gang £
Unswitched	10	13·20	25·00
Switched	,,	20·10	37·40
Switched with neon indicator	,,	27·80	52·70

13 amp socket outlet to B.S. 1363: 1967; with white moulded plastic coverplate and box for surface mounting

Unswitched	,,	9·60	17·10
Switched	,,	12·90	22·60
Switched with neon indicator	,,	19·50	35·90

13 amp socket outlet to B.S. 1363: 1967; with white plastic coverplate and galvanized steel box to B.S. 4662: 1970 for flush mounting

Switched	,,	16·10	27·50
Switched with neon indicator	,,	22·80	40·70

Socket outlet as last but with matt chrome steel coverplate

Switched	,,	25·10	42·40
Switched with neon indicator	,,	31·80	55·80

CONNECTION UNITS

Moulded plastic connection unit to B.S. 816: 1952, with white moulded plastic box and coverplate for surface mounting

	Unit	Switched £	Unswitched £
Standard	10	22·60	20·00
With neon indicator	,,	29·20	—
With flue outlet	,,	23·90	21·30
With flex outlet and neon indicator	,,	30·50	

Connection unit as last but with galvanized steel box for flush mounting

Standard	,,	23·30	20·80
With neon indicator	,,	30·00	
With flex outlet	,,	24·60	21·90
With flex outlet and neon indicator	,,	31·20	

Galvanized pressed steel pattern connection unit with matt chrome or satin brass finish coverplate with plastic inserts and galvanized steel box for flush mounting

Standard	,,	31·20	28·10
With neon indicator	,,	37·80	
With flex outlet	,,	33·10	28·90
With flex outlet and neon indicator	,,	39·70	

Connection unit as above but for surface mounting

Standard	,,	29·30	27·70
With neon indicator	,,	36·00	
With flex outlet	,,	30·50	28·80
With flex outlet and neon indicator	,,	37·10	

Telephone outlet plate, moulded plastic complete with box

	Unit	White finish Single outlet £	Double outlet £	Bronze or chrome finish Single outlet £	Double outlet £
Flush mounted	10	11·20	12·00	19·40	20·60
Surface mounted	,,	14·00	14·80	22·20	23·40

Electrical Installations − Market Prices of Materials 255

FITTINGS AND ACCESSORIES

CONNECTION UNITS continued

		Single Outlet	Double Outlet	Isolated T.V./F.M. Single Outlet	Double Outlet
T.V. co-axial socket outlet, moulded plastic, white finish	Unit	£	£	£	£
Flush mounted	10	21·30	28·70	34·20	46·60
Surface mounted	,,	24·10	31·50	37·00	49·40

CEILING ROSE

		2 terminals & earth	2 terminals, loop & earth	2 terminals loop, earth & strain	Plug in type
		£	£	£	£
Moulded plastic ceiling rose for mounting over conduit box	Unit 10	4·90	5·20	5·90	11·80

LAMPHOLDER

		£
White moulded plastic pendant with non-rising terminals and cord grip	,,	8·20

BATTEN HOLDERS

White moulded plastic to B.S. 52 (1963)

		Straight	Angled	Adjustable
		£	£	£
3 terminals	,,	13·40	15·60	14·00

SHAVER SUPPLY UNITS

		Surface type with white moulded plastic box	Flush type with galvanized steel box	Surface type with shower light and dual voltage
Single voltage supply unit, with white moulded plastic faceplate, unswitched, (not suitable for bathrooms) over conduit box	Unit 10	£ 5·25	£ 5·28	£ —
Dual voltage supply unit, with white moulded plastic coverplate and switches with neon indicator lamp	,,	14·83	16·40	24·38

COOKER CONTROL UNITS

45 amp cooker control unit to B.S. 4177: 1967 comprising 45 amp D.P. main switch and a 13 amp switched socket outlet with white painted steel box and coverplate with moulded plastic insert; suitable for surface mounting

		£
Standard	Unit 10	41·20
With neon indicator	,,	56·90

Cooker control unit as last but with galvanized steel box for flush mounting

| Standard | ,, | 45·70 |
| With neon indicator | ,, | 61·60 |

45 amp cooker control unit to B.S. 4177: 1967 comprising 45 amp D. main switch and a 13 amp switched socket output; white moulded plastic coverplate and box; suitable for surface mounting

Standard	,,	40·60
With neon indicator	,,	53·15
Connector unit; moulded white plastic cover and block and galvanized steel box	,,	15·92

FITTINGS AND ACCESSORIES

CONSUMER UNITS

	Unit	2 way £	3 way £	4 way £	6 way £	8 way £	12 way £
60 amp 240 volt D.P./A.C. switched, all insulated with moulded plastic case and cover.							
Fitted rewirable fuses.	No.	4·45	5·89	7·29	9·29	11·81	
Fitted cartridge fuses.	,,	4·96	6·65	8·30	10·80	13·82	
Fitted miniature circuit breakers.	,,	8·76	12·35	15·98	22·19	29·02	
60 amp 240 volt D.P./A.C. switched, metal clad unit with enamelled steel case and moulded plastic fuse cover.							
Fitted rewirable fuses.	,,	5·35	7·40	8·49	10·58	12·84	
Fitted cartridge fuses.	,,	5·85	8·15	9·49	12·10	14·86	
Fitted miniature circuit breakers.	,,	9·65	13·85	17·09	23·49	30·05	
100 amp 240 volt ditto							
Fitted rewirable fuses.	,,				12·81	15·23	23·21
Fitted cartridge fuses.	,,				14·32	17·24	26·24
Fitted miniature circuit breakers.	,,				25·71	32·44	49·02

FLUORESCENT LIGHT FITTINGS

	Unit	600 mm Single 20 W £	600 mm Twin 20 W £	1200 mm Single 40 W £	1200 mm Twin 40 W £	1500 mm Single 65 W £
Batten type; surface mounting complete with tube	No.	6·49	10·89	8·43	15·65	9·93
Ditto with prismatic diffuser.	,,	9·55	15·35	12·98	23·44	14·95

		1500 mm Twin 65 W £	1800 mm Single 75 W £	1800 mm Twin 75 W £	2400 mm Single 100 W £	2400 mm Twin 100 W £
Batten type.	,,	18·63	11·48	20·39	15·73	26·74
Ditto with prismatic diffuser.	,,	28·37	17·74	31·86	23·82	40·90

	Unit	300 × 1200 mm 40 W – 1 lamp £	300 × 1200 mm 40 W – 2 lamp £	300 × 1800 mm 75 W – 1 lamp £
Modular type; recessed mounting with opal diffuser and tubes	No.	33·27	40·71	40·78

		300 × 1800 mm 75 W – 2 lamp £	600 × 600 mm 40 W – 2 lamp £	600 × 600 mm 40 W – 3 lamp £
	,,	49·90	52·79	58·30

		600 × 1200 mm 40 W – 3 lamp £	600 × 1200 mm 40 W – 4 lamp £	600 × 1800 mm 75 W – 2 lamp £
	,,	60·16	67·46	69·32

		600 × 1800 mm 75 W – 3 lamp £	600 × 1800 mm 75 W – 4 lamp £	
	,,	76·24	85·40	

Electrical Installations – Market Prices of Materials

FITTINGS AND ACCESSORIES

FLUORESCENT LIGHT FITTINGS *continued*

		600 mm single 20 W £	600 mm twin 20 W £	1200 mm single 40 W £	1200 mm twin 40 W £
Corrosion resistant; G.R.P. body, gasket sealed, acrylic diffuser, complete with tubes to F.P. 54 & B.S. 4533-2-2	Unit No.	22·01	27·22	23·63	29·05
		1500 mm single 65 W £ 22·54	*1500 mm twin 65 W £* 36·34	*1800 mm single 75 W £* 27·76	*1800 mm twin 75 W £* 42·97
Flameproof to I.P. 65 B.S. 229 (1957) & B.S. 889 (1965) complete with tubes	,,	*600 mm single 40 W £* 90·32	*600 mm twin 40 W £* 125·58	*1200 mm single 40 W £* 90·65	*1200 mm twin 40 W £* 141·06
		1500 mm single 65 W £ 96·69	*1500 mm twin 65 W £* 142·67	*1800 mm single 75 W £* 101·59	*1800 mm twin 75 W £* 149·61
Adjustable brackets	,,	£ 2·26			

		Round pattern 4 W £	Square pattern 4 W £	Rectangular pattern 4 W £
Self-contained non-maintained emergency fitting, 3 hour duration complete with glass diffuser and tubes	,,	43·78	44·53	43·78

		Exit £	Emergency Exit £	Fire Exit £
Exit sign, normal and non-maintained type, surface mounting, complete with tube, to B.S. 5266 Size 430 × 195 × 120 mm . .	,,	70·22	70·22	70·22

ELECTRICAL INSTALLATIONS

Constants of Labour

Constants of labour are given for the major items of work for which prices are given in 'Prices for Measured Work'. 'Prices for Measured Work' have been based on the quantities of materials required at the rates given in 'Market Prices of Materials' plus 10% to cover overheads and 5% for profit, and the 'Constants' of labour priced at a man hour rate of £6·92, this rate including an allowance of 40% to cover site and head office overheads and preliminary items and 5% for profit.

Reference to the 'Constants' should assist the reader

(1) To compare the prices given to those used in his own organization.
(2) To calculate the effect of changes in wage rates.
(3) To calculate analogous prices for work similar to but differing in detail from the examples given.

SWITCHGEAR AND DISTRIBUTION

BUSBAR CHAMBERS SHEET STEEL CASE FIXED TO BRICKWORK
man hours each

	100 amp	200 amp	400 amp	600 amp	800 amp	1200 amp
1524 mm section	3·00	3·25	3·50	4·00	4·25	4·75

DISTRIBUTION BOARDS (NON FULLY SHROUDED) SHEET STEEL CASE; FIXED AND CONNECTED
man hours each

	2 way	3 way	4 way	6 way	8 way	10 way
20 amp S.P. & N.			1·10	1·35	1·60	1·85
T.P. & N.		1·60	1·90	2·40	2·80	3·20
30 amp S.P. & N.			1·25	1·50	1·75	2·00
T.P. & N.		1·65	1·75	2·40	2·80	3·30
60 amp S.P. & N.	1·15	1·45	2·00	2·15	2·30	
T.P. & N.	1·50	1·75	2·00	2·30	2·75	3·00
100 amp T.P. & N.	1·90	2·40	2·75	3·25	3·50	3·75
200 amp T.P. & N.	3·00	3·75	4·75	5·75		
300 amp T.P. & N.	3·00	3·75	4·75	5·75		

	12 way	14 way	16 way	18 way	20 way	24 way
20 amp S.P. & N.	2·10	2·35	2·60	3·00		
T.P. & N.	3·80	4·40	4·80	5·20	5·50	6·00
30 amp S.P. & N.	2·25					
T.P. & N.	3·80	4·40	4·80	5·20	5·50	
60 amp T.P. & N.	3·80					
100 amp T.P. & N.	4·00					

SWITCHES AND ISOLATORS; FIXED AND CONNECTED
man hours each

	20 amp	30 amp	60 amp	100 amp
D.P.	1·00	1·00	1·50	1·50
T.P.	1·25	1·25	1·75	1·75
T.P. & N.	1·25	1·25	1·75	1·75

SWITCHFUSES; FIXED AND CONNECTED
man hours each

	20 amp	30 amp	60 amp	100 amp
S.P. & N.	1·00	1·25	1·50	1·75
D.P.	1·00	1·25	1·50	1·75
T.P.	1·50	1·75	2·00	2·50
T.P. & N.	1·75	2·00	2·25	2·50

SWITCHGEAR AND DISTRIBUTION

FUSESWITCHES; FIXED AND CONNECTED
man hours each

	60 amp	100 amp	200 amp	300 amp	400 amp	600 amp	800 amp
S.P. & N.	1·50	1·75	2·25				
D.P.	1·50	1·75	2·25				
T.P.	1·75	2·25	2·50	2·75	3·00	3·75	4·50
T.P. & N.	1·75	2·25	2·50	2·75	3·00	3·75	4·50
T.P.S. & N.			2·75		3·75	4·50	4·75

262   *Electrical Installations – Constants of Labour*

STEEL CONDUIT

HEAVY GAUGE SCREWED STEEL CONDUIT AND FITTINGS SURFACE FIXED WITH SADDLES
man hours per 100 m

	16 mm	20 mm	25 mm	32 mm	40 mm	50 mm
Lighting and power circuits	50·00	50·00	55·00	60·00	70·00	115·00

CONDUIT BOXES – CAST IRON ADAPTABLE
man hours per 100

Square pattern	150 × 150 × 75 mm 91·70	150 × 150 × 100 mm 91·70	225 × 225 × 75 mm 100·00	300 × 300 × 100 mm 105·00		
Rectangular pattern	150 × 150 × 100 mm 91·70	150 × 100 × 50 mm 91·70	150 × 100 × 75 mm 91·70	225 × 150 × 75 mm 98·30	225 × 150 × 100 mm 100·00	300 × 150 × 75 mm 105·00

CONDUIT BOXES – SHEET STEEL
man hours per 100

Square pattern	50 × 50 × 37·5 mm 88·30	75 × 75 × 37·5 mm 88·30	75 × 75 × 50 mm 88·30	75 × 75 × 75 mm 90·00	100 × 100 × 50 mm 90·00	150 × 150 × 50 mm 90·00
	150 × 150 × 75 mm 91·70	150 × 150 × 100 mm 91·70	225 × 225 × 50 mm 98·30	225 × 225 × 100 mm 100·00	300 × 300 × 100 mm 105·00	
Rectangular pattern	100 × 75 × 37·5 mm 88·30	100 × 75 × 50 mm 88·30	150 × 75 × 50 mm 88·30	150 × 75 × 75 mm 90·00	150 × 100 × 75 mm 90·00	225 × 75 × 50 mm 90·00
	225 × 150 × 75 mm 91·70	225 × 150 × 100 mm 91·70	300 × 150 × 50 mm 98·30	300 × 150 × 75 mm 100·00	300 × 225 × 100 mm 105·00	

JUNCTION BOXES, WEATHERPROOF SURFACE MOUNTED AND CONNECTED
man hours each

Box with terminal block	*10 way* 1·33	*20 way* 1·97	*40 way* 3·34	*60 way* 4·69	*100 way* 7·70	*150 way* 10·72
	200 way 14·73	*250 way* 18·00	*300 way* 21·50	*350 way* 22·50	*400 way* 28·50	*450 way* 34·50

CABLE TRUNKING

STEEL CABLE TRUNKING FIXED WITH PLUGS AND SCREWS
man hours per 100 m

	50 × 50 mm	75 × 50 mm	75 × 75 mm	100 × 50 mm	100 × 75 mm	100 × 100 mm	150 × 75 mm	150 × 100 mm	150 × 150 mm
Single compartment	42·00	50·00	54·00	54·00	67·00	75·00	88·00	96·00	108·00
Double compartment	45·00	55·00	60·00	60·00	75·00	83·00	95·00	105·00	120·00

FITTINGS
man hours per 100

	50 × 50 mm	75 × 50 mm	75 × 75 mm	100 × 50 mm	100 × 75 mm	100 × 100 mm	150 × 75 mm	150 × 100 mm	150 × 150 mm
90° angle single compartment	50·00	54·00	62·00	62·00	67·00	67·00	75·00	79·00	83·00
Tee	71·00	71·00	75·00	75·00	83·00	83·00	92·00	92·00	100·00
Flanged connection	8·00	8·00	12·00	12·00	12·00	15·00	15·00	16·00	16·00
Sealed end	8·00	8·00	8·00	8·00	12·00	12·00	12·00	12·00	12·00

PVC TRUNKING WITH CLIP ON LID; FIXED
man hours per 100 m

	50 × 50 mm	75 × 50 mm	75 × 75 mm	100 × 50 mm	100 × 75 mm	100 × 100 mm	150 × 75 mm	150 × 100 mm	150 × 150 mm
Single compartment	25·00	25·00	25·00	30·00	35·00	35·00	40·00	40·00	45·00

FITTINGS
man hours per 100

	50 × 50 mm	75 × 50 mm	75 × 75 mm	100 × 50 mm	100 × 75 mm	100 × 100 mm	150 × 75 mm	150 × 100 mm	150 × 150 mm
Cross over	18·00	18·00	18·00	20·00	20·00	25·00	25·00	25·00	25·00
Stop end	5·00	5·00	5·00	5·00	5·00	8·00	8·00	8·00	8·00
Flanged coupling	20·00	20·00	20·00	30·00	30·00	30·00	40·00	40·00	40·00
Internal coupling	6·00	6·00	6·00	8·00	8·00	8·00			
External coupling	6·00	6·00	6·00	8·00	8·00	8·00	10·00	10·00	10·00
Angle	10·00	10·00	10·00	10·00	15·00	15·00	15·00	20·00	20·00
Tee	15·00	15·00	15·00	20·00	20·00	20·00	25·00	25·00	25·00
Dividing strip	5·00	5·00	7·00						

Electrical Installations – Constants of Labour

CABLE TRAY

LIGHT DUTY STEEL TRAY FIXED TO CONCRETE
man hours per 100 m

Width

Tray	50 mm	75 mm	100 mm	150 mm	200 mm	225 mm
	65·00	65·00	70·00	70·00	75·00	75·00

	300 mm	375 mm	450 mm	600 mm	750 mm	900 mm
	90·00	90·00	125·00	150·00	165·00	185·00

TRAY FITTINGS
man hours per 100

	50 mm	75 mm	100 mm	150 mm	200 mm	225 mm
Bend	30·00	35·00	35·00	40·00	50·00	50·00
Tee	40·00	45·00	50·00	50·00	65·00	75·00

	300 mm	375 mm	450 mm	600 mm	750 mm	900 mm
Bend	65·00	75·00	90·00	90·00	100·00	100·00
Tee	80·00	85·00	100·00	125·00	125·00	125·00

Electrical Installations – Constants of Labour 265

LADDER RACK

STEEL LADDER RACK FIXED TO CONCRETE
man hours per 100 m

	150 mm	200 mm	Width 300 mm	400 mm	600 mm	1000 mm
Ladder rack	75·00	75·00	100·00	125·00	150·00	175·00

LADDER RACK FITTINGS
man hours per 100

Bend, radius	30·00	40·00	60·00	100·00	120·00	120·00
Bend, outside radius		30·00	40·00	100·00	120·00	
Tee junction		40·00	60·00	100·00	120·00	140·00
Cross junction		50·00	70·00	120·00	140·00	
Riser	40·00	60·00	70·00	100·00	120·00	120·00
End connection	50·00	70·00	70·00	120·00	120·00	120·00
Universal coupling	50·00	70·00	70·00	120·00	120·00	
Cantilever arm	30·00					
Wall bracket	30·00					

CABLE

SINGLE CORE PVC INSULATED CABLE
man hours per 100 m

	1·00 mm²	1·50 mm²	2·50 mm²	4 mm²	6 mm²	10 mm²
450/750 volt grade drawn into conduit	2·00	2·50	3·00	3·50	4·00	6·00

	16 mm²	25 mm²	35 mm²	50 mm²	70 mm²	95 mm²
	8·00	12·50	17·50	25·50	36·00	49·00

	1·00 mm²	1·50 mm²	2·50 mm²	4 mm²	6 mm²	10 mm²
450/750 volt grade laid in trunking	1·50	2·00	2·50	3·00	3·50	5·00

	16 mm²	25 mm²	35 mm²	50 mm²	70 mm²	95 mm²
	7·00	10·00	16·00	24·00	30·00	40·00

	120 mm²	150 mm²	185 mm²	240 mm²	300 mm²	400 mm²	500 mm²	630 mm²
450/750 volt grade drawn into ducts	30·00	31·50	33·50	37·00	42·00	46·00	52·00	60·00

ARMOURED PVC INSULATED AND SHEATHED CABLE FIXED WITH CLIPS
man hours per 100 m

600/1000 volt grade

	1·5 mm²	2·5 mm²	4 mm²	6 mm²	10 mm²	16 mm²
2 core	12·00	12·00	17·00	17·00	25·00	33·00
3 core	17·00	17·00	25·00	25·00	37·00	47·00
4 core	17·00	17·00	25·00	25·00	37·00	47·00
5 core	20·00	25·00	33·00	50·00		
7 core	20·00	25·00	33·00	50·00		
8 core	20·00	25·00	33·00			
10 core	25·00	33·00	45·00			
12 core	25·00	33·00	45·00			
14 core	33·00	45·00	50·00			
22 core	45·00	50·00	60·00			
37 core	60·00	70·00				
48 core	70·00	70·00				

MINERAL INSULATED COPPER SHEATHED CABLE FIXED WITH SADDLES
man hours per 100 m heavy duty

600 volt grade	2L 1·00	2L 1·50	2L 2·50	2L 4	3L 1·00
Bare	18·00	21·00	30·00	41·00	23·00
PVC sheathed	22·00	26·00	35·00	45·00	28·00

	3L 1·50	4L 1	4L 1·50	7L 1	7L 1·50
Bare	28·00	26·00	33·00	40·00	50·00
PVC sheathed	33·00	31·00	38·00	45·00	56·00

Electrical Installations – Constants of Labour 267

CABLE

MINERAL INSULATED COPPER SHEATHED CABLE FIXED WITH SADDLES
man hours per 100 m heavy duty

1000 volt grade	1H 6	1H 10	1H 16	1H 25	1H 35
Bare	30·00	40·00	55·00	73·00	78·00
PVC sheathed	33·00	45·00	60·00	80·00	85·00
	1H 50	1H 70	1H 95	1H 120	1H 150
Bare	83·00	100·00	112·00	125·00	150·00
PVC sheathed	90·00	106·00	120·00	133·00	158·00
	4H 1·5	4H 2·5	4H 4	4H 6	4H 10
Bare	51·00	65·00	70·00	75·00	83·00
PVC sheathed	56·00	70·00	76·00	83·00	100·00
	4H 16	7H 1·5	7H 2·5	12H 2·5	19H 1·5
Bare	100·00	73·00	80·00	91·00	100·00
PVC sheathed	120·00	81·00	88·00	101·00	110·00

TERMINATIONS FOR MINERAL INSULATED COPPER SHEATHED CABLE
man hours per 100

Light duty 600 volt grade	2L 1·0	2L 1·5	2L 2·5	2L 4	3L 1·0
Low temperature to 135° Celsius	24·00	26·00	28·00	30·00	26·00
Low temperature with PVC shroud	28·00	30·00	32·00	34·00	30·00
	3L 1·5	4L 1·0	4L 1·5	7L 1·0	7L 1·5
Low temperature to 135° Celsius	28·00	28·00	30·00	30·00	35·00
Low temperature with PVC shroud	32·00	32·00	34·00	34·00	40·00

Heavy duty 1000 volt grade	1H 6	1H 10	1H 16	1H 25	1H 35
Low temperature to 135° Celsius	24·00	27·00	30·00	35·00	36·00
Low temperature with PVC shroud	28·00	31·00	34·00	37·00	40·00
	1H 50	1H 70	1H 95	1H 120	1H 150
Low temperature to 135° Celsius	40·00	44·00	48·00	54·00	56·00
Low temperature with PVC shroud	44·00	48·00	52·00	56·00	60·00

	2H 1·5	2H 2·5	2H 4	2H 6	2H 10	2H 16	2H 25
Low temperature to 135° Celsius	33·00	33·00	37·00	37·00	37·00	40·00	40·00
Low temperature with PVC shroud	37·00	37·00	41·00	41·00	41·00	44·00	44·00
	3H 1·5	3H 2·5	3H 4	3H 6	3H 10	3H 16	3H 25
Low temperature to 135° Celsius	33·00	37·00	39·00	39·00	39·00	42·00	42·00
Low temperature with PVC shroud	37·00	41·00	43·00	43·00	43·00	46·00	46·00
	4H 1·5	4H 2·5	4H 4	4H 6	4H 10	4H 16	4H 25
Low temperature to 135° Celsius	35·00	39·00	40·00	40·00	40·00	44·00	44·00
Low temperature with PVC shroud	39·00	43·00	44·00	44·00	44·00	48·00	48·00
	7H 1·5	7H 2·5	12H 2·5	19H 1·5			
Low temperature to 135° Celsius	50·00	50·00	70·00	116·00			
Low temperature with PVC shroud	54·00	54·00	74·00	120·00			

FITTINGS AND ACCESSORIES

LIGHTING SWITCHES
man hours each

	1 gang	2 gang	3 gang	4 gang	6 gang	9 gang	10 gang	12 gang		
5 amp metal clad surface mounted switch fixed and connected 1 and 2 way	0·50	0·70	1·00	1·17	1·50	2·50	2·80	3·00		
As last but weatherproof 1 and 2 way	0·50	0·70	1·00	1·17						

	1 gang	2 gang	3 gang	4 gang	6 gang	8 gang	9 gang	12 gang	18 gang	24 gang
5 amp modular type switch with steel or plastic coverplates and steel box, fixed and connected 1 and 2 way	0·50	0·70	1·00	1·17	1·50	2·20	2·50	3·00	3·60	4·60

Ceiling switches, moulded white plastic fixed and connected

5 amp 1 way	5 amp 2 way	15 amp 1 way	30 amp 1 way
0·25	0·30	0·40	0·40

SOCKET OUTLETS
man hours each

13 amp socket outlet, switched or unswitched, flush or surface; fixed and connected .

1 gang	2 gang
0·45	0·45

CONNECTION UNITS
man hours each

Moulded plastic pattern with steel box; fixed and connected . . . . 0·34
Telephone outlet plate with box; fixed and connected single or double outlet . 0·30

SHAVER SUPPLY UNITS
man hours each

Single voltage, flush or surface mounted; fixed and connected . . . 0·45
Dual voltage, ditto . . . . 0·50

LAMPHOLDER
man hours each

Lightning pendant fitting comprising rose, lampholder and 3-core flex fixed and connected . . . . 0·50

COOKER CONTROL UNITS
man hours each

45 amp with 13 amp socket outlet; moulded plastic or steel cover plate and box, flush or surface; fixed and connected . . . . 0·33

Electrical Installations – Constants of Labour

FITTINGS AND ACCESSORIES

CONSUMER UNITS
man hours each

	2 way	3 way	4 way	6 way	8 way
All insulated moulded plastic, DP/AC switch; fixed and connected	1·25	1·50	1·50	1·83	2·00

	6 way	8 way	12 way
Metalclad, DP/AC switch; fixed and connected	1·83	2·00	3·00

IMMERSION HEATERS
man hours each

3 kW up to 915 mm long fitted and connected . . . . 0·75
Thermostat . . . . 0·25

LIGHT FITTINGS

FLUORESCENT FITTINGS FIXED AND CONNECTED INCLUDING TUBES
man hours each

	600 mm single or twin	1200 mm single or twin	1500 mm single or twin	1800 mm single or twin	2400 mm single or twin		
Batten type with diffuser, surface mounted	0·50	0·65	0·75	1·00	1·25		

	300 × 1200 mm	300 × 1800 mm	600 × 600 mm	600 × 1200 mm	600 × 1800 mm 2 tubes	600 × 1800 mm 3 tubes	600 × 1800 mm 4 tubes
Modular type with diffuser, recessed mounting	1·50	1·75	1·50	1·75	2·00	2·15	2·25

	600 mm single or twin	1200 mm single or twin	1500 mm single or twin	1800 mm single or twin
Corrosion resistant batten type	0·50	0·65	0·75	1·00
Flameproof type	1·00	1·25	1·50	1·75

Emergency fittings Non-maintained unit 3 hour duration	0·50
Exit signs	0·50

ELECTRICAL INSTALLATIONS

Prices for Measured Work

These prices are intended to apply to new work in the London area and include allowances for all overhead charges, preliminary items and profit. The prices are for reasonable quantities of work and the user should make suitable adjustments if the quantities are especially small or especially large. Adjustments may also be required for locality (e.g. outside London) and for the market conditions (e.g. volume of work on hand or on offer) at the time of use.

Prices are given in metric terms. They are based on an inclusive labour rate of £6·92 per man hour and on the prices of materials given in 'Market Prices of Materials' or as indicated.

The labour element includes allowances of 40% to cover site and head office overheads and preliminary items and 5% for profit.

The material element includes allowances of 10% to cover overheads and 5% for profit.

The labour element, inclusive of overheads and profit, in any particular item can be ascertained by deducting the amount which appears in italics below the measured price from the measured price. The amount shown in italics represents the value of material content contained in the measured price including relevant allowances for waste, overheads and profit.

For details of the inclusive labour rate and assumptions made in calculating 'Prices for Measured Work' see 'Directions'.

Prices do not allow for any charges in respect of V.A.T.

Electrical Installations – Prices for Measured Work

SWITCHGEAR AND DISTRIBUTION

BUSBAR CHAMBERS
(fixed to brickwork)

		100 amp £	200 amp £	400 amp £	600 amp £	800 amp £	1200 amp £
Sheet steel case 1524 mm long	Unit No.	118·90 *98·15*	162·50 *140·00*	202·45 *178·22*	308·25 *280·57*	446·00 *416·59*	603·00 *570·15*

DISTRIBUTION BOARDS
(sheet steel case, fully shrouded, fixed including connections)

415/440 volt		*4* way £	*6* way £	*8* way £	*10* way £	*12* way £	*14* way £	*16* way £
20 amp H.R.C. fuse								
S.P. & N.	Unit No.	54·30 *47·57*	66·85 *58·19*	79·10 *68·71*	92·95 *80·84*	107·00 *93·08*	136·00 *120·51*	150·00 *132·86*
T.P. & N.	„	94·30 *83·91*	113·40 *100·28*	132·75 *116·84*	159·65 *140·96*	183·00 *159·54*	254·00 *228·53*	333·00 *303·46*
30 amp H.R.C. fuse								
S.P. & N.	„	58·65 *50·68*	75·35 *65·64*	94·30 *82·88*	110·60 *97·45*	130·75 *115·85*		
T.P. & N.	„	120·70 *107·53*	155·75 *139·14*	188·70 *169·31*	223·85 *201·71*	259·30 *233·02*		

		2 way £	*3* way £	*4* way £	*6* way £	*8* way £	*10* way £
60 amp H.R.C. fuse							
T.P. & N.	„	140·75 *129·68*	152·00 *137·50*	166·25 *148·95*	220·00 *199·20*	269·75 *245·52*	371·00 *344·83*
100 amp H.R.C. fuse							
T.P. & N.	„	259·50 *246·38*	279·00 *262·32*	306·25 *287·20*	385·60 *363·13*	440·00 *415·77*	

Electrical Installations – Prices for Measured Work

SWITCHGEAR AND DISTRIBUTION

DISTRIBUTION BOARDS *continued*
(sheet steel case, non fully shrouded, rewirable, fixed including connections)

415/440 volt		4 way £	6 way £	8 way £	10 way £	12 way £	14 way £	16 way £	18 way £
20 amp ways S.P. & N..	Unit No.	30·80 23·19	38·55 29·21	44·77 33·70	52·80 40·00	58·35 43·82	72·00 55·73	86·00 68·09	98·75 78·00

		3 way £	4 way £	6 way £	8 way £	10 way £	12 way £
T.P. & N..	,,	45·34 34·27	54·20 41·04	68·95 52·34	85·90 66·52	99·00 76·85	115·00 88·61

		14 way £	16 way £	18 way £	20 way £	24 way £
T.P. & N..	,,	181·15 150·70	218·00 184·22	235·50 194·49	258·00 219·73	284·00 242·52

		4 way £	6 way £	8 way £	10 way £	12 way £
30 amp ways S.P. & N..	,,	36·20 27·55	42·65 32·26	48·80 36·68	63·50 49·68	71·85 56·28

		3 way £	4 way £	6 way £	8 way £	10 way £
T.P. & N..	,,	55·65 44·22	64·60 52·47	87·40 70·79	105·60 86·29	125·80 103·00

		12 way £	14 way £	16 way £	18 way £	20 way £
T.P. & N..	,,	146·00 119·75	221·00 190·47	246·00 212·45	263·00 226·80	286·00 247·97

		2 way £	3 way £	4 way £	6 way £	8 way £	10 way £	12 way £
60 amp ways S.P. & N..	,,	49·65 41·68	59·40 49·36	68·80 54·98	92·80 77·90	111·50 95·60		
T.P. & N..	,,	80·20 69·84	105·80 93·68	120·85 107·00	171·00 155·09	219·00 199·53	253·00 232·65	303·00 277·14
100 amp ways T.P. & N..	,,	122·30 109·16	161·00 144·24	209·00 190·11	274·00 251·43	370·00 346·14	441·00 415·00	511·00 483·24
200 amp ways (H.R.C.) T.P. & N..	,,	411·30 390·50	495·00 469·50	672·00 639·50	863·00 823·67			
300 amp ways T.P. & N..	,,	511·00 490·40	645·00 619·15	872·00 839·22	1162·00 1122·40			

SWITCHGEAR AND DISTRIBUTION

SWITCHES AND ISOLATORS
(415/440 volt fixed to brickwork including connections)

Cast iron enclosure	Unit	20 amp £	30 amp £	60 amp £	100 amp £
D.P.	No.	19·70	21·40	34·60	49·80
		12·77	*14·46*	*24·23*	*39·42*
T.P.	,,	23·60	27·00	41·10	56·90
		14·96	*18·35*	*30·68*	*46·48*
T.P. & N.	,,	25·30	28·85	44·00	61·50
		16·66	*20·20*	*33·58*	*49·41*
Sheet steel case					
D.P.	,,	20·25	23·08	34·10	45·60
		13·32	*16·16*	*23·72*	*35·23*
T.P.	,,	23·00	26·10	40·27	53·80
		14·37	*17·44*	*28·16*	*41·68*
T.P. & N	,,	23·80	27·00	42·66	57·05
		15·15	*18·39*	*30·55*	*44·95*

SWITCHFUSES
(415/440 volt, fixed to brickwork including connections)

Cast iron enclosure	Unit	20 amp £	30 amp £	60 amp £	100 amp £
S.P. & N.	No.	22·70	26·00	42·00	66·50
		15·78	*19·06*	*31·59*	*54·41*
D.P.	,,	23·90	27·44	45·50	71·00
		16·99	*20·52*	*35·10*	*58·94*
T.P.	,,	28·00	34·05	60·85	90·40
		19·36	*25·40*	*48·75*	*78·27*
T.P. & N.	,,	29·70	36·00	63·00	93·60
		21·06	*27·37*	*50·90*	*81·52*
Sheet steel case					
S.P. & N.	,,	22·00	27·00	40·00	56·90
		15·13	*18·33*	*29·45*	*44·56*
D.P.	,,	23·00	28·50	42·70	63·80
		16·04	*19·83*	*32·33*	*51·70*
T.P.	,,	28·45	35·45	54·30	84·20
		18·05	*23·33*	*40·43*	*66·89*
T.P. & N.	,,	31·20	38·70	58·35	87·25
		19·10	*24·86*	*42·77*	*69·95*

SWITCHGEAR AND DISTRIBUTION

FUSE SWITCHES
(415/440 volt, fixed and connected)

Sheet steel case	Unit	60 amp £	100 amp £	200 amp £	300 amp £	400 amp £	600 amp £	800 amp £
S.P. & N.	No.	64·70	87·40	121·50				
		54·31	*75·31*	*105·84*				
D.P.	,,	65·00	87·40	121·50				
		54·39	*75·31*	*105·84*				
T.P.	,,	69·50	93·80	127·00	213·00	300·00	494·00	549·00
		57·44	*78·27*	*109·70*	*194·32*	*279·00*	*467·80*	*518·10*
T.P. & N.	,,	75·60	101·00	133·60	221·00	306·00	504·00	559·00
		63·49	*85·27*	*116·27*	*201·70*	*285·40*	*478·00*	*528·15*
T.P.S. & N.	,,			158·50		347·00	603·00	664·00
				139·50		*321·10*	*572·00*	*631·00*

CONTROL GEAR

ALL INSULATED COMPLETE CONSUMER UNITS
(surface mounting pattern and fixing to any surface including connections)

60 amp 240 volt D.P./A.C. switched, all insulated with moulded plastic case and cover	Unit	2 way £	3 way £	4 way £	6 way £	8 way £
Fitted rewirable fuses	No.	13·85	17·25	18·90	23·45	27·55
		5·21	*6·87*	*8·49*	*10·80*	*13·71*
Fitted cartridge fuses	,,	14·45	18·13	20·00	25·20	29·90
		5·80	*7·75*	*9·66*	*12·54*	*16·03*
Fitted miniature circuit breakers	,,	18·85	24·70	28·90	37·85	47·45
		10·19	*14·33*	*18·53*	*25·20*	*33·59*

METAL CLAD CONSUMER UNITS
(flush or surface mounting pattern and fixing to any surface including connections)

60 amp 240 volt D.P./A.C. switched, metal clad with moulded plastic fuse cover	Unit	2 way £	3 way £	4 way £	6 way £	8 way £	12 way £
Fitted rewirable fuses	No.	14·90	19·00	20·25	24·95	28·75	
		6·25	*8·62*	*9·88*	*12·29*	*14·90*	
Fitted cartridge fuses	,,	15·50	19·85	21·40	26·70	31·10	
		6·83	*9·48*	*11·03*	*14·04*	*17·23*	
Fitted miniature circuit breakers	,,	19·85	26·45	30·20	39·85	48·60	
		11·22	*16·07*	*19·81*	*27·20*	*34·78*	
100 amp 240 volt D.P./A.C. ditto							
Fitted rewirable fuses	,,				27·50	31·50	47·65
					14·86	*17·66*	*26·88*
Fitted cartridge fuses	,,				29·27	33·80	51·15
					16·61	*19·98*	*30·38*
Fitted miniature circuit breakers	,,				42·40	51·38	77·45
					29·76	*37·54*	*56·69*

Electrical Installations – Prices for Measured Work 277

STEEL CONDUIT

HEAVY GAUGE SCREWED, WELDED STEEL CONDUIT
(including standard pattern boxes and fittings, surface fixed with and including spacer bar type saddles)

	Unit	16 mm £	20 mm £	25 mm £	32 mm £	40 mm £	50 mm £
Black enamelled	m	4·57	4·75	5·53	6·54	7·89	12·90
		1·11	*1·29*	*1·72*	*2·39*	*3·05*	*4·96*
Galvanized	„	5·06	5·31	6·28	7·27	8·83	13·86
		1·60	*1·85*	*2·47*	*3·12*	*3·99*	*5·90*

CONDUIT FITTINGS AND COVERS

Although the rates for conduit include an allowance for fittings, as required by the Standard Method of Measurement, a range of prices is given below for fittings which may be used in other conditions.

	Unit	16 mm £	20 mm £	25 mm £	32 mm £
Back outlet box					
Black enamelled	No.	2·12	2·12	2·55	
		0·95	*0·95*	*1·37*	
Galvanized	„	2·42	2·42	2·97	
		1·24	*1·24*	*1·80*	
Terminal box					
Black enamelled	„	1·98	1·98	2·18	3·48
		0·81	*0·81*	*0·98*	*2·31*
Galvanized	„	2·22	2·22	2·47	4·04
		1·05	*1·05*	*1·30*	*2·86*
Terminal and back outlet box					
Black enamelled	„		2·34	2·75	
			1·17	*1·58*	
Galvanized	„		2·65	3·11	
			1·48	*1·94*	
Angle box					
Black enamelled	„	2·05	2·05	2·27	3·81
		0·88	*0·88*	*1·09*	*2·63*
Galvanized	„	2·31	2·31	2·60	4·44
		1·13	*1·13*	*1·42*	*3·26*
Through box					
Black enamelled	„	2·05	2·05	2·27	3·81
		0·88	*0·88*	*1·09*	*2·63*
Galvanized	„	2·31	2·31	2·60	4·44
		1·13	*1·13*	*1·42*	*3·26*
Through way and back outlet box					
Black enamelled	„		2·41	3·08	
			1·24	*1·90*	
Galvanized	„		2·76	3·63	
			1·59	*2·46*	

CONDUIT BOXES

GREY CAST IRON ADAPTABLE BOXES
(with heavy covers fixed to brickwork)

Square pattern	Unit	150 × 150 × 75 mm £	150 × 150 × 100 mm £	225 × 225 × 75 mm £	300 × 300 × 100 mm £
Black	No.	12·44	14·60	20·10	39·55
		6·09	*8·25*	*13·17*	*32·28*
Galvanized	,,	15·20	18·08	25·50	50·35
		8·85	*11·73*	*18·56*	*43·08*

Rectangular pattern		150 × 75 × 50 mm £	150 × 100 × 50 mm £	150 × 100 × 75 mm £	225 × 150 × 75 mm £
Black	,,	9·90	11·82	12·35	17·30
		3·56	*5·47*	*6·01*	*10·49*
Galvanized	,,	11·30	14·10	15·65	21·70
		4·95	*7·76*	*9·31*	*14·88*

		225 × 150 × 100 mm £	300 × 150 × 75 mm £
Black	,,	21·75	26·60
		14·85	*19·35*
Galvanized	,,	27·50	36·10
		20·59	*28·82*

Electrical Installations – Prices for Measured Work

CONDUIT BOXES

SHEET STEEL
(standard gauge)

		50 × 50 × 37·5 mm £	75 × 75 × 37·5 mm £	75 × 75 × 50 mm £	75 × 75 × 75 mm £
Square pattern	Unit				
Black	No.	6·78	6·90	6·92	7·26
		0·67	*0·79*	*0·81*	*1·03*
Galvanized	,,	7·05	7·22	7·25	7·67
		0·94	*1·11*	*1·14*	*1·44*

		100 × 100 × 50 mm £	150 × 150 × 50 mm £	150 × 150 × 75 mm £	150 × 150 × 100 mm £
Black	,,	7·15	7·52	7·90	8·60
		0·82	*1·29*	*1·55*	*2·28*
Galvanized	,,	7·57	8·10	8·60	9·67
		1·34	*1·87*	*2·26*	*3·32*

		225 × 225 × 50 mm £	225 × 225 × 100 mm £	300 × 300 × 100 mm £	
Black	,,	9·50	10·35	13·25	
		2·70	*3·43*	*5·98*	
Galvanized	,,	10·70	11·90	16·00	
		3·90	*4·98*	*8·71*	

		100 × 75 × 37·5 mm £	100 × 75 × 50 mm £	150 × 75 × 50 mm £	150 × 75 × 75 mm £
Rectangular pattern					
Black	,,	6·92	6·94	7·05	7·36
		0·81	*0·83*	*0·94*	*1·13*
Galvanized	,,	7·25	7·30	7·46	7·88
		1·14	*1·19*	*1·35*	*1·65*

		150 × 100 × 75 mm £	225 × 75 × 50 mm £	225 × 150 × 75 mm £	225 × 50 × 100 mm £
Black	,,	7·58	7·63	8·72	9·06
		1·35	*1·40*	*2·37*	*2·71*
Galvanized	,,	8·16	8·26	9·82	10·32
		1·93	*2·03*	*3·47*	*3·97*

		300 × 150 × 50 mm £	300 × 150 × 75 mm £	300 × 225 × 100 mm £	
Black	,,	9·78	9·97	11·71	
		2·98	*3·05*	*4·44*	
Galvanized	,,	11·12	11·38	13·74	
		4·32	*4·46*	*6·47*	

Electrical Installations – Prices for Measured Work

JUNCTION BOXES

WEATHERPROOF JUNCTION BOXES

Sheet steel enclosure, galvanized finish with rail mounted plastic terminal blocks, gland plates and gaskets, side hung door with padlock; fixed and connected	Unit	10 way £	20 way £	40 way £	60 way £	100 way £	150 way £
	No	40·55	47·70	69·25	84·00	128·00	162·20
		31·36	34·05	46·12	51·51	74·51	88·01

		200 way £	250 way £	300 way £	350 way £	400 way £	450 way £
	,,	222·50	260·20	298·00	318·50	373·50	428·00
		120·52	135·62	149·33	162·77	176·21	189·65

Electrical Installations – Prices for Measured Work

CABLE TRUNKING

STEEL CABLE TRUNKING GALVANIZED OR STOVE ENAMELLED
(fixed with plugs and screws)

	Unit	50 × 50 mm £	75 × 50 mm £	75 × 75 mm £	100 × 50 mm £	100 × 75 mm £	100 × 100 mm £	150 × 50 mm £	150 × 100 mm £	150 × 150 mm £
Single compartment	m	5·48	6·29	6·84	6·98	8·03	9·02	10·97	12·60	14·40
		2·57	*2·83*	*3·10*	*3·24*	*3·39*	*3·83*	*4·88*	*5·92*	*6·91*
Flanged connector	No.	1·41	1·42	1·42	1·76	1·76	1·81	1·81	2·04	2·43
		0·30	*0·31*	*0·31*	*0·38*	*0·38*	*0·38*	*0·43*	*0·43*	*0·66*
Sealing end	,,	1·41	1·42	1·42	1·76	1·76	1·81	1·81	2·04	2·43
		0·30	*0·31*	*0·31*	*0·38*	*0·38*	*0·38*	*0·43*	*0·43*	*0·66*
Double compartment	m	6·47	7·41	8·16	8·20	9·54	10·74	11·60	14·55	17·25
		3·36	*3·60*	*4·01*	*4·05*	*4·35*	*5·00*	*5·84*	*7·28*	*8·96*
Flanged connector	No.	1·41	1·42	1·42	1·76	1·76	1·81	1·81	2·04	2·43
		0·30	*0·31*	*0·31*	*0·38*	*0·38*	*0·38*	*0·43*	*0·43*	*0·66*
Sealing end	,,	1·41	1·42	1·42	1·76	1·76	1·81	1·81	2·04	2·43
		0·30	*0·31*	*0·31*	*0·38*	*0·38*	*0·38*	*0·43*	*0·43*	*0·66*
Triple compartment	m		8·90	9·85	9·75	11·16	12·60	13·10	16·40	19·45
			4·39	*5·00*	*4·90*	*5·28*	*6·17*	*6·63*	*8·43*	*10·45*
Flanged connector	No.		1·42	1·42	1·76	1·76	1·81	1·81	2·04	2·43
			0·31	*0·31*	*0·38*	*0·38*	*0·38*	*0·43*	*0·43*	*0·66*
Sealing end	,,		1·42	1·42	1·76	1·76	1·81	1·81	2·04	2·43
			0·31	*0·31*	*0·38*	*0·38*	*0·38*	*0·43*	*0·43*	*0·66*
Four compartment	m				10·90	12·45	14·15	14·20	17·85	21·25
					5·73	*6·23*	*7·35*	*7·39*	*9·56*	*11·92*
Flanged connector	No.				1·76	1·76	1·81	1·81	2·04	2·43
					0·38	*0·38*	*0·38*	*0·43*	*0·43*	*0·66*
Sealing end	,,				1·76	1·76	1·81	1·81	2·04	2·43
					0·38	*0·38*	*0·38*	*0·43*	*0·43*	*0·66*

	Unit	225 × 100 mm £	225 × 150 mm £	225 × 225 mm £	300 × 75 mm £	300 × 100 mm £	300 × 150 mm £	300 × 225 mm £	300 × 300 mm £
Single compartment	m	15·94	17·16	20·16	17·73	19·87	20·50	22·45	24·00
		8·19	*9·41*	*11·86*	*9·43*	*11·22*	*11·86*	*13·11*	*14·32*

CABLE TRUNKING

EXTRA FOR TRUNKING FITTINGS

	Unit	50 × 50 mm £	75 × 50 mm £	75 × 75 mm £	100 × 50 mm £	100 × 75 mm £	100 × 100 mm £	150 × 50 mm £	150 × 75 mm £	150 × 100 mm £	150 × 150 mm £	225 × 75 mm £
90° bend												
Single compartment	No.	5·38	6·00	6·55	6·75	7·15	7·47	8·00	8·64	9·35	10·12	11·77
		1·92	2·19	2·40	2·46	2·51	2·63	2·81	3·45	3·81	4·24	5·89
Double compartment	,,	5·78	6·43	6·79	7·12	7·49	7·83	8·41	9·02	9·64	10·41	
		2·32	2·62	2·64	2·83	2·85	2·99	3·22	3·88	4·10	4·53	
Triple compartment	,,		6·80	7·20	7·45	7·84	8·18	8·74	9·38	10·00	10·78	
			2·99	3·05	3·16	3·20	3·34	3·55	4·19	4·46	4·90	
Four compartment	,,				7·84	8·29	8·57	9·08	9·72	10·37	11·14	
					3·55	3·65	3·73	3·89	4·53	4·83	5·26	

	Unit	225 × 100 mm £	225 × 150 mm £	225 × 225 mm £	300 × 75 mm £	300 × 100 mm £	300 × 150 mm £	300 × 225 mm £	300 × 300 mm £
Single compartment	,,	12·86	14·30	15·57	15·59	15·75	16·96	18·40	21·77
		6·98	7·73	9·00	8·32	8·48	9·00	10·45	13·12

	Unit	50 × 50 mm £	75 × 50 mm £	75 × 75 mm £	100 × 50 mm £	100 × 75 mm £	100 × 100 mm £	150 × 50 mm £	150 × 75 mm £	150 × 100 mm £	150 × 150 mm £	225 × 75 mm £
Tee piece												
Single compartment	,,	7·12	7·47	7·85	8·10	8·87	8·96	9·84	10·60	11·14	12·26	14·69
		2·21	2·56	2·66	2·91	2·99	3·08	3·27	4·03	4·57	4·99	7·42
Double compartment	,,	7·90	8·29	8·60	8·85	9·60	9·70	10·60	11·35	11·83	12·92	
		2·99	3·38	3·41	3·66	3·73	3·81	4·03	4·78	5·26	5·65	
Triple compartment	,,		9·00	9·30	9·60	10·36	10·40	11·33	12·06	12·58	13·67	
			4·10	4·11	4·42	4·48	4·53	4·76	5·49	6·01	6·40	
Four compartment	,,				10·38	11·11	11·17	12·08	12·81	13·32	14·43	
					5·19	5·23	5·29	5·51	6·24	6·75	7·16	

	Unit	225 × 100 mm £	225 × 150 mm £	225 × 225 mm £	300 × 75 mm £	300 × 100 mm £	300 × 150 mm £	300 × 225 mm £	300 × 300 mm £
Single compartment	,,	16·97	18·85	20·65	20·34	20·90	22·11	24·40	28·00
		9·70	10·89	12·69	11·69	12·25	12·77	14·71	17·63

	Unit	50 × 50 mm £	75 × 50 mm £	75 × 75 mm £	100 × 50 mm £	100 × 75 mm £	100 × 100 mm £	150 × 50 mm £	150 × 75 mm £	150 × 100 mm £	150 × 150 mm £	225 × 75 mm £
Cross horizontal												
Single compartment	,,	8·16	8·95	9·75	10·13	10·97	11·24	12·14	12·52	12·96	14·00	18·00
		2·55	3·34	3·87	4·25	4·40	4·67	4·87	5·25	5·69	6·03	10·04
Double compartment	,,	8·90	9·70	10·42	10·90	11·68	11·96	12·92	13·28	13·60	14·64	
		3·29	4·10	4·54	5·01	5·11	5·39	5·65	6·01	6·32	6·68	
Triple compartment	,,		10·46	11·18	11·67	12·40	12·70	13·63	14·00	14·29	15·38	
			4·85	5·30	5·79	5·84	6·13	6·36	6·73	7·02	7·42	
Four compartment	,,				12·40	13·17	13·48	14·37	14·60	15·03	16·11	
					6·51	6·60	6·91	7·10	7·33	7·76	8·15	

	Unit	225 × 100 mm £	225 × 150 mm £	225 × 225 mm £	300 × 75 mm £	300 × 100 mm £	300 × 150 mm £	300 × 225 mm £	300 × 300 mm £
Single compartment	,,	20·10	22·00	23·95	23·40	23·85	24·65	26·75	31·15
		12·13	13·34	15·30	14·38	14·83	15·30	17·40	20·77

Electrical Installations – Prices for Measured Work

CABLE TRUNKING

PVC TRUNKING WITH CLIP ON LID
(grey finish fixed to brickwork or concrete)

	Unit	50 × 50 mm £	75 × 50 mm £	75 × 75 mm £	100 × 50 mm £	100 × 75 mm £	100 × 100 mm £	150 × 75 mm £	150 × 100 mm £	150 × 150 mm £
Single compartment	m	4·18 *2·45*	4·80 *3·08*	5·70 *3·98*	7·16 *5·08*	8·06 *5·64*	8·93 *6·51*	13·70 *10·94*	16·45 *13·69*	17·38 *14·27*
Extra for trunking fittings										
Cross over	No.	7·83 *6·58*	8·46 *7·21*	9·07 *7·82*	13·50 *12·10*	14·85 *13·48*	15·20 *13·48*	17·45 *15·71*	21·40 *19·68*	26·55 *24·82*
Stop end	,,	0·63 *0·28*	0·74 *0·39*	0·89 *0·54*	1·11 *0·76*	1·56 *1·21*	1·76 *1·21*	3·73 *3·19*	4·50 *3·92*	6·00 *5·44*
Flanged coupling	,,	3·03 *1·65*	3·25 *1·87*	3·75 *2·38*	4·78 *2·70*	5·45 *3·36*	5·45 *3·36*	6·70 *3·92*	7·65 *4·87*	8·95 *6·17*
Internal coupling	,,	1·02 *0·58*	1·09 *0·67*	1·09 *0·67*	1·54 *0·99*	1·77 *1·22*	1·77 *1·22*			
External coupling	,,	1·17 *0·75*	1·34 *0·92*	1·55 *1·13*	2·42 *1·87*	2·89 *2·34*	2·89 *2·34*	4·14 *3·45*	4·95 *4·27*	6·00 *5·29*
Angle/top cover	,,	1·95 *1·25*	2·55 *1·86*	2·97 *2·28*	5·25 *4·57*	7·55 *6·51*	7·55 *6·51*	11·75 *10·71*	14·05 *12·68*	20·00 *18·58*
Angle/external cover	,,	1·95 *1·25*	4·04 *3·35*	5·11 *4·42*	5·67 *4·98*	9·90 *8·87*	9·90 *8·87*	11·30 *10·27*	13·58 *12·20*	20·00 *18·58*
Angle/internal cover	,,	1·95 *1·25*	4·04 *3·35*	5·11 *4·42*	5·67 *4·98*	9·90 *8·87*	9·90 *8·87*	11·30 *10·27*	13·58 *12·20*	20·00 *18·58*
Tee/top cover	,,	3·03 *1·99*	4·02 *2·98*	4·39 *3·35*	6·75 *5·39*	9·07 *7·69*	9·07 *7·69*	14·95 *13·22*	18·70 *16·96*	23·00 *21·29*
Tee/external cover	,,	5·53 *4·49*	5·94 *4·90*	6·87 *5·83*	8·30 *6·93*	10·60 *9·24*	10·60 *9·24*	14·95 *13·22*	18·70 *16·96*	23·00 *21·29*
Tee/internal cover	,,	5·53 *4·49*	5·94 *4·90*	6·87 *5·83*	8·30 *6·93*	10·60 *9·24*	10·60 *9·24*	14·95 *13·22*	18·70 *16·96*	23·00 *21·29*
Dividing strip per 1·8 m length	,,	2·28 *1·93*	2·90 *2·56*	3·75 *3·26*						

CABLE TRAY

STANDARD GALVANIZED MILD STEEL LIGHT DUTY CABLE TRAY
(fixed to concrete 3 m above f.f.l.)

	Unit	50 mm £	75 mm £	100 mm £	150 mm £	200 mm £	225 mm £
Length	m	6·46 *1·96*	6·65 *2·15*	7·35 *2·51*	8·36 *3·52*	9·03 *3·84*	10·10 *4·91*
Ditto with return flange	,,			10·67 *5·13*	11·84 *6·30*		14·35 *7·42*
Bend	No.	3·60 *1·52*	4·00 *1·57*	4·11 *1·69*	4·96 *2·19*	6·13 *2·67*	6·67 *3·21*
Ditto with return flange	,,			13·68 *10·91*	17·10 *13·98*		20·30 *16·47*
Tee	,,	5·07 *2·30*	5·47 *2·36*	6·00 *2·53*	6·75 *3·29*	8·51 *4·01*	10·00 *4·83*
Ditto with return flange	,,			19·50 *16·07*	21·60 *17·78*		26·70 *21·48*

	Unit	300 mm £	375 mm £	450 mm £	600 mm £	750 mm £	900 mm £
Length	m	12·70 *6·46*	14·10 *7·86*	17·70 *9·05*	25·75 *15·35*	31·75 *20·32*	37·20 *24·42*
Ditto with return flange	,,	15·70 *8·78*		22·50 *13·86*	27·10 *16·71*	31·35 *19·23*	
Bend	No.	9·25 *4·75*	12·15 *6·96*	14·50 *8·28*	19·15 *12·91*	26·70 *19·80*	35·65 *28·71*
Ditto with return flange	,,	23·80 *18·98*		38·10 *31·16*	51·40 *43·45*	62·70 *54·03*	
Tee	,,	12·67 *7·13*	16·33 *10·45*	19·35 *12·43*	28·05 *19·38*	38·33 *29·68*	51·70 *43·07*
Ditto with return flange	,,	32·35 *23·70*		48·70 *39·34*	65·45 *55·41*	79·80 *69·42*	
Coupling piece	,,	2·25 *1·21*					

LADDER RACK

LADDER RACK
(galvanized finish fixed to brickwork or concrete)

	Unit	150 mm £	200 mm £	width 300 mm £	400 mm £	600 mm £	1000 mm £
Straight	m	10·87 *5·68*	11·15 *5·95*	13·00 *6·09*	14·87 *6·22*	17·30 *6·90*	23·60 *11·47*
Bend (radius)	No.	15·75 *13·66*	16·80 *14·04*	19·10 *14·93*	22·15 *15·21*	25·70 *17·38*	42·90 *34·58*
Bend (outside radius)	,,		24·30 *22·21*	29·20 *26·40*	33·30 *29·15*	39·00 *32·06*	61·60 *53·30*
T-Junction	,,		26·40 *23·60*	28·10 *23·93*	31·20 *24·26*	33·00 *24·71*	48·20 *38·53*
X-Junction	,,		34·75 *31·29*	36·90 *32·06*	40·65 *32·37*	43·20 *33·52*	
Riser	,,	16·95 *14·18*	19·30 *15·12*	20·33 *15·49*	22·90 *15·97*	25·05 *16·77*	27·65 *19·36*
Cantilever bracket	,,	5·03 *1·57*	6·53 *1·69*	7·50 *2·64*	11·70 *3·38*	12·60 *4·31*	20·90 *12·58*
Wall bracket	,,	5·76 *2·30*	7·20 *2·37*	7·50 *2·68*	11·10 *2·80*	11·40 *3·11*	
End connector	,,	3·04 *0·96*					
Universal coupling	,,	3·97 *1·89*					

CABLE

SINGLE CORE PVC INSULATED CABLE

		1·0 mm²	1·5 mm²	2·5 mm²	4 mm²	6 mm²	10 mm²	16 mm²	25 mm²	35 mm²
Drawn into conduit	Unit	£	£	£	£	£	£	£	£	£
450/750 volt grade	m	0·18	0·22	0·29	0·39	0·49	0·78	1·10	1·78	2·48
		0·04	*0·05*	*0·08*	*0·15*	*0·21*	*0·36*	*0·55*	*0·91*	*1·27*

		50 mm²	70 mm²	95 mm²
		£	£	£
Ditto	„	3·48	4·93	6·84
		1·72	*2·44*	*3·45*

		1·0 mm²	1·5 mm²	2·5 mm²	4 mm²	6 mm²	10 mm²	16 mm²	25 mm²	35 mm²
		£	£	£	£	£	£	£	£	£
Laid in trunking										
450/750 volt grade	„	0·14	0·19	0·25	0·36	0·45	0·71	1·03	1·60	2·38
		0·04	*0·05*	*0·08*	*0·15*	*0·21*	*0·36*	*0·55*	*0·91*	*1·27*

		50 mm²	70 mm²	95 mm²	120 mm²	150 mm²	185 mm²	240 mm²	300 mm²	400 mm²
		£	£	£	£	£	£	£	£	£
Ditto	„	3·38	4·52	6·22	7·90	9·16	10·75	13·37	15·93	22·60
		1·72	*2·44*	*3·45*	*4·78*	*5·91*	*7·29*	*9·56*	*11·92*	*18·23*

		400 mm²	630 mm²
		£	£
Ditto	„	26·75	33·80
		22·25	*29·00*

		120 mm²	150 mm²	185 mm²	240 mm²	300 mm²	400 mm²	500 mm²	630 mm²
		£	£	£	£	£	£	£	£
Drawn into ducts									
450/750 volt grade	„	6·86	8·09	9·60	12·12	14·80	21·40	25·85	33·12
		4·78	*5·91*	*7·28*	*9·56*	*11·92*	*18·23*	*22·25*	*28·97*

PVC INSULATED AND SHEATHED COPPER

		1·0 mm²	1·5 mm²	2·5 mm²	4 mm²	6 mm²	10 mm²	16 mm²
Fixed to timber								
450/750 volt grade	Unit	£	£	£	£	£	£	£
Single core	m	0·55	0·58	0·67	0·80	0·90	1·21	1·77
		0·09	*0·11*	*0·15*	*0·25*	*0·32*	*0·52*	*0·72*
Flat twin core	„	0·62	0·71	0·82	1·06	1·35	2·09	2·76
		0·15	*0·19*	*0·27*	*0·47*	*0·65*	*0·95*	*1·54*
Twin core and E.C.C.	„	0·63	0·71	0·82	1·09	1·40	2·18	2·89
		0·16	*0·19*	*0·27*	*0·50*	*0·71*	*1·14*	*1·71*
Flat three core	„	0·73	0·82	1·02	1·39	1·98	2·75	3·77
		0·21	*0·27*	*0·43*	*0·70*	*0·94*	*1·57*	*2·45*
Three core and E.C.C.	„	0·77	0·85	1·06	1·44	2·17	2·88	4·06
		0·25	*0·30*	*0·47*	*0·75*	*1·13*	*1·84*	*2·88*

Electrical Installations – Prices for Measured Work

CABLE

ARMOURED PVC INSULATED AND SHEATHED CABLE 600–1000 VOLT GRADE

Fixed with clips	Unit	1.5 mm² £	2.5 mm² £	4 mm² £	6 mm² £	10 mm² £	16 mm² £
2 core	m	1.37 / *0.54*	1.52 / *0.69*	2.14 / *0.96*	2.38 / *1.20*	3.60 / *1.89*	4.50 / *2.23*
3 core	,,	1.82 / *0.64*	2.00 / *0.82*	2.90 / *1.17*	3.32 / *1.59*	4.91 / *2.35*	6.34 / *3.09*
4 core	,,	1.90 / *0.72*	2.12 / *0.94*	3.27 / *1.54*	3.67 / *1.94*	5.46 / *2.90*	7.40 / *4.15*
5 core	,,	2.37 / *0.99*	3.04 / *1.31*	4.50 / *2.22*	6.44 / *2.98*		
7 core	,,	2.58 / *1.20*	3.32 / *1.59*	4.95 / *2.67*	7.27 / *3.81*		
8 core	,,	2.89 / *1.51*	3.85 / *2.12*	5.56 / *3.28*			
10 core	,,	3.52 / *1.79*	4.69 / *2.41*	7.30 / *4.20*			
12 core	,,	3.68 / *1.95*	4.96 / *2.68*	7.77 / *4.66*			
14 core	,,	4.90 / *2.61*	6.60 / *3.50*	8.90 / *5.44*			
22 core	,,	6.82 / *3.71*	8.74 / *5.28*	12.50 / *8.35*			
37 core	,,	9.60 / *5.44*	12.30 / *7.46*				
48 core	,,	11.38 / *6.54*	14.50 / *9.65*				

TERMINATION FOR ARMOURED PVC INSULATED AND SHEATHED CABLE 600–1000 VOLT GRADE

Cable gland including brass locknut and PVC shroud	Unit	1.5 mm² £	2.5 mm² £	4 mm² £	6 mm² £	10 mm² £	16 mm² £
2 core	No.	4.88 / *1.07*	4.88 / *1.07*	5.17 / *1.36*	5.70 / *1.55*	7.08 / *2.24*	8.12 / *2.24*
3 core	,,	4.88 / *1.07*	5.17 / *1.36*	5.17 / *1.36*	5.70 / *1.55*	7.08 / *2.24*	8.12 / *2.24*
4 core	,,	5.57 / *1.07*	5.86 / *1.36*	7.09 / *1.55*	7.09 / *1.55*	9.85 / *2.24*	11.24 / *2.24*
5 core	,,	6.55 / *1.36*	6.55 / *1.36*	9.50 / *1.55*			
7 core	,,	6.90 / *1.36*	7.10 / *1.55*	10.54 / *2.24*			
8 core	,,	8.63 / *1.36*	8.82 / *1.55*	10.90 / *2.24*			
10 core	,,	10.20 / *1.55*	10.90 / *2.24*	12.60 / *2.24*			

CABLE

MINERAL INSULATED COPPER SHEATHED CABLE
(fixed to brickwork or concrete with copper clips or saddles)

		2L 1·0 £	2L 1·5 £	2L 2·5 £	2L 4 £	3L 1·0 £
Light duty 600 volt grade	Unit					
Bare	m	2·64	2·80	3·02	4·13	3·32
		0·56	0·72	0·94	1·29	0·69
PVC sheathed	„	2·84	3·01	3·24	4·51	3·05
		0·63	0·80	1·03	1·40	0·77

		3L 1·5 £	4L 1·0 £	4L 1·5 £	7L 1·0 £	7L 1·5 £
Light duty 600 volt grade						
Bare	„	3·37	3·10	3·70	3·97	4·95
		0·87	0·82	1·07	1·20	1·49
PVC sheathed	„	3·45	3·53	3·79	4·42	5·49
		0·96	0·90	1·16	1·31	1·61

		1H 6 £	1H 10 £	1H 16 £	1H 25 £	1H 35 £
Heavy duty 1000 volt grade						
Bare	„	3·69	3·96	5·44	7·36	8·45
		0·92	1·19	1·63	2·31	3·05

		1H 50 £	1H 70 £	1H 95 £	1H 120 £	1H 150 £
Heavy duty 1000 volt grade						
Bare	„	9·62	12·00	14·40	16·72	20·37
		3·88	5·06	6·65	8·07	9·99

		1H 6 £	1H 10 £	1H 16 £	1H 25 £	1H 35 £
Heavy duty 1000 volt grade						
PVC sheathed	„	4·00	4·60	5·90	8·00	9·08
		1·00	1·29	1·76	2·44	3·20

		1H 50 £	1H 70 £	1H 95 £	1H 120 £	1H 150 £
Heavy duty 1000 volt grade						
PVC sheathed	„	10·28	12·60	15·23	17·60	21·25
		4·05	5·25	6·93	8·38	10·32

		2H 1·5 £	2H 2·5 £	2H 4 £	2H 6 £	2H 10 £	2H 16 £	2H 25 £
Heavy duty 1000 volt grade	Unit							
Bare . . .	m	3·66	4·43	5·77	6·65	7·90	9·86	12·40
		1·03	1·25	1·62	2·08	2·72	3·91	5·47
PVC sheathed . .	„	4·26	5·05	6·33	7·36	8·64	10·75	13·40
		1·15	1·38	1·76	2·17	2·90	4·12	5·78

		3H 1·5 £	3H 2·5 £	3H 4 £	3H 6 £	3H 10 £	3H 16 £	3H 25 £
Heavy duty 1000 volt grade								
Bare	„	4·12	5·33	6·41	7·14	8·75	10·95	14·65
		1·14	1·45	1·84	2·30	3·35	4·73	7·20
PVC sheathed . .	„	4·60	5·94	7·05	7·80	9·40	12·65	15·55
		1·27	1·58	1·99	2·46	3·54	5·02	7·53

		4H 1·5 £	4H 2·5 £	4H 4 £	4H 6 £	4H 10 £	4H 16 £	4H 25 £
Heavy duty 1000 volt grade								
Bare	„	4·95	6·30	7·05	8·10	9·90	12·95	17·10
		1·42	1·80	2·20	2·92	4·18	6·04	8·82
PVC sheathed . .	„	5·43	6·80	7·65	8·85	11·30	14·70	18·30
		1·55	1·94	2·37	3·10	4·39	6·38	9·29

Electrical Installations – Prices for Measured Work

CABLE

MINERAL INSULATED COPPER SHEATHED CABLE continued
(fixed to brickwork or concrete with copper clips or saddles)

	Unit	7H 1·5 £	7H 2·5 £	12H 2·5 £	19H 1·5 £
Heavy duty 1000 volt grade					
Bare	m	7·00	8·19	10·95	14·05
		1·95	*2·65*	*4·64*	*7·13*
PVC sheathed	,,	7·70	8·90	11·90	15·05
		2·10	*2·82*	*4·93*	*7·43*

CABLE

TERMINATION FOR MINERAL INSULATED COPPER SHEATHED CABLE

		2L 1·0	2L 1·5	2L 2·5	2L 4	3L 1·0
Light duty 600 volt grade	Unit	£	£	£	£	£
Low temperature to 105 °C	No.	2·31	2·45	2·59	2·73	2·45
		0·65	*0·65*	*0·65*	*0·65*	*0·65*
Low temperature with PVC shroud	„	2·47	2·61	2·75	2·89	2·61
		0·81	*0·81*	*0·81*	*0·81*	*0·81*

		3L 1·5	4L 1·0	4L 1·5	7L 1·0	7L 1·5
Light duty 600 volt grade		£	£	£	£	£
Low temperature to 105 °C	„	2·59	2·59	2·73	3·33	3·67
		0·65	*0·65*	*0·65*	*1·25*	*1·25*
Low temperature with PVC shroud	„	2·75	2·75	2·89	3·53	3·87
		0·81	*0·81*	*0·81*	*1·45*	*1·45*

		1H 6	1H 10	1H 16	1H 25	1H 35	1H 50	1H 70
Heavy duty 1000 volt grade	Unit	£	£	£	£	£	£	£
Low temperature to 105°C	No.	2·31	2·52	2·73	3·07	3·14	4·02	4·29
		0·65	*0·65*	*0·65*	*0·65*	*0·65*	*1·25*	*1·25*
Low temperature with PVC shroud	„	2·47	2·68	2·89	3·23	3·30	4·22	4·49
		0·81	*0·81*	*0·81*	*0·81*	*0·81*	*1·45*	*1·45*

		1H 95	1H 120	1H 150	2H 1·5	2H 2·5	2H 4	2H 6
Heavy duty 1000 volt grade		£	£	£	£	£	£	£
Low temperature to 105 °C	„	4·57	6·28	6·42	2·94	2·94	3·22	3·22
		1·25	*2·54*	*2·54*	*0·66*	*0·66*	*0·66*	*0·66*
Low temperature with PVC shroud	„	4·77	6·53	6·67	3·10	3·10	3·38	3·38
		1·45	*2·79*	*2·79*	*0·82*	*0·82*	*0·82*	*0·82*

		2H 10	2H 16	2H 25	3H 1·5	3H 2·5	3H 4	3H 6
Heavy duty 1000 volt grade		£	£	£	£	£	£	£
Low temperature to 105 °C	„	3·81	4·01	4·31	2·94	3·22	3·24	3·95
		1·25	*1·25*	*1·54*	*0·66*	*0·66*	*0·66*	*1·25*
Low temperature with PVC shroud	„	4·01	4·22	4·56	3·10	3·38	3·52	4·15
		1·45	*1·45*	*1·79*	*0·82*	*0·82*	*0·82*	*1·45*

		3H 10	3H 16	3H 25	4H 1·5	4H 2·5	4H 4	4H 6
Heavy duty 1000 volt grade		£	£	£	£	£	£	£
Low temperature to 105 °C	„	3·95	4·16	8·16	3·07	3·24	4·02	4·02
		1·25	*1·25*	*5·25*	*0·65*	*0·66*	*1·25*	*1·25*
Low temperature with PVC shroud	„	4·15	4·36	8·55	3·23	3·52	4·22	4·22
		1·45	*1·45*	*5·64*	*0·81*	*0·82*	*1·45*	*1·45*

		4H 10	4H 16	4H 25	7H 1·5	7H 2·5	12H 2·5	19H 1·5
Heavy duty 1000 volt grade		£	£	£	£	£	£	£
Low temperature to 105 °C	„	4·02	5·44	8·29	4·71	4·71	7·24	13·28
		1·25	*2·54*	*5·25*	*1·25*	*1·25*	*2·54*	*5·25*
Low temperature with PVC shroud	„	4·22	5·83	8·68	4·91	4·91	7·63	13·67
		1·45	*2·79*	*5·64*	*1·45*	*1·45*	*2·79*	*5·64*

FITTINGS AND ACCESSORIES

LIGHTING SWITCHES

5 amp metalclad surface mounted switch including box, fixed and connected	Unit	1 gang £	2 gang £	3 gang £	4 gang £	6 gang £
One way	No.	5·87	8·95	13·05	16·25	22·60
		2·41	4·11	6·13	8·15	12·24
Two way	,,	6·20	9·60	14·00	17·60	24·60
		2·74	4·76	7·10	9·46	14·18

		8 gang	9 gang	10 gang	12 gang
One way	,,	31·50	35·60	39·75	45·15
		16·29	18·31	20·36	24·41
Two way	,,	34·10	38·50	43·00	49·00
		18·87	21·22	23·61	28·30

Weatherproof 5 amp metalclad surface mounted A.C./D.C. switch, fixed and connected		1 gang £	2 gang £	3 gang £	4 gang £
One way	,,	8·60	14·70	21·55	27·20
		5·12	9·86	14·62	19·08
Two way	,,	8·90	15·35	22·50	28·50
		5·44	10·51	15·59	20·37

Modular type 5 amp switch comprising galvanized steel box and switch mounting harness, switch units and bronze or satin chrome cover plate; assembled, fixed and connected to wiring	Unit	1 gang £	2 gang £	3 gang £	4 gang £	6 gang £	8 gang £	9 gang £	12 gang £
One way	No.	6·00	7·90	11·70	13·35	18·50	24·30	28·30	33·25
		2·54	3·03	4·77	5·26	8·12	9·09	11·01	12·49
Two way	,,	6·20	8·25	12·25	14·10	19·60	25·80	30·00	35·50
		2·73	3·40	5·34	5·99	9·24	10·58	12·69	14·73

FITTINGS AND ACCESSORIES

SOCKET OUTLETS

13 amp metalclad socket outlet to B.S. 1363: 1967; galvanized steel box and coverplate with white plastic inserts; surface mounting; fixed and connected to wiring

	Unit	1 gang £	2 gang £
Unswitched	No.	4·70	6·10
		1·59	2·96
Switched	,,	5·50	7·50
		2·40	4·39
Switched with neon indicator	,,	6·40	9·25
		3·28	6·16

13 amp socket outlet to B.S. 1363: 1967; with white moulded plastic coverplate including fixing box to brickwork or concrete and connecting to wiring

		Surface type with moulded plastic box		Flush type with galvanized steel box	
	Unit	1 gang £	2 gang £	1 gang £	2 gang £
Unswitched	No.	4·30	5·15		
		1·18	2·04		
Switched	,,	4·65	5·80	5·05	6·35
		1·56	2·68	1·93	3·25
Switched with neon indicator	,,	5·45	7·35	5·80	8·00
		2·32	4·22	2·70	4·77

13 amp socket outlet to B.S. 1363: 1967; with matt chrome steel coverplate and galvanized steel box to B.S. 4662: 1970; for flush mounting; fixed to brickwork or concrete and connected to wiring

	Unit	1 gang £	2 gang £
Switched	No.	6·10	8·10
		2·97	4·97
Switched with neon indicator	,,	6·85	9·60
		3·74	6·81

Electrical Installations – Prices for Measured Work

FITTINGS AND ACCESSORIES

CONNECTION UNITS

		Switched £	Unswitched £
Moulded pattern unit to B.S. 816: 1952, white moulded plastic box and coverplate for surface mounting fixed to brickwork or concrete and connected to wiring	Unit		
Standard fused	No.	5·80 / 2·68	5·50 / 2·39
Fused with neon indicator	,,	6·55 / 3·44	
Fused with flex outlet	,,	5·95 / 2·83	5·65 / 3·59
Fused with flex outlet and neon indicator	,,	6·70 / 3·59	
Moulded pattern unit to B.S. 816: 1952 white moulded plastic coverplate and galvanized steel box for flush mounting; fixed to brickwork or concrete and connected to wiring			
Standard fused	,,	5·90 / 2·76	5·60 / 2·47
Fused with neon indicator	,,	6·65 / 3·52	
Fused with flex outlet	,,	6·00 / 2·91	5·70 / 2·60
Fused with flex outlet and neon indicator	,,	6·80 / 3·67	
Galvanized pressed steel pattern with matt chrome or satin finish, white moulded plastic inserts, galvanized steel box for mounting fixed to brickwork or concrete and connected to wiring			
Standard fused	,,	6·80 / 3·67	6·40 / 3·31
Fused with neon indicator	,,	7·55 / 4·44	
Fused with flex outlet	,,	7·00 / 3·89	6·50 / 3·41
Fused with flex outlet and neon indicator	,,	7·75 / 4·65	

		Single outlet £	Double outlet £
Telephone outlet (moulded plastic plate with box fixed and connected)			
Flush mounted	,,	3·44 / 1·36	3·54 / 1·46
Surface mounted	,,	3·76 / 1·68	3·86 / 1·78
As above but with bronze or satin chrome-plate			
Flush mounted	,,	4·39 / 2·31	4·53 / 2·45
Surface mounted	,,	4·71 / 2·63	4·85 / 2·77

		Single outlet £	Double outlet £	Isolated T.V./F.M. single outlet £	Isolated T.V./F.M. double outlet £
T.V. co-axial socket outlet, moulded plastic with box fixed and connected	Unit				
Flush mounted	No.	4·60 / 2·53	5·45 / 3·38	6·10 / 4·02	7·55 / 5·45
Surface mounted	,,	4·92 / 2·85	5·77 / 3·70	6·42 / 4·34	7·87 / 5·77

Electrical Installations – Prices for Measured Work
FITTINGS AND ACCESSORIES

CEILING ROSE

		Rose with 2 loop terminals earth and strain	Rose with 2 loop terminals and earth	Rose with 2 terminals and earth	Plug-in type with 2 A plug
White moulded plastic surface mounted to conduit box and connected	Unit No.	£ 3·86 0·75	£ 3·78 0·67	£ 3·75 0·64	£ 4·54 1·43

LAMPHOLDER

Lighting pendant fitting comprising ceiling rose, lamp holder and 3-core flex; fixed and connected to wiring	Unit No.	Rose with 3 terminals, earth and strain £ 5·20 1·74

BATTEN HOLDER

		Straight pattern	Angle pattern	Adjustable pattern
White moulded plastic, 3 terminals to B.S. 52 (1963) fixed to conduit box and connected	Unit No.	£ 4·73 1·62	£ 4·98 1·87	£ 4·80 1·69

SHAVER UNITS

		Surface type with white moulded plastic box	Flush type with galvanized steel box
Single voltage supply unit, with white moulded plastic faceplate, unswitched, fixed to brickwork or concrete and connected to wiring	Unit No.	£ 9·24 6·13	£ 9·28 6·17
Dual voltage supply unit with white moulded plastic faceplate and switches fixed to brickwork or concrete and connected to wiring with neon indicator	„	20·66 17·20	20·39 16·93
With light unit	„	31·70 28·23	

FITTINGS AND ACCESSORIES

COOKER CONTROL UNITS

45 amp cooker control unit to B.S. 4177: 1967; comprising 45 amp D.P. main switch and a 13 amp switched socket outlet, steel box and coverplate with moulded plastic inserts; fixed to brickwork or concrete and connected to wiring

	Unit	Surface mounted £	Flush mounted £
Standard	No.	7·95	8·45
		4·82	5·35
With neon indicator	,,	9·75	10·30
		6·64	7·18

Cooker control unit as last but moulded plastic box and coverplate; fixed to brickwork or concrete and connected to wiring

Standard	,,	7·80
		4·70
With neon indicator	,,	9·30
		6·18

Connector unit moulded plastic cover and block, galvanized steel box for flush mounting fixed to brickwork or concrete and connected to wiring

	,,	5·00
		1·90

CONSUMER UNITS

60 amp 240 volt D.P./A.C. switched all insulated unit with moulded plastic case and cover for surface mounting fixed to brickwork or concrete and connected to wiring

		2 way £	3 way £	4 way £	6 way £	8 way £
Fitted rewirable fuses	,,	13·85	17·25	18·90	23·45	27·55
		5·21	6·87	8·49	10·80	13·71
Fitted cartridge fuses	,,	14·45	18·15	20·00	25·20	30·00
		5·80	7·75	9·66	12·54	16·03
Fitted miniature circuit breakers	,,	18·85	24·70	29·00	37·85	47·45
		10·19	14·33	18·53	25·20	33·59

60 amp 240 volt D.P./A.C. switched metalclad unit with enamelled steel case and moulded plastic fuse cover fixed to brickwork or concrete and connected

Fitted rewirable fuses	,,	14·90	19·00	20·25	24·95	28·75
		6·25	8·62	9·88	12·29	14·90
Fitted cartridge fuses	,,	15·50	19·85	21·40	26·70	31·10
		6·83	9·48	11·03	14·04	17·23
Fitted miniature circuit breakers	,,	19·90	26·45	30·20	39·85	48·60
		11·22	16·07	19·81	27·20	34·78

100 amp 240 volt D.P./A.C. switched metalclad unit with enamelled steel case and moulded plastic fuse cover fixed to brickwork or concrete and connected to wiring

		6 way £	8 way £	12 way £
Fitted miniature circuit breackers	,,	27·50	31·50	47·65
		14·86	17·66	26·88

Electrical Installations – Prices for Measured Work

FITTINGS AND ACCESSORIES

IMMERSION HEATERS

	Unit	Up to 760 mm long £	760 to 915 mm long £
3 kW immersion heater fitted and connected	No.	13·40 8·19	14·50 9·31
Thermostat	,,	5·85 4·12	5·85 4·12

FLUORESCENT FITTINGS

		600 mm Single 20 W £	600 mm Twin 20 W £	1200 mm Single 40 W £	1200 mm Twin 40 W £
Batten type; surface mounted fixed and connected and tubes fitted	Unit No.	11·05 7·57	16·10 12·65	14·30 9·81	22·65 18·15
As last but with prismatic diffuser	,,	14·55 11·10	21·25 17·80	19·55 15·06	31·65 27·14

		1500 mm Single 65 W £	1500 mm Twin 65 W £	1800 mm Single 75 W £	1800 mm Twin 75 W £	2400 mm Single 100 W £	2400 mm Twin 100 W £
Batten type as before	,,	16·75 11·54	26·75 21·59	20·25 13·33	30·55 23·62	26·90 18·24	39·60 30·95
As last but with prismatic diffuser	,,	22·55 17·34	38·00 32·84	27·50 20·56	43·80 36·87	36·25 27·58	56·00 47·31

		300 × 1200 mm 40 W – 1 lamp £	300 × 1200 mm 40 W – 2 lamp £	300 × 1800 mm 75 W – 1 lamp £	300 × 1800 mm 75 W – 2 lamp £
Modular type recessed mounting with opal diffuser, fixed and connected and tubes fitted	Unit No.	48·90 38·50	57·50 47·09	59·30 47·17	69·80 57·70

		600 × 600 mm 40 W – 2 lamp £	600 × 600 mm 40 W – 3 lamp £	600 × 1200 mm 40 W – 3 lamp £	600 × 1200 mm 40 W – 4 lamp £
	,,	71·40 61·04	77·90 67·50	81·70 69·55	90·10 77·99

		600 × 1800 mm 75 W – 2 lamp £	600 × 1800 mm 75 W – 3 lamp £	600 × 1800 mm 75 W – 4 lamp £
	,,	94·00 80·13	103·00 88·13	114·30 98·71

LIGHT FITTINGS

FLUORESCENT FITTINGS

	Unit No.	600 mm Single 20 W £	600 mm Twin 20 W £	1200 mm Single 40 W £	1200 mm Twin 40 W £	1500 mm Single 65 W £	1500 mm Twin 65 W £	1800 mm Single 75 W £	1800 mm Twin 75 W £
Corrosion resistant, G.R.P. body, gasket sealed with acrylic diffuser fixed and connected and tubes fitted	,,	28·95 25·49	35·00 31·51	31·85 27·36	38·10 33·62	31·30 26·10	47·25 42·04	39·06 32·13	56·60 49·70
Flameproof to I.P.G.S., B.S. 229 (1957) and B.S. 899 (1965) fixed and connected and lamps fitted	,,	111·25 104·32	152·00 145·04	113·35 104·70	171·60 162·92	122·00 111·68	175·15 164·78	129·50 117·34	185·00 172·80

EMERGENCY FITTINGS

	Unit No.	Round pattern 4 W £	Square pattern 4 W £	Rectangular pattern 4 W £
Self-contained non-maintained 3 hour duration complete with glass diffuser, fixed and connected	,,	54·10 50·64	55·00 51·50	54·10 50·64
Exit sign maintained type to B.S. 5266 surface mounted fixed and connected 430 × 195 × 120 mm	,,	84·65 81·17		

PART THREE

Approximate Estimating

Direction, *page 301*
Mechanical Installations, *page 303*
Lift and Escalator Installations, *page 315*
Electrical Installations, *page 316*
Cost Indices, *page 324*
Elemental Cost Plan, *page 325*

APPROXIMATE ESTIMATING

Directions

Prices given in this section are average prices for typical buildings and are exclusive of fees for professional services. The prices are based upon the total floor area of all storeys measured between the external walls and without deduction for internal walls, except in the case of local authority flats which are based upon the areas as defined in M.O.H.L.G. (now the Department of the Environment) circular No. 1/68.

Although the prices have been carefully considered in the light of recent tenders for each element and changes in specification requirements, average prices can never provide more than a rough guide to the probable cost. In general no attempt has been made to elaborate the description of the services as the lack of detail should serve to emphasize that these can only be average prices for typical requirements and that such prices can vary considerably depending upon a number of factors.

This is particularly true of housing where the services costs per dwelling can vary very little with size. Most dwellings will have one bath, one sink and one or at most two W.C.s and washbasins. Furthermore heating, if provided to the Department of the Environment minimum standard, is required only to the living rooms, dining and circulation areas and kitchen; the cost of heating these areas will vary very little irrespective of the number of persons per dwelling.

The standards and costs of most of the building types contained in this section are controlled by the central government and it can be assumed unless otherwise stated that rates per square metre are average rates taken from a number of tenders which reflect these standards. The exception is office building, where average standards of office buildings constructed for owner occupation, as opposed to rented accommodation, have been assumed.

The rates per square metre should not therefore be used indiscriminately; each case must be assessed on its merits.

The cost of heating can probably vary more than any of the other services because it depends upon such a variety of circumstances. Among these are: type of heating (e.g. impelled warm air, central boiler with radiators), type of fuel, geographical location, orientation of the building, ventilation rate, fabric of the external envelope, proportion of the building to be heated, internal temperature standard, the heat gains from occupants and equipment and the period of the day during which heat is required. Each of these should be considered if an accurate estimate is to be obtained. However heating estimates will continue to be called for well before many of these factors can be determined and it is in these circumstances that the prices in this section can form a guide.

Where however more detail is or becomes available estimating by means of priced approximate quantities is always more accurate than by using rates based on a square metre basis and the information in this section should be useful in these circumstances for services. Rates are given for boiler plant of varying capacity, distribution pipe-work outside plant rooms, valves, radiators, hot and cold water services; ductwork, air conditioning, fire fighting services, electrical sub-station equipment, heavy power and general electrical installation, external lighting and lightning protection. These rates are based on the 'Prices for Measured Work' section but allow for incidentals which would be measured separately in a Bill of Quantities. This detail has also been used in the Cost Plan for the services to a typical office block.

Approximate costs are also given for central heating installations for varying sizes of

DIRECTIONS

homes. The prices do not include for incidental builders' work nor for profit and attendance by a main contractor where the work is executed as a sub-contract; they do however include for profit and overheads for the services contractor and for $2\frac{1}{2}\%$ cash discount for the main contractor. Capital contributions to statutory authorities and public undertakings and the cost of work carried out by them have been excluded.

Despite the increasing proportion of mechanical and electrical services incorporated in buildings there are no published cost indices available for general use. The editors have compiled a table of cost indices for both services and these are incorporated at the end of this section. Clearly such indices can only give a very general picture of the pattern of costs over the last few years.

Approximate Estimating

MECHANICAL INSTALLATIONS

BASIC SERVICES – PRICES PER SQUARE METRE

Approximate prices per m² of floor area of the following mechanical services:	Secondary Schools £	University arts and administration buildings £	Offices £	Local authority flats £	Local authority houses £	Non-teaching hospitals £
Heating installation	32·00	34·50	24·50	23·25	18·00	28·00
Hot water services	6·00	7·00	4·00	4·75	3·75	9·50
Cold water services	6·00	10·75	4·00	5·50	4·00	9·00
Ventilation	4·00		16·50	1·75		
Ancillary services	3·30			1·20	1·20	40·00
Total approximate cost of above services	51·30	52·25	49·00	36·45	26·95	86·50

Notes

University arts and administration buildings: The prices exclude the cost of all special services, e.g. laboratory services and the like.

The total price of mechanical services in science teaching buildings, including special services but excluding lifts, is likely to be in the region of

Physics buildings . . . . £ 54·00 per m²
Chemistry buildings . . . . £104·00 per m²
Biological science buildings . . . . £ 88·00 per m²

Local authority flats and houses: The prices are based upon tenders for large estates and assume, on average, dwellings accommodating 4 persons. The price for heating assumes impelled warm air serving the whole dwelling in the case of flats but serving only the living and dining rooms, kitchen and circulation areas in the case of houses. Solid embedded electric underfloor heating to flats would cost about £10·75 per m².

Hospitals: Prices exclude the heat source.

Hot and cold water services: Prices exclude the cost of all sanitary fittings.

Ancillary services: For housing and schools this covers gas installation; for hospitals it covers piped mechanical services, steam and condense, fire fighting, mechanical ventilation and equipment.

AIR CONDITIONING INSTALLATION

Approximate costs of air conditioning installations for two specimen office blocks with floor areas of 6000 m² and 15,000 m² in four and eight storeys respectively. The types of installation selected are the variable air volume and the induction system.

The costs allow for all plant and equipment, distribution ductwork, pipework for heating, chilled and cooling water, automatic controls, fire detection systems and all associated electrical work.

	Floor area 6,000 m² £	15,000 m² £
Variable air volume system, per m²	109·00	83·00
Induction system, per m²	97·00	70·00

Air conditioning installations vary considerably according to the type of plant selected, the ancillary services chosen, the methods of heating and cooling, the sophistication of automatic controls, the requirements for fire protection, the type of fuel available for heating and many other considerations. No two buildings will have precisely the same requirements.

The following information has been compiled to indicate the average cost of a number of different design solutions. Brief specification notes are provided to enable the user to make his own adjustments to the cost of individual elements which take into account his own design criteria. It has been assumed that the building is double glazed, has better than average insulation and a window to wall ratio not exceeding 50%.

MECHANICAL INSTALLATIONS

AIR CONDITIONING INSTALLATION continued

VARIABLE AIR VOLUME SYSTEM

	Office block of 6,000 m²		Office block of 15,000 m²		
Elements	Cost of element £	Cost of element per m² of floor area £	Cost of element £	Cost of element per m² of floor area £	Specification Ref. Page
Boilers					
Plant and instruments	16,000	2·67	24,300	1·62	A
Flue	6,100	1·02	8,100	0·54	B
Water treatment	6,100	1·02	10,800	0·72	C
Gas installation	2,500	0·42	2,700	0·18	D
Space heating					
Distribution pipework	17,300	2·88	33,300	2·22	E
Convectors and/or radiators	3,250	0·54	8,100	0·54	F
Heating to batteries	34,500	5·75	59,400	3·96	H
Chilled water to batteries	25,900	4,32	45,000	3·00	J
Condenser cooling water Distribution pipework	12,200	2·03	19,800	1·32	K
Cooling plant					
Chillers	54,700	9·12	88,200	5·88	L
Cooling towers	6,100	1·02	9,000	0·60	M
Automatic controls	47,900	7·98	72,000	4·80	N
Ductwork					
Supply	232,500	38·75	506,000	33·73	P
Extract	49,000	8·17	106,000	7·07	P
Air conditioning plant					
Heating batteries	15,800	2·63	26·100	1·74	Q
Humidifiers & cooling batteries	28,100	4·68	46,800	3·12	Q
Fans and filters	43,200	7·20	70,200	4·68	R
Sound attenuation	23,000	3·83	49,500	3·30	S
Fire detection	13,700	2·28	23,400	1·56	T
Electrical work in connection	16,200	2·70	28,800	1·92	V
Totals	654,050	109·01	1,237,500	82·50	
		say 109·00		say 83·00	

MECHANICAL INSTALLATIONS

AIR CONDITIONING INSTALLATION continued

INDUCTION SYSTEM

Office block of 6,000 m²   *Office block of 15,000 m²*

Elements	Cost of element £	Cost of element per m² of floor area £	Cost of element £	Cost of element per m² of floor area £	Specification Ref. Page
Boilers					
Plant and instruments	16,000	2·67	24,300	1·62	A
Flue	6,100	1·02	9,000	0·60	B
Water treatment	6,100	1·02	10,800	0·72	C
Gas installation	2,500	0·42	2,500	0·17	D
Space heating					
Distribution pipework	17,300	2·88	33,300	2·22	E
Convectors and/or radiators	3,250	0·54	8,100	0·54	F
Induction units	59,500	9·92	149,400	9·96	G
Heating to batteries	65,500	10·92	102,500	6·83	H
Chilled water					
Distribution pipework	84,600	14·10	146,700	9·78	J
Condenser cooling water					
Distribution pipework	12,200	2·03	17,100	1·14	K
Cooling plant					
Chillers	42,500	7·08	70,200	4·68	L
Cooling towers	5,500	0·92	8,100	0·54	M
Automatic controls	52,500	8·75	90,000	6·00	N
Ductwork					
Supply	68,000	11·33	117,000	7·80	P
Extract	34,250	5·71	70,200	4·68	P
Air conditioning plant					
Heating batteries	11,800	1·97	20,700	1·38	Q
Humidifiers & cooling batteries	20,600	3·43	30,600	2·04	Q
Fans and filters	29,500	4·92	51,300	3·42	R
Sound attenuation	15,800	2·63	38,700	2·58	S
Fire detection	11,200	1·87	18,900	1·26	T
Electrical work in connection	14,750	2·46	23,400	1·56	V
Totals	579,450	96·57	1,042,800	69·52	
		say 97·00		say 70·00	

MECHANICAL INSTALLATIONS

AIR CONDITIONING INSTALLATIONS

Brief Specification Notes

Ref.
A **Boilers**
  Plant and instruments: Three gas-fired boilers each of approximately 250 and 580 kW capacity for the two buildings respectively; together with burners, pumps, direct-mounted instruments, feed and expansion tanks. Normal standby facilities are included.
B Flue: Mild steel insulated in boiler house, internal lining to vertical builders' stack.
C Water treatment: Chemical dosage equipment.
D **Gas installation:** Pipework internal to building, meter, solenoid values.
  Space heating
E Distribution pipework: **Pipework from boilers to terminal equipment, all valves, fittings and supports, insulation.**
F Convector and/or radiators: Panel radiators, or natural convectors in circulation areas and staircases.
G **Induction units:** High velocity units suitable for four-pipe system utilizing ducted fresh air.
H **Heating to air heater batteries:** Distribution pipework to batteries and including valves, fittings and supports, insulation.
J **Chilled water to batteries and induction units:** Distribution pipework including valves, fittings and supports, insulation.
K **Condenser cooling water:** Distribution pipework, valves, fittings and supports, insulation.
  Cooling plant
L Chillers: Centrifugal chiller units of approximately 190 and 470 tons total capacity for the two buildings respectively, including mountings and supports, insulation and pumps. Normal standby facilities are included.
M Cooling towers: Forced or induced draught fans, roof-mounted cooling towers with supports.
N **Automatic controls:** Pneumatic controls including motorized valves, all thermostats, control panels, actuators, interconnecting wiring and tubing.
  Ductwork
P Supply and extract: Galvanized mild steel ductwork, fittings and supports, mixing boxes (for induction system), dampers, grilles and diffusers, insulation.
  Air conditioning plant
Q Heating and cooling batteries: Humidifiers, batteries and casing and connections.
R Fans and filters: Centrifugal and axial flow fans with casings and connections, and automatic roll type filters.
S Sound attenuation: Silencers and duct lining (short lengths only).
T **Fire detection:** Heat detectors, smoke detectors, gas detectors, control panel, interconnecting wiring (excluding other fire protection services not directly associated with air conditioning installation).
V **Electrical work in connection:** Electrical supplies to control panels and mechanical plant, mechanical services distribution board.

Approximate Estimating

MECHANICAL INSTALLATIONS

BOILER PLANT

Low pressure hot water boiler plant of welded construction comprising 2 No boilers with instruments and boiler mountings and burners for town or natural gas, combustion chamber reverse gas flow principle combustion including delivery and commissioning (*p.c.* as shown) 2 No heating circulating pumps (*p.c.* as shown) 2 No hot water service supply pumps (*p.c.* as shown) 2 No hot water service circulation pumps (*p.c.* as shown). Boiler control panel and thermostatic controls for heating and hot water service circuits, weather compensator, automatic controls, pipework and fittings including valves to boilers and pumps and to hot water service calorifiers with all necessary lagging together with feed and expansion tank and cold feed and vent pipes.

	Total cost of plant £
Total boiler capacity 275 kW Boilers *p.c.* £2000·00 each Heating pumps *p.c.* £1250·00 each Hot water service primary pumps *p.c.* £650·00 each Hot water service secondary pumps *p.c.* £825·00 each	40,000·00
Total boiler capacity 1100 kW Boilers *p.c.* £5500·00 each Heating pumps *p.c.* £1250·00 each Hot water service primary pumps *p.c.* £750·00 each Hot water service secondary pumps *p.c.* £1000·00 each	44,000·00
Total boiler capacity 2250 kW Boilers *p.c.* £8100·00 each Heating pumps *p.c.* £1250·00 Hot water service primary pumps *p.c.* £1050·00 each Hot water service secondary pumps *p.c.* £1000·00 each	67,200·00

Low pressure hot water boiler plant of welded construction comprising 2 No boilers and combined calorifiers with instruments and boiler mounting and burners for natural gas, combustion chamber reverse gas flow principle combustion including delivery and commissioning (*p.c.* as shown) 2 No heating circulating pumps, (*p.c.* as shown) 2 No hot water service circulation pumps (*p.c.* as shown). Boiler control panel and thermostatic controls for heating and hot water service circuits, weather compensator, automatic controls, all interconnecting pipework and fittings including valves to boilers and pumps with all necessary lagging together with feed and expansion tank and cold feed and vent pipes.

Total boiler capacity 275 kW Boilers *p.c.* £2500·00 each Heating pumps *p.c.* £1250·00 each Hot water service secondary pumps *p.c.* £850·00 each	47,500·00
Total boiler capacity 550 kW Boilers *p.c.* £3200·00 each Heating pumps *p.c.* £1250·00 each Hot water service secondary pumps *p.c.* £1000·00 each	48,750·00
Total boiler capacity 1100 kW Boilers *p.c.* £5600·00 each Heating pumps *p.c.* £1250·00 each Hot water service secondary pumps *p.c.* £1050·00 each	51,000·00

MECHANICAL INSTALLATIONS

BOILER PLANT

Low pressure hot water boiler plant of welded construction comprising 2 No boilers with instruments and boiler mountings and burners for 35 second oil including delivery and commissioning (*p.c.* as shown) 2 No heating circulating pumps (*p.c.* as shown) 2 No hot water service supply pumps (*p.c.* as shown) 2 No hot water service circulating pumps (*p.c.* as shown) 2 No oil storage tanks (*p.c.* as shown). Boiler control panel and thermostatic controls for heating and hot water service circuits, weather compensator, automatic controls all interconnecting pipework and fittings including valves to boilers and pumps with all necessary lagging together with feed and expansion tank and cold feed and vent pipes.

	Total cost of plant £
Total boiler capacity 275 kW Boilers *p.c.* £1600·00 each Heating pumps *p.c.* £1250·00 each Hot water service primary pumps *p.c.* £650·00 each Hot water service secondary pumps *p.c.* £900·00 each Oil storage tank *p.c.* £750·00 each	46,900·00
Total boiler capacity 1100 kW Boilers *p.c.* £3800·00 each Heating pumps *p.c.* £1250·00 each Hot water service primary pumps *p.c.* £825·00 each Hot water service secondary pumps *p.c.* £1000·00 each Oil storage tank *p.c.* £1500·00 each	60,400·00
Total boiler capacity 2250 kW Boilers *p.c.* £5500·00 each Heating pumps *p.c.* £1250·00 each Hot water service primary pumps *p.c.* £1100·00 each Hot water service secondary pumps *p.c.* £1000·00 each Oil storage tank *p.c.* £2500·00 each	75,000·00

Low pressure hot water boiler plant of welded construction comprising 2 No boilers and combined calorifiers with instruments and boiler mountings and burners for 35 second oil including delivery and commissioning (*p.c.* as shown) 2 No heating circulation pumps (*p.c.* as shown) 2 No hot water service circulating pumps (*p.c.* as shown) 2 No oil storage tanks (*p.c.* as shown). Boiler control panel and thermostatic controls for heating and hot water service circuits, weather compensator, automatic controls all interconnecting pipework and fittings including valves to boilers and pumps and to hot water service calorifiers with feed and expansion tank and cold feed and vent pipes.

Total boiler capacity 275 kW Boilers *p.c.* £2200·00 each Heating pumps *p.c.* £1250·00 each Hot water service secondary pumps *p.c.* £900·00 each Oil storage tank *p.c.* £750·00 each	54,500·00
Total boiler capacity 550 kW Boilers *p.c.* £2800·00 each Heating pumps *p.c.* £1250·00 each Hot water service secondary pumps *p.c.* £1100·00 each Oil storage tank *p.c.* £1150·00 each	56,750·00
Total boiler capacity 1100 kW Boilers *p.c.* £5200·00 each Heating pumps *p.c.* £1250·00 each Hot water service secondary pumps *p.c.* £1000·00 each Oil storage tank *p.c.* £1500·00 each	67,200·00

Approximate Estimating

MECHANICAL INSTALLATIONS

DISTRIBUTION PIPEWORK HEATING OUTSIDE PLANT ROOM

Mild steel tube to B.S. 1387 with joints in the running length allowance for waste, fittings and brackets assuming average runs.

	Cost per metre Screwed £	Welded £
Black Medium Weight		
15 mm	13·50	14·90
20 mm	7·20	8·40
25 mm	7·65	8·90
32 mm	8·15	9·20
40 mm	8·60	9·60
50 mm	17·90	20·80
65 mm	21·40	25·20
80 mm	24·50	24·30
100 mm	21·12	23·70
125 mm	25·70	28·30
150 mm	48·90	40·00
Black Heavy Weight		
15 mm	13·90	15·20
20 mm	7·60	8·90
25 mm	8·10	9·90
32 mm	8·80	10·40
40 mm	9·25	10·50
50 mm	18·60	21·90
65 mm	22·30	26·70
80 mm	24·70	26·70
100 mm	22·90	23·70
125 mm	28·00	38·40
150 mm	48·40	46·20

Approximate Estimating
MECHANICAL INSTALLATIONS

VALVES

Bronze globe valve with renewable disc joint to pipework with screwed joints.

		£
15 mm	No.	9·50
20 mm	No.	12·50
25 mm	No.	16·30
32 mm	No.	21·50
40 mm	No.	27·20
50 mm	No.	38·50

Malleable iron ball type isolating gas valve with flanged ends to B.S. table D including bolted joints.

80 mm	No.	105·00
100 mm	No.	255·00

Bronze wedge disc non-rising stem gate valve with screwed joints.

20 mm	No.	13·10
25 mm	No.	16·10
32 mm	No.	21·50
40 mm	No.	27·00
50 mm	No.	37·00

Gate valve as above but with flanged ends to B.S. table F including bolted joints.

50 mm	No.	80·00
76/80 mm	No.	140·00

Cast iron gate valve with non-rising stem with flanged ends to B.S. table D including bolted joints.

50 mm	No.	50·00
80 mm	No.	58·00
100 mm	No.	75·00
150 mm	No.	125·00

Bronze equilibrium ball valve, hydraulic working pressure 10 bar, with flanged end to B.S. table D including bolted joint.

50 mm	No.	260·00
76/80 mm	No.	340·00

Bronze straight pattern radiator valve wheelhead or lockshield matt finish and screwed joints to tube.

15 mm	No.	9·00
20 mm	No.	12·00
25 mm	No.	15·00

Bronze angle pattern radiator valve wheelhead or lockshield matt finish and screwed joints to tube.

15 mm	No.	8·00

For additional types of valves see 'Prices for Measured Work' section.

MECHANICAL INSTALLATIONS

RADIATORS

Pressed steel panel type radiator prime finish including brackets and fixing, taking down once for painting by others. (cost per m² of heating surface)

		305 mm high £	432 mm high £	584 mm high £	740 mm high £
Single Panel	m²	20·80	17·60	18·40	17·70
Double Panel	m²	16·70	14·20	14·50	15·00

For floor or wall mounted individual convector and unit heaters see 'Prices for Measured Work' section.

HOT WATER SERVICE

Light gauge copper tube to B.S. 2871 part 1 table X with joints as described including allowance for waste, fittings and brackets assuming average runs.

Cost per metre £

Capillary Joints
- 15 mm . . . . 8·50
- 22 mm . . . . 5·90
- 28 mm . . . . 7·30
- 35 mm . . . . 7·75
- 42 mm . . . . 11·90
- 54 mm . . . . 13·25

Compression Joints
- 15 mm . . . . 7·70
- 22 mm . . . . 5·80
- 28 mm . . . . 7·70
- 35 mm . . . . 8·35
- 42 mm . . . . 13·40
- 54 mm . . . . 13·00

Bronze Welded Joints
- 76 mm . . . . 46·30

CALORIFIERS AND PUMPS

Calorifiers and pumps have not been detailed in this section, for sizes and types see 'Prices for Measured Work' section.

COLD WATER SERVICE

Light gauge copper tube to B.S. 2871 part 1 table X with joints as described including allowance for waste, fittings and brackets assuming average runs.

Cost per metre £

Capillary Joints
- 15 mm . . . . 8·20
- 22 mm . . . . 7·00
- 28 mm . . . . 8·60
- 35 mm . . . . 8·80
- 42 mm . . . . 13·00
- 54 mm . . . . 13·80

Compression Joints
- 15 mm . . . . 9·45
- 22 mm . . . . 7·35
- 28 mm . . . . 8·80
- 35 mm . . . . 9·30
- 42 mm . . . . 14·20
- 54 mm . . . . 13·60

Bronze Welded Joints
- 76 mm . . . . 48·40

MECHANICAL INSTALLATIONS

THERMAL INSULATION

Asbestos free calcium silicate sections 25 mm thick fixed with aluminium bands at 450 mm intervals to pipework of the nominal sizes shown, including all pipe fittings, flanges and valves.

Nominal size mm	Cost per metre £
15	5·00
20/22	4·95
25/28	5·20
40/42	5·50
50/54	7·50
76/80	10·30
100/108	10·90
150/159	13·80

PUMPS

Pumps have not been detailed in this section, for sizes and types see 'Prices for Measured Work' section.

TANKS

Pressed steel sectional tank of standard construction, weather-proof cover, one 460 mm manhole with hinged lid and one 150 mm cowl ventilator. An allowance has been made for pipe connections. Finished one coat bituminous primer. Erected on prepared base at ground level or on independent prepared structures up to 15 m high.

Capacity	Cost per unit £
25,500 litres	4800·00
83,500 litres	9425·00
480,000 litres	33100·00
1,025,000 litres	59500·00

AIR CONDITIONING AND VENTILATION

To calculate the weight of ductwork multiply the duct length (measured on the centre line overall fittings) by their respective girths and then applying the appropriate sheeting weights to the superficial areas without making any allowance for the additional sheeting in joints, seams, welts, waste, etc. The rates below allow for ductwork and for all other labour and material in fabrication fittings, supports and jointing to equipment, stop and capped ends, elbows, bends, diminishing and transition pieces, regular and reducing couplings, branch diffuser and 'snap on' grille connections, ties, 'Ys', crossover spigots, etc., turning vanes, damper access doors and openings, handholes, test holes and covers, blanking plates, flanges, stiffeners tie rods and all supports and brackets fixed to concrete or brickwork.

Rectangular low velocity galvanized mild steel ductwork in accordance with table 1 of the HVCA DW 141 (metric).
£3000 per tonne
Rectangular high velocity galvanized mild steel ductwork in accordance with table 2 of the HVCA DW 141 (metric)
£3500 per tonne

Approximate Estimating

MECHANICAL INSTALLATIONS

AIR CONDITIONING AND VENTILATION *continued*

Package air handling unit with a duty of 0·94 m³/sec against 250 N/m² resistance. Comprising mixing box, automatic roll fitter, spray coil humidifier with 8 row cooling coil, 2 row reheat coil and centrifugal fan complete with anti-vibration mountings and flexible connections . . . .
As above but with a duty of 2·10 m³/sec against 500 N/m² resistance . . .
As above but with a duty of 5·10 m³/sec against 500 N/m² resistance . . .
As above but with a duty of 9·10 m³/sec against 500 N/m² resistance . . .
See *'Prices for Measured Work' section for fans and roof extract units.*

Cost per unit
£
3950·00
6250·00
9025·00
15200·00

Chilled water installation comprising two refrigeration compressors, cooling towers, pumps, pipework valves and fittings, controls, thermostatic controls, insulation, anti-vibration mountings and starters

Cost per ton of cooling capacity
£
390–480

Recommended air conditioning loads for various applications:

Computer rooms	. 10 m² of floor area per ton
Restaurants	. 16 m² of floor area per ton
Banks (main area)	. 22 m² of floor area per ton
Large Office Buildings (exterior zone)	. 25 m² of floor area per ton
Supermarkets	. 30 m² of floor area per ton
Large Office Block (interior zone)	. 32 m² of floor area per ton
Small Office Block (interior zone)	. 35 m² of floor area per ton

FIRE FIGHTING SERVICES

100 mm dry riser main including dry riser breeches horizontal inlet with 2 No. 64 mm male instantaneous coupling and landing valve. Complete with padlock and leather strap.

£

Price per landing . . . . 370·00
Hose reel pedestal mounted type with 37 metres of hose including approximately 15 metres of pipework:
Price per hose reel . . . . 280·00
Sprinkler installation including sprinkler head and all associated pipework, valve sets, booster pumps and water storage:
Price per sprinkler head . . . . 85·00
Recommended maximum area coverage per sprinkler head:
Extra light hazard, 21 m² of floor area.
Ordinary hazard, 12 m² of floor area.
Extra high hazard, 9 m² of floor area.

MECHANICAL INSTALLATIONS

DOMESTIC CENTRAL HEATING

Solid fuel central heating installation comprising either open fire room heater or boiler, fuel storage, pump, small or microbore distribution pipework, steel radiators, valves, expansion tank, room thermostat, and all insulation:

	£
Heating for 3 rooms comprising 2 radiators plus hot water service	840–970
Heating for 4 rooms comprising 3 radiators plus hot water service	950–1120
Heating for 5 rooms comprising 4 radiators plus hot water service	1120–1270
Heating for 6 rooms comprising 5 radiators plus hot water service	1230–1350
Heating for 7 rooms comprising 6 radiators plus hot water service	1350–1400
Heating for 7 rooms comprising 7 radiators plus hot water service	1450–1650

Gas fired central heating installation comprising boiler, pump small or microbore distribution pipework, steel radiators, valves, expansion tank, room thermostat and all insulation:

Heating for 3 rooms comprising 3 radiators plus hot water service	925–1230
Heating for 4 rooms comprising 4 radiators plus hot water service	1220–1350
Heating for 5 rooms comprising 5 radiators plus hot water service	1350–1450
Heating for 6 rooms comprising 6 radiators plus hot water service	1425–1525
Heating for 7 rooms comprising 7 radiators plus hot water service	1525–1850

Oil fired central heating installation comprising oil storage tank and supports, pump, room thermostats, small or microbore distribution pipework, steel radiators, valves and insulation together with hot water cylinder for hot water supply:

Heating for 7 rooms comprising 7 radiators and boiler plus hot water service . .	1600–1850
Heating for 10 rooms comprising 10 radiators and boiler plus hot water service .	1530–2020

Approximate Estimating

LIFT AND ESCALATOR INSTALLATIONS

LIFT INSTALLATIONS

The cost of lift installations will vary depending upon a variety of circumstances. The following prices assume a floor to floor height of 3 metres and standard finishes to cars and gates.

Passenger lifts
Electrically operated two speed general purpose 10 person lift serving 10 levels with directional collective controls and a speed of 1 metre per second . . . . £37,000 per lift
    Add to above for
        Bottom motor room . . . . £1,000 per lift
        Extra levels served . . . . £1,750 per level
        Increased speed of travel from 1 to 1·50 metres per second . . £1,000 per lift
        Increased speed of travel from 1 to 2·50 metres per second . . £17,000 per lift
        Enhanced finish to car . . . . £2,000 per car
As above but 20 person lift with a speed of 1·5 metres per second . £43,000 per lift
    Add to above for
        Bottom motor room . . . . £1,000 per lift
        Extra levels served . . . . £1,750 per level
        Increased speed of travel from 1·5 to 2·5 metres per second . £16,000 per lift
        Enhanced finish to car . . . . £2,000 per car
For any floor bypassed add 45% of the cost of an extra level served.

Goods lifts
Electrically operated two speed general purpose goods lift serving 5 levels, to take 500 kg load; inter manually operated shutters and automatic push button control and a speed of 0·25 metres per second £22,500 per lift
    Add to above for
        Extra levels served . . . . £1,250 per level
        Increased capacity up to 2000 kg . . . . . £5000 per lift
        Increased speed of travel from 0·25 to 0·50 metres per second . £500 per lift
Electrically operated heavy duty goods lift serving 5 levels, to take 1500 kg load with manually operated doors and automatic push button control and a speed of 0·25 metres per second . . . £22,500 per lift
    Add to above for
        Increased capacity up to 3000 kg . . . . . £7,000 per lift
        Extra levels served . . . . £1250 per level
        Through car . . . . £1,000 per lift
Oil hydraulic operated heavy duty goods lift serving 4 levels, to take 500 kg load with manually operated shutters and automatic push button control and a speed of 0·25 metres per second . . . £23,000 per lift
    Add to above for
        Increased capacity up to 1000 kg . . . . . £1,000 per lift
        Increased capacity up to 1500 kg . . . . . £2,000 per lift
        Increased capacity up to 2000 kg . . . . . £2,500 per lift

ESCALATOR INSTALLATIONS

35° pitch escalator with a rise of 3·50 metres and with opaque balustrades
    600 mm tread . . . . £26,900·00 per escalator
    810 mm tread . . . . £27,500·00 per escalator
    1000 mm tread . . . . £29,900·00 per escalator
    Extra for glass balustrades . . . . £800 per escalator

ELECTRICAL INSTALLATIONS

BASIC SERVICES – PRICES PER SQUARE METRE

Approximate prices per m² of floor area for the following electrical services:	Secondary Schools £	University arts and administration buildings £	Offices £	Local authority flats £	Local authority houses £	Non-teaching hospitals £
Lighting	4·30	6·80	5·60	4·00	3·40	6·15
Power	3·20	7·20	5·50	3·55	3·15	7·60
Mains and switchgear	5·00	5·75	3·70	1·80	1·10	6·15
Lighting fittings	6·20	8·75	8·20	1·80	1·70	9·10
Ancillary equipment	5·30	8·00	8·00	0·65	0·65	11·00
Total approximate cost of above services	24·00	36·50	31·00	11·80	10·00	40·00

Notes:
The price of electrical services in science teaching buildings, including special services is likely to be:

Physics buildings . . . . £40·00 per m²
Chemistry buildings . . . . £42·00 per m²
Biological science buildings . . . . £45·00 per m²

Local authority flats and houses: The prices are average prices assuming dwellings accommodating 2 to 6 persons.

Ancillary equipment: This includes, where appropriate, telephone outlets, TV and radio aerial outlets, fire-alarm installation, connections to mechanical services, staff call systems and public address equipment.

Approximate Estimating

ELECTRICAL INSTALLATIONS

LIGHTING AND POWER

Approximate prices for wiring of lighting and power points complete, including accessories and socket outlets with plugs but excluding lighting fittings, consumer control units shown separately.

Consumer control units
8-way 60 amp S.P. & N. surface mounted insulated consumer control units fitted with miniature circuit breakers including 2 m long 32 mm screwed welded conduit with three runs of 16 mm² PVC cables ready for final connections by the supply authority .   *Each*   78·00
As above but 100 amp metal cased consumer control unit and 25 mm² PVC cables   .   90·00

Lighting circuits   *Per point*
Wired in PVC insulated and PVC sheathed cable in flats and houses installed in cavi-   £
ties and roof space protected, where buried, by light gauge conduit . . . . 19·50
As above but in commercial property . . . . 23·50
Wired in PVC insulated cable in screwed welded conduit in flats and homes . . 33·50
As above but in commercial property . . . . 42·00
As above but in industrial property . . . . 48·00
Wired in M.I.C.C. cable in flats and houses . . . . 38·00
As above but in commercial property . . . . 36·50
As above but in industrial property . . . . 47·50

Single 13 amp switched socket outlets
Wired in PVC insulated and PVC sheathed cable in flats and houses on a ring main circuit protected, where buried, by light gauge conduit . . . . . 23·00
As above but in commercial property . . . . 27·25
Wired in PVC insulated cable in screwed welded conduit throughout on a ring main circuit in flats and homes . . . . 35·50
As above but in commercial property . . . . 36·75
As above but in industrial property . . . . 45·50
Wired in M.I.C.C. cable on a ring main circuit in flats and houses . . . . 36·25
As above but in commercial property . . . . 41·50
As above but in industrial property . . . . 53·50

Cooker control circuits
30 amp circuit including unit, wired in PVC insulated and PVC sheathed cable, protected where buried, by conduit . . . . 45·00
As above but wired in PVC insulated cable in screwed welded conduit . . . 55·75
As above but wired in M.I.C.C. cable . . . . 76·00

ELECTRICAL INSTALLATIONS

ELECTRIC HEATING

Electric underfloor heating by heating cable
　The approximate cost of solid embedded underfloor heating including distribution fuseboard, time switches and thermostats is £10·75 per m² of floor area based on 145 watts per m².

Storage heaters
　Storage heater with built in thermostat, temperature controller and safety cut-out.

Capacity kW	Supply and installation cost of storage heating including wiring Unit £
2·000 (P.C. £75·00)	130·00
2·625 (P.C. £85·00)	140·00
3·375 (P.C. £100·00)	160·00

Note

Cost of wiring is based on 10 metre run of circuit including a 25 amp double pole switch with metal box, share of distribution fuse board and 4 mm² PVC cable drawn into 20 mm conduit.
For wiring with PVC insulated and PVC sheathed cable the above costs may be reduced by £6·00 in each case.

Approximate Estimating 319

ELECTRICAL INSTALLATIONS

HEAVY POWER AND ANCILLARY INSTALLATIONS

Indoor substation equipment
The installation cost of equipment for an indoor substation is governed by various factors e.g. the duty it is required to perform, local electricity authority requirements, earthing needs depending on the soil resistivity of the ground in the vicinity of the substation and the actual location of the site, etc.
An accurate cost of proposed substation equipment can be estimated only when full design details and all relevant information is available. However for budgetary purposes cost figures given below in respect of typical equipment utilised in a modern substation may serve as a helpful guide.
Three phase 11 kV 250 MVA manually operated five panel metalclad switchboard including kWH meter, maximum demand indicator with associated voltage and current transformers, interconnecting cables, cable boxes and cable glands comprising:
Two incoming 400 amp oil circuit breakers complete with one set of busbars.
Two outgoing 400 amp oil circuit breakers each complete with one set of busbars, overcurrent and earth fault protection.
One bus section and metering panel, one 400 amp oil circuit breaker and two sets of busbars.
Three phase and neutral 415 volt 26 MVA manually operated two tier, nineteen circuit seven cubicle metalclad switchboard including cable boxes and cable glands comprising:
Two 1250 amp incoming air circuit breakers each complete with one set of busbars, one voltmeter, one ammeter and thermomagnetic overload relay.
One 1250 amp bus section complete with two sets of busbars.
Ten 200 amp T.P. & N. fuse switches complete with HBC fuse-links.
Six 600 amp T.P. & N. fuse switches complete with HBC fuse-links.
Two 800 kVA 11 kV/433 volt 50 h.z. delta/star double wound naturally cooled oil immersed core type transformers with off load tap changers.

Cost per kVA
£
Equipment . . . . 30·00
Installation including testing and commissioning . . . . . 7·00

Note
Copper used in the manufacture of transformers is based on a price of £1100·00 per tonne.
Above costs do not include for first filling of oil in the oil circuit breaker tanks.
No allowance has been made in the above costs for earthing system beyond the earth terminals provided on the equipment by manufacturers.
Installation cost is exclusive of builder's work.
Low voltage distribution fuseboards.
500 volt distribution fuseboard of sheet steel all welded fabricated construction with hinged and lockable door complete with fuse holders and HBC fuse links fixed to brickwork or concrete and connected.

No. of ways	20 amp T.P. & N. £	30 amp T.P. & N. £	60 amp T.P. & N. £	100 amp T.P. & N. £
4	58·00	94·00	138·00	235·00
8	90·00	155·00	235·00	
12	125·00	215·00	325·00	

Approximate Estimating
ELECTRICAL INSTALLATIONS

HEAVY POWER AND ANCILLARY INSTALLATIONS continued

RISING MAIN BUSBAR TRUNKING

Four pole mild steel sheet rising main busbar trunking with copper busbars, complete with tap off points, fire barriers at 3·5 m intervals, one thrust block, one expansion joint assembly and one cable entry chamber with glands for incoming cables and including tap off cable clamps, interconnections, internal wiring, shields, phase buttons, labels and external earth tape; cost of installation based on a length of 60 m long rising main.

	Per metre £
200 amp rising main	70·00
300 amp rising main	80·00
400 amp rising main	94·00
600 amp rising main	125·00

SWITCHBOARD SAFETY MATTING

	Per square metre £
6 mm thick matting tested to 11,000 volts	14·50
10 mm thick matting tested to 15,000 volts	22·00

	Each £
Electric shock treatment instructions	
Printed on paper	0·75
Printed on cardboard	1·50
Printed on plasticized card	2·30

Low voltage distribution cables, PVC, SWA, PVC
600/1000 volt grade four core l.v. distribution cable with stranded copper conductors, PVC insulated, PVC tape bedded, single steel wire armoured and PVC sheathed to B.S. 6346/69 including clamps and supports, fixed to brickwork or concrete.

Conductor cross-sectional area mm^2	per metre £	per termination including gland £
25	7·00	5·25
35	8·00	6·50
50	10·00	8·00
70	12·50	10·30
95	15·50	13·75
120	18·75	15·60
150	22·00	19·00
185	26·50	22·80
240	34·00	29·00

Note
Items of copper in the above cables are based on copper market price of £1100·00 per tonne. Cost of cables is based on ordering quantities between 200 and 1000 metres.

Low voltage power circuits
Three phase four wire circuit feeding an individual load, wired with PVC insulated cable drawn in heavy gauge black enamelled screwed welded conduit including standard associated fittings and flexible PVC sheathed metallic conduit, not exceeding one metre long, fixed to brickwork or concrete with distance saddles in surface work per 10 metre run.

Approximate Estimating 321

ELECTRICAL INSTALLATIONS

HEAVY POWER AND ANCILLARY INSTALLATIONS *continued*

Cable size mm^2	Conduit size mm	£
1·5	20	50·50
2·5	20	53·00
4	20	58·50
6	25	77·00
10	25	90·00
16	32	124·00

Three phase four wire circuit feeding an individual load item, wired with M.I.C.C./PVC cable including terminations, glands and flexible PVC sheathed metallic conduit not exceeding 1 metre long and through conduit box with connectors, fixed with PVC covered clips, per 10 metre run.

Cable size mm^2	£
1·5	67·50
2·5	81·50
4	92·00
6	108·00
10	134·00
16	175·00

Fluorescent lighting fittings
  Fluorescent lighting fitting complete with 1 metre long drop rods, three core heat resistant flexible cable, connector block, control gear, tubes, conduit boxes, supports and fixings, fixed to brickwork or concrete and connected excluding cost of wiring (in case of recessed fitting opening in false ceiling provided by ceiling contractor)

	£
4' single 40 watt surface type with metal reflector	28·00
6' single 75 watt surface type with metal reflector	36·00
8' single 100 watt surface type with metal reflector	46·00
6' twin 75 watt surface type with plastic diffuser	53·00
6' twin 75 watt recessed type with opal diffuser	82·00

Air handling fluorescent fitting
  6' twin 75 watt recessed type air handling fitting complete with main body, reflector louvre, three core heat resistant flexible cable, connector blocks, control gear, tubes, conduit boxes and drop rods not exceeding 1 metre long, fixed to brickwork or concrete and connected.
  Installed cost excluding wiring, fixing and levelling of main body to suspended ceiling, supplying and fixing of flow plates or any other mechanical items, e.g. air boxes etc., as part of air flow system.

£
per fitting 95·00

Emergency lighting circuits
  Wired in 2·5 mm^2 PVC insulated cable drawn in 20 mm heavy gauge black enamelled screwed welded conduit including standard associated fittings, fixed to brickwork or concrete with spacer bar saddles in surface work and connected.
  Installed cost based on an average circuit run of 10 metres per point including lighting fitting and share of distribution board, battery charger and automatic changeover control gear.

£
per point 146·00

Fire alarm circuits
  Breakglass contacts and alarm bells wired in four wire system with 1·5 mm^2 PVC insulated cable drawn in 20 mm heavy gauge black enamelled screwed welded conduit including standard associated fittings, fixed to brickwork or concrete with spacer bar saddles in surface work and connected including share of alarm indicator panel, relays, battery and trickle charger.

Approximate Estimating

ELECTRICAL INSTALLATIONS

HEAVY POWER AND GENERAL continued

Installed cost per pair of breakglass contact and alarm bell based on an average distance of 30 metres between contacts or bells.

£
320·00

Note:
For inclusion of smoke detectors allow £120·00 per unit.
For inclusion of heat detectors allow £40·00 per unit.

Clock circuits
  D.C. clock circuit wired with 2 × 1·5 mm^2 single core PVC insulated cable drawn in 20 mm heavy gauge black enamelled screwed welded conduit including standard associated fittings, clock connector and share of battery unit fixed to brickwork or concrete with spacer bar saddles in surface work.
  Cost based on an average circuit run of 30 metres per clock point (excluding provisions for clocks).

£
per clock point   155·00

  A.C. mains clock circuit wired as above but cost based on an average circuit run of 4 metres per clock point.

£
20·00

Power and telephone common underfloor ducts.
  225 × 25 mm three compartment underfloor steel cable duct including associated fittings and complete with power and telephone outlet boxes, laid in screed, excluding builder's work.
  Installed cost based on an approximate run of 2 metres of ducting per 10 m^2 of floor area

£
per metre   17·10

Floor box outlets, twin socket outlets with plug tops and twin telephone outlets
  Installed cost based on an average of one box per 5 metres of ducting

per box   42·00

Boiler plant automatic control panels
  For budgetary purposes installation costs of boiler plant automatic control panels and associated wiring are given below. These costs relate to standard equipment normally used for the electrification of boiler plants associated with office blocks and various buildings. The equipment consists of metal encased panel fitted with automatic controllers for motors and motorized valves, isolators for incoming supply and individual load items, indicating instruments and all necessary control and main wiring in M.I.C.C./PVC cable fixed to cable tray with PVC covered clips.

Total capacity with two boilers kW	Installed cost £
275	5450·00
1100	8600·00
2250	12200·00

Note:
Above boilers are standard low pressure hot water boilers using natural gas for combustion wiring and connections only, to mechanical services.
  The cost of electrical connections to mechanical services equipment will obviously vary depending on the type of building and complexity of the equipment but an allowance of £3·10 per m^2 of floor area should be a useful guide.

Approximate Estimating

ELECTRICAL INSTALLATIONS

EXTERNAL LIGHTING

Estate road lighting.
Post type road lighting lantern 80 watt mbf/u complete with 4·5 m high column with hinged lockable door, control gear and cut-out including all internal wiring, interconnections and earthing, fed by 2·5 mm² two core PVC SWA PVC underground cable.
Approximate installed price per metre road length (based on 300 metres run) including time switch but excluding all trench and builder's work
    Columns erected on same side of road at 30 m intervals . . . . £9·00
    Columns erected on both sides of road at 60 m intervals in staggered formation . £11·00
Bollard lighting
Bollard lighting fitting 50 watt mbf/u including controlgear, all internal wiring, interconnections, earthing and 10 metres of 2·5 mm² two core PVC SWA PVC underground cable
    Approximate installed price excluding trench work and builder's work . . . £110·00
Outdoor flood lighting
Wall mounted outdoor flood light fitting complete with tungsten halogen lamp, mounting bracket, wire guard and all internal wiring; fixed to brickwork or concrete and connected
    Installed price 500 watt . . . . £48·00 each
    Installed price 1000 watt . . . . £60·00 each
Pedestal mounted outdoor flood light fitting complete with 1000 watt mbf/u lamp, mounting bracket, control gear, contained in weatherproof steel box, all internal wiring, interconnections and earthing; fixed to brickwork or concrete and connected
    Approximate installed price excluding builder's work . . . . . £330·00 each

LIGHTNING PROTECTION

Lightning conductor tape with standard associated clips fixed to brickwork or concrete surfaces

	Per metre £
25 mm × 3 mm aluminium tape	3·10
25 mm × 3 mm aluminium tape PVC covered	3·75
25 mm × 3 mm copper tape	5·60
25 mm × 3 mm copper tape PVC covered	6·40

	Each £
16 mm diameter steel cored copper bond earth rod 2·44 metres long including bronze coupling, driving stud, connection clamp and 'Denso' protected joint, driving earth rod into ground and making connections	50·00
Gunmetal test clamp and connecting	8·50
Concrete inspection pit comprising surround and lid . . . (P.C.)	7·50

Note:
Due to a large variation in the earthing requirements for different projects, installation costs of special items, e.g. air terminals, earthplates and various types of tape fixings, have not been included.
Installation cost of copper tape is based on copper market price of £1100·00 per tonne.

STANDBY DIESEL GENERATING SET

440 volt three phase four wire 50 Hz packaged standby skid mounted diesel generating set complete with radio and television suppressors, fuel tank with associated piping, fresh air inlet and exhaust ducts, control panel, mains failure relay, starting battery with charger, all internal wiring, interconnections, earthing and labels.

Budgetary price of installation including connecting, testing and commissioning, but excluding builder's work	Per kVA £
50 to 150 kVA rating	130·00
200 to 500 kVA rating	110·00
600 to 1000 kVA rating	125·00

COST INDICES

The following tables reflect the major changes in cost to contractors but do not necessarily reflect changes in tender levels. In addition to changes in labour and materials costs tenders are affected by other factors such as the degree of competition in the particular industry and area where the work is to be carried out, the availability of labour and the general economic situation. This has meant in recent years that when there has been an abundance of work tender levels have often increased at a greater rate than can be accounted for by increases in basic labour and material costs and, conversely, when there is a shortage of work this has often resulted in keener tenders. Allowances for these factors are impossible to assess on a general basis and can only be based on experience and knowledge of the particular circumstances.

In compiling the tables the cost of labour has been calculated on the basis of a notional gang as set out elsewhere in the book. The proportion of labour to materials has been assumed as follows:

Mechanical Services   30:70
Electrical Services   50:50

Mechanical Services

January 1965 = 100

Year	First quarter	Second quarter	Third quarter	Fourth quarter
1965	103	105	105	105
1966	110	110	112	113
1967	116	116	118	118
1968	124	124	125	125
1969	130	130	131	131
1970	137	138	139	142
1971	147	150	150	153
1972	162	165	167	169
1973	171	177	183	195
1974	213	232	244	252
1975	267	277	289	299
1976	307	324	341	356
1977	364	371	376	383
1978	401	410	417	443
1979	457	468	497	520
1980	548			

Electrical Services

January 1965 = 100

Year	First quarter	Second quarter	Third quarter	Fourth quarter
1965	102	103	104	104
1966	104	111	112	115
1967	118	123	123	124
1968	132	132	136	137
1969	137	137	142	146
1970	146	150	153	160
1971	165	166	166	168
1972	178	181	184	187
1973	189	200	207	214
1974	235	249	259	263
1975	294	299	316	327
1976	351	368	377	382
1977	402	407	408	411
1978	448	474	478	486
1979	528	537	542	569
1980	624			

Approximate Estimating 325

ELEMENTAL COST PLAN

The Elemental Cost Plan which follows is an example of how priced approximate quantities may be used to compile a detailed estimate for the mechanical and electrical service, to a building using a known or assumed specification.

The rates used in the pricing of the Cost Plan are taken from the *Approximate Estimating* or *Prices for Measured Work* sections. In the latter case page references have been given to assist the reader.

This example is for a four storey office block having a floor area of 2400 m² (measured within the external walls and over all internal walls and partitions) and a car park of 500 m² forming part of the ground floor.

The mechanical services include, heating, hot and cold water installations, gas and fire fighting installations.

The electrical services include, sub-mains, lighting and power, connections to mechanical services, telephone conduit and fire alarms.

The resultant sums have been used in the elemental cost plan for the complete office block which appears in the companion volume of *Spons Architects' and Builders' Price Book.*

Element	Total cost of element £	Cost of element per m² of floor area £
MECHANICAL SERVICES		
Cold water installation		
Serving 24 No lavatory basins, 4 No sinks and 24 No w.c. suites.		
Rising main		
Light gauge copper tube to B.S. 2871 part 1, table X, with capillary joints, and fittings,		
15 mm diameter tube and fittings, 16 m @ £8·20	131·20	
Light gauge copper tube to B.S. 2871 part 1, table X, with bronze welded joints and fittings,		
76 mm diameter tube and fittings, 25 m @ £48·40	1210·00	
76/80 mm flanged gate valve, 2 No @ £125·00	250·00	
76/80 mm flanged gunmetal equilibrium ball valve, 1 No @ £340·00	340·00	
Pressed steel sectional storage tank, 83,500 litres capacity 1 No @ £9050·00	9050·00	
25 mm thick asbestos free calcium silicate sectional insulation to 15 mm diameter pipework,		
16 m @ £5·00	80·00	
Continued	11061·20	— —

ELEMENTAL COST PLAN

Element		Total cost of element £		Cost of element per m^2 of floor area £
	Continued	11061·20	—	—

MECHANICAL SERVICES continued
25 mm thick asbestos free calcium silicate sectional insulation to 76 mm diameter pipework,
25 m @ £10·30 . . . . 257·50   11318·70   4·72

Cold water down service
Light gauge copper tube to B.S. 2871, part 1, table X, with capillary joints and fittings,
15 mm diameter, 80 m @ £8·20 . . . . 656·00
22 mm diameter, 56 m @ £7·00 . . . . 392·00
28 mm diameter, 35 m @ £8·60 . . . . 301·00
35 mm diameter, 45 m @ £8·80 . . . . 396·00
42 mm diameter, 20 m @ £13·00 . . . . 260·00
54 mm diameter, 62 m @ £13·80 . . . . 855·60
Light gauge copper tube to B.S. 2871, part 1, table X, with bronze welded joints and fittings,
76 mm diameter, 15 m @ £48·40 . . . . 726·00
Bronze gate valve with screwed joints,
20 mm, 10 No @ £13·10 . . . . . . 131·00
25 mm, 8 No @ £16·10 . . . . . . 128·80
50 mm, 6 No @ £37·00 . . . . . . 222·00
76/80 mm flanged gate valve, 1 No @ £140·00 . . 140·00
25 mm thick asbestos free calcium silicate sectional insulation to:
15 mm diameter pipework and fittings, 80 m @ £5·00. 400·00
22 mm diameter pipework and fittings, 56 m @ £4·95. 277·20
28 mm diameter pipework and fittings, 35 m @ £5·20. 182·00
35 mm diameter pipework and fittings, 45 m @ £5·50. 247·50
42 mm diameter pipework and fittings, 20 m @ £5·50. 110·00
54 mm diameter pipework and fittings, 62 m @ £7·50. 465·00   5890·10   2·45

Hot water installation
Serving 24 No lavatory basins and 4 No sinks,
Light gauge copper tube to B.S. 2871, part 1, table X, with capillary joints and fittings,
15 mm diameter tube and fittings, 50 m @ £8·50 . 425·00
22 mm diameter tube and fittings, 35 m @ £5·90 . 206·50
28 mm diameter tube and fittings, 20 m @ £7·30 . 146·00
35 mm diameter tube and fittings, 25 m @ £7·75 . 193·75
54 mm diameter tube and fittings, 30 m @ £13·25 . 397·50
Bronze gate valve with screwed ends,
20 mm, 6 No @ £13·10 . . . . . . 78·60

Continued   1447·35   17208·80   7·17

Approximate Estimating

ELEMENTAL COST PLAN

Element		Total cost of element £	Cost of element per m² of floor area £
Continued	1447·35	17208·80	7·17

MECHANICAL SERVICES *continued*

25 mm, 4 No @ £16·10	64·40	
50 mm, 4 No @ £37·00	148·00	
25 mm thick asbestos free calcium silicate section to:		
15 mm diameter pipework and fittings, 30 m @ £5·00.	150·00	
22 mm diameter pipework and fittings, 35 m @ £4·95.	173·25	
28 mm diameter pipework and fittings, 20 m @ £5·20.	104·00	
35 mm diameter pipework and fittings, 25 m @ £5·50.	137·50	
54 mm diameter pipework and fittings, 30 m @ £7·50.	225·00	2449·50	1·02

See 'Heating installation' for combined boiler calorifiers

Fire fighting installation
Dry riser installation
Serving three floors.
100 mm diameter steel riser with breeches with 2 No
64 mm instantaneous coupling and landing valve, 2 No
@ £1500·00 . . . . 3000·00 3000·00 1·25

Hose reel installation
Serving four floors.
Hose reel pedestal mounted type with 37 m of hose and
approximately 15 m of pipework, No. 8 @ £280·00 . 2240·00 2240·00 0·93

Gas installation
Serving 2 No boilers.
Black heavy weight mild steel tube to B.S. 1387 with
welded joints and fittings,
  80 mm diameter tube and fittings, 30 m @ £26·70 . 801·00
  100 mm diameter tube and fittings, 15 m @ £23·70 . 355·50
Gas valves with flanged inlet tube,
  80 mm, 2 No @ £105·00 . . . . . . 210·00
  100 mm, 1 No @ £255·00 . . . . . 255·00 1621·50 0·68

Heating installation
Low pressure hot water boiler plant having a total capacity of 1100 kW comprising 2 No boilers and combined calorifiers with instruments and boiler mountings, burners for natural gas, 2 No heating circulating pumps, 2 No hot water service circulating pumps, boiler control panel and thermostatic control for heating and hot water circuits, weather compensator, automatic controls, all interconnecting pipework and fittings including valves to boilers and pumps, with all necessary lagging together with feed and expansion tank and cold feed and vent pipes, 1 No £51,000·00 . . . . . . 51000·00

Continued	51000·00	26519·80	11·05

ELEMENTAL COST PLAN

Element	Total cost of element £	Cost of element per m² of floor area £
Continued 51000·00	26519·80	11·05

MECHANICAL SERVICES *continued*
Black medium weight mild steel tube to B.S. 1387 with screwed joints and fittings,

15 mm diameter tube and fittings, 120 m @ £13·50	1620·00		
20 mm diameter tube and fittings, 200 m @ £7·20	1440·00		
25 mm diameter tube and fittings, 140 m @ £7·65	1071·00		
32 mm diameter tube and fittings, 100 m @ £8·15	815·00		
50 mm diameter tube and fittings, 240 m @ £17·90	4296·00		
80 mm diameter tube and fittings, 90 m @ £24·50	2205·00		
100 mm diameter tube and fittings, 100 m @ £21·12	2112·00		

Bronze globe valve with screwed joints,

15 mm, 70 No @ £9·50	665·00
20 mm, 20 No @ £12·50	250·00
32 mm, 10 No @ £21·50	215·00
50 mm, 35 No @ £38·50	1347·50

Cast iron gate valve, flanged,

80 mm, 8 No @ £58·00	464·00
100 mm, 6 No @ £75·00	450·00

711 mm high continuous sill line natural convector front panel, 560 m @ £22·73 (p. 191) . . . . . 12728·80
Corners, 30 No @ £16·17 (p. 191) . . . 485·10
Valve boxes, 72 No @ £12·71 (p. 191) . . 915·12
33 mm double element finned copper tube, 520 m @ £34·32 (p. 191) . . . . 17846·40

25 mm thick asbestos free calcium silicate sectional insulation to:

15 mm diameter pipework and fittings, 120 m @ £5·00	600·00
20 mm diameter pipework and fittings, 200 m @ £4·95	990·00
25 mm diameter pipework and fittings, 140 m @ £5·20	728·00
32 mm diameter pipework and fittings, 100 m @ £5·50	550·00
50 mm diameter pipework and fittings, 240 m @ £7·50	1800·00
80 mm diameter pipework and fittings, 90 m @ £10·30	927·00
100 mm diameter pipework and fittings, 100 m @ £10·90	1090·00

	106610·92	44·42
	133130·72	55·47

Allow for marking of holes, testing and commissioning, protection of the work, record drawings and identification equipment,
add 1½% . . . . 1996·96   1996·96   0·83

	135127·68	56·30
Add for general contractors' discount, 1/39	3464·81	1·44

TOTAL COST OF MECHANICAL SERVICES . 138592·49

TOTAL COST PER m² OF FLOOR AREA . .   57·74

Approximate Estimating

ELEMENTAL COST PLAN

Element	Total cost of element		Cost of element per m^2 of floor area
ELECTRICAL SERVICES	£	£	£
Lighting installation			
20 mm black enamelled conduit, 1150 m @ £4·75 (p. 277)	5462·50		
20 mm galvanized conduit, 100 m @ £5·31 (p. 277)	531·00		
100 × 75 mm steel cable trunking, 140 m @ £8·03 (p. 281)	1124·20		
Angles 12 No @ £7·15 (p. 282)	85·80		
Tees 12 No @ £8·87 (p. 282)	106·44		
Ends 30 No @ £1·76 (p. 281)	52·80		
1·5 mm² single core P.V.C. insulated cable,			
In conduit, 4000 m @ £0·22 (p. 286)	880·00		
In trunking, 4000 m @ £0·19 (p. 286)	760·00		
One gang, one way 5 amp switch, 40 No @ £5·87 (p. 291)	234·80		
One gang, two way, 5 amp switch, 8 No @ £6·20 (p. 291)	49·60		
4 ft single 40 watt fluorescent lighting fitting complete with tube, 60 No @ £28·00	1680·00		
6 ft twin 75 watt recessed modular fitting complete with tubes, 250 No @ £82·00	20500·00		
Emergency lighting circuits, 16 points @ £146·00	2336·00	33803·14	14·08
Power installation			
20 mm black enamelled conduit, 300 m @ £4·75 (p. 277)	1425·00		
20 mm galvanized conduit, 40 m @ £5·31 (p. 277)	212·40		
2·5 mm² single core, PVC insulated cable,			
In conduit, 1500 m @ £0·29 (p. 286)	435·00		
In trunking, 3000 m @ £0·25 (p. 286)	750·00		
13 amp single switch socket outlet with plug top, 40 No @ £6·35 (p. 292)	254·00		
13 amp switch connection unit with neon indicator, 2 No @ £6·55 (p. 293)	13·10		
Budget estimate for conduit, cables and control gear for Kitchen equipment	3000·00	6089·50	2·54
Connections to mechanical services			
2400 m² @ £3·10	7440·00	7440·00	3·10
Clock installation			
A.C. mains clock installation, 8 No @ £20·00	160·00		
240 volt electric clock and fused connector (p.c. £30·00), 8 No @ £36·00	288·00	448·00	0·19
Continued		47780·64	19·91

ELEMENTAL COST PLAN

Element	Total cost of element £	Cost of element per m² of floor area £
Continued		47780·64	19·91

ELECTRICAL SERVICES *continued*

Telephone and power under floor duct installation

	£		
225 × 25 mm three compartment underfloor steel cable duct, 500 m @ £17·10 .	8550·00		
Twin socket outlets with plug tops and twin telephone outlets, 100 No @ £42·00 .	4200·00	12750·00	5·31

Fire alarm installation

Contact and alarm bells, 9 pr. @ £320·00 .	2880·00	2880·00	1·20

Sub-mains installation

Budget estimate for supply and installation of sub-station equipment @ approximately £37·00 per kVA. Floor area of 2400 m² at approximately 0·077 kVA per m² . . . . 6837·60
12 way metal clad S.P. & N. sub-distribution board complete with M.C.B.s, 8 No @ £77·45 (p. 276) . . 619·60
100 amp T.P. & N. switch fuse, 1 No @ £101·00 (p. 275) 101·00
200 amp T.P. & N. switch fuse, 2 No @ £158·50 (p. 275) 317·00
50 mm² four core PVC, S.W.A., PVC cable, 60 m @ £10·00 . . . . 600·00
Terminations, 18 No @ £8·00 . . . . . 144·40
70 mm² four core PVC, S.W.A., PVC cable, 120 m @ £12·50 . . . . 1500·00
Terminations, 4 No @ £10·30 . . . . . 41·20   10160·40   4·23

                                                                                          73571·04   30·65

Allow for marking of holes, testing and commissioning, protection of the work, record drawings and identification equipment, add 1½% . . . . 1103·57   1103·57   0·46

                                                                                          74674·61   31·11

Add for general contractor's discount, 1/39 . . . 1914·73   1914·73   0·80
   TOTAL COST OF ELECTRICAL SERVICES   76589·34

   TOTAL COST PER m² OF FLOOR AREA . .   31·91

PART FOUR

Daywork

Heating and Ventilating Industry, *page* 333
Electrical Industry, *page* 337

When work is carried out in connection with a contract which cannot be valued in any other way it is usual to assess the value on a cost basis with suitable allowances to cover overheads and profit. The basis of costing is a matter for agreement between the parties concerned but definitions of prime cost for the Heating and Ventilating and Electrical Industries have been prepared and published jointly by the Royal Institution of Chartered Surveyors and the appropriate bodies of the industries concerned for the convenience of those who wish to use them. These documents are reproduced on the following pages by kind permission of the publishers.

HEATING AND VENTILATING INDUSTRY

Definition of Prime Cost of daywork carried out under a heating, ventilating, air conditioning, refrigeration, pipework and/or domestic engineering contract

This Definition of Prime Cost is published by the Royal Institution of Chartered Surveyors and the Heating and Ventilating Contractors' Association for convenience, and for use by people who choose to use it. Members of the Heating and Ventilating Contractors' Association are not in any way debarred from defining Prime Cost and rendering accounts for work carried out on that basis in any way they choose. Building owners are advised to reach agreement with contractors on the Definition of Prime Cost to be used prior to entering into a contract or sub-contract.

SECTION 1: APPLICATION

1.1 This Definition provides a basis for the valuation of daywork executed under such heating, ventilating, air conditioning, refrigeration, pipework and/or domestic engineering contracts as provide for its use.

1.2 It is not applicable in any other circumstances, such as jobbing or other work carried out as a separate or main contract nor in the case of daywork executed after a date of practical completion.

1.3 The terms 'contract' and 'contractor' herein shall be read as 'sub-contract' and 'sub-contractor' as applicable.

SECTION 2: COMPOSITION OF TOTAL CHARGES

2.1 The Prime Cost of daywork comprises the sum of the following costs:

(*a*) Labour as defined in Section 3.
(*b*) Materials and goods as defined in Section 4.
(*c*) Plant as defined in Section 5.

2.2 Incidental costs, overheads and profit as defined in Section 6, as provided in the contract and expressed therein as percentage adjustments, are applicable to each of 2.1 (*a*)–(*c*).

SECTION 3: LABOUR

3.1 The standard wage rates, emoluments and expenses referred to below and the standard working hours referred to in 3.2 are those laid down for the time being in the rules or decisions or agreements of the Joint Conciliation Committee of the Heating, Ventilating and Domestic Engineering Industry applicable to the works (or those of such other body as may be appropriate) and to the grade of operative concerned at the time when and the area where the daywork is executed.

HEATING AND VENTILATING INDUSTRY

3.2 Hourly base rates for labour are computed by dividing the annual prime cost of labour, based upon the standard working hours and as defined in 3.4, by the number of standard working hours per annum. See example.

3.3 The hourly rates computed in accordance with 3.2 shall be applied in respect of the time spent by operatives directly engaged on daywork, including those operating mechanical plant and transport and erecting and dismantling other plant (unless otherwise expressly provided in the contract) and handling and distributing the materials and goods used in the daywork.

3.4 The annual prime cost of labour comprises the following:

(a) Standard weekly earnings (i.e. the standard working week as determined at the appropriate rate for the operative concerned).
(b) Any supplemental payments.
(c) Any guaranteed minimum payments (unless included in Section 6.1 (a)–(p)).
(d) Merit money.
(e) Differentials or extra payments in respect of skill, responsibility, discomfort, inconvenience or risk (excluding those in respect of supervisory responsibility – see 3.5).
(f) Payments in respect of public holidays.
(g) Any amounts which may become payable by the contractor to or in respect of operatives arising from the rules etc. referred to in 3.1 which are not provided for in 3.4 (a)–(f) nor in Section 6.1 (a)–(p).
(h) Employer's contributions to the Annual Holiday with Pay and Welfare Benefits Scheme or payments in lieu thereof.
(i) Employer's National Insurance contributions as applicable to 3.4 (a)–(h).
(j) Any contribution, levy or tax imposed by Statute, payable by the contractor in his capacity as an employer.

3.5 Differentials or extra payments in respect of supervisory responsibility are excluded from the annual prime cost (see Section 6). The time of principals, staff, foremen, charge hands and the like when working manually is admissible under this Section at the rates for the appropriate grades.

SECTION 4: MATERIALS AND GOODS

4.1 The prime cost of materials and goods obtained specifically for the daywork is the invoice cost after deducting all trade discounts and any portion of cash discounts in excess of 5%.

4.2 The prime cost of all other materials and goods used in the daywork is based upon the current market prices plus any appropriate handling charges.

4.3 The prime cost referred to in 4.1 and 4.2 includes the cost of delivery to site.

4.4 Any Value Added Tax which is treated, or is capable of being treated, as input tax (as defined by the Finance Act 1972, or any re-enactment or amendment thereof or substitution therefor) by the contractor is excluded.

SECTION 5: PLANT

5.1 Unless otherwise stated in the contract, the prime cost of plant comprises the cost of the following:

(a) use or hire of mechanically-operated plant and transport for the time employed on and/or provided or retained for the daywork;

HEATING AND VENTILATING INDUSTRY

(b) use of non-mechanical plant (excluding non-mechanical hand tools) for the time employed on and/or provided or retained for the daywork;
(c) transport to and from site and erection and dismantling where applicable.

5.2. The use of non-mechanical hand tools and of erected scaffolding, staging, trestles or the like is excluded (see Section 6), unless specifically retained for the daywork.

SECTION 6: INCIDENTAL COSTS, OVERHEADS AND PROFIT

6.1 The percentage adjustments provided in the contract which are applicable to each of the totals of Sections 3, 4 and 5 comprise the following:

(a) Head office charges.
(b) Site staff including site supervision.
(c) The additional cost of overtime (other than that referred to in 6.2).
(d) Time lost due to inclement weather.
(e) The additional cost of bonuses and all other incentive payments in excess of any included in 3.4.
(f) Apprentices' study time.
(g) Fares and travelling allowances.
(h) Country, lodging and periodic allowances.
(i) Sick pay or insurances in respect thereof, other than as included in 3.4.
(j) Third party and employers' liability insurance.
(k) Liability in respect of redundancy payments to employees:
(l) Employer's National Insurance contributions not included in 3.4.
(m) Use and maintenance of non-mechanical hand tools.
(n) Use of erected scaffolding, staging, trestles or the like (but see 5.2).
(o) Use of tarpaulins, protective clothing, artificial lighting, safety and welfare facilities, storage and the like that may be available on site.
(p) Any variation to basic rates required by the contractor in cases where the contract provides for the use of a specified schedule of basic plant charges (to the extent that no other provision is made for such variation – see 5.1).
(q) In the case of a sub-contract which provides that the subcontractor shall allow a cash discount, such provision as is necessary for the allowance of the prescribed rate of discount.
(r) All other liabilities and obligations whatsoever not specifically referred to in this Section nor chargeable under any other Section.
(s) Profit.

6.2. The additional cost of overtime where specifically ordered by the Architect/Supervising Officer shall only be chargeable in the terms of a prior written agreement between the parties.

Daywork
HEATING AND VENTILATING INDUSTRY

Example of calculation of typical standard hourly base rate (as defined in Section 3) for Advanced Fitter (qualified gas and arc) and Mate employed in the Heating, Ventilating, Air Conditioning, Piping and Domestic Engineering Industry under NJIC Rules based upon rates ruling at 4 February 1980.

	Rate £	Advanced Fitter (qual. gas & arc) £	Rate £	Mate £
Standard Weekly Earnings – 48 weeks × 38 hours	2·26	4,122·24	1·64	2,991·36
Welding Supplement (gas and arc) – 48 weeks × 38 hours	0·20	364·80	—	—
Merit Money and other variables as applicable		*		*
		4,487·04		2,991·36
Employer's National Insurance Contribution at 13·5%		605·75		403·83
		5,092·79		3,395·19
CITB Annual Levy		60·00		6·00
Weekly Holiday Credit and Welfare Stamp – 48 weeks	9·83	471·84	8·08	387·84
Annual Labour Cost as defined in Section 3		£5,624·63		£3,789·03
Hourly base rates as defined in Section 3, clause 3.2.	$\dfrac{£5,624·63}{1763·2} =$	£3·19	$\dfrac{£3,789·03}{1763·2} =$	£2·15

Note: (1) Standard working hours per annum calculated as follows:

52 weeks at 38 hours	=	1976
Less:		
4 weeks holiday at 38 hours	=	152
8 days Public Holidays at (average) 7·6 hours per day	=	60·8
		212·8
		1763·2

(2) Where applicable, Merit Money and other variables (e.g. Daily Abnormal Conditions Money), which attract Employer's National Insurance Contribution, should be included at *.

It should be noted that all labour costs incurred by the Contractor in his capacity as an Employer, other than those contained in the hourly base rate, as defined under Section 3, are to be taken into account under Section 6.

(3) The above example is for the convenience of users only and does not form part of the Definition; all the basic costs are subject to re-examination according to the time when and the area where the daywork is executed.

ELECTRICAL INDUSTRY

Definition of Prime Cost of daywork carried out under an electrical contract

This Definition of Prime Cost is published by The Royal Institution of Chartered Surveyors and The Electrical Contractors' Association for convenience, and for use by people who choose to use it. Building owners and/or members of The Electrical Contractors' Association are not in any way debarred from defining Prime Cost in any way they choose and rendering accounts for work carried out on that basis. Building owners are advised to reach agreement with contractors on the Definition of Prime Cost to be used prior to entering into a contract or sub-contract.

This Definition applies solely to daywork carried out under and incidental to an Electrical Main Contract or Sub-Contract. It does not cover:
(*i*) jobbing work, or
(*ii*) daywork ordered by the Architect/Supervising Officer to be carried out after the date of commencement of the Defects Liability Period, which may be the subject of separate agreement.

SECTION 1: LABOUR

(*i*) The amount of wages at the standard graded wage rates applicable when the day works are carried out and such extra payments as are fixed by the Joint Industry Board for the Electrical Contracting Industry, or where other trades are concerned, in accordance with the rules and awards of the recognized wage-fixing bodies of such trades in force in the area in which the work is carried out. Rates above that of an approved electrician will only be allowed after prior agreement with the Architect/Supervising Officer.

(*ii*) Any payment made at the sole discretion of the contractor of more than the standard wage rate to any approved electrician selected on the grounds of responsibility, which is called for in carrying out any daywork operation, and for which payment in accordance with the aforementioned rules and awards is previously authorized, may be included in the amounts of wages, otherwise such payments shall be deemed to be included in the overheads as Section (4).

(*iii*) The time of principals, foremen or other supervising grades at the standard time rates for the trades practised when actually working with their hands, unless previously otherwise authorized.

(*iv*) The additional cost of overtime, where specifically ordered or subsequently sanctioned in writing by the Architect/Supervising Officer to be worked on daywork.

SECTION 2: MATERIALS

(*i*) The cost of materials, including delivery to the site.
(*ii*) Materials supplied from the Contractor's or sub-trader's stock – at current prices plus justifiable charges for handling and delivery to the site.
(*iii*) Materials manufactured by the Contractor or sub-trader at the trade price or in the absence thereof at a price to be agreed.

The cost of materials referred to in paragraphs (*i*) and (*ii*) above is the cost to the Contractor or sub-trader less all trade discounts, but including all discounts for cash not exceeding 5%.

ELECTRICAL INDUSTRY

SECTION 3: PLANT

(*i*) Use or hire of mechanically-operated plant and transport for the time engaged in dayworks.

(*ii*) The delivery, erection, use or hire charges for non-mechanical plant (excluding hand tools), specially provided for daywork operation for such time as the Architect Supervising Officer considers reasonable.

(*iii*) Should hired plant be operated by the Electrical Contractor's labour, such time would be charged under Section (1) (*i*).

SECTION 4: OVERHEADS

Where the above Definition of Prime Cost is used, Overheads as defined below and Profit may be dealt with by means of percentages on the totals of Prime Cost in each of the Sections (1), (2) and (3) above at the rates stated in the Contract Bills or Specifications as the case may be subject always to the terms of the Contract or Sub-Contract.

(*i*) Head Office charges.
(*ii*) Site supervision and site staff.
(*iii*) Additional payments other than those allowed under Section (1) (*ii*).
(*iv*) Additional cost of overtime other than that allowed under Section (1) (*iv*).
(*v*) Time lost due to inclement weather.
(*vi*) Apprentices' study time.
(*vii*) Employer's contribution to National Insurances, including graduated pensions.
(*viii*) Selective Employment Tax and any other similar tax or imposition.
(*ix*) Payment for annual and public holidays.
(*x*) Fares and time allowances for travelling except for 'in lieu payment for travelling time' as defined in the working Rule Agreement which is reimbursable under Section (1) (*i*).
(*xi*) Lodging, country and periodic leave allowances.
(*xii*) Safety and Welfare facilities.
(*xiii*) Third Party, employer's liability and other necessary insurances.
(*xiv*) Welfare contributions, i.e. sickness and other payments or insurance in respect thereof.
(*xv*) Obligations under the Redundancy Payments Act 1965.
(*xvi*) Use, repair and sharpening of small tools.
(*xvii*) All non-mechanically-operated plant and other equipment, protective clothing artificial lighting (for own use), storage facilities and the like that may be in general use on the site.
(*xviii*) All other liabilities and obligations whatsoever.
(*xix*) In the case of a Sub-Contract for Electrical Installation work which is the subject of a Prime Cost item in a contract for building work, an allowance of a cash discount of $2\frac{1}{2}\%$ to the main contractor.

PART FIVE

Fees for Professional Services

Extracts from the scales of fees for consulting engineers follow; these are reproduced by kind permission of The Association of Consulting Engineers, Alliance House, 12, Caxton Street, London, SW1H 0QL. The full scale is not reproduced here and the extracts are given for guidance only; the full scale should be consulted before concluding any agreement. It should be noted that this scale of fees is not necessarily applicable to work carried out for Government and Local Authorities.

In recent years there has been an increasing tendency for Bills of Quantities to be prepared for mechanical and electrical services. Where this is done by Chartered Quantity Surveyors the fees will be in accordance with the appropriate scale of fees issued by The Royal Institution of Chartered Surveyors, 12 Great George Street, London, SW1P 3AD.

CONSULTING ENGINEERS' FEES
CONDITIONS OF ENGAGEMENT – DECEMBER 1970 PRINTING

1. PREAMBLE

The following paragraphs describe the scope of professional services provided by the Consulting Engineer and give general advice about his appointment.

The Association of Consulting Engineers has drawn up standard Conditions of Engagement to form the basis of the agreement between the Client and the Consulting Engineer, for five different types of appointment as hereafter described. Each of the standard Conditions of Engagement is preceded by a recommended Memorandum of Agreement. These two documents, taken together, constitute the recommended form of Agreement in each case.

When the standard Conditions of Engagement are not so used, it is in the interests of both the Client and the Consulting Engineer that there should be an exchange of letters defining the duties which the Consulting Engineer is to perform and the terms of payment.

All fees and charges set out in this Agreement and schedule are exclusive of Value Added Tax, the amount of which, at the rate and in the manner prescribed by law, shall be paid by the Client to the Consulting Engineer. Where Value Added Tax is chargeable on Disbursements and Out-of-pocket Expenses, this will be based upon the VAT-exclusive cost of such outgoings.

2. REPORTS AND ADVISORY WORK

For reports, and for advisory work, the services required from the Consulting Engineer will usually comprise one or more of the following:

(*a*) Investigating and advising on a project and submitting a Report thereon. The Consulting Engineer may be asked to examine alternatives; review all technical aspects; make an economic appraisal of costs and benefits; draw conclusions and make recommendations.

CONSULTING ENGINEERS' FEES

(b) Inspecting an existing structure (e.g. a reservoir or a building or an installation) and reporting thereon. If the client requires continuing advice on maintenance or operation of an existing project, the Consulting Engineer may be appointed to make periodic visits.

(c) Making a special investigation of an engineering problem and reporting thereon.

(d) Making valuations of plant and undertakings.

Payment for the services provided under Items (a) to (d) above should normally be on a time basis, as described in Section 10. It is further recommended that the conditions governing an appointment under Item (a) above should be based on:

Schedule 1—Conditions of Engagement for Report and Advisory Work.

3. DESIGN AND SUPERVISION OF CONSTRUCTION

When the Client has approved the proposals recommended in the Report (see Section 2 (a) above) and has decided to proceed with the construction of engineering works or the installation of plant, it is normal practice for the Consulting Engineer who prepared or assisted in the preparation of the Report to be appointed for the subsequent stages of the work or the relevant part thereof. The Consulting Engineer will assist the Client in obtaining the requisite approval, then prepare the designs and tender documents to enable competitive tenders to be obtained or orders to be placed, and will be responsible for the technical control and administration of the construction of the Works.

It is recommended that the agreement for this type of appointment should be based on the most appropriate of the following:

Schedule 2: Conditions of Engagement for the design and supervision of Civil Mechanical and Electrical Works.

Schedule 3: Conditions of Engagement for the design and supervision of Structural Engineering work in Buildings and other Structures, where an Architect is appointed by the Client.

Schedule 4: Conditions of Engagement for the design and supervision of Engineering Systems (formerly referred to as 'Engineering Services') in Buildings and other Projects.

4. VARIATION IN TERMS OF PAYMENT FOR DESIGN AND SUPERVISION OF CONSTRUCTION

The Association of Consulting Engineers believes that, when the circumstances are normal and the design work is of average complexity, competent and responsible engineering services cannot be provided at a level of remuneration lower than that represented by the scales of percentage charges shown in the standard A.C.E. Conditions of Engagement. Variation in these terms may, however, be necessary in special circumstances. For example, the following would justify an increase in the charge for services:

(a) The design work is of an unusually complex character.

(b) The Works are to be constructed by means of an abnormally large number of separate contracts.

(c) A substantial proportion of the project involves alterations or additions to existing structures, plant or services.

(d) The completion of the project is retarded through circumstances over which the Consulting Engineer has no control—e.g. by the Client in order to spread the aggregate cost over an extended period. If the project is to be built in stages, the definition of 'Works' in the standard A.C.E. Conditions of Engagement should apply separately to each stage.

The following would justify a reduction in the charge for services:

(e) The design work is unusually simple or repetitive in character.

(f) The fact that part of the normal services set out in the applicable standard A.C.E. Conditions of Engagement has already been provided by the Consulting Engineer (e.g. included in a project report, as described in Section 2) and has been paid for by the Client. In such a case, an appropriate reduction in the charge for services should be agreed between the Client and the Consulting Engineer to take account only of those earlier services, directly relevant to the Works and not to alternative schemes, which would otherwise have to be performed as part of the normal services.

6. PARTIAL SERVICES

When the Client wishes to appoint the Consulting Engineer for partial services only, and not for both design and supervision of construction, it is important that both parties recognize the limitation which such an appointment places upon the professional responsibility of the Consulting Engineer who cannot be held liable for matters that are outside his control and knowledge.

The terms of reference for the appointment should be carefully drawn up and the relevant A.C.E. Conditions of Engagement should be adapted to suit the scope of services required.

Professional charges for partial services are usually best calculated on the time basis described in Section 10, but may, in suitable cases, be a commensurate part of the percentage charge for normal services shown in the standard Conditions of Engagement.

Examples of partial service are described in the two succeeding Sections 7 and 8.

7. INSPECTION SERVICES

The inspection of materials and plant during manufacture or on site is usually required during the construction stage of a project. Consequently, the arrangements for this service are described in the standard Conditions of Engagement referred to in Section 3. If, however, the Client wishes to engage the Consulting Engineer to provide only inspection services, the professional charges of the Consulting Engineer may be either a percentage of the cost of the materials to be inspected, or on a time basis as described in Section 10.

8. INDUSTRIALIZED BUILDING

A Client who has decided to employ a building contractor for the design and construction of an Industrialized Building may require the services of a Consulting Engineer for advice on the selection of a contractor, for checking the design and for supervising construction. It is recommended that the agreement for this type of appointment should be based on:

Schedule 5: Conditions of Engagement for Structural Engineering services in connection with Industrialized Building.

CONSULTING ENGINEERS' FEES

9. ACTING AS ARBITRATOR, UMPIRE OR EXPERT WITNESS

A Consulting Engineer may be appointed to act as Arbitrator or Umpire, or be required to attend as an Expert Witness at Parliamentary Committees, Courts of Law, Arbitrations or Official Inquiries.

Payment for any of these services should be on the basis of a lump sum retainer plus time charges and expenses as described in Section 10, not less than three hours per day being chargeable for attendance, however short, either before or after a mid-day adjournment.

10. PAYMENT ON A TIME BASIS

When it is impossible to estimate in advance the duration and extent of the Consulting Engineer's services, neither a lump sum payment alone nor a percentage of the estimated construction cost would normally be a fair basis of remuneration. The most satisfactory and equitable method of payment in these cases makes allowance for the actual time occupied in providing the services required, and comprises the following elements, as applicable:

10.1 A charge in the form of hourly rate(s) for the services of a Partner or a Consultant of the firm. The hourly rate will depend upon his standing, the nature of the work and any special circumstances. Alternatively a lump sum fee may be charged instead of the said hourly rate(s).

10.2 A charge which covers technical and supporting staff salary and other payroll costs actually incurred by the Consulting Engineer, together with a fair proportion of his overhead costs, plus an element of profit. This charge is most conveniently calculated by applying a multiplier to the salary cost and then adding the net amount of other payroll costs. For technical and supporting staff working in or based on the Consulting Engineer's office, a multiplier of 2·5 is normally applicable. The major part of the multiplier is attributable to the Consulting Engineer's overheads which may include, *inter alia*, the following indirect costs and expenses:

(*a*) rent, rates and other expenses of upkeep of his office, its furnishings, equipment and supplies;
(*b*) insurance premiums other than those recovered in the payroll cost;
(*c*) administrative, accounting, secretarial and financing costs;
(*d*) the expense of keeping abreast of advances in engineering;
(*e*) the expense of preliminary arrangements for new or prospective projects;
(*f*) loss of productive time of technical staff between assignments.

10.3 A charge for use of a computer or other special equipment.

10.4 Disbursements made by the Consulting Engineer which can be directly identified with the work.

When calculating amounts chargeable on a time basis, a Consulting Engineer is entitled to include time spent by Partners, Consultants and technical and supporting staff in travelling in connection with the performance of the services. The time spent by secretarial staff or by staff engaged on general accountancy or administration duties in the Consulting Engineer's office is not chargeable unless otherwise agreed.

Fees for Professional Services

CONSULTING ENGINEERS' FEES

11. APPOINTMENTS OUTSIDE THE UNITED KINGDOM

The standard A.C.E. Conditions of Engagement are suitable for appointments in the United Kingdom. For work outside the United Kingdom the arrangement and wording of the clauses will in general be found suitable, but modifications and supplementary clauses may be needed to suit the circumstances and locality of the work.

GENERAL CONDITIONS AND SCALE OF CHARGES FOR REPORT AND ADVISORY WORK CARRIED OUT UNDER SCHEDULE 1

1. DEFINITIONS

'Salary Cost'
means the annual salary including bonuses of any person employed by the Consulting Engineer, divided by 1600 (being deemed to be the average annual total of effective working hours of an employee) and multiplied by the number of working hours spent by such person in performing any of the services in respect of which payment under this Agreement is to be made to the Consulting Engineer upon the basis of Salary Cost. For the purposes of this definition the annual salary of a person employed by the Consulting Engineer for a period less than a full year shall be calculated pro rata to such person's salary (including bonuses) for such lesser period.

'Other Payroll Cost'
means the annual amount of all contributions and payments made by the Consulting Engineer on behalf of or in respect of a person employed by him for staff pension and life assurance schemes, and also for National Health Insurance, Graduated Pension Fund, Selective Employment Tax and for any other tax, charge, levy, impost or payment of any kind whatsoever which the Consulting Engineer at any time during the performance of this Agreement is obliged by law to make on behalf of or in respect of such person, divided by 1600 (being deemed to be the average annual total of effective working hours of an employee) and multiplied by the number of working hours spent by such person in performing any of the services in respect of which payment under this Agreement is to be made to the Consulting Engineer upon the basis of Other Payroll Cost. For the purposes of this definition the annual amount of all contributions and payments made by the Consulting Engineer on behalf of or in respect of a person employed by him for a period less than a full year shall be calculated pro rata to the amount of such contributions and payments for such lesser period.

Obligations of Consulting Engineers

6. NORMAL SERVICES

The services to be provided by the Consulting Engineer shall comprise

(*a*) all or any of the services stated in the Appendix to the Memorandum of Agreement and

(*b*) advising the Client as to the need for the Client to be provided with additional services in accordance with Clause 7.

CONSULTING ENGINEERS' FEES

7. ADDITIONAL SERVICES NOT INCLUDED IN NORMAL SERVICES

7.1 As services additional to those specified in Clause 6, the Consulting Engineer shall, if so requested by the Client, provide any of the services specified in Clause 7.2 and provide or take all reasonable steps to arrange for the provision of any of the services specified in Clause 7.3.

7.2 (*a*) Carrying out work consequent upon a decision by the Client to seek parliamentary powers.

(*b*) Carrying out work in connection with any application by the Client for any order, sanction, licence, permit or other consent, approval or authorization necessary to enable the Task to proceed.

(*c*) Carrying out work arising from the failure of the Client to award a contract in due time.

(*d*) Carrying out work consequent upon any assignment of a contract by the Contractor or upon the failure of the Contractor properly to perform any contract or upon delay by the Client in fulfilling his obligations under Clause 8 or in taking any other step necessary for the due performance of the Task.

(*e*) Advising the Client upon and carrying out work following the taking of any step in or towards any litigation or arbitration relating to the Task.

(*f*) Carrying out work in conjunction with others employed to provide any of the services specified in Clause 7.3.

7.3 (*a*) Specialist technical advice on any abnormal aspects of the Task.

(*b*) Architectural, legal, financial and other professional services.

(*c*) Services in connection with the valuation, purchase, sale or leasing of lands and the obtaining of wayleaves.

(*d*) The carrying out of marine, air and land surveys, and the making of model tests or special investigations.

7.4 The Consulting Engineer shall obtain the prior agreement of the Client to the arrangements which he proposes to make on the Client's behalf for the provision of any of the services specified in Clause 7.3. The Client shall be responsible to any person or persons providing such services for the cost thereof.

Obligations of Client

8. INFORMATION TO BE SUPPLIED TO THE CONSULTING ENGINEER

8.1 The Client shall supply to the Consulting Engineer without charge and within a reasonable time all necessary and relevant data and information in the possession of the Client and shall give such assistance as shall reasonably be required by the Consulting Engineer in the performance of the Task.

8.2 The Client shall give his decision on all sketches, drawings, reports, recommendations, tender documents and other matters properly referred to him for decision by the Consulting Engineer in such reasonable time as not to delay or disrupt the performance by the Consulting Engineer of the Task.

9. PAYMENT FOR SERVICES

9.1 In respect of services provided by the Consulting Engineer under Clauses 6 and 7, the Client shall pay the Consulting Engineer

CONSULTING ENGINEERS' FEES

(a) For technical and supporting staff working in or based on the Consulting Engineer's Office: Up to a maximum of Salary Cost times 2·5, plus Other Payroll Cost.
(b) For technical and supporting staff, working as field staff, in or based on any field office established in pursuance of Clause 10: Salary Cost times the multiplier specified in Clause 4 (a) of the Memorandum of Agreement plus Other Payroll Cost.
(c) The fee specified in Clause 4 (b) of the Memorandum of Agreement, which shall be deemed to cover the services of Partners and Consultants of the firm but not their expenses which are reimbursable separately under Clause 11 (c).
(d) A reasonable charge for the use of a computer or other special equipment which charge shall, if possible, be agreed between the Client and the Consulting Engineer before the work is put in hand.
9.2 Time spent by technical and supporting staff in connection with the use of a computer or other special equipment, including the development and writing of programmes and the operation of the computer in trial and final runs, and time spent by technical and supporting staff in travelling in connection with the Task, shall be chargeable.
9.3 Unless otherwise agreed between the Client and the Consulting Engineer, the Consulting Engineer shall not be entitled to any payment in respect of time spent by secretarial staff or by staff engaged on general accountancy or administration duties in the Consulting Engineer's office.

10. PAYMENT FOR FIELD STAFF FACILITIES

The Client shall be responsible for the cost of providing such field office accommodation, furniture, telephones, equipment and transport as shall be reasonably necessary for the use of field staff, and for the reasonable running costs of such necessary field office accommodation and other facilities, including those of stationery, telephone calls, telegrams and postage. Unless otherwise agreed between the Client and the Consulting Engineer, the Consulting Engineer shall arrange for the provision of field office accommodation and facilities for the use of field staff.

11. DISBURSEMENTS

The Client shall reimburse the Consulting Engineer in respect of all the Consulting Engineer's disbursements properly made in connection with:
(a) Printing, reproduction and purchase of all documents, drawings, maps and records.
(b) Telegrams and telephone calls other than local.
(c) Travelling, hotel expenses and other similar disbursements.
(d) Advertising for tenders and for field staff.
(e) The provision of additional services to the Client pursuant to Clause 7.4.

12. PAYMENT FOLLOWING TERMINATION OR SUSPENSION BY THE CLIENT

12.1 Upon a termination or suspension by the Client in pursuance of Clause 2.3, the Client shall pay to the Consulting Engineer the sums specified in (a), (b) and (c) of this sub-clause (less the amount of payments previously made to the Consulting Engineer under the terms of this Agreement).

CONSULTING ENGINEERS' FEES

(a) All amounts due to the Consulting Engineer on a time basis in accordance with Clause 9 in respect of services rendered up to the date of termination or suspension together with a sum calculated in accordance with Clause 9 in respect of time worked by the Consulting Engineer's staff in complying with Clause 2.6.

(b) A fair and reasonable proportion of any lump sum specified in Clause 4(b) of the Memorandum of Agreement. In the assessment of such proportion, the services carried out by the Consulting Engineer up to the date of termination or suspension and in pursuance of Clause 2.6 shall be compared with a reasonable assessment of the services which the Consulting Engineer would have carried out but for the termination or suspension.

(c) Amounts due to the Consulting Engineer under any other clauses of this Agreement.

12.2 In any case in which the Client has required the Consulting Engineer to suspend the carrying out of the Consulting Engineer's services in pursuance of the power conferred by Clause 2.3, the Client may, at any time within the period of 12 months from the date of his requirement in writing to the Consulting Engineer to suspend the carrying out of the Consulting Engineer's services, require the Consulting Engineer in writing to resume the performance of such services. In such event the Consulting Engineer shall within a reasonable time of receipt by him of the Client's said requirement in writing resume the performance of his services in accordance with this Agreement. Upon such a resumption, the amount of any payment made to the Consulting Engineer under Clause 12.1(b) shall rank as a payment made on account of the total sum payable to the Consulting Engineer under this Agreement, but no adjustment shall be made of any other sum paid or payable to the Consulting Engineer upon suspension.

12.3 If the Consulting Engineer shall need to perform any additional services in connection with the resumption of his services in accordance with Clause 12.2, the Client shall pay the Consulting Engineer in respect of the performance of such additional services in accordance with Clause 9, and any appropriate reimbursements in accordance with Clause 11.

13. PAYMENT FOLLOWING TERMINATION BY THE CONSULTING ENGINEER

Upon a termination by the Consulting Engineer in pursuance of Clause 2.5, the Client shall pay to the Consulting Engineer the sums specified in Clause 12.1(a), (b) and (c) (less the amount of payments previously made to the Consulting Engineer under the terms of this Agreement). Upon payment of such sums, the Consulting Engineer shall deliver to the Client such completed drawings and other similar documents relevant to the Task as are in his possession. The Consulting Engineer shall be permitted to retain copies of any documents so delivered to the Client.

GENERAL CONDITIONS AND SCALE OF CHARGES FOR THE DESIGN AND SUPERVISION OF CIVIL, MECHANICAL AND ELECTRICAL WORKS

1. DEFINITIONS

'Salary Cost'
means the annual salary including bonuses of any person employed by the Consulting

CONSULTING ENGINEERS' FEES

Engineer, divided by 1600 (being deemed to be the average annual total of effective working hours of an employee) and multiplied by the number of working hours spent by such person in performing any of the services in respect of which payment under this Agreement is to be made to the Consulting Engineer upon the basis of Salary Cost. For the purposes of this definition the annual salary of a person employed by the Consulting Engineer for a period less than a full year shall be calculated pro rata to such person's salary (including bonuses) for such lesser period.

'Other Payroll Cost'
means the annual amount of all contributions and payments made by the Consulting Engineer on behalf of or in respect of a person employed by him for staff pension and life assurance schemes and also for National Health Insurance, Graduated Pension Fund, Selective Employment Tax and for any other tax, charge, levy, impost or payment of any kind whatsoever which the Consulting Engineer at any time during the performance of this Agreement is obliged by law to make on behalf of or in respect of such person, divided by 1600 (being deemed to be the average annual total of effective working hours of an employee) and multiplied by the number of working hours spent by such person in performing any of the services in respect of which payment under this Agreement is to be made to the Consulting Engineer upon the basis of Other Payroll Cost. For the purposes of this definition the annual amount of all contributions and payments made by the Consulting Engineer on behalf of or in respect of a person employed by him for a period less than a full year shall be calculated pro rata to the amount of such contributions and payments for such lesser period.

6. NORMAL SERVICES

6.1 Design Stage I
The services to be provided by the Consulting Engineer at this stage shall comprise all or any of the following as may be necessary in the particular case:

(*a*) Investigating data and information relevant to the Works which are reasonably accessible to the Consulting Engineer, and considering any reports relating to the Works which have either been previously prepared by the Consulting Engineer or else prepared by others and made available to the Consulting Engineer by the Client.

(*b*) Making any normal topographical survey of the proposed site of the Works which may be necessary to supplement the topographical information already available to the Consulting Engineer.

(*c*) Advising the Client on the need to carry out any geotechnical investigations which may be necessary to supplement the geotechnical information already available to the Consulting Engineer, arranging for such investigations when authorized by the Client, certifying the amount of any payments to be made by the Client to the persons or firms carrying out such investigations under the Consulting Engineer's direction, and advising the Client on the results of such investigations.

(*d*) Advising the Client on the need for arrangements to be made, in accordance with Clause 7, for the carrying out of special surveys, special investigations or model tests, and advising the Client of the results of any such surveys, investigations or tests carried out.

(*e*) Consulting any Architect appointed by the Client in connection with the architectural treatment of the Works.

CONSULTING ENGINEERS' FEES

(*f*) Preparing such documents as are reasonably necessary to enable the Client to consider the Consulting Engineer's general proposals for the construction of the Works in the light of the investigations carried out by him at this stage, and to enable the Client to apply for approval in principle of the execution of the Works in accordance with such proposals.

6.2 Design Stage II

The services to be provided by the Consulting Engineer at this stage shall comprise all or any of the following as may be necessary in the particular case:

(*a*) Preparing designs and tender drawings in connection with the Works.

(*b*) Advising as to the appropriate conditions of contract to be incorporated in any contract to be made between the Client and a Contractor.

(*c*) Preparing such specifications, schedules and bills of quantities as may be necessary to enable the Client to obtain tenders or otherwise award a contract for carrying out the Works.

(*d*) Advising the Client as to the suitability for carrying out the Works of persons and firms tendering and further as to the relative merits of tenders, prices and estimates received for carrying out the Works.

As soon as the Consulting Engineer shall have submitted advice to the Client upon tenders, his services at this stage shall be complete.

6.3 Construction Stage

The Consulting Engineer shall not accept any tender in respect of the Works unless the Client gives him instructions in writing to do so, and any acceptance so made by the Consulting Engineer on the instructions of the Client shall be on behalf of the Client. The services to be provided by the Consulting Engineer at this stage shall include all or any of the following as may be necessary in the particular case:

(*a*) Advising on the preparation of formal contract documents relating to accepted tenders for carrying out the Works or any part thereof.

(*b*) Inspecting and testing during manufacture and installation such electrical and mechanical materials, machinery and plant supplied for incorporation in the Works as are usually inspected and tested by Consulting Engineers, and arranging and witnessing acceptance tests.

(*c*) Advising the Client on the need for special inspection or testing other than that referred to in sub-clause (*b*).

(*d*) Advising the Client on the appointment of site-staff in accordance with Clause 8.

(*e*) Preparing bar bending schedules and any further designs and drawings which may be necessary.

(*f*) Examining the Contractor's proposals.

(*g*) Making such visits to site as the Consulting Engineer shall consider necessary to satisfy himself as to the performance of any site-staff appointed pursuant to Clause 8, and to satisfy himself that the Works are executed generally according to contract and otherwise in accordance with good engineering practice.

CONSULTING ENGINEERS' FEES

(*h*) Giving all necessary instructions to the Contractor, provided that the Consulting Engineer shall not without the prior approval of the Client give any instructions which in the opinion of the Consulting Engineer are likely substantially to increase the cost of the Works unless it is not in the circumstances practicable for the Consulting Engineer to obtain such prior approval.

(*i*) Issuing certificates for payment to the Contractor.

(*j*) Performing any services which the Consulting Engineer may be required to carry out under any contract for the execution of the Works, including where appropriate the supervision of any specified tests and of the commisioning of the Works, provided that the Consulting Engineer may decline to perform any services specified in a contract the terms of which have not initially been approved by the Consulting Engineer.

(*k*) Delivering to the Client on the completion of the Works such records and manufacturers' manuals as are reasonably necessary to enable the Client to operate and maintain the Works.

(*l*) Deciding any dispute or difference arising between the Client and the Contractor and submitted to the Consulting Engineer for his decision, provided that this service shall not extend to advising the Client following the taking of any step in or towards any arbitration or litigation in connection with the Works.

6.4 General

Without prejudice to the preceding provisions of this clause, the Consulting Engineer shall from time to time as may be necessary advise the Client as to the need for the Client to be provided with additional services in accordance with Clause 7.

7. ADDITIONAL SERVICES NOT INCLUDED IN NORMAL SERVICES

7.1 As services additional to those specified in Clause 6, the Consulting Engineer shall, if so requested by the Client, provide any of the services specified in Clause 7.2 and provide or take all reasonable steps to arrange for the provision of any of the services specified in Clause 7.3.

7.2 (*a*) Preparing any report or additional contract documents required for consideration of proposals for the carrying out of alternative works.

(*b*) Carrying out work consequent upon a decision by the Client to seek parliamentary powers.

(*c*) Carrying out work in connection with any application already made by the Client for any order, sanction, licence, permit or other consent, approval or authorization necessary to enable the Works to proceed.

(*d*) Carrying out work arising from the failure of the Client to award a contract in due time.

(*e*) Preparing details for shop fabrication of ductwork, metal or plastic frameworks.

(*f*) Carrying out work consequent upon any assignment of a contract by the Contractor or upon the failure of the Contractor properly to perform any contract or upon delay by the Client in fulfilling his obligations under Clause 9 or in taking any other step necessary for the due performance of the Works.

(*g*) Advising the Client and carrying out work following the taking of any step in or towards any litigation or arbitration relating to the Works.

CONSULTING ENGINEERS' FEES

(*h*) Carrying out work in conjunction with others employed to provide any of the services specified in Clause 7.3.
(*i*) Carrying out such other additional services, if any, as are specified in Clause 8 of the Memorandum of Agreement.
7.3 (*a*) Specialist technical advice on any abnormal aspects of the Works.
(*b*) Architectural, legal, financial and other professional services.
(*c*) Services in connection with the valuation, purchase, sale or leasing of lands and the obtaining of wayleaves.
(*d*) The carrying out of marine and air surveys, and land surveys other than those referred to in Clause 6, and the making of model tests or special investigations.
(*e*) The carrying out of special inspections or tests advised by the Consulting Engineer under sub-clause 6.3 (*c*).
7.4 The Consulting Engineer shall obtain the prior agreement of the Client to the arrangements which he proposes to make on the Client's behalf for the provision of any of the services specified in Clause 7.3. The Client shall be responsible to any person or persons providing such services for the cost thereof.

8. SUPERVISION ON SITE

8.1 If in the opinion of the Consulting Engineer the nature of the Works, including the carrying out of any geotechnical investigation pursuant to Clause 6.1, warrants full-time or part-time engineering supervision on site, the Client shall not object to the appointment of such suitably qualified technical and clerical site-staff as the Consulting Engineer shall consider reasonably necessary to enable such supervision to be carried out.
8.2 Persons appointed pursuant to the previous sub-clause shall be employed either by the Consulting Engineer or, if the Client and the Consulting Engineer shall so agree, by the Client directly, provided that the Client shall not employ any person as a member of the site-staff unless the Consulting Engineer has first selected or approved such person as suitable for employment.
8.3 The terms of service of all site-staff to be employed by the Consulting Engineer shall be subject to the approval of the Client, which approval shall not be unreasonably withheld.
8.4 The Client shall procure that the contracts of employment of site-staff employed by the Client shall stipulate that the person employed shall in no circumstances take or act upon instructions other than those of the Consulting Engineer.
8.5 Where any of the services specified in Clause 6.3 are performed by site-staff employed by the Client, the Consulting Engineer shall not be responsible for any failure on the part of such staff properly to comply with any instructions given by the Consulting Engineer.

Obligations of Client

9. INFORMATION TO BE SUPPLIED TO THE CONSULTING ENGINEER

9.1 The Client shall supply to the Consulting Engineer, without charge and within a reasonable time, all necessary and relevant data and information in the possession of the Client and shall give such assistance as shall reasonably be required by the Consulting Engineer in the performance of his services under this Agreement.
9.2 The Client shall give his decision on all sketches, drawings, reports, recommenda-

Fees for Professional Services 351

CONSULTING ENGINEERS' FEES

tions, tender documents and other matters properly referred to him for decision by the Consulting Engineer in such reasonable time as not to delay or disrupt the performance by the Consulting Engineer of his services under this Agreement.

10. PAYMENT FOR NORMAL SERVICES

10.1 Payment depending upon the actual cost of the Works.
10.1.1 The sum payable by the Client to the Consulting Engineer for his services under Clause 6 shall be calculated as follows:
(*a*) The Works shall first be classified into one or more of the following classes as shall be appropriate:

Class 1: Structural work in reinforced concrete, prestressed concrete steel and other metals.
Class 2: Buildings including engineering systems associated with buildings, but excluding Class 1 work.
Class 3: Civil engineering including geotechnical investigation, but excluding Class 1 and Class 2 work.
Class 4: Mechanical and electrical plant and equipment.

(*b*) The cost of each relevant class of work shall next be calculated, and
(*c*) The sum payable by the Client to the Consulting Engineer shall then be calculated and shall be an amount or the sum of the amounts calculated in respect of each relevant class of work in accordance with the Scales of Charges set out in Clause 10.1.3 where the cost of the class of work is not less than £10,000, or in accordance with the Scale of Charges in Clause 11 or otherwise as may be agreed between the Client and the Consulting Engineer, where the cost of the class of work is less than £10,000.
10.1.2 The cost of work shall be calculated in accordance with Clause 19. Where the Works have been classified in accordance with Clause 10.1.1 into more than one class, there shall be attributed to each class an appropriate portion of any 'General or Preliminary Items' included in the total cost of the Works, so that the total cost of all classes of work shall equal the total cost of the Works. In the classification of the Works, Class 1 work shall be taken to include concrete, reinforcement, prestressing tendons and anchorages, formwork, inserts and all labours.
10.1.3 The Scales of Charges (suitable for work of average complexity) referred to in Clause 10.1.1 are as follows:

Cost of class of work		Class 1 work	Charge for Class 2 or Class 3 or Class 4 work
On the first	£ 10,000	15%	11%
On the next	£ 15,000	13%	9%
On the next	£ 25,000	$11\frac{1}{2}$%	$7\frac{1}{2}$%
On the next	£ 50,000	$9\frac{3}{4}$%	$6\frac{1}{2}$%
On the next	£ 100,000	$8\frac{1}{4}$%	6%
On the next	£ 300,000	$7\frac{1}{4}$%	$5\frac{1}{2}$%
On the next	£ 500,000	$6\frac{1}{2}$%	5%

CONSULTING ENGINEERS' FEES

Cost of class of work		Class 1 work	Charge for Class 2 or Class 3 or Class 4 work
On the next	£1,000,000	$6\frac{1}{4}\%$	$4\frac{1}{2}\%$
On the next	£2,000,000	6%	$4\frac{1}{4}\%$
On the next	£4,000,000	$5\frac{3}{4}\%$	4%
On the remainder		$5\frac{1}{4}\%$	$3\frac{3}{4}\%$

The charge for Class 1 work can also be calculated conveniently from the appropriate line of the following table:

	Cost of Class 1 work	Charge	
From	£ 10,000 to £ 25,000	£ 1,500 + 13%	of balance over £ 10,000
	£ 25,000 to £ 50,000	£ 3,450 + $11\frac{1}{2}\%$	of balance over £ 25,000
	£ 50,000 to £ 100,000	£ 6,325 + $9\frac{3}{4}\%$	of balance over £ 50,000
	£ 100,000 to £ 200,000	£ 11,200 + $8\frac{1}{4}\%$	of balance over £ 100,000
	£ 200,000 to £ 500,000	£ 19,450 + $7\frac{1}{4}\%$	of balance over £ 200,000
	£ 500,000 to £1,000,000	£ 41,200 + $6\frac{1}{2}\%$	of balance over £ 500,000
	£1,000,000 to £2,000,000	£ 73,700 + $6\frac{1}{4}\%$	of balance over £1,000,000
	£2,000,000 to £4,000,000	£136,200 + 6%	of balance over £2,000,000
	£4,000,000 to £8,000,000	£256,200 + $5\frac{3}{4}\%$	of balance over £4,000,000
Over	£8,000,000	£486,200 + $5\frac{1}{4}\%$	of balance over £8,000,000

The charge for Class 2, or Class 3 or Class 4 work can also be calculated conveniently from the appropriate line of the following table:

	Cost of Class 2 or Class 3 or Class 4 work	Charge	
From	£ 10,000 to £ 25,000	£ 1,100 + 9%	of balance over £ 10,000
	£ 25,000 to £ 50,000	£ 2,450 + $7\frac{1}{2}\%$	of balance over £ 25,000
	£ 50,000 to £ 100,000	£ 4,325 + $6\frac{1}{2}\%$	of balance over £ 50,000
	£ 100,000 to £ 200,000	£ 7,575 + 6%	of balance over £ 100,000
	£ 200,000 to £ 500,000	£ 13,575 + $5\frac{1}{2}\%$	of balance over £ 200,000
	£ 500,000 to £1,000,000	£ 30,075 + 5%	of balance over £ 500,000
	£1,000,000 to £2,000,000	£ 55,075 + $4\frac{1}{2}\%$	of balance over £1,000,000
	£2,000,000 to £4,000,000	£100,075 + $4\frac{1}{4}\%$	of balance over £2,000,000
	£4,000,000 to £8,000,000	£185,075 + 4%	of balance over £4,000,000
Over	£8,000,000	£345,075 + $3\frac{3}{4}\%$	of balance over £8,000,000

10.1.4 If the Client decides to have the Works constructed in more than one phase and as a consequence the services which it may be necessary for the Consulting Engineer to perform under Clause 6 have to be undertaken by the Consulting Engineer separately in respect of each phase, then the provisions of this payment clause shall apply separately to each phase and as if the expression 'the Works' as used in this clause meant, in the case of each phase, the work comprised in that phase.

CONSULTING ENGINEERS' FEES

10.2 Payment of a fixed sum
The sum payable by the Client to the Consulting Engineer for his services under Clause 6 shall be the sum specified in Clause 5 (*a*) of the Memorandum of Agreement, provided that the Consulting Engineer shall, in addition to the said sum, be paid in accordance with Clause 11.2 for any services of the kind specified in Clause 6.3 (*l*) which it is necessary for him to provide.

10.3 Payment on the basis of Salary Cost times multiplier, plus Other Payroll Cost plus fee.

10.3.1 In respect of services provided by the Consulting Engineer under the following Clauses:

Clause 6	Normal services
Clause 7	Additional services
Clause 12	Computer, etc.
Clause 13.3	Site Visits
Clause 15	Alterations, etc.
Clause 16	Works damaged, etc.

the Client shall pay the Consulting Engineer
(*a*) Technical and supporting staff Salary Cost times the multiplier specified in Clause 5 (*b*) of the Memorandum of Agreement, plus Other Payroll Cost.
(*b*) The fee specified in Clause 5 (*c*) of the Memorandum of Agreement, which shall be deemed to cover the services of Partners and Consultants of the firm but not their expenses which are reimbursable separately under Clause 14 (*c*).
(*c*) Any charge for the use of a computer or other special equipment payable under Clause 12 (*d*).

10.3.2 Time spent by technical and supporting staff in travelling in connection with the Works shall be chargeable on the above basis.

10.3.3 Unless agreed between the Client and the Consulting Engineer, the Consulting Engineer shall not be entitled to any payment in respect of time spent by secretarial staff or by staff engaged on general accountancy or administration duties in the Consulting Engineer's office.

10.3.4 The Consulting Engineer shall submit to the Client at the time of submission of the monthly accounts referred to in Clause 20 such supporting data as may be agreed between the Client and the Consulting Engineer.

11. PAYMENT FOR ADDITIONAL SERVICES

11.1 In respect of additional services provided by the Consulting Engineer under Clause 7, the Client shall, subject to Clause 10.3, pay the Consulting Engineer in accordance with the Scale of Charges set out in Clause 11.2.

11.2 Scale of Charges:
(*a*) Partners and Consultants: At the hourly rate or rates specified in Clause 5 (*d*) of the Memorandum of Agreement.
(*b*) Technical and supporting staff: Up to a maximum Salary Cost times 2·5, plus Other Payroll Cost.
(*c*) Time spent by Partners, Consultants, technical and supporting staff in travelling in connection with the Works shall be chargeable on the above basis.

CONSULTING ENGINEERS' FEES

(*d*) Unless otherwise agreed between the Client and the Consulting Engineer, the Consulting Engineer shall not be entitled to any payment in respect of time spent by secretarial staff or by staff engaged on general accountancy or administration duties in the Consulting Engineer's office.

12. PAYMENT FOR USE OF COMPUTER OR OTHER SPECIAL EQUIPMENT

Where the Client has agreed to pay the Consulting Engineer

(*a*) in accordance with Clause 10.3 and the Consulting Engineer decides to use a computer or other special equipment in the carrying out of any of his services, or

(*b*) for his services under Clause 6 in accordance with Clause 10.1 or 10.2 and the Consulting Engineer decides to use a computer or other special equipment in carrying out any additional services in accordance with Clause 7 or is expressly required by the Client to use a computer or other special equipment in the carrying out of his services under Clause 6,

the Client shall, unless otherwise agreed between the Client and the Consulting Engineer, pay the Consulting Engineer

(*c*) for the time spent in connection with the use of a computer or other special equipment, including the development and writing of programmes and the operation of the computer in trial and final runs, in accordance with Clause 10.3 when applicable and otherwise in accordance with the Scale of Charges set out in Clause 11.2, and

(*d*) a reasonable charge for the use of the computer or other special equipment, which charge shall, if possible, be agreed between the Client and the Consulting Engineer before the work is put in hand.

13. PAYMENT FOR SITE SUPERVISION

13.1 In addition to any other payment to be made by the Client to the Consulting Engineer under this Agreement, the Client shall

(*a*) reimburse the Consulting Engineer in respect of all salary and wage payments made by the Consulting Engineer to site-staff employed by the Consulting Engineer pursuant to Clause 8 and in respect of all other expenditure incurred by the Consulting Engineer in connection with the selection, engagement and employment of site-staff, and

(*b*) pay to the Consulting Engineer a sum calculated at 7 per cent of the amounts payable to the Consulting Engineer under the preceding sub-clause in respect of head office overhead costs incurred on the site-staff administration,

provided that in lieu of payments under (*a*) and (*b*) above the Client and the Consulting Engineer may agree upon inclusive monthly or other rates to be paid by the Client to the Consulting Engineer for each member of site-staff employed by the Consulting Engineer.

13.2 The Client shall also in all cases be responsible for the cost of providing such local office accommodation, furniture, telephones, equipment and transport as shall be reasonably necessary for the use of site-staff appointed pursuant to Clause 8, and for the reasonable running costs of such necessary local office accommodation and other facilities, including those of stationery, telephone calls, telegrams and postage. Unless otherwise agreed betweeen the Client and the Consulting Engineer, the Consulting Engineer shall arrange, whether through the Contractor or otherwise, for the provision of local office accommodation and facilities for the use of site-staff.

CONSULTING ENGINEERS' FEES

13.3 In cases where the Consulting Engineer has thought it proper that site-staff should not be appointed, or where the necessary site-staff is not available at site due to sickness or any other cause, the Consulting Engineer shall, subject to Clause 10.3, be paid in accordance with the Scale of Charges set out in Clause 11.2 for site visits which would have been unnecessary but for the absence or non-availability of site-staff.

14. DISBURSEMENTS

The Client shall in all cases reimburse the Consulting Engineer in respect of all the Consulting Engineer's disbursements properly made in connection with:

(a) Printing, reproduction and purchase of all documents, drawings, maps and records.
(b) Telegrams and telephone calls other than local.
(c) Travelling, hotel expenses and other similar disbursements.
(d) Advertising for tenders and for site-staff.
(e) The provision of additional services to the Client pursuant to Clause 7.4.

The Client may, however, by agreement between himself and the Consulting Engineer make to the Consulting Engineer a lump sum payment or payment of a sum calculated as a percentage charge on the cost of the Works in satisfaction of his liability to the Consulting Engineer in respect of the Consulting Engineer's disbursements.

15. PAYMENT FOR ALTERATION OR MODIFICATION TO DESIGN

If after the completion by the Consulting Engineer of his services under Clause 6.1, or where the Client has agreed to make payment to the Consulting Engineer in accordance with Clause 10.2 at any time after the date of this Agreement, any design whether completed or in progress or any specification, drawing or other document prepared in whole or in part by the Consulting Engineer shall require to be modified or revised by reason of instructions received by the Consulting Engineer from the Client, or by reason of circumstances which could not reasonably have been foreseen, then the Client shall make additional payment to the Consulting Engineer for making any necessary modifications or revisions and for any consequential reproduction of documents. Subject to Clause 10.3, and unless otherwise agreed between the Client and the Consulting Engineer, the additional sum to be paid to the Consulting Engineer shall be calculated in accordance with the Scale of Charges set out in Clause 11.2, and shall also include any appropriate reimbursements in accordance with Clause 14.

16. PAYMENT WHEN WORKS ARE DAMAGED OR DESTROYED

If at any time before completion of the Works any part of the Works or any materials, plant or equipment whether incorporated in the Works or not shall be damaged or destroyed, the Client shall make additional payment to the Consulting Engineer in respect of any expenses incurred or additional work required to be carried out by the Consulting Engineer as a result of such damage or destruction. Subject to Clause 10.3, the amount of such additional payment shall be calculated in accordance with the Scale of Charges set out in Clause 11.2, and shall also include any appropriate reimbursements in accordance with Clause 14.

CONSULTING ENGINEERS' FEES

17. PAYMENT FOLLOWING TERMINATION OR SUSPENSION BY THE CLIENT

17.1 Upon a termination or suspension by the Client in pursuance of Clause 2.3, the Client shall pay to the Consulting Engineer the sums specified in (*a*), (*b*) and (*c*) of this sub-clause (less the amount of payments previously made to the Consulting Engineer under the terms of this Agreement).

(*a*) A fair and reasonable proportion of the sum which would have been payable to the Consulting Engineer under Clause 10 if no such termination or suspension had taken place. In the assessment of such proportion, the services carried out by the Consulting Engineer up to the date of termination or suspension and in pursuance of Clause 2.6 shall be compared with a reasonable assessment of the services which the Consulting Engineer would have carried out but for the termination or suspension. In any case in which it is necessary to assess the payment to be made to the Consulting Engineer in accordance with this sub-clause by reference to the cost of the Works, then to the extent that such cost is not known the assessment shall be made upon the basis of the Consulting Engineer's best estimates of cost.

(*b*) Amounts due to the Consulting Engineer under any other clauses of this Agreement.

(*c*) A disruption charge equal to one sixth of the difference between the sum which would have been payable to the Consulting Engineer under Clause 10, but for the termination or suspension, and the sum payable under (*a*) above.

17.2 In any case in which the Client has required the Consulting Engineer to suspend the carrying out of the Consulting Engineer's services in pursuance of the power conferred by Clause 2.3, the Client may, at any time within the period of 12 months from the date of his requirement in writing to the Consulting Engineer to suspend the carrying out of the Consulting Engineer's services, require the Consulting Engineer in writing to resume the performance of such services. In such event

(*a*) the Consulting Engineer shall within a reasonable time of receipt by him of the Client's said requirement in writing resume the performance of his services in accordance with this Agreement, the payment made under Clause 17.1(*a*) ranking as payment on account towards the total sum payable to the Consulting Engineer under Clause 10, but

(*b*) notwithstanding such resumption the Consulting Engineer shall be entitled to retain or receive as an additional payment due in accordance with this Agreement the disruption charge referred to in Clause 17.1(*c*).

17.3 If the Consulting Engineer shall need to perform any additional services in connection with the resumption of his services in accordance with Clause 17.2 the Client shall pay the Consulting Engineer in respect of the performance of such additional services in accordance as the case may be with Clause 10.3 or the Scale of Charges set out in Clause 11, and any appropriate reimbursements in accordance with Clause 14.

18. PAYMENT FOLLOWING TERMINATION BY THE CONSULTING ENGINEER

Upon a termination by the Consulting Engineer in pursuance of Clause 2.5, the Client shall pay to the Consulting Engineer the sums specified in Clause 17.1(*a*) and (*b*) (less the amount of payments previously made to the Consulting Engineer under the terms of this Agreement). Upon payment of such sums, the Consulting Engineer shall deliver to the

CONSULTING ENGINEERS' FEES

Client such completed drawings, specifications and other similar documents relevant to the Works as are in his possession. The Consulting Engineer shall be permitted to retain copies of any documents so delivered to the Client.

19. COST OF THE WORKS

19.1 The cost of the Works or any part thereof shall be deemed to include:

(*a*) The cost to the Client of the Works however incurred, including any payments (before deduction of any liquidated damages or penalties payable by the Contractor to the Client) made by the Client to the Contractor by way of bonus, incentive or ex-gratia payments, or in settlement of claims.
(*b*) A fair valuation of any labour, materials, manufactured goods, machinery or other facilities provided by the Client, and of the full benefit accruing to the Contractor from the use of construction plant and equipment belonging to the Client which the Client has required to be used in the execution of the Works.
(*c*) The market value, as if purchased new, of any second-hand materials, manufactured goods and machinery incorporated in the Works.
(*d*) The cost of geotechnical investigations (Clause 6.1 (*c*)).
(*e*) A fair proportion of the total cost to the Client of any work in connection with the provision or diversion of public utilities systems which is carried out, other than by the Contractor, under arrangements made by the Consulting Engineer. The said fair probation shall be assessed with reference to the costs incurred by the Consulting Engineer in making such arrangements.

19.2 The cost of the Works shall not include:

(*a*) Administration expenses incurred by the Client.
(*b*) Costs incurred by the Client under this Agreement.
(*c*) Interest on capital during construction, and the cost of raising moneys required for carrying out the construction of the Works.
(*d*) Cost of land and wayleaves.

GENERAL CONDITIONS AND SCALE OF CHARGES FOR THE DESIGN AND SUPERVISION OF ENGINEERING SYSTEMS IN BUILDINGS AND OTHER PROJECTS CARRIED OUT UNDER SCHEDULE 4

1. DEFINITIONS

'Salary Cost'
means the annual salary including bonuses of any person employed by the Consulting Engineer, divided by 1600 (being deemed to be the average annual total of effective working hours of an employee) and multiplied by the number of working hours spent by such person in performing any of the services in respect of which payment under this Agreement is to be made to the Consulting Engineer upon the basis of Salary Cost. For the purposes of this definition the annual salary of a person employed by the Consulting Engineer for a period less than a full year shall be calculated pro rata to such person's salary (including bonuses) for such lesser period.

CONSULTING ENGINEERS' FEES

'Other Payroll Cost'
means the annual amount of all contributions and payments made by the Consulting Engineer on behalf of or in respect of a person employed by him for staff pension and life assurance schemes and also for National Health Insurance, Graduated Pension Fund, Selective Employment Tax and for any other tax, charge, levy, impost or payment of any kind whatsoever which the Consulting Engineer at any time during the performance of this Agreement is obliged by law to make on behalf of or in respect of such person, divided by 1600 (being deemed to be the average annual total of effective working hours of an employee) and multiplied by the number of working hours spent by such person in performing any of the services in respect of which payment under this Agreement is to be made to the Consulting Engineer upon the basis of Other Payroll Cost. For the purposes of this definition the annual amount of all contributions and payments made by the Consulting Engineer on behalf of or in respect of a person employed by him for a period less than a full year shall be calculated pro rata to the amount of such contributions and payments for such lesser period.

'Tender Drawings'
means drawings prepared by the Consulting Engineer in sufficient detail to enable those tendering to interpret correctly the design for the Works and to submit competitive tenders for the execution of the Works.

'Detail Drawings'
means drawings, additional to the Tender Drawings, specially prepared by the Consulting Engineer at the request of the Architect or the Client.

'Co-ordination Drawings'
means drawings prepared by the Consulting Engineer or the Contractor, showing the inter-relation of two or more engineering systems.

'Installation Drawings'
means drawings prepared by the Contractor for approval by the Consulting Engineer, showing details of the Contractor's proposals for the execution of the Works.

'Builder's Work Drawings'
means drawings normally prepared by the Contractor for approval by the Consulting Engineer, showing details of work of a structural nature which is required to be carried out by a builder or other party to facilitate the execution of the Works.

'Record Drawings'
means drawings normally prepared by the Contractor for approval by the Consulting Engineer, showing clearly the general scheme and details of the Works as completed.

Obligations of Consulting Engineer

6. NORMAL SERVICES

6.1 Design Stage I
The services to be provided by the Consulting Engineer at this stage shall comprise all or any of the following as may be necessary in the particular case:

(*a*) Investigating data and information relating to the Project and relevant to the Works which are reasonably accessible to the Consulting Engineer, and considering any reports

CONSULTING ENGINEERS' FEES

relating to the works which have either been prepared by the Consulting Engineer or else prepared by others and made available to the Consulting Engineer by the Client.
(*b*) Advising the Client on the need for arrangements to be made, in accordance with Clause 7, for the carrying out of special surveys, special investigations or model tests, and advising the Client of the results of any such surveys, investigations or tests carried out.
(*c*) Consulting any local or other authorities on matters of principle in connection with the design of the Works.
(*d*) Consulting the Architect or any other professional adviser appointed by the Client in connection with the Project.
(*e*) Providing sufficient preliminary information and approximate estimates (based on unit volume, unit surface area or similar bases of estimation) regarding the Works to enable the Client or the Architect to prepare architectural sketch plans and budget estimates for the Project.

6.2 Design Stage II
The services to be provided by the Consulting Engineer at this stage shall comprise all or any of the following as may be necessary in the particular case:

(*a*) Preparing designs, Tender Drawings and specifications for the Works in such detail as may be necessary to enable competitive tenders for the execution of the Works to be obtained.
(*b*) Providing outline information as to plant rooms, chimneys, air-conditioning and ventilation ducts, main service ducts and other similar elements incorporated in the building structure, and providing information as to the approximate weights of items of heavy plant and equipment which are to be incorporated in the Works.
(*c*) Advising on conditions of contract relevant to the Works and forms of tender and invitations to tender as they relate to the Works.
(*d*) Advising the Client as to the suitability for carrying out the Works of persons and firms tendering, and further as to the relative merits of tenders, prices and estimates received for carrying out the Works.

As soon as the Consulting Engineer shall have submitted advice to the Client upon tenders, his services at this stage shall be complete.

6.3 Installation Stage
The Consulting Engineer shall not accept any tender in respect of the Works unless the Client gives him instructions in writing to do so, and any acceptance so made by the Consulting Engineer on the instructions of the Client shall be on behalf of the Client. The services to be provided by the Consulting Engineer at this stage shall include all or any of the following as may be necessary in the particular case:

(*a*) Advising on the preparation of formal contract documents relating to accepted tenders for carrying out the Works or any part thereof.
(*b*) Advising the Client on the appointment of site-staff in accordance with Clause 10.
(*c*) Providing the Contractor with such further information as is necessary in the opinion of the Consulting Engineer, to enable the Installation Drawings to be prepared.
(*d*) Examining the Contractor's proposals.
(*e*) Making such visits to site as the Consulting Engineer shall consider necessary to satisfy himself as to the performance of any site-staff appointed pursuant to Clause 10,

CONSULTING ENGINEERS' FEES

and to satisfy himself that the Works are executed generally according to his designs and specifications and otherwise in accordance with good engineering practice.

(*f*) Giving all necessary instructions to the Contractor, provided that the Consulting Engineer shall not without the prior approval of the Client give any instructions which are in his opinion likely substantially to increase the cost of the Works unless it is not in the circumstances practicable for the Consulting Engineer to obtain such prior approval.

(*g*) Advising the Client or the Architect as to the need to vary any part of the Project for a reason or reasons relating to the Works.

(*h*) Approving the Contractor's commissioning procedures and performance tests, and inspecting the Works on completion.

(*i*) Advising on interim valuations, issuing certificates for payment to the Contractor where appropriate and advising on the settlement of the Contractor's final accounts.

(*j*) Performing any services which the Consulting Engineer may be required to carry out under any document which he has prepared relating to the Works. The Consulting Engineer may decline to perform any services specified in a contract the terms of which have not initially been expressly approved by him in writing.

(*k*) Delivering to the Client on the completion of the Works copies of Record Drawings, the Contractor's operating instructions, and, where appropriate, certificates of works tests.

(*l*) Assisting in settling any dispute or difference which may arise between the Client and the Contractor, provided that this service shall not extend to advising the Client following the taking of any step in or towards any arbitration or litigation in connection with the Works.

6.4 General

Without prejudice to the preceding provisions of this clause, the Consulting Engineer shall from time to time as may be necessary advise the Client as to the need for the Client to be provided with additional services in accordance with Clause 7.

7. ADDITIONAL SERVICES NOT INCLUDED IN NORMAL SERVICES

7.1 As services additional to those specified in Clause 6, the Consulting Engineer shall, if so requested by the Client, provide any of the services specified in Clause 7.2 and provide or take all reasonable steps to arrange for the provision of any of the services specified in Clause 7.3.

7.2 (*a*) Preparing any report or additional contract documents required for consideration of proposals for the carrying out of alternative works.

(*b*) Carrying out work consequent upon a decision by the Client to seek parliamentary powers.

(*c*) Carrying out work in connection with any application already made by the Client for any order, sanction, licence, permit or other consent, approval or authorization necessary to enable the Works to proceed.

(*d*) Carrying out work arising from the failure of the Client to award a contract in due time.

(*e*) Carrying out special cost investigations or detailed valuations, including estimates or cost analyses based on measurement or forming an element of a cost planning service.

(*f*) Carrying out surveys, including surveys of existing installations.

CONSULTING ENGINEERS' FEES

(g) Negotiating and arranging for the provision or diversion of utility services.
(h) Negotiating any contract or sub-contract with a Contractor selected otherwise than by competitive tendering, including checking and agreeing quantities and nett costs of materials and labour, arithmetical checking and agreeing added percentages to cover overheads and profit.
(i) Preparing Detail Drawings, Co-ordination Drawings, Builder's Work Drawings, Record Drawings or any detailed Schedules.
(j) Preparing details for shop fabrication of ductwork, metal or plastic frameworks.
(k) Checking and advising upon any part of the Project not designed by the Consulting Engineer.
(l) Inspecting or witnessing the testing of materials or machinery during manufacture.
(m) Carrying out commissioning procedures or performance tests.
(n) Carrying out work consequent upon any assignment of the contract by the Contractor or upon the failure of the Contractor properly to perform any contract or upon delay by the Client in fulfilling his obligations under Clause 11 or in taking any other step necessary for the due performance of the Works.
(o) Providing manuals and other documents describing the design, operation and maintenance of the Works.
(p) Advising the Client and carrying out work following the taking of any step in or towards any litigation or arbitration relating to the Works.
(q) Carrying out work in conjunction with others employed to provide any of the services specified in Clause 7.3.
(r) Carrying out such other additional services, if any, as are specified in Clause 6 of the Memorandum of Agreement or made necessary by RIBA 'Plan of Work' or other special procedures required by the Client or the Architect.

7.3 (a) Specialist technical advice on any abnormal aspects of the Works.
The services provided by the Consulting Engineer under Clause 6 will include the provision of all expert technical advice and skills which are normally required for the class of work for which the Consulting Engineer's services are engaged.

(b) Legal, financial, architectural, structural and other professional services.
(c) Services in connection with the valuation, purchase, sale or leasing of lands and the obtaining of wayleaves.
(d) Investigation of the nature and strength of existing works and the making of model tests or special investigations.

7.4 The Consulting Engineer shall obtain the prior agreement of the Client to the arrangements which he proposes to make on the Client's behalf for the provision of any of the services specified in Clause 7.3. The Client shall be responsible to any person or persons providing such services for the cost thereof.

8. BILLS OF QUANTITIES

8.1 The Consulting Engineer shall advise the Client on the need for preparing Bills of Quantities in respect of the works before the invitation of tenders therefor.

8.2 If the Client so requests, the Consulting Engineer shall

(a) prepare detailed Bills of Quantities for the Works if the design of the Project is sufficiently advanced to enable him so to do, or

CONSULTING ENGINEERS' FEES

(b) prepare Bills of Approximate Quantities for the Works, if the design of the Project is not sufficiently advanced to enable him to prepare detailed Bills of Quantities.

8.3 In any case in which the Consulting Engineer has prepared Bills of Approximate Quantities, he shall subsequently correct the quantities and reprice the Bills in consultation with the Contractor.

9. VARIATIONS

9.1 The Consulting Engineer shall initiate all necessary variation orders in connection with the Works.

9.2 The Consulting Engineer shall measure or assess the extent of all variations in the Works and shall negotiate, and if possible agree with the Contractor, the value thereof. The Consulting Engineer shall also check all relevant entries in the Contractor's interim and final accounts.

9.3 If, however, the Consulting Engineer and the Contractor agree to remeasure the Works on completion for the purposes of Clause 8.3, Clause 9.2 shall not apply.

9.4 The Consulting Engineer shall check and approve the Contractor's assessments of the value of all fluctuations (increases and/or decreases) in the cost of labour and materials.

10. SUPERVISION ON SITE

10.1 If in the opinion of the Consulting Engineer the nature of the Works warrants full-time or part-time engineering supervision on site, the Client shall not object to the appointment of such suitably qualified technical and clerical site-staff as the Consulting Engineer shall consider reasonably necessary to enable such supervision to be carried out.

10.2 Persons appointed pursuant to the previous sub-clause shall be employed either by the Consulting Engineer or, if the Client and the Consulting Engineer shall so agree, by the Client directly, provided that the Client shall not employ any person as a member of the site-staff unless the Consulting Engineer has first selected or approved such person as suitable for employment.

10.3 The terms of service of all site-staff to be employed by the Consulting Engineer shall be subject to the approval of the Client, which approval shall not be unreasonably withheld.

10.4 The Client shall procure that the contracts of employment of site-staff employed by the Client shall stipulate that the person employed shall in no circumstances take or act upon instructions other than those of the Consulting Engineer.

10.5 Where any of the services specified in Clause 6.3 are performed by site-staff employed by the Client, the Consulting Engineer shall not be responsible for any failure on the part of such staff properly to comply with any instructions given by the Consulting Engineer.

Obligations of Client

11. INFORMATION TO BE SUPPLIED TO THE CONSULTING ENGINEER

11.1 The Client shall supply to the Consulting Engineer, without charge and within a reasonable time, all necessary and relevant data and information in the possession of the

CONSULTING ENGINEERS' FEES

Client and shall give such assistance as shall reasonably be required by the Consulting Engineer in the performance of his services under this Agreement. The information to be provided by the Client to the Consulting Engineer shall include:

(*a*) All such drawings as may be necessary to make the Client's or the Architect's requirements clear, including plans and sections of all buildings (to a scale of not less than 1 to 100) and essential details (to a scale of not less than 1 to 25) together with site plans (to a scale of not less than 1/1,250) and levels.

(*b*) Copies of all contract documents, variation orders and supporting documents relating to those parts of the Project which are relevant to the Works.

11.2 The Client shall give his decision on all sketches, drawings, reports, recommendations, tender documents and other matters properly referred to him for decision by the Consulting Engineer in such reasonable time as not to delay or disrupt the performance by the Consulting Engineer of his services under this Agreement.

12. PAYMENT FOR NORMAL SERVICES

12.1 The sum payable by the Client to the Consulting Engineer for his services under Clause 6 shall be calculated in accordance with the Scale of Charges set out in Clause 12.2 where the cost of the Works is not less than £10,000, or in accordance with the Scale of Charges set out in Clause 13.2, or otherwise as may be agreed between the Client and the Consulting Engineer, where the cost of the Works is less than £10,000.

12.2 The Scale of Charges (suitable for work of average complexity) referred to in Clause 12.1 is as follows:

Cost of the Works		Charge
On the first	£ 10,000	11%
On the next	£ 15,000	9%
On the next	£ 25,000	8%
On the next	£ 50,000	$7\frac{1}{2}$%
On the next	£100,000	7%
On the next	£800,000	$6\frac{3}{4}$%
On the remainder		$6\frac{1}{2}$%

The charge can also be calculated conveniently from the appropriate line of the following table:

Cost of the Works	Charge
From £ 10,000 to £ 25,000	£ 1,100 + 9% of balance over £ 10,000
£ 25,000 to £ 50,000	£ 2,450 + 8% of balance over £ 25,000
£ 50,000 to £ 100,000	£ 4,450 + $7\frac{1}{2}$% of balance over £ 50,000
£ 100,000 to £ 200,000	£ 8,200 + 7% of balance over £ 100,000
£ 200,000 to £1,000,000	£15,200 + $6\frac{3}{4}$% of balance over £ 200,000
Over £1,000,000	£69,200 + $6\frac{1}{2}$% of balance over £1,000,000

CONSULTING ENGINEERS' FEES

12.3 If the Client decides to have the Works constructed in more than one phase and as a consequence the services which it may be necessary for the Consulting Engineer to perform under Clause 6 have to be undertaken by the Consulting Engineer separately in respect of each phase, then the provisions of this payment clause shall apply separately to each phase and as if the expression 'the Works' as used in this clause meant, in the case of each phase, the work comprised in that phase.

13. PAYMENT FOR ADDITIONAL SERVICES

13.1 In respect of additional services provided by the Consulting Engineer under Clause 7, the Client shall pay the Consulting Engineer in accordance with the Scale of Charges set out in the next sub-clause.

13.2 Scale of Charges:

(*a*) Partners and Consultants: At the hourly rate or rates specified in Clause 4 of the Memorandum of Agreement.

(*b*) Technical and supporting staff: Up to a maximum of Salary Cost times 2·5, plus Other Payroll Cost.

(*c*) Time spent by Partners, Consultants, technical and supporting staff in travelling in connection with the Works shall be chargeable on the above basis.

(*d*) Unless otherwise agreed between the Client and the Consulting Engineer, the Consulting Engineer shall not be entitled to any payment in respect of time spent by secretarial staff or by staff engaged on general accountancy or administration duties in the Consulting Engineer's office.

14. PAYMENT FOR BILLS OF QUANTITIES

14.1 For services provided under Clause 8, the Consulting Engineer shall be paid an additional amount according to the procedure adopted, as follows:

(*a*) Where Bills of Quantities are prepared, either by the Consulting Engineer or by a Quantity Surveyor appointed by the Client to provide services in respect of the Works, the Consulting Engineer shall be paid for providing the additional information required for preparing the Bill of Quantities, at the following rates:

	Detailed Bills of Quantities	Bills of Approximate Quantities
On the cost of the Works in the Bills		
On the first £500,000	1%	$\frac{1}{2}$%
On the excess thereafter	$\frac{3}{4}$%	$\frac{3}{8}$%

(*b*) For preparing Bills of Quantities as described in Clauses 8.2(*a*) and 8.2(*b*), the Consulting Engineer shall be paid at the following rates:

	Detailed Bills of Quantities	Bills of Approximate Quantities
On the cost of the Works in the Bills and subject to Clause 14.1(*c*)		
On the first £500,000	$1\frac{1}{2}$%	$\frac{3}{4}$%
On the excess thereafter	$1\frac{1}{4}$%	$\frac{5}{8}$%

CONSULTING ENGINEERS' FEES

In the case of Approximate Bills of Quantities the Consulting Engineer shall be paid a further amount of $1\frac{1}{2}\%$ on the cost of the Works, for correcting and repricing the Bills.

(c) On any Agreements concluded after 8 March 1977 the amount payable to the Consulting Engineer shall be increased by $12\frac{1}{2}\%$ of the sum as calculated in Clause 14.1(b).

14.2 Payments made in accordance with either Clause 14.1 or Clause 14.2 shall be calculated on the total cost of the works in the Bills, including all provisional and prime cost items therein.

14.3 Unless and until the Bills are fully and accurately priced, the cost of the works in the Bills shall be the total of the tender recommended for acceptance or, where no tender has been received, shall be the Consulting Engineer's best estimate of the said cost.

15. PAYMENT FOR VARIATIONS

15.1 (a) For his services under Clause 9, the Consulting Engineer shall be paid an additional amount of $2\frac{1}{2}\%$ on the value of additional works and of fluctuations (increases and/or decreases) in the cost of labour and materials, and $1\frac{1}{2}\%$ on the value of omitted works.

(b) On any Agreements concluded after 8 March 1977, the amount payable to the Consulting Engineer shall be increased by $12\frac{1}{2}\%$ of the sum as calculated in Clause 15.1(a).

15.2 Where measurement of the completed Works is carried out for the purposes of Clause 8.3, the preceding sub-clauses shall not apply.

16. PAYMENT FOR USE OF COMPUTER OR OTHER SPECIAL EQUIPMENT

Where the Consulting Engineer decides to use a computer or other special equipment in carrying out any additional services in accordance with Clause 7 or is expressly required by the Client to use a computer or other special equipment in the carrying out of his services under Clause 6, the Client shall, unless otherwise agreed between the Client and the Consulting Engineer, pay the Consulting Engineer

(a) for the time spent in connection with the use of a computer or other special equipment, including the development and writing of programmes and the operation of the computer in trial and final runs, in accordance with the Scale of Charges set out in Clause 13.2, and

(b) a reasonable charge for the use of the computer or other special equipment, which charge shall, if possible, be agreed between the Client and the Consulting Engineer before the work is put in hand.

17. PAYMENT FOR SITE SUPERVISION

17.1 In addition to any other payment to be made by the Client to the Consulting Engineer under this Agreement, the Client shall

(a) reimburse the Consulting Engineer in respect of all salary and wage payments made by the Consulting Engineer to site-staff employed by the Consulting Engineer pursuant to Clause 10 and in respect of all other expenditure incurred by the Consulting Engineer in connection with the selection, engagement and employment of site-staff, and

CONSULTING ENGINEERS' FEES

(b) pay to the Consulting Engineer a sum calculated at 7 per cent of the amounts payable to the Consulting Engineer under the preceding sub-clause in respect of head office overhead costs incurred on site-staff administration.

provided that in lieu of payments under (a) and (b) above the Client and the Consulting Engineer may agree upon inclusive monthly or other rates to be paid by the Client to the Consulting Engineer for each member of site-staff employed by the Consulting Engineer.

17.2 The Client shall also in all cases be responsible for the cost of providing such local office accommodation, furniture, telephones, equipment and transport as shall be reasonably necessary for the use of site-staff appointed pursuant to Clause 10, and for the reasonable running costs of such necessary local office accommodation and other facilities, including those of stationery, telephone calls, telegrams and postage. Unless otherwise agreed between the Client and the Consulting Engineer, the Consulting Engineer shall arrange, whether through the Contractor or otherwise, for the provision of local office accommodation and facilities for the use of site-staff.

17.3 In cases where the Consulting Engineer has thought it proper that site staff should not be appointed, or where the necessary site-staff is not available at site due to sickness or any other cause, the Consulting Engineer shall be paid in accordance with the Scale of Charges set out in Clause 13.2 for site visits which would have been unnecessary but for the absence or non-availability of site-staff.

18. DISBURSEMENTS

The Client shall in all cases reimburse the Consulting Engineer in respect of all the Consulting Engineer's disbursements properly made in connection with:

(a) Printing, reproduction and purchase of all documents, drawings, maps and records.
(b) Telegrams and telephone calls other than local.
(c) Travelling, hotel expenses and other similar disbursements.
(d) Advertising for tenders and for site-staff.
(e) The provision of additional services to the Client pursuant to Clause 7.4.

The Client may, however, by agreement between himself and the Consulting Engineer make to the Consulting Engineer a lump sum payment or payment of a sum calculated as a percentage charge on the cost of the Works in satisfaction of his liability to the Consulting Engineer in respect of the Consulting Engineer's disbursements.

19. PAYMENT FOR ALTERATION OR MODIFICATION TO DESIGN

If after the completion by the Consulting Engineer of his services under Clause 6.1 any design whether completed or in progress or any specification, drawing or other document prepared in whole or in part by the Consulting Engineer shall require to be modified or revised by reason of instructions received by the Consulting Engineer from the Client or from the Architect, or by reason of circumstances which could not reasonably have been foreseen, then the Client shall make additional payment to the Consulting Engineer for making any necessary modifications or revisions and for any consequential reproduction of documents. Unless otherwise agreed between the Client and the Consulting Engineer, the additional sum to be paid to the Consulting Engineer shall be calculated in accordance

CONSULTING ENGINEERS' FEES

with the Scale of Charges set out in Clause 13.2, and shall also include any appropriate reimbursements in accordance with Clause 18.

20. PAYMENT WHEN WORKS ARE DAMAGED OR DESTROYED

If at any time before completion of the Works any part of the Works or any materials plant or equipment whether incorporated in the Works or not shall be damaged or destroyed, the Client shall make additional payment to the Consulting Engineer in respect of any expenses incurred or additional work required to be carried out by the Consulting Engineer as a result of such damage or destruction. The amount of such additional payment shall be calculated in accordance with the Scale of Charges set out in Clause 13.2, and shall also include any appropriate reimbursements in accordance with Clause 18.

21. PAYMENT FOLLOWING TERMINATION OR SUSPENSION BY THE CLIENT

21.1 Upon a termination or suspension by the Client in pursuance of Clause 2.3, the Client shall pay to the Consulting Engineer the sums specified in (*a*), (*b*) and (*c*) of this sub-clause (less the amount of payments previously made to the Consulting Engineer under the terms of this Agreement).

(*a*) A fair and reasonable proportion of the sum which would have been payable to the Consulting Engineer under Clause 12 if no such termination or suspension had taken place. In the assessment of such proportion, the services carried out by the Consulting Engineer up to the date of termination or suspension and in pursuance of Clause 2.6 shall be compared with a reasonable assessment of the services which the Consulting Engineer would have carried out but for the termination or suspension. In any case in which it is necessary to assess the payment to be made to the Consulting Engineer in accordance with this sub-clause by reference to the cost of the Works, then to the extent that such cost is not known the assessment shall be made upon the basis of the Consulting Engineer's best estimates of cost.

(*b*) Amounts due to the Consulting Engineer under any other clauses of this Agreement.

(*c*) A disruption charge equal to one sixth of the difference between the sum which would have been payable to the Consulting Engineer under Clause 12, but for the termination or suspension, and the sum payable under (*a*) above.

21.2 In any case in which the Client has required the Consulting Engineer to suspend the carrying out of the Consulting Engineer's services in pursuance of the power conferred by Clause 2.3, the Client may, at any time within the period of 12 months from the date of his requirement in writing to the Consulting Engineer to suspend the carrying out of the Consulting Engineer's services, require the Consulting Engineer in writing to resume the performance of such services. In such event

(*a*) the Consulting Engineer shall within a reasonable time of receipt by him of the Client's said requirement in writing resume the performance of his services in accordance with this Agreement, the payment made under Clause 21.1(*a*) ranking as payment on account towards the total sum payable to the Consulting Engineer under Clause 12, but

(*b*) notwithstanding such resumption the Consulting Engineer shall be entitled to retain

CONSULTING ENGINEERS' FEES

or receive as an additional payment due in accordance with this Agreement the disruption charge referred to in Clause 21.1 (c).

21.3 If the Consulting Engineer shall need to perform any additional services in connection with the resumption of his services in accordance with Clause 21.2, the Client shall pay the Consulting Engineer in respect of the performance of such additional services in accordance with the Scale of Charges set out in Clause 13.2, and any appropriate reimbursements in accordance with Clause 18.

22. PAYMENT FOLLOWING TERMINATION BY THE CONSULTING ENGINEER

Upon a termination by the Consulting Engineer in pursuance of Clause 2.5, the Client shall pay to the Consulting Engineer the sums specified in Clause 21.1(a) and (b) (less the amount of payments previously made to the Consulting Engineer under the terms of this Agreement). Upon payment of such sums, the Consulting Engineer shall deliver to the Client such completed drawings, specifications and other similar documents relevant to the Works as are in his possession. The Consulting Engineer shall be permitted to retain copies of any documents so delivered to the Client.

23. COST OF THE WORKS

23.1 The cost of the Works or any part thereof shall be deemed to include:

(a) The cost to the Client of the Works however incurred, including any payments (before deduction of any liquidated damages or penalties payable by the Contractor to the Client) made by the Client to the Contractor by way of bonus, incentive or ex-gratia payments, or in settlement of claims.

(b) Where the Works are carried out as a sub-contract or sub-contracts awarded under a main contract, the allowances made in the main contract to cover attendance and profit relating to the Works, together with the cost of items of builder's work required in connection with the Works, and a part of the cost of the preliminary and general items included in the main contract being the proportion that the cost of the sub-contract or sub-contracts bears to the total cost of the main contract.

(c) A fair proportion of the cost of any chimneys, and any air-conditioning or ventilation ducts and their insulation, forming part of the building structure. The said fair proportion shall be assessed with reference to the Consulting Engineer's estimate of the cost of independent construction.

(d) A fair valuation of any labour, materials, manufactured goods, machinery or other facilities provided by the Client, and of the full benefit accruing to the Contractor from the use of construction plant and equipment belonging to the Client which the Client has required to be used in the execution of the Works.

(e) The market value, as if purchased new, of any second-hand materials, manufactured goods and machinery incorporated in the Works.

23.2 The cost of the Works shall not include:

(a) Administration expenses incurred by the Client.
(b) Costs incurred by the Client under this Agreement.
(c) Interest on capital during construction, and the cost of raising moneys required for carrying out the construction of the Works.
(d) Cost of land and wayleaves.

PART SIX

Large Industrial Projects

Directions, *page* 371
The Contract, *page* 373
Labour, *page* 383
Estimating, *page* 437

LARGE INDUSTRIAL PROJECTS

Directions

GENERALLY

Great care should be taken in applying the material in this section to work associated with buildings and vice versa as so many of the factors contributing to cost and price differ between the two groups of engineering work.

In addition to these directions, the reader is referred to the introductions preceding the sub-sections in Labour and Estimating. In particular, the introduction to the Estimating sub-section describes the methods of estimating normally adopted for large industrial projects.

CONTRACTS

This sub-section introduces the legal contracts usually encountered in large industrial projects. The first part deals with contracts for construction that can include engineering and procurement whilst the second part highlights those matters of particular relevance in contracts where the process design is supplied by the Contractor.

LABOUR

This sub-section gives a typical site agreement for oil and petro-chemical work and incorporates The Mechanical Construction Engineering Agreement negotiated by the Engineering Employers' Federation for other work. The rates of pay for oil and petro-chemical work are set out on page 392 whilst those for other work are in the text of the extracts on page 433.

Two typical calculations of actual cost per man hour are given in the Estimating Sub-section on page 483.

ESTIMATING

The costs given in the Cost Guide sub-section were current in April 1980 for British manufacture and can be used in the preparation of a budget estimate for a substantial United Kingdom grass roots' petro-chemical project in the £3,500,000–£10,000,000 range. However the costs given are not suitable for the preparation of a definitive estimate. Land costs are excluded. The costs are not based on any particular location.

Materials and components costs do not vary much from one location to another and as labour is given in terms of hours rather than cost, labour rates appropriate to the location required can be applied. All costs including those for sub-contracted items are net.

The abbreviations used in the Cost Guide sub-section are explained on page 446.

The cost analysis of a typical plant is included to show how the cost code list items may be logically assembled to achieve a total cost or price. The figures given are not definitive and the relationships between them should not be applied without adjustment to a project under consideration. Nevertheless, they are based on actual projects and can be considered as giving reasonable general guidances.

DIRECTIONS

The cost indices can be used to calculate general cost and price trends in the industry but it should be kept in mind that items do not alter in price consistently and the costs of at least the major commodity elements should if possible be substantiated by current quotations.

LARGE INDUSTRIAL PROJECTS

The Contract

THE TYPE OF CONTRACT

In determining the type of contract to be used for a project, two basic considerations have to be taken into account, namely:

(a) the precise nature and scope of work to be undertaken by the contractor and his responsibilities in connection with it, and
(b) the basis of remuneration.

The extent of the contractor's design responsibilities must be ascertained. In some cases he will be responsible for all aspects of the design (including the process design) whilst in others his design responsibilities will be limited to that of engineering and detailing using the process design supplied by the customer. If the scope of work includes civil and building construction the question of civil and building design will also need to be considered.

The extent of material and equipment to be supplied by the contractor is also a major factor. In many cases the contractor will supply all materials and equipment for the project but it is not uncommon for the customer by way of separate contracts, to supply specialist items of equipment and apparatus 'free issue' to the contractor to erect.

The construction work will normally be carried out by the contractor using specialist sub-contract trades as necessary. Here again however, it is not uncommon for the construction work to be divided into separate contracts using the services of a main civil contractor and a main mechanical erection contractor, particularly where the project is extensive and complex.

The major principles to be determined governing basis of remuneration are whether the contract is to be on a lump sum or a cost reimbursement basis. If the scope of the work can be fully defined, a lump sum contract will be more appropriate and will enable the customer to obtain competitive tenders.

If the scope of the work cannot be fully ascertained, some form of cost reimbursement contract will have to be drawn up. The lump sum contract differs widely from the cost reimbursement. The general features of these two different contracts can be summarized as follows:

LUMP SUM

Subject always to the limits of responsibility attaching to the contractor in relation to the scope of his work and duties, the commercial risk in the performance of the contract will rest with the contractor. Having been supplied with the full technical details of the project and all other relevant information, the contractor will be able to submit a tender which if accepted will be binding between the parties. The contractor will also be deemed to have priced the risks and liabilities set forth in the terms and conditions of the contract and to have taken due account of them in his tender.

COST REIMBURSEMENT PLUS FEE

In this form of transaction the commercial risk will largely rest with the customer. The element of the unknown cannot be priced by the contractor and the price of the project to

THE CONTRACT

the customer will ultimately be based on what it costs. It follows therefore that a large degree of control must be exercised and the customer's involvement throughout the project will be very marked. Cost estimates are normally prepared by the contractor at the outset and thereafter cost control procedures will be implemented. Many contracts of this nature will contain a cost savings bonus which will enable the contractor and the customer to share in any savings made to the definitive estimate, but any overrun in the cost will be borne by the customer. The basic advantage of the cost reimbursement plus fee contract is that it enables the project to start without delay. At some stage however the design will be 'frozen' and the scope of supply defined, at which stage the customer and the contractor will often re-negotiate the contract on a lump sum basis. Often a Target Programme Concept is used where the contractor receives a bonus for completion before the agreed target date but is liable to a penalty should he overrun the target date.

THE PREPARATION OF THE CONTRACT DOCUMENTS

Having established the basic principles the contract documents can be drawn up. Drafts will be prepared by either the customer or the contractor; frequently the customer will look to the contractor to carry out this service as the task will be a familiar one to him and the process of preparing and agreeing the drafts can be concluded more quickly. Nevertheless, this process can often be protracted as there can be a number of contingencies, such as planning approvals, financial appropriation and site availability, to be overcome before the documents can be finalised and signed. Early work on a project will often proceed, however, against a letter of intent signed by the customer in which the following may be expressed:

(a) Authority to the contractor to proceed with certain works with a limit to expenditure, except as otherwise subsequently authorised.

(b) The basis of payment for the work executed.

(c) Provisions about terminating the work if the contract is not signed within a specified period.

(d) Exclusion of liability for damages suffered by either party in the event that the project does not proceed beyond that work authorized in the letter of intent.

In competitive lump sum tendering the contract documents will normally be prepared in detail by or on behalf of the customer for the purpose of inviting tenders and thereafter the concluding of a contract will normally be achieved with the minimum delay. The number of documents comprising the contract documents vary from one project to another but the following may be regarded as typical:

Form of Agreement
General Conditions of Contract
Special Conditions of Contract
Specification(s)
Drawings
Bills of Quantities (where the project includes building and civil construction)
Price Variation Formulae (unless the contract is on a fixed price basis)
Preliminary Programme Bar Chart

THE ADMINISTRATION OF THE CONTRACT

It is desirable on most contracts for a co-ordination procedure to be prepared and agreed by both parties, setting out in detail the lines of communication, systems and procedures

THE CONTRACT

to be adopted. This can sometimes present a complex network of activities, designed to ensure that the project is executed in an organized manner. The co-ordination procedure does not normally form part of the contract documents as it is subject to many (unwritten) revisions during the life of the project.

Both the customer and the contractor will normally appoint a project manager to head their respective teams and in most cases the customer's project manager will be 'The Engineer' as defined in the contract documents. In this capacity certain powers and discretions will be vested in him, including the power to vary the work, order additional work, certify monies due to the contractor and approve designs and drawings. The role of The Engineer may in many respects be compared with that of The Architect in a building contract.

On certain large projects the customer may engage the services of a consulting engineer to act in a professional capacity for him, either on the entire project or for certain parts of it, and in such cases the consulting engineer's functions will be even more similar to those of an architect. The customer may further engage the services of an independent firm of quantity surveyors to handle cost control and many other financial aspects of the project and in such cases the quantity surveyor's powers, duties and responsibilities will be set forth in the terms of the contract and will generally follow those associated with the quantity surveyor in a building contract.

THE FUNDAMENTAL PROVISIONS OF A LUMP SUM CONTRACT

The fundamental provisions of a lump sum contract are as follows:

(*a*) Definition of Works – this is a clear definition of the work to be undertaken by the contractor ('The Works') supported by the specification(s), drawings and other relevant data which will be incorporated into the form part of the contract.

(*b*) The Contract Price – this may be expressed as one total sum or may be divided into two or more sums, covering different sections of the works, or applying to specific services to be supplied by the contractor. Part of the contract price may be ascertainable by reference to rates; for example plant hire rates.

(*c*) Date for Completion – completion will normally be defined as the date when the works have been constructed and passed certain specified mechanical and electrical tests. A taking-over certificate will be issued by the customer from the date of which the Defects Liability Period will commence. Completion and taking-over of the works can be in stages.

(*d*) Variations – the customer has the right through The Engineer to alter the works, including incorporating additional work or omitting work already specified. The contractor is paid for any additional work so ordered by agreement on the extra prices for it or the contract may contain pre-agreed daywork rates or other cost reimbursement terms on which the cost of variations are to be based. Omission of work reduces the contract price. Limits are usually placed on the extent of variations permitted unless otherwise agreed by the contractor. The contract cannot be varied by omitting the whole of the Works or duplicating the Works. (These comments on variations do not apply to Process Contracts.)

(*e*) Extension of Time – provisions are normally included entitling the contractor to an extension of time for completion as a result of variations or by reason of any delay due to causes beyond the contractor's control. The period of extensions will be agreed between the contractor and The Engineer.

THE CONTRACT

(*f*) Damages for Delay in Completion – damages will be 'ascertained and liquidated' and the amount specified in the contract. This sum may be expressed as a weekly or daily sum and can be limited in total to a specified percentage of the contract price – normally not exceeding 10%. Such damages will be in full satisfaction for the delay and the customer will be entitled to deduct the amount of such damages from monies due to the contractor. The contractor's liability for such damages will commence from the Date for Completion (as stated in the contract or as extended) and will continue until the works are completed, or the limit of damages has been reached, whichever first occurs. Where a project is subject to sectional completion damages would apply to the sectional completion accordingly.

(*g*) Payment – the contract normally provides for the contract price to be paid in monthly instalments against the contractor's invoices or The Engineer's certificate. The customer is entitled to retain from each instalment the amount specified in the contract for retention. One-half of the retention money will be released to the contractor on taking over the plant and the balance at the end of the Defects Liability Period. Retention money will normally be between 5% and 10% of the money due.

(*h*) Defects Liability Period – this can be otherwise called the 'maintenance period' or 'guarantee period'. Normally the period is for twelve months (or in the case of associated building work, six months) commencing from the effective date of the taking-over certificate. The contractor will be responsible for rectifying at his own cost during this period, any part of the works designed, supplied or constructed by him which have become defective, otherwise than through maloperation of the plant by the customer or any other causes for which the contractor is not responsible. The contractor's liability under these provisions will be in lieu of any condition or warranty as to the suitability of the works implied by law.

The key clauses in the contract can include the following:

(*a*) Secrecy – The contractor undertakes to keep secret and confidential all information supplied by the customer relating to the process and know-how and the customer's commercial operations, (such information to be restricted to those of the contractor's employees who have need of it in connection with the project). The undertaking will normally exclude information which is part of the public domain or which subsequently becomes part of the public domain through no fault of the contractor. A secrecy agreement is often signed separately from and in advance of the contract and it can contain detailed provisions with regard to the disclosure of information.

(*b*) Patent and Other Protected Rights – The contractor indemnifies the customer against any infringement or alleged infringement of any letters patent, registered design, trade mark or copyright in respect of designs, methods and equipment supplied by the contractor. The contractor will (unless otherwise provided) have charge of all litigation arising from it.

The customer for his part indemnifies the contractor in similar form in respect of any processes, designs, methods and equipment supplied by the customer for use by the contractor for the purpose of contracting the works.

(*c*) Errors in Drawings, Designs and Other Data – The contractor accepts responsibility for errors, discrepancies or omissions in drawings and other data supplied by him, unless arising from faulty information supplied by the customer. The customer accepts responsibility for his own faulty designs, drawings or other information supplied to the contractor.

THE CONTRACT

(*d*) Vesting – All equipment, materials and plant delivered to the construction site for permanent incorporation in the works shall vest in and become the property of the customer. The contractor is precluded from removing such equipment, materials and plant except where they do not conform to or are surplus to the requirements of the contract and need to be modified or replaced. For such purposes property in the same re-vests with the contractor.

The contractor is further precluded from removing from the site any temporary facilities, construction plant and the like until the item or items in question have served their purpose.

(*e*) On-Site and Off-Site Inspection – The authorized representatives of the customer have the right to carry out inspection at such places where equipment to be comprised in the works is being manufactured and witness tests on such equipment. The same rights as to inspection shall similarly apply during the course of construction of the works on site. Such inspection and witnessing of tests is to be made after prior arrangement with and in the presence of the contractor.

If the works are within or adjacent to existing process manufacturing or production facilities, the contractor will be bound to comply with the customer's safety and security regulations at any time in force.

(*f*) Care of the Works – The works remain at the risk of the contractor until handed over and accepted by the customer. All loss or damage to the works is to be replaced or repaired by the contractor at his own expense so that the works are handed over in all respects in accordance with the contract, save in the case of loss or damage arising from:

(*a*) any act, omission or negligence on the part of the customer, his servants and agents (not being the Contractor).

(*b*) normal 'excepted' risks such as:
War, hostilities, revolution, insurrection and radioactive contamination.

(*g*) Insurance – An obligation is placed upon the contractor to insure the works against loss or damage arising from any act, omission or negligence on the part of the contractor. Such insurance will normally include sub-contractors and the customer as joint assureds, and extend to cover the parties in respect of:

(*a*) any damage caused to the works whilst the contractor or his sub-contractors are on site for the purpose of rectifying defects during the Defects Liability Period.

(*b*) any damage occurring during the Defects Liability Period from a cause existing prior to handing over.

The terms of the insurance and the insurers will be to the satisfaction of the customer. The contractor is to provide to the customer evidence of premium payments and the customer will have the right to effect the insurance if the contractor shall fail to do so, and deduct the premium costs from monies due to the contractor.

The contract will also normally include arrangements for insuring the works against loss or damage by fire, explosion and other risks of a kindred nature and for third party insurances.

(*h*) Mechanical Tests on Completion – On substantial completion of the erection of the plant (or any self-contained or significant section of it) the contractor shall carry out such hydraulic, pneumatic, electrical and other mechanical tests to demonstrate that the works have been constructed in accordance with the specifications and sound engineering principles but without actually putting the works on-stream.

THE CONTRACT

The customer shall have the right to attend and witness such tests. The contractor is obligated to make such adjustments and modifications to the works as may be necessary at his own expense in the event of failure to meet any test requirement, and to carry out further tests until satisfactory results are achieved.

Once the mechanical tests have been satisfactorily proved, the Taking Over Certificate is issued by the customer, and the works are considered to have been constructed in accordance with the contract. Minor construction faults will not prejudice the issuing of the certificate; such faults, however, can be identified in a schedule and their rectification should be carried out by the contractor before leaving the site.

(i) Termination of Contract – The customer will have the right to terminate the contract if the contractor effectively abandons it or persistently fails to perform and further rights will be vested in the customer including the right to employ others to complete the works and to the free use of the contractor's plant and equipment on the site for the purpose of so completing. No further payments will become due to the contractor until the works are so completed and any extra costs incurred by the customer will be for the contractor's account. The contractor will have similar rights to terminate if the customer fails to make payments in accordance with the contract and in such circumstances the contractor will retain his rights at law to recover the outstanding sums.

(j) Consequential Losses – By any express term of the contract, the contractor will not be liable for any loss of profit or trading revenue, loss of production, loss of contracts or other 'consequential' losses arising through any breach or default on the part of the contractor, save and except to the extent that such losses are pre-ascertained and included in the Liquidated Damages stated in the contract, in respect of delay in completion.

STANDARD FORMS OF CONTRACT

Whilst the types of contract used for the engineering, supply and construction of industrial projects vary considerably in detail their overall concept lies within a familiar pattern. Over the years certain standard forms of contract have been developed by various professional institutions and trade associations which through practice and custom in the engineering and construction industry have become firmly established. It is however rare that a standard set of conditions can adequately meet the specific requirements of a particular project, or of a customer, and it is normal practice therefore to amend a standard form as necessary and further to supplement it with additional conditions.

Many of the larger industrial customer concerns have their own standard forms of contract orientated to their own particular policy and requirements, the major ones are listed below.

THE INSTITUTION OF MECHANICAL ENGINEERS
THE INSTITUTION OF ELECTRICAL ENGINEERS
THE ASSOCIATION OF CONSULTING ENGINEERS

Model Form A	(Home Erection)
Model Form B1	(Delivery Free on Board or carriage, insurance and freight)
Model Form B2	(Delivery Free on Board, carriage, insurance, freight or free on rail with Supervision of Erection)
Model Form B3	(Delivery to and Erection on Site – Export)
Model Form C	(Sale of Electrical and Mechanical Goods other than Electric Cables (Home without Erection)

THE CONTRACT

THE INSTITUTION OF ELECTRICAL ENGINEERS—ALONE

Model Form E (Home Cable Contracts with installations)

THE INSTITUTION OF CHEMICAL ENGINEERS

Model Form of Condition of Contract for Process Plants suitable for lump sum and reimbursable contracts in the United Kingdom. This form is specially designed for projects where the contractor is, inter alia, responsible for the process design of the plant. A process plant contract has certain characteristics which are introduced in the next section.

THE INSTITUTION OF CIVIL ENGINEERS
THE ASSOCIATION OF CONSULTING ENGINEERS
THE FEDERATION OF CIVIL ENGINEERING CONTRACTORS

Conditions of Contract for use in conjunction with works of Civil Engineering Construction, commonly referred to as I.C.E. Conditions.

FÉDÉRATION INTERNATIONALE DES INGÉNIEURS-CONSEILS

Conditions of Contract for use on international contracts where a number of contractors may be working on the same complex.

THE PROCESS PLANT CONTRACT

For the purpose of these comments the process plant contract means a contract under which the contractor supplies the process design and in consequence is responsible for the process performance of the plant under design conditions. The wide variability of process plants in their technical and performance criteria requires the need for each particular contract to be purpose-made. It is common practice to refer to such contracts as 'turn-key' projects, but in order to correctly apply this label, the following criteria must be met:

(a) the process design is supplied by the contractor.
(b) the scope of the work is fully defined before the contract is signed.
(c) the contract price is a lump sum.
(d) the contractor is contracted to undertake the entire project, using as necessary, his own sub-contractors.
(e) the contractor has absolute control of the project throughout.

The Fundamental Provisions of a Process Plant contract include:

(a) Rights of Process – Rights in the process remain with the contractor. The process design (know-how) and the right to operate the process is supplied and granted by the contractor under licence to the customer for the purposes of the project only.

(b) Limitation on the use of the process – The customer is restricted from using the contractor's process design, drawings and other data except for the purpose of operating and maintaining the process plant as designed. Subsequent modification to increase capacity or any duplication of the plant or of any significant features of the plant are to be subject to the contractor's agreement and additional payment.

(c) Development and improvement to the process design – The contractor undertakes to provide the customer, without additional payment (unless provided expressly to the

THE CONTRACT

contrary) all improvements developed by the contractor to the process over a specified period. This provision is normally to be found in export contracts only.

(*d*) Variations – The customer has no power to instruct the contractor to alter the plant. Variations are to be subject to the contractor's approval and additional payment. Approval by the contractor is subject to withdrawal or modification by the contractor of performance guarantees, warranties and other contractual obligations which may be affected by such variations. This is a fundamental difference from the engineering and construction contracts introduced above.

(*e*) Mechanical Testing – As with the engineering and construction contract, the works will be subject to mechanical tests on completion of construction but in the process plant contract, fulfilment of these tests does not constitute the same degree of finality of the contractor's obligations as is the case with engineering and construction contracts.

A statement to the effect that the works have been constructed prima facie in accordance with the contract will probably be signed by the customer and the contractor and from the date of such statement the Defects Liability Period may operate. This date more often than not will be used for the purposes of establishing any liability which may attach to the contractor in respect of liquidated damages for delay in completion.

(*f*) Preparation for Start-Up and Start-Up Performance Tests – The contract will provide for the respective responsibilities of the contractor and the customer during the final phase of the project – usually referred to as 'commissioning' – during which preparations for and the start-up of the works takes place. The customer will have obligations to supply the necessary operating labour and supervisors, routine maintenance services, feedstock and utilities of the correct quantity and quality.

The contract will normally provide a time limit within which the performance guarantees are to be proved, and failure by the contractor to meet this requirement will render him liable to the customer for the payment of liquidated damages for all guarantees not met, provided such failure is due to causes for which the contractor is responsible.

The contractor's responsibilities and liabilities in respect of the performance guarantees shall be discharged:

(*a*) when the performance guarantees have been met,

(*b*) upon the payment by the contractor of liquidated damages for failure to achieve the guarantees within the time specified in the contract in the circumstances referred to above and

(*c*) if through causes beyond the control of the contractor a performance test is not held within a specified period of time,

whichever shall first occur.

(*g*) Liquidated Damages – In addition to any conditions as to the payment by the contractor of liquidated damages for delay in completion, the process plant contract will normally provide for the payment of liquidated damages for non-performance, that is to say, the failure of the plant under design conditions, to meet the performance guarantees. Such damages will be ascertained by calculating over a given period of time the additional costs which the customer will incur by way of increased operating costs in raw material and utilities' consumptions or loss of product capacity and/or quality and the amount so ascertained will be stated in the contract on a cost by unit basis. The damages will be limited to a total sum (normally expressed as a percentage of the contract price) and the contractor's liability will be limited to such sum.

THE CONTRACT

A process plant contract will also contain similar provisions to those listed above as fundamental and key clauses in engineering and construction contracts appropriately modified.

The contractor for his part will retain full control and responsibility over all operations during this phase and will provide specialist engineers to supervise all commissioning and start-up activities. The contractor may, under the terms of the contract, be required to train the customer's personnel in the operation and maintenance of the works in accordance with the contractor's operating and maintenance manuals.

When the contractor is satisfied that the works are operating in accordance with the design conditions a performance test will be carried out under the supervision of the contractor, in order to demonstrate that the works comply in all respects with the specifications and meet the performance guarantees given by the contractor.

Performance guarantees may include all or any of the following:

Plant Capacity
Product Quality
Utilities Consumption (power, steam, etc.)
Feed-stock Consumption (raw materials, operating chemicals, etc.)
Effluent (quantity and quality)

The criteria against which plant performance is assessed will be set forth in the specifications or some other document comprised in the contract and will include the duration of the test, methods of analysis, measurement, tolerance factors, laboratory tests and all relevant matters relating to performance.

If the works fail to meet the performance guarantees through causes for which the contractor is responsible, the contractor shall at his own expense, carry out such work of re-design and make such modifications to the works as he considers necessary. To enable the contractor to comply with these obligations, the customer will be obliged to shut down the works (or the affected part) and make it available to the contractor. Further performance tests shall be repeated.

LARGE INDUSTRIAL PROJECTS

Labour

INTRODUCTION

The pay and working conditions of operatives constructing large industrial projects are established after negotiation. Generally, all contractors for oil and petrol-chemical projects working at a particular site, negotiate collectively individual site agreements whilst those contractors for other work in the mechanical construction industry (such as chemical and process plant construction, outside steelwork erection, pipework installation, steelworks, smelter and power station construction, boiler erection, turbine and electrical plant installation, platework and tank and pressure vessel fabrication) many of whom are members of The Engineering Employers' Federation subscribe to The Mechanical Construction Engineering Agreement negotiated nationally by that body. Contractors who are not members of the Federation adopt their own wages structure which is, however, generally based on the national agreement. Some oil and petro-chemical work customers take direct interest in the agreements for their sites and incorporate a requirement to apply them in their contracts. There is a separate agreement relating to Engineering Construction Hook-up work on Oil and Gas Production Platform Sites.

OIL AND PETRO-CHEMICAL SITE AGREEMENTS

Generally speaking, the terms and conditions of employment for employees on oil and petro-chemical projects are governed by Site Agreements. These Agreements are negotiated at local level with the full-time trade union officials and the contractor(s) concerned. Site Agreements vary depending upon area, local practices and client wishes etc. More recently the trend has been towards a 40-hour week without overtime working except under extreme emergencies. On these sites a two tier wages structure operates, e.g. a basic rate plus a productivity allowance or a measured incentive bonus scheme. On other projects, where overtime is worked, a consolidated basic rate applies. The object of having Site Agreements is to avoid, where possible, employees of contractors having substantial differences in the average take home pay packet for work people carrying out similar work.

It is customary for each operative to be issued with a copy of the Agreement for the site on which he works.

THE ENGINEERING EMPLOYERS' FEDERATION NATIONAL AGREEMENT

The Engineering Employers' Federation has negotiated a national agreement for mechanical construction engineering site workers with different trades unions or groups of trades unions. The agreement defines a normal 40-hour week (overtime being arranged at site level) and national minimum rates of basic pay, to which are added productivity allowances or bonus incentive payments based on measurements of productivity negotiated at company or site level, to supplement the national time rate where deemed necessary.

383

AGREEMENTS (O.C.P.C.A.)

The National Agreement replaces the previous National Sector Agreements (The Outside Steelwork Erection and the Steam Generating Plant Erection Agreements and the Steel Plate Erection Code).

The petro-chemical site agreements tend to follow, as a matter of policy, the general framework of the E.E.F. agreement except with regard to the rate and level of remuneration.

A TYPICAL SITE AGREEMENT FOR OIL AND CHEMICAL PLANT CONSTRUCTORS' WORK

The typical site agreement which follows relates to a 40 hour week site with a basic rate plus a productivity allowance. It must be appreciated that the conditions shown will vary from area to area, these matters being taken into account and resolved during the process of negotiations with the union officials.

Monetary rates have only been included in the following text where such rates can be considered as having an application wider than one geographical location. In other cases the reader is referred to the italicized notes where general rate guidance is given.

SECTION 1: TITLE

Standard Site Agreement for the mechanical, electrical and instrumentation construction work to be undertaken by Contractors for (*name of Client*).

PROCEDURE

SECTION 2: PREAMBLE

2.1 This Agreement regulates the wages and defines the conditions of employment which will apply to all hourly paid chargehands, skilled craftsmen and labourers engaged on mechanical, electrical and instrumentation construction work and is made between the contractors listed below and other contractors and sub-contractors subsequently appointed by the clients or the main contractors (herein referred to as the Employers) and the Trade Unions signatory to this Agreement.

2.2 The acceptance of this entire Agreement is a condition of employment. No variations or amendment to these conditions will be sought or entertained by any of the parties during the operation of this Agreement, otherwise than in accordance with Section 5 of this Agreement.

2.3 Both the Employers and the Trade Unions accept that this Agreement is binding in honour upon them, but both parties expressly agree that it is not intended to constitute a legally enforceable agreement.

SECTION 3: SCOPE

3.1 This Agreement covers all aspects of mechanical, electrical and instrumentation construction work to be undertaken by Contractors for the British Petroleum Company and its associates and other Clients at the locations specified in Section 1.

3.2 The scope includes:

(*a*) Construction of new capital projects including modifications and extensions to existing plants but excludes maintenance and work placed by local factory management associated with existing plant.

(b) Testing and pre-commissioning work up to the point where the contractor has fulfilled his obligations for the construction of the plant.
(c) Temporary steel-framed and pre-cast concrete buildings.
(d) Tankage within the process area.
(e) Power station mechanical, electrical and instrumentation work.

3.3 The scope excludes:
(a) Building and civil engineering operations covered by the Working Rule Agreements of the National Joint Council for the Building Industry and the Civil Engineering Construction Conciliation Board.
(b) Catering staff of canteens.
(c) Off-site tank erection in the tank farm.
(d) Erection of temporary site buildings of non steel-framed construction.
(e) Insulation, painting and associated work.

SECTION 4: PRIOR AND COLLATERAL AGREEMENTS

This Agreement shall as of the date of commencement hereof cancel and supersede all prior agreements between the parties hereto, and shall constitute the entire agreement between the parties hereto, relating to men employed on the site and no other stipulation or understanding shall have effect unless it is agreed in writing and signed by all parties hereto subsequent to the date of this Agreement.

SECTION 5: DURATION

This Agreement will come into effect on 1 June 1974 and will remain in effect indefinitely subject to the following:

(a) Three months' notice to be given in writing by either party if termination or revision is required.

(b) The period of notice for termination or revision may be waived by mutual agreement.

SECTION 6: PRODUCTIVITY AND INTERCHANGEABILITY

6.1 It is recognized that incentive bonus schemes are precluded by the nature of the work. The wage rates and conditions of employment contained in this Agreement have been established by productivity bargaining between the Trade Unions and the Employers concerned and the allowance prescribed in Section 16 is conditional upon the observance of the following:

6.2 It is agreed that every effort will be made to achieve maximum productivity and, in order to achieve this, the following code of behaviour will be strictly observed: Every effort will be made to eliminate wasteful practices, restrictions or customs likely to impede maximum productivity.

6.3 It is further agreed that every craft union signatory to this Agreement will have equal right to submit and the Employers to accept for employment any member with the necessary skills for a particular job regardless of past practices or demarcation agreements.

6.4 In order to achieve this maximum productivity the most efficient form of working will be accepted by the Trade Unions at all times including, but not limited to, the following:

AGREEMENTS (O.C.P.C.A.)

6.4.1 All modern aids to production provided by Employers will be used by each trade.

6.4.2 Pumps, motors, compressors and other equipment, which require no special erecting or rigging, will be handled by the basic craft at ground level and up to approximately two feet above base elevation.

6.4.3 The lifting of switch gear equipment and transformers up to one ton in weight fitted with simple hooks or eye bolts will be handled by the basic craft, including simple slinging, provided it is a simple task requiring no specialized rigging work.

6.4.4 All craftsmen will use pull-lifts and the like for simple positioning and final lining-up of pipe-work and plant provided that it is a simple task requiring no specialized rigging work.

6.4.5 Each trade will be required to fabricate on site and install work associated with its own craft when instructed to do so by management, e.g. brackets, supports, cable trays and trunking.

6.4.6 Each trade may erect and dismantle a two stage scaffold of the Easy-Fix type. Three stage Easy-Fix scaffolding pre-erected by scaffolders may be placed and used by craftsmen to progress their own work. All tubular scaffolding of conventional type will be erected and dismantled by scaffolders.

6.4.7 Valve packing will be carried out by any available capable craftsmen.

6.4.8 Millwrights and pipefitters in composite gangs engaged on testing will be completely interchangeable on the first flange.

6.4.9 All craftsmen are to undertake own fixing to concrete and brickwork e.g. 'redheads' required by normal construction provided that they are suitably instructed in accordance with the appropriate regulations in the use of percussive tools of the Hilti and Spitmatic type.

6.4.10 Welders will grind and clean their own welds and assist in final alignment of butts with pipe-fitters. They will assist in the handling of welding cables.

6.4.11 Spot priming and painting of identification marks on materials are to be undertaken by any available employee.

6.4.12 Lamp replacement is to be undertaken by any available employee.

6.4.13 When insufficient electricians are available for pulling cables in trenches gangs may be augmented by labourers for this purpose.

6.5 In the event of a temporarily unbalanced labour force occurring as a result of absence caused by non-attendance, sickness, leave or shortage of labour, or any other abnormal job condition, the Trade Unions agree to interchangeability between crafts during the progress of the work in order to maintain production. Such interchangeability will also apply when inclement weather precludes outdoor working.

6.6 Any difficulties arising from the above will be dealt with through negotiating procedure.

SECTION 7: NEGOTIATING PROCEDURE

7.1 In the event of any dispute arising which cannot be quickly settled on site, there will be no stoppage of work until the full resources of negotiating procedure have been utilized. Questions arising on the site regarding the Agreement will be dealt with as follows:

Stage 1 The matter at issue will be taken up on the site between the shop steward and foreman and if there is failure to agree the matter will be referred to the employer's site industrial relations officer by either party.

AGREEMENTS (O.C.P.C.A.)

Stage 2 If agreement is not reached the matter will be referred to the trade union full time official by either party and a meeting held on site or at a mutually convenient place within seven days of the matter being reported.

Stage 3 Failing settlement at Stage 2 either party may refer the matter to a central conference to be held within seven days at a place suitable to all parties. The conference will be composed of representatives of the Employers and full time officials of the Trade Unions signatory to the Agreement.

Stage 4 In the event of the foregoing procedure being exhausted without settlement, the trade union or the employer will have the right to refer the matter at issue to a conference of representatives of the Employers and Trade Union national officials for settlement. This conference, if requested by either side is to take place within fourteen days of application.

7.2 Disputes at Stage 2 concerning a sub-contractor are to be dealt with in conjunction with the main contractor. The main contractor will conduct cases at Stages 3 and 4 on behalf of the sub-contractor.

7.3 Any claim referred to Stage 4 will be led at Stage 4 by the individual trade union sponsoring the claim if it so desires.

SECTION 8: PROCEDURE FOR ASSESSMENT OF ABNORMAL CONDITIONS AND THE AWARD OF PAYMENTS

8.1 The only condition payments prescribed by this Agreement are for Height, Confined Space and for Working Outside Defined Normal Construction Areas. However, it is recognized that from time to time conditions may exist on site which are abnormal to the generality of new construction work and cannot be removed. The following procedure is provided for the negotiation of payments for work in such abnormal conditions.

8.1.1 Claims for abnormal condition payments, submitted for negotiation at Stage 2, are to be reported without delay to the chairman of the site managers' committee, who, in collaboration with the local full-time trade union official(s) involved in the claim is to appoint an inspection panel comprising two independent site managers and two local full-time officials also, whenever possible, independent.

8.1.2 The panel is to carry out an inspection of the work area concerned for the purpose of determining the facts of the case. When the panel is able to agree unanimously that an abnormal condition payment is warranted, it may recommend that an immediate payment not exceeding 10p per half-shift or part thereof, for work in the relevant condition, should be awarded at Stage 2. Such payment is subject to ratification, revision or withdrawal at Stage 3.

8.1.3 In those cases when the panel is unable to reach agreement, the claim is to be referred to a Stage 3 meeting to be convened within five working days of the inspection, or as mutually convenient. The claim is then to be dealt with by negotiation on the basis of the facts as reported by the inspection panel.

WORKING CONDITIONS

SECTION 9: WORKING HOURS

9.1 The normal hours of work will be 40 hours worked on five days (Monday–Friday inclusive) comprising eight hours per day.

AGREEMENTS (O.C.P.C.A.)

9.2 It is agreed by both Employers and Trade Unions that these normal hours of work will be strictly observed, and no overtime wil be worked except in exceptional circumstances and without prejudice to the normal hours of work on the site.

9.3 Employees will be required to attend their work regularly during normal working hours as stated above. Unauthorized lateness or absenteeism will render employees liable to discharge after prior warning. Warnings will be given in writing with copies to the employee's shop steward and full-time trade union officials.

SECTION 10: OVERTIME

10.1 In exceptional circumstances when overtime is required to be worked to meet the contingencies of construction and commissioning, the number of men required will be decided by management. Transport will be provided by management if public transport is not readily available. Men working overtime will be advised of transport arrangements prior to commencing overtime work.

10.2 A full day will be worked before overtime is reckoned with the following exceptions:

(*a*) Time lost through sickness certified to the satisfaction of the employer.

(*b*) Laying off on account of working throughout the previous night.

(*c*) Absence with leave or enforced idleness.

10.3 Overtime rates will be calculated on the basic hourly rate prescribed in Section 15.

10.4 For the first five days of the week overtime worked before or after normal working hours up to midnight will be paid at the rate of time-and-a-half and if continued after midnight, at the rate of double-time until 8.0 a.m. or normal day-work starting time.

10.5 If having completed the normal working hours of any day and having left the site an employee is requested to resume work, he will be paid from the time of recommencing work as if he had worked continuously. Whilst the interval between ceasing and recommencing work will not be paid for, the period of overtime concerned will provide for a guaranteed payment of not less than three hours at premium time.

10.6 An employee sent home between midnight and 2 a.m. will be paid double-time from midnight until the time he is sent home and will be paid at the basic hourly rate from the time he is sent home until starting time on the same morning. An employee sent home after 2 a.m. will be paid double-time from midnight until the time he is sent home and will be paid time-and-a-half from the time he is sent home until 8 a.m. or the normal day-work starting time.

10.7 An employee required to work through the recognized midday meal break will be paid at the rate of time-and-a-half and an alternative meal interval (for which no payment will be made) granted.

10.8 Time worked on Saturday will be paid at the rate of time-and-a-half until 4 p.m.

10.9 All hours worked between 4 p.m. on Saturday and starting time on Monday will be paid at the rate of double-time.

10.10 In the interest of safety and the maintenance of production an employee who, after completing his normal working hours for the day, continues to work beyond 2 a.m. will not be permitted or required to continue working beyond normal starting time of the succeeding day shift. An employee precluded from continuing to work as stated above will be paid for the normal hours at his appropriate basic hourly rate for the day on which he was precluded from working.

AGREEMENTS (O.C.P.C.A.)

SECTION 11: NIGHTWORK

11.1 Nightwork will be calculated on the basic hourly rate prescribed in Section 15.

11.2 Nightwork conditions will apply when men other than daywork men work throughout the night for not less than three consecutive nights. The nightwork week will consist of 40 hours which may comprise either 4 nights of 10 hours or 5 nights of 8 hours, whichever arrangement is agreed for the site as a whole. (Where four nights of ten hours are operated no premium time is payable for any of the ten hours.)

11.3 Hours worked on nightwork will be paid at the rate of time-and-one-third.

11.4 Hours worked before or after the full nightwork has been worked will be paid at time-and-two-thirds of the basic hourly rate. Employees on nightwork will receive double-time for all hours worked between 4 p.m. on Saturday and normal starting time on Monday.

11.5 Daywork men who worked during the day beyond the midday meal break and are required to go on nightwork the same night will be paid overtime rates for the night's work in accordance with Section 10.

11.6 Nightwork men required to change over to daywork will be warned at least 24 hours in advance of the necessity for such a change of shift.

SECTION 12: SHIFT-WORKING

12.1 **Preamble**
Shift-working may be introduced temporarily to cover pre-commissioning and commissioning and when it is essential for technical reasons subject to the following conditions:

(a) The period of shift-working is not less than one week.

(b) The introduction of shift-working does not constitute a condition of employment to the men involved, except where men are recruited specifically for shift-working.

(c) The men required for shift-working have been given a minimum of 24 hours notice.

(d) Existing employees have first been offered the shift-working opportunities prior to further recruitment to fill remaining vacancies.

(e) There has been prior consultation at Stage 3 between the parties to this Agreement.

12.2 **Three shift working** (5 days per week – 3 gangs).

12.2.1 Three shifts, each of eight hours, will be worked to cover 24 hours per day, five days per week, from Monday to Friday. Shift hours will be as follows:

Morning	6.0 a.m. to 2.0 p.m.
Afternoon	2.0 p.m. to 10.0 p.m.
Night	10.0 p.m. to 6.0 a.m.

The work force which worked Morning Shift in Week 1 will change to Afternoons in Week 2, and to Nights in Week 3, and so on.

12.2.2 Three shift working will be paid at the following rate calculated on the basic hourly rate prescribed in Section 15 as follows:
6 a.m. Monday to 6 a.m. Saturday – Time & $\frac{1}{3}$ + P. & I.A. + I.W.W.A.

12.2.3 No overtime will be worked except in exceptional circumstances. In such circumstances, overtime hours worked by three shift men will be paid at the basic rate plus one-half, except between 2.0 p.m. on Saturday and 6.0 a.m. on Monday, when they will be paid for at double-time.

12.2.4 Each 8 hour shift includes a 30 minute paid meal-break.

12.3 **Rotating shifts** (7 days per week – 4 gangs).

AGREEMENTS (O.C.P.C.A.)

12.3.1 Shift hours will be as follows:

Morning 6.0 a.m. to 2 p.m.
Afternoon 2.0 p.m. to 10.0 p.m.
Evening 10.0 p.m. to 6.0 a.m.

Four gangs will be required to cover the 168 hour week, giving an average of 42 hours per week per gang. In order that no man will be required to work more than an average of 40 normal hours per week he will take one morning shift off work in each 4 week period. The actual rotation of the shifts will be by mutual arrangement between employer and employees such that a full cycle is completed in each four week period.

12.3.2 Rotating shift work will be paid at the following rates calculated on the basic hourly rate prescribed in Section 15 as follows:

6 a.m. Monday to 6 a.m. Saturday – Time & $\frac{1}{3}$ + P. & I.A. + I.W.W.A. 6 a.m. Saturday to 2 p.m. Saturday – Time & $\frac{1}{2}$ + P. & I.A. + I.W.W.A. 2 p.m. Saturday to 6 a.m. Monday – Double Time + P. & I.A. + I.W.W.A.

12.3.3 No overtime will be worked except in exceptional circumstances. In such circumstances overtime worked by rotating shift men will be paid for at the basic rate plus one-half, except between 2.0 p.m. on Saturday and 6.0 a.m. on Monday, when they will be paid for at double-time.

12.3.4 Each 8 hour shift includes a 30 minute paid meal break.

SECTION 13: GUARANTEED MINIMUM EARNINGS

13.1 Employees who have been continuously employed on site by the employer for not less than four weeks will be guaranteed employment for five days in each normal pay week. In the event of work not being available for the whole or part of the five days, employees covered by the guarantee will be assured earnings at the basic hourly rate prescribed in Section 15 for 40 hours.

13.2 This guarantee is subject to the following conditions:

(*a*) That the employees are capable of, available for and willing to perform satisfactorily, during the period of guarantee, the work associated with their usual occupation, or suitable alternative work where their usual work is not available.

(*b*) In the case of a recognised holiday the period of guarantee will be reduced proportionately.

(*c*) In the event of dislocation of production as a result of industrial dispute on site, the operation of the guarantee will be automatically suspended.

(*d*) In computing the assured earnings referred to above, premium time on overtime worked will be ignored.

SECTION 14: PAYMENT FOR TIME LOST OWING TO INCLEMENT WEATHER

14.1 **Preamble**

14.1.1 The intention of this Section is that, subject always to safety considerations, employees will be expected to work in the weather conditions normally encountered on outside construction sites in the United Kingdom. The employers will provide protective clothing and suitable alternative work such as is available if it is unsafe to carry out normal duties.

14.1.2 Employees will co-operate by working under temporary and permanent covers by undertaking suitable alternative work in accordance with Clause 6.2 of this Agreement.

14.2 Inclement weather working allowance (IWWA)

14.2.1 An allowance will be paid for all clocked hours subject to compliance with the terms of this section as follows:
(a) 32½p per hour for hourly paid chargehands and skilled craftsmen.
(b) 24½p per hour for labourers.
The allowance will not be enhanced for overtime.

14.2.2 The allowance will not be paid for the following:

(a) Travelling time.
(b) Guaranteed week payments.
(c) Holiday credits and statutory holidays.

14.3 Unless otherwise instructed by the employer, an employee will present himself for work on the site and will there remain available for work throughout the normal working hours.

14.3.1 Employees are not expected to continue working in the open in sustained severe, inclement weather, but the employer will decide when weather conditions justify the cessation or resumption of work and if some or all men will work at a particular time.

14.3.2 The following are tasks which will be carried out in all weather conditions:
Off-loading materials.
Supply of materials and services to men working under cover.
Proceeding to alternative work including under cover work.
Working under permanent or temporary cover.
Placing temporary covers over work so that it may be continued when directed.
Dealing with emergencies and in the interests of safety.

14.3.3 *Temporary shelters at the place of work*
Any place provided for carrying out work or taking shelter during periods of intermittent inclement weather such as temporary fabrication shops, elephant shelters and other temporary covered areas on the job site may be used as temporary shelter as directed by the employer but such places will not be provided with heating or seating.

Mess huts, canteens, changing rooms and other forms of permanent site establishment are not to be considered as places of temporary shelter.

14.3.4 Upon the onset of inclement weather an employee may take cover at or adjacent to his place of work in temporary shelter as defined in 14.3.3 above. From the time of taking cover for a period of up to 30 minutes there shall be no loss of earnings provided that the employee holds himself available to resume work immediately upon instruction.

14.3.5 When the period for the use of temporary shelters has elapsed, employees will be returned to cabins, unless the inclement weather has sufficiently improved to allow normal working to be resumed.

14.3.6 *Alternative work*
When because of inclement weather no work is available temporarily for an employee in his own occupation, he must accept any suitable alternative work.
Employees for whom work cannot be found will not interfere with those employees who are at work.

14.3.7 *Payment for time spent in the cabins*
Employees who have been returned to the cabins upon the expiry of 30 minutes in

AGREEMENTS (O.C.P.C.A.)

temporary shelters will be paid at the basic hourly rate prescribed in Section 15, and enhanced in the case of nightwork by the appropriate differential prescribed for normal working hours plus the inclement weather working allowance in 14.1 above provided that they remain available for work throughout the prescribed working hours.

14.3.8 If an employee, during the normal working hours of any day (or night), fails to keep himself available he will not be entitled to payment in accordance with 14.3.7 above for any portion of such day (or night) except such hours as he has actually held himself available for work.

WAGES, PRODUCTIVITY ALLOWANCE PROFICIENCY AND RESPONSIBILITY PAYMENTS

SECTION 15: BASIC HOURLY RATE

15.1 (a) Hourly paid chargehands   235p per hour
(b) Skilled craftsmen   220p per hour
(c) Labourers   171p per hour

15.2 It is agreed that the above basic hourly rates include all condition payments, including previous *ad hoc* arrangements of any kind, except for fixed condition payments as provided for in Sections 29, 30 and 31 of this Agreement.

15.3 *Overtime, Nightwork and Shiftworking premia*
Premia for overtime (Section 10), nightwork (Section 11) and shiftworking (Section 12) are calculated on the above basic hourly rates except in the case of proficiency payments (Section 17) which are added to the basic hourly rate for the purpose of the calculation.

15.4 Under this agreement the above rates are effective from 1 January 1980. The basic hourly rates may then be reviewed on 1 January 1981 and thereafter at agreed intervals.

SECTION 16: PRODUCTIVITY AND INTERCHANGEABILITY ALLOWANCE (P. & I.A.)

16.1 This allowance will be payable in addition to the hourly rates in 15.1 above, subject to the implementation of and full compliance with Section 6 of this Agreement.

16.2 The allowance will be at the rate of:

(a) 65p per hour for hourly paid chargehands and skilled craftsmen.

(b) 48.75p per hour for labourers.

16.3 The allowance will be paid for all hours of productive work on the site, and will not be enhanced for overtime. The allowance will alternatively cease to be paid if Section 6 is not complied with.

16.4 The allowance will not be paid for:

(a) Time lost due to inclement weather.

(b) Travelling time.

(c) Guaranteed week payments.

(d) Holiday credits and statutory holidays.

SECTION 17: PROFICIENCY PAYMENTS

17.1 Proficiency payments will be made as follows:

17.1.1 To craftsmen employed as welders and qualified by test in electric arc or gas

AGREEMENTS (O.C.P.C.A.)

welding to API/ASME code or Lloyds Class 1 standard on all metals and alloys, regardless of the type of welding, electrode or filler rod used . . . 5p per hour.

17.1.2 To labourers employed as drivers in such cases where the employer requires that the driver should hold a current driving licence for the class of vehicle involved . . . 2½p per hour.

17.2 The above payments will be enhanced for overtime.

SECTION 18: DUTIES OF WORKING CHARGEHANDS

Hourly paid skilled chargehands are employed to supervise production but occasions will arise when they may have to use the tools of their trade to assist the employees under their supervision. In such cases no objections will be raised by the remainder of the workforce to such chargehands using the tools of their trade in the performance of their duties.

SECTION 19: PAY WEEK

19.1 *Payment*
Wages will be distributed on the Thursday following the end of pay week, in the employer's time.

19.2 *Pay queries*
Written pay queries are to be handed into the Site Office on arrival at work in the morning and will be answered in writing or verbally by lunch time on the following day.

19.3 *Advances of Pay*
Applications for advances of pay (subs), will be dealt with only during the midday meal break.

HOLIDAYS, LODGING AND RADIUS ALLOWANCES

SECTION 20: ANNUAL HOLIDAYS

20.1 It is the intention that every workman will have four weeks and two days holiday. Three weeks of these holidays may be arranged in accordance with the needs of the job by mutual agreement between the employer and employee, between 1 May and 31 October. A fourth week of holiday will be taken during the Christmas/New Year period.

The balance of two days is to be taken on dates of managements' choice after consultation with trade union representatives. The two days shall not be taken as single days but will be attached to other holiday periods whether statutory or annual holidays.

20.2 There will be credited to every worker in respect of such holidays for each full week's work performed a sum calculated so as to provide holiday pay for each full week of holiday entitlement of 40 hours at the basic hourly rate plus one third. The sum credited will be:

	Chargehands	Craftsmen	Labourers
Hourly Rate	28·9636p	27·1148p	21·0756p
Weekly Rate	£11·58	£10·85	£8·43

20.3 Employees working 40 hours, normal and/or overtime, in a seven-day week (Monday to Sunday) are entitled to full holiday credit for that week. When less then 40 hours (Monday to Sunday) are worked, the credit is reduced proportionately.

20.4 A man absent on authorized leave or through sickness or accident certified to the satisfaction of the employer will be credited with holiday credits for normal working hours so lost up to a maximum of twelve weeks in any one year of his employment.

AGREEMENTS (O.C.P.C.A.)

20.5 Accrued holiday credits will only be paid when the holiday is taken or the employment is terminated.

SECTION 21: STATUTORY HOLIDAYS

21.1 There will be eight statutory holidays on which the site will be closed. The dates of such holidays will be posted at the site at the commencement of the project.

21.2 Payment to men on holiday on each of these eight days will be at the appropriate basic hourly rate prescribed in Section 15 plus one-third, for the eight hours constituting the normal working day.

21.3 Employees are required to work the full normal hours of the working day immediately prior to a statutory holiday and to return to work by the first available means of transport on the working day following. The disciplinary procedure for bad timekeeping (See Clause 28.1) will apply to those men who fail to do so. See also Clause 23.8 with regard to the entitlement to radius allowance and Clause 24.4 with regard to the entitlement to lodging allowance. See also sub-Clause 25.1.3 with regard to the payment of travelling time to men returning from periodic leave.

21.4 A worker who is redundant within 14 days of the occurrence of a Statutory Holiday or a week-end to which a Statutory Holiday is attached will be paid for that holiday in accordance with Clause 21.2.

SECTION 22: PAYMENT FOR WORK DONE ON STATUTORY HOLIDAYS

22.1 Payment will be made at double-time of the basic hourly rate for all hours worked on the eight days of the statutory holidays.

22.2 In addition a day off in lieu will be taken at a convenient date later, for which payment will be made at the basic hourly rate for hours constituting a normal working day.

SECTION 23: RADIUS ALLOWANCE

23.1 Radius allowance will be paid to any man who travels daily between his home and a site over 1½ miles away. This allowance is compensation for travelling time and fares and will, therefore, only be paid in respect of days worked.

23.2 The scales of radius allowance payable are as follows with effect from 14 May 1979.

Distance in miles one way	Daily Travel Allowance Scale 1	Scale 2
0 to 1½	Nil	Nil
1½ to 5	£1·25	£0·84
5 to 8	£1·73	£1·15
8 to 11	£2·30	£1·53
11 to 14	£2·87	£1·91
14 to 17	£3·55	£2·37
17 to 20	£4·14	£2·75
20 to 25	£4·65	£3·10
25 to 30	£5·05	£3·35
30 to 35	£5·39	£3·60
Over 35	£5·72	£3·83

Scale 2 allowances are payable in respect of distances for which suitable transport is provided.

AGREEMENTS (O.C.P.C.A.)

23.3 All distances in each scale will be measured on a straight line basis, but exceptional cases (e.g. circuitous routes due to natural barriers) where this is considered to be inappropriate may be raised with the employer in accordance with the agreed procedure.

23.4 Any man who, by arrangement with the employer, uses both provided and other transport and travels a total distance of over one and a half miles each way daily will be entitled to an extra payment of 10p per day in addition to any allowance(s) payable under Scale 1 or 2, or both, provided the total allowance does not exceed £5.72 per day (plus the extra payment of 10p per day).

23.5 Any man receiving a total radius allowance, however made up, equal to lodging allowance may, by arrangement with the employer move into lodgings and be paid lodging allowance in accordance with Section 24 of this Agreement instead of radius allowance.

23.6 Radius allowance is not payable to any man who is in receipt of lodging allowance, but special consideration will be given where lodgings are not available close to the site.

23.7 Where a man opts to travel by his own means he will receive radius allowance under Scale 1 and will not be eligible to use provided transport. If he decides to use provided transport he will receive payment in accordance with Scale 2. If he subsequently decides to travel by his own means he will continue to receive radius payment under Scale 2 until he is able to nominate another employee to fill his place in the provided transport.

23.8 Any man who is absent during working hours or leaves the site early or returns late at holiday times without good cause or without permission is liable to forfeit the radius allowance to which he would otherwise be entitled for the day or days concerned.

SECTION 24: LODGING ALLOWANCE

24.1 Lodging allowance can be paid only on the completion of an 'Application for Lodging Allowance' form (see Appendix 'B'). Men necessarily living away from home in lodgings as agreed by the employer will receive £5·80 per day or £40·60 per week lodging allowance with effect from 14 May 1979.

24.2 Lodging allowance is not paid during absence but sympathetic consideration is given to employees who are sick in lodgings and submit the necessary medical certificates. This consideration will not extend beyond 14 days' absence for any one period.

24.3 Men normally in receipt of lodging allowance who are sick at home may, on production of satisfactory medical evidence and relevant receipts, claim the lodging retainer for periods up to 14 days.

24.4 Lodging allowance will continue during authorized absence on week-end leave, and statutory holidays detailed in Section 21 provided that the employee returns to work by the first available means of transport on Monday or the day following statutory holidays. A man absent without permission on either the working day preceding or the working day following periodic week-end leave will forfeit lodging allowance for the working day of absence only. A man absent without permission both the preceding and the following working days will forfeit lodging allowance for the whole of the period.

24.5 Any man who is absent during working hours on a normal working day (Monday to Friday) without good cause, or without permission, will forfeit the lodging allowance payable for that day.

24.6 No employees in receipt of lodging allowance will be entitled to payment of radius allowance but special consideration will be given in the case of men unable to obtain

AGREEMENTS (O.C.P.C.A.)

lodgings near the site. Such consideration, however, will not extend beyond the payment provided in the appropriate scale in Section 23 for distances not exceeding 8 miles.

24.7 Employees absent from their lodgings on annual holiday will be paid a lodging retainer of £1·80 per day, £12·60 per week with effect from 14 May 1979.

SECTION 25: TRAVELLING ON COMMENCEMENT AND TERMINATION OF EMPLOYMENT AND PERIODIC LEAVE APPLICABLE TO MEN IN LODGINGS

25.1 The following provisions will apply to men brought into the district at the wish of the employer and necessarily living away from the place in which they normally reside.

25.1.1 *Travel payment*
Fares and travelling time will be payable between the site and the man's home or port of disembarkation in the United Kingdom.

25.1.2 *Commencement of employment applicable to men in lodgings*
Men will receive a free travel warrant or fares for a second-class railway journey and be paid at the appropriate basic hourly rate prescribed in Section 15 for the time actually spent in travelling to the site on commencement of employment, except where the distance exceeds 100 miles, when the allowance will be one hour paid at the aforesaid rate for each 25 miles of the journey to the site.

25.1.3 *Periodic leave for men in lodgings.*
Men will be granted periodic leave at the rate of one every calendar month and will receive a free travel warrant for a return second-class journey up to a maximum of twelve in any one year of employment. Variations in the calendar month period may be made to meet exigencies of the particular job or to coincide with statutory or annual holidays.

Provided that a man returns to work by the first available means of transport on Monday or the day following a statutory holiday he will be paid at the appropriate basic hourly rate prescribed in Section 15 for the time actually spent on the journey back to the site except where the distance exceeds 100 miles, when the allowance will be one hour paid at the aforesaid rate for each 25 miles of the journey to the site.

25.1.4 *Measurement of distance*
Distance will be measured in a straight line for the purpose of sub-clauses 25.1.2 and 25.1.3 hereof.

25.2 This section will not apply to a man who discharges himself voluntarily (other than for compassionate reasons accepted by the employer) or who is discharged for misconduct.

TERMINATION OF EMPLOYMENT

SECTION 26: PERIODS OF NOTICE

26.1 During the first four weeks of employment either the man or the employer may terminate the employment by tendering two hours' notice to expire at the end of the normal working hours of any day.

26.2 After completion of four weeks employment and less than 104 weeks, a man will be paid 40 hours' pay at the appropriate total hourly rate, in lieu of 40 normal hours notice of termination of employment.

Notification to the man of termination under this clause will be effective from the Friday of the current week.

A man wishing to terminate his employment will be required to give 40 normal working hours' notice expiring on a Friday but consideration will be given to men having urgent grounds for leaving.

26.3 After 104 weeks' employment, notice will be given in accordance with the appropriate statutory requirements, and is to expire on a Friday.

SECTION 27: REDUNDANCY AND TRANSFER

27.1 On termination of employment, because of redundancy, as defined in Section 81 of the Employment Protection (Consolidation) Act 1978, a job severance payment will be made for each full week of continuous employment up to a maximum of 103 weeks at the following weekly rates:

(a) Skilled craftsmen and skilled chargehands £2·30
(b) Labourers £1·725

27.2 Payments on termination of employment because of redundancy after exactly 104 weeks' continuous employment will be made in accordance with the Employment Protection (Consolidation) Act.

27.3 Employees who leave voluntarily or who are dismissed for reasons other than redundancy are not entitled to payments. Time in employment includes notice periods under the Employment Protection (Consolidation) Act or periods for which payments in lieu of notice are made.

27.4 An employee who accepts transfer with the same employer to another site is not entitled to a job severance payment, as continuity of employment is maintained for the purpose of the Employment Protection (Consolidation) Act. If subsequently such an employee is made redundant before the 104 weeks of employment are completed, there will be an entitlement to accrued job severance pay for all sites to which this standard scheme has been applied.

27.5 An employee cannot be entitled to both a job severance payment and payment under the Employment Protection (Consolidation) Act. It follows that if a job severance payment is made under the above provisions and the employee subsequently becomes entitled to the statutory payment under the Employment Protection (Consolidation) Act, then the redundancy payment should be reduced by the amount of the job severance payment previously made in respect of the same employment.

27.6 All payments made on redundancy, other than under the Employment Protection (Consolidation) Act, are normally taxable, but by arrangement with the Inland Revenue dated 1 November 1978, payments made strictly in accordance with this standard scheme will not be subject to tax.

27.7 A worker who is redundant within fourteen days prior to a statutory holiday will receive payment for that holiday in accordance with Section 21 of this Agreement.

SECTION 28: UNSATISFACTORY WORK AND MISCONDUCT

28.1 **Dismissal after prior warning**

28.1.1 An employee may be dismissed for unsatisfactory work or bad timekeeping after prior warning and the terms of Section 26 will apply.

28.1.2 Warning for poor performance under this clause will be given orally in the presence of the shop steward. A written copy of second warnings will be given to the employee concerned, the shop steward in attendance and the full-time trade union offi-

AGREEMENTS (O.C.P.C.A.)

cial. If further poor performance occurs a third and final warning will be issued as above. If further poor performance occurs after a third warning the man will be dismissed.

28.1.3 Offences for which termination under this clause is appropriate include, but are not limited to:

bad timekeeping,
lack of capability,
bad workmanship or performance,
failure to comply with rules and regulations,
entering the canteen at other than published meal breaks,
disorderly conduct, such as verbal abuse to the canteen staff.

Note
Bad timekeeping includes:
absence without permission,
clocking in late,
clocking out early,
clocking in and out again during normal working hours without permission,
poor attendance.

28.2 Dismissal for gross misconduct

28.2.1 In the case of gross misconduct an employee may be summarily dismissed at any time.

The kind of acts (or omissions) which constitute gross misconduct include, but are not limited to:

fighting,
threatening behaviour,
physical violence,
stealing,
wilful damage or destruction of property,
possession of or under the influence of alcohol or drugs during working hours,
fake booking of work,
clocking offences,
clocking-in or out the time-card of another employee,
leaving the work area for any reason without clocking out,
smoking in certain specified no-smoking areas,
serious breaches of safety regulations or company rules,
gross negligence or recklessness.

28.3 The level of management responsible for discipline will be the employer's agent or the agent's deputy.

28.4 Where applicable, these disciplinary procedures also cover offences committed anywhere on the client's property and on transport and facilities provided by the employer.

28·5 Arbitration Panels

28.5.1 If an employee wishes to challenge the reasons for his summary dismissal he may, within a time limit of 5 days, through his shop steward and full-time official, request suspension without pay pending the decision of an Arbitration Panel. Claims for suspension made initially by shop stewards are subject to confirmation by the full-time official. It is the responsibility of the shop steward to report at the earliest op-

Large Industrial Projects – Labour

AGREEMENTS (O.C.P.C.A.)

portunity claims for suspension to the full-time official, who will confirm or withdraw the claim as he deems most appropriate.

28.5.2 Claims for re-instatement or re-employment will be referred to the next Arbitration Panel which will sit periodically or as required. The Panel will comprise two local trade union officials, and two employers' representatives, none of whom will be parties concerned in the case.

28.5.3 The Panel will elect its own chairman and will be conducted informally as follows:

(*a*) Considers written evidence.
(*b*) Site manager or his representative presents the case for the dismissal, calling witnesses as necessary.
(*c*) The dismissed man's representative presents the case for reinstatement or re-employment as appropriate and calls witnesses as necessary.
(*d*) Panel calls for further evidence and re-examines witnesses as necessary.
(*e*) Panel considers its decision in private.
(*f*) Panel gives its formal decision to both parties and issues written communique.

28.5.4 The decision of the Panel will be within the following categories:

(*a*) Confirmation of dismissal.
(*b*) Re-engagement on a specified date with or without continuity of employment.
(*c*) Full re-instatement.

28.5.5 The decision of the Panel is binding on both parties.

FIXED CONDITION PAYMENTS

SECTION 29: HEIGHT MONEY

29.1 Payment will be made for working at heights as follows:

Men working at	Payment per hour
30 ft. and over but less than 50 ft.	1p
50 ft. and over but less than 75 ft.	1·5p
75 ft. and over but less than 100 ft.	3p
100 ft. and over but less than 150 ft.	4·5p
150 ft. and over but less than 200 ft.	9p
200 ft. and over but less than 250 ft.	13·5p
250 ft. and over but less than 300 ft.	18p

Thereafter 4·5p for each additional 50 ft.

29.2 These scales do not apply when work is being carried out on a completed covered floor (part of a permanent structure) with protective rail or walls, unless such work is over the side of the floor or over an opening in the floor.

SECTION 30: CONFINED SPACE MONEY

30.1 Men working in wholly enclosed process columns or vessels in which the working space is restricted by internal fittings, will be paid an allowance in accordance with the following table:

AGREEMENTS (O.C.P.C.A.)

Internal diameter of shell	Payment per hour
8 ft. and over	2·5p
5 ft. to 8 ft.	3p
3 ft. to 5 ft.	4p
Under 3 ft.	7·5p

The allowance to be paid will be fixed according to the internal diameter of the shell with due consideration to the arrangement of the internal fittings.

30.2 Men working in specific sections of water tube boilers where access to areas in and around the boiler is severely restricted will receive payments in accordance with Clause 30.1.

30.3 Men working INSIDE PIPES will be paid the following confined space allowance:

Internal diameter of pipe	per hour
Under 3 ft.	7·5p
3 ft. and over but under 5 ft.	4p
5 ft. and over but under 8 ft.	3p

SECTION 31: SUPPLEMENTARY PAYMENT FOR WORKING OUTSIDE DEFINED NORMAL CONSTRUCTION AREAS

31.1 This Agreement principally covers the construction of new capital projects in a 'green field' environment. However, during the course of a project it may be necessary for new construction labour to work outside defined normal construction areas.

31.2 It is therefore an additional requirement of this Agreement that new construction workers should be prepared to work outside defined normal construction areas.

31.3 In consideration of this requirement and of all conditions directly associated with such areas, wherever they may arise, a supplementary allowance of 10p per hour will be paid.

31.4 The allowance will be paid for all hours of productive work on the site and will not be enhanced for overtime.

31.5 The allowance will not be paid for:

(*a*) Time lost due to inclement weather.
(*b*) Travelling time.
(*c*) Guaranteed week payments.
(*d*) Holiday credits and statutory holidays.

MEAL AND TEA BREAKS

SECTION 32: MEAL AND TEA BREAKS

32.1 The midday meal break, for which no payment will be made, will be of 45 minutes' duration. This period may be reduced to 30 minutes when site conditions allow.

32.2 Morning tea will be available and taken under cover in the work area of the individual work crews during periods (not exceeding ten minutes) suitable to minimum disruption of work.

32.3 Where overtime of two hours or more in any one normal working day (Monday to

AGREEMENTS (O.C.P.C.A.)

Friday) is worked a ten-minute paid tea break will be granted at normal finishing time, provided it is taken in the work area of the individual work crews.
32.4 No other tea breaks will be permitted.

REGULATIONS

SECTION 33: REGULATIONS

All employees must comply with the Statutory Regulations and the Employers' and Clients' regulations for safety and good order (non-smoking etc.) made either for the construction site or area and notified on the site notice boards.

SECTION 34: SAFETY HELMETS

The wearing of safety helmets is a condition of employment for employees engaged on the site. These will be supplied by the employer. Should the attitude of any employee be contrary to the spirit and intention of this Section, which is for the purpose of safe working, then the Trade Union will fully support the employer in his non-employment or dismissal of such an offender.

SECTION 35: CONTRACTORS' BADGES

All employees will be issued with contractors' badges or identity cards which shall remain the property of the employer and be returned to the time-keeper's office on the termination of employment. The badge will be worn in a position such that the employee can be easily identified at all times as a contractor's employee.

SECTION 36: PROTECTIVE CLOTHING

36.1 *Free Issue*

36.1.1 There will be an entitlement on first recruitment to a free issue of one pair safety boots, one donkey jacket and two sets of overalls. Thereafter replacement will be made at cost price as required.
36.1.2 A new employee, joining site for the first time and whose employment is expected to last for a period in excess of four weeks is to receive a free issue of protective clothing on commencement.
36.1.3 A new employee who receives an issue of protective clothing as in 36.1.2 above, but leaves his employment for any reason before the four week period has elapsed is not to receive a further free issue from any employer on site with whom he may subsequently take up employment.
36.1.4 Short term employees, i.e., for periods of less than four weeks, will not be entitled to free issue of protective clothing.
36.1.5 Any employee of one contractor who following termination immediately takes up employment with another contractor on site, having already received a free issue of protective clothing, will not receive another issue upon re-employment.
36.1.6 A new employee on the site, having had previous employment on site with either the same or another contractor, in the course of which employment he received a free issue of protective clothing, will not be entitled to a second issue unless a period of at least one year has elapsed since the previous issue.

AGREEMENTS (O.C.P.C.A.)

36.1.7 Employees are expected to use and to take reasonable care of protective clothing issued.

36.1.8 Overalls, donkey jackets and protective clothing of the same standard are to be issued to all employees.

36.2 Additional Clothing

Other protective clothing required e.g. oilskins, leggings, rubber boots and gloves, will be provided by the employer and will be worn when required.

Any clothing so issued to employees at any time will remain the property of the employer. It is the employee's responsibility to take reasonable care of the clothing.

SECTION 37: TRADE UNION FACILITIES

37.1 General

37.1.1 The Trade Unions signatory to this Agreement are recognized as the negotiating Trade Unions for the categories of employees covered by this Agreement.

37.1.2 Employers will encourage all their employees covered by this Agreement to be paid up members of the signatory Trade Unions.

37.2 Shop Stewards

37.2.1 The Employers recognize the right of employees to elect a shop steward in accordance with the rules of their trade unions and to act on their behalf in accordance with the terms of the negotiating procedure for this Agreement.

37.2.2 A shop steward will not represent employees of any employer other than his own.

37.2.3 The name of the shop steward, and the categories of employees he represents will be notified to the employer by the trade union concerned at the time of his election.

37.2.4 Shop stewards will be subject to the control of their trade union and will act in support of this Agreement.

37.2.5 A shop steward will be afforded reasonable agreed facilities for dealing with questions arising between the employer and employees he represents. He will obtain permission from his foreman or supervisor before leaving his place of work and such permission will not be unreasonably withheld. Shop stewards' hourly earnings will be maintained whilst carrying out their duties with the above permission.

37.3 Agreed Facilities

37.3.1 In order that shop stewards can carry out their job efficiently in the interests of their members, the Trade Unions and the Employers, it is agreed that they should be afforded access to telephone facilities and notice board.

37.3.2 Off-site accommodation for meetings of shop stewards and their meetings with local officials, should also be provided allowing for the necessary privacy appropriate to such occasions.

37.3.3 Shop stewards will be granted time off with pay for attending appropriate training courses, by prior arrangement with the employer.

37.3.4 On each contract there may be a senior shop steward for the purpose of dealing with general matters affecting all trade unions on the particular contract.

37.4 Shop Stewards Committee

37.4.1 It is recognized as desirable that a Shop Stewards Committee comprising properly accredited shop stewards should be formed for the purpose of co-ordination.

37.4.2 Facilities will be granted for weekly meetings of the Shop Stewards' Committee,

AGREEMENTS (O.C.P.C.A.)

for which payment will be made, during the last hour of a working day to be decided in consultation with the employer. The number of shop stewards and the representation will be determined by agreement between the Employers and the local officials of the Trade Unions.

37.5 Time off for Trade Union Duties and Activities – Code of Practice

37.5.1 **Preamble**
It is agreed between the parties that the spirit and intent of this Code of Practice in keeping with statutory provisions, is to aid and improve the conduct of Industrial Relations.

37.5.2 *Introduction*
Sections 27 and 28 of the Employment Protection (Consolidation) Act provide for time off to be permitted by an employer for those employees who are officials or members of recognized trade unions to enable them to engage in trade union duties and activities.

37.5.3 *Official of an Independent Trade Union*
An official of an independent trade union recognized by the employer will be permitted reasonable paid time off during working hours for the purpose of:

(*a*) Carrying out those duties which are concerned with Industrial Relations between the employer and employees, and which do not conflict with the terms of this Agreement.

(*b*) Undergoing training relevant to their Industrial Relations duties, which is approved by the T.U.C. or their own trade union.

Note
Under this Section, the employee will be paid as if he had worked during the period when the time off was granted.

37.5.4 *A member of an Independent Trade Union*
A member of an independent trade union, recognized by the employer will be permitted reasonable time off during his working hours for trade union activities such as:
(*a*) Taking part in trade union elections.
(*b*) Attendance at meetings called to discuss subjects which have been agreed by the employer and trade union officials as being of particular urgency.

Note
Under this Section, there is no obligation for the employer to pay the employee for time off for union activities.

37.5.5 *General Matters*
(*a*) It is agreed between the parties, that the official who seeks time off will ensure that site management is informed as far as possible in advance. At the same time, the official should indicate the nature of the business for which time off is required, the intended location and expected period of absence.

(*b*) The trade unions shall inform site management in writing of the appointment or resignation of officials.

(*c*) Where it is necessary to hold meetings of members during working hours, the union or its official should seek to agree the arrangement with site management as far in advance as possible, and any such arrangement will have full regard for the effect upon productivity.

(*d*) When meetings of members take place, it is agreed to leave at work such members

AGREEMENTS (E.E.F.)

as are essential for safety reasons. The parties also agree that such meetings may be deferred on those occasions when site management is able to provide evidence of the serious effects which would arise from the meeting being held at the proposed time.

(e) Union officials and union members should not unduly or unnecessarily prolong the time they are absent from work on union duties or activities.

(f) Management and unions have a responsibility to use agreed procedures to resolve problems constructively and avoid industrial action. Time off is arranged for this purpose.

(g) Time off is not granted in cases where officials and/or members are engaged in industrial action.

(h) If a dispute arises from the application of this code of practice the existing machinery for the avoidance of disputes shall be fully utilized.

SECTION 38: SAFETY MEETINGS

38.1 Safety meetings will be held weekly on each main site or area.

38.2 The employer's representatives will be the safety officer and the industrial relations officer and every fourth week the site manager will attend.

38.3 The employees will have two, or on larger sites three, representatives with the right to co-opt additional representatives where necessary.

38.4 Minutes of the meetings will be kept and the distribution will include the site manager, and all shop stewards in the particular area.

SICKNESS, ACCIDENT AND LIFE INSURANCE

SECTION 39: SICKNESS, ACCIDENT AND LIFE INSURANCE SCHEME

Sickness, accident and life insurance cover will be provided by the Employers through an insurance policy for all those employed under this Agreement. Details of this cover are contained in a separate explanatory booklet to be issued to each employee.

SECTION 40: PAID BEREAVEMENT LEAVE

40.1 Upon application supported by evidence satisfactory to the Employer up to three days' bereavement leave may be granted upon the death of wife, a child, a parent or a parent-in-law.

40.2 For each day of bereavement leave so granted eight hours at the basic hourly rate will be paid.

THE MECHANICAL CONSTRUCTION ENGINEERING AGREEMENT MADE BY THE ENGINEERING EMPLOYERS' FEDERATION

The Mechanical Construction Engineering Agreement replaces and consolidates the National Sector Agreements. The Agreement dated 26 March 1975 is reproduced here in full and is followed by the Supplementary National Agreement of 30 May 1979.

SCOPE AND SIGNATORIES

0.01 This Agreement shall be known as the Mechanical Construction Engineering

AGREEMENTS (E.E.F.)

Agreement. It is made by the Engineering Employers' Federation on behalf of all member firms in its Sites Group Organization with the following trades unions:

Amalgamated Union of Engineering Workers (Constructional Section)
Amalgamated Union of Engineering Workers (Engineering Section)
The Amalgamated Society of Boilermakers, Shipwrights, Blacksmiths and Structural Workers
The Electrical, Electronic, Telecommunication and Plumbing Union
The National Union of Sheet Metal Workers, Coppersmiths and Heating and Domestic Engineers
The Transport and General Workers' Union
The General and Municipal Workers' Union

0.02 The Agreement covers engineering construction, installation, and erection activities undertaken by all Sites Group member firms on outside construction sites. Relevant work would include the following activities:

Steam Generating Plant Erection
Turbine Generator and Power Plant Installation
Outside Steelwork Erection and Bridgebuilding
Pipework and Process Plant Installation
Steel Plate and Tank Erection
The Installation of Pumps, Compressors, Switch Gear and other components and sub assemblies where associated with activities of the above nature on outside construction sites.

0.03 This Agreement supersedes all previously existing versions of the Outside Steelwork Erection and Steam Generating Plant Erection Agreements made between the Federation and the above signatory Unions. It also replaces the provisions of the Steel Plate Erection Code.

0.04 It is accepted that all federated firms who operate under the provisions of the O.S.E. and S.G.P.E. Agreements at the date of signature of this Agreement shall with effect from 1 April 1974 operate under the provisions of the Mechanical Construction Engineering Agreement.

0.05 This Agreement shall provide the basis of terms and conditions of employment for workers in federated firms who are employed on construction, erection or installation work on outside sites. This Agreement shall not apply to out-workers temporarily working on site and deployed from engineering establishments, except that such workers should overall receive terms and conditions not less favourable than those provided by this Agreement.

0.06 Any questions concerning the interpretation or application of this Agreement shall be dealt with under the Avoidance of Disputes Provisions contained in Section VIII (Procedure).

SECTION I: WORKING CONDITIONS

(a) **Working Week**

1.01 The normal working week shall consist of 40 hours which shall be worked on 5 days (Monday to Friday).

1.02 Where there is a practice of recognizing tea breaks the privilege must not be abused.

AGREEMENTS (E.E.F.)

1.03 Workers shall not leave the site during working hours, but where breaks are recognized, arrangements shall be made for refreshments to be brought round to the men.

1.04 As regards the mid-day meal, the time and duration shall be arranged between the foreman and the workers.

(b) **Overtime Working**

1.05 A full day shall be worked before overtime is reckoned, with the following exceptions: lost time through sickness certified to the satisfaction of the employers; lying off on account of working all the previous night; absence with leave or enforced idleness.

1.06 Overtime worked either before or after the normal working hours shall be paid in respect of the first 2 hours at the rate of time-and-one-third and thereafter at the rate of time-and-half.

1.07 All hours worked on a Saturday shall be paid for at the rate of time-and-half.

1.08 All hours worked between midnight Saturday and midnight Sunday shall be paid at the rate of double time.

1.09 A workman working through his meal break due to abnormal circumstances shall be paid at overtime rates unless an equivalent period is allowed.

1.10 A workman sent home between midnight and 2.00 a.m. shall be paid time-and-half for the hours worked after midnight and shall receive an allowance of plain time for working hours between the time when he is sent home and 6.00 a.m.

1.11 A workman sent home after 2.00 a.m. shall be paid time-and-half for the hours worked after midnight and receive an allowance of time-and-half for working hours between the time he is sent home and 6.00 a.m.

1.12 Where to meet exceptional circumstances a dayshift workman, having completed his normal dayshift working hours, is required to continue working throughout the night and into the normal dayshift hours of the following day, he shall be paid for the hours worked into that dayshift at the rate of time-and-half.

1.13 Workmen who, before dayshift stopping time, are notified to start at any time between midnight and the dayshift starting time the same morning shall be paid for hours worked between those periods at the rate of time-and-third the first two hours and time-and-half thereafter.

1.14 In the event of a workman being called upon to return to work after having ceased work and gone home for the day, overtime shall commence and be paid for from the time of restarting at the rate payable for that hour as though he had worked continuously.

1.15 A workman required to return to work and not so notified until after he has ceased work and gone home for the day shall be guaranteed payment equivalent to three hours at the appropriate overtime rate for the period from the time he restarts until the time he finishes work.

1.16 Any payments due to the worker in connection with the same recall under clauses 1.10 and 1.11 above shall be taken into consideration in the computation of the sum payable under this guarantee.

1.17 Payment for overtime shall be calculated on dayshift rates.

(c) **Nightshift**

1.18 Nightshift is where men, other than dayshift men, work throughout the night for not less than three consecutive nights.

1.19 A full nightshift shall consist of 40 working hours worked on four or five nights with one or two unpaid breaks for meals each night to be mutually arranged. Time lost

AGREEMENTS (E.E.F.)

through sickness certified to the satisfaction of the employers; lying off on account of working all the previous day; absence with leave or enforced idleness shall not be taken into account.

1.20 Nightshift shall be paid at the rate of time-and-one-third for all hours worked. Hours worked after the full night has been worked shall be paid at the rate of time-and-half.

1.21 Nightshift men shall receive double time for all hours worked between midnight Saturday and midnight Sunday.

1.22 Payment for nightshift and for overtime on nightshift shall be calculated on dayshift rates.

(d) **Double Dayshift and/or Three Shift System**

1.23 The normal shift week shall consist of five shifts of $7\frac{1}{2}$ hours worked from Monday to Friday, i.e. $37\frac{1}{2}$ working hours. Each shift shall have a half hour break for a meal. The following division of the day is recommended subject to any modification which local considerations may render necessary:

Monday to Friday – first shift 6.00 a.m.–2.00 p.m.
Monday to Friday – second shift 2.00 p.m.–10.00 p.m.
Monday to Friday – third shift 10.00 p.m.–6.00 a.m.

1.24 For a normal shift week of $37\frac{1}{2}$ hours, 44 hours at the time rate of the worker concerned shall be paid for the first and the second shifts and 48 hours at the time rate of the worker concerned shall be paid for the third shift.

1.25 In the case of a workman carrying on in place of any workman on the succeeding shift the former shall be paid for the extra hours so worked at the shift rate of his shift or the succeeding shift, whichever is the higher. Where for any other reason a shift worker works beyond the normal hours of his shift, overtime premiums in accordance with the overtime section of this Agreement, calculated on the time rate, shall apply.

1.26 Overtime hours worked on a Saturday by workers on all shifts shall be paid for at time-and-half, and any shifts worked on Sunday shall be paid for at double time up to midnight.

SECTION II: PRODUCTIVITY PRINCIPLES

2.01 The Unions and the Federation unreservedly agree that there is an urgent and continuing necessity, in the interests of those employed in the engineering industry on outside sites and of the community as a whole, for productive resources and the manpower of the industry to be deployed and used more efficiently.

2.02 For its part, the Federation, through its constituent Sites Groups, undertakes to have its members continue their efforts to initiate improvements in productivity, such initiatives to be based on the fullest consultation with the workers concerned and their representatives.

2.03 Accordingly, the Unions individually and collectively accept and undertake to ensure that at all levels their committees, officials and members accept all appropriate and recognized techniques for analysing or evaluating methods of site erection and installation work (including method study, work measurement and job evaluation) as ways in which the task of improving efficiency and wages can be tackled effectively at site level. The Unions also undertake to co-operate fully in the elimination of impediments to the efficient utilisation of labour and equipment which cause site labour costs to be higher than they should be.

AGREEMENTS (E.E.F.)

2.04 The following are some examples of impediments to the efficient utilization of labour, the elimination of which would fall to be dealt with under clause 2.03 above.

(*i*) Inappropriate and uneconomic manning and use of labour.
(*ii*) Lack of flexibility in the deployment of labour.
(*iii*) Resistance in principle to the introduction of shift working on the basis of nationally agreed conditions.
(*iv*) Resistance to the planned use of all working hours.
(*v*) Non-productive and lost time.

2.05 The parties agree that improvements in pay and conditions within the framework of National Agreements may be made by individual companies, provided that there is a measured increase in labour productivity or efficiency to which the efforts of the workers concerned have contributed.

2.06 Increased payments so earned may be achieved through jointly approved or established incentive schemes governed by recognized control methods, or, where they already exist, by fixed lieu payments closely related to improved productive efficiency.

2.07 Agreements reached between recognized representatives of both parties, at whatever level, should be honourably adhered to and changes to such agreements should not be considered except through constitutional methods.

2.08 Every attempt will be made to reach agreement on improved productivity, pay and conditions through domestic discussions but, if agreement cannot be reached, the parties accept that the Procedure in Section VIII of this Agreement shall be honoured.

SECTION III: HOLIDAYS WITH PAY

Part A: General

3.01 Each worker shall receive 28 days holiday per annum as follows

20 days of annual holiday
8 Bank or other days

The 8 Bank or other days shall include the following:
In England, Wales, and Northern Ireland
CHRISTMAS DAY, NEW YEAR'S DAY, and either EASTER MONDAY or GOOD FRIDAY
In Scotland – NEW YEAR'S DAY

3.02 To comply with this Agreement it is necessary that every worker shall take his full holiday entitlement. Where a worker is required to work on any recognized holiday or part thereof an equivalent holiday in lieu shall be made available for him at another mutually agreed period.

Part B: Arrangements for Taking Holidays

3.03 The above holidays shall be arranged locally by mutual agreement in accordance with the usual practice for fixing holidays in the firm or district concerned. Whilst it is desirable that the holidays of site workers should generally conform with those observed by local custom, this shall not preclude other mutually acceptable arrangements being made in exceptional circumstances such as when the majority of workers are not local men.

AGREEMENTS (E.E.F.)

3.04 It is agreed that a week of holiday should be taken in the winter period, to be attached to the Christmas holiday and so arranged as to provide a continuous holiday to include Christmas Day to New Year's Day, and the three days provided under the Agreement of 6 June 1974 should be used for this purpose. If, after the fullest consultation and the exhaustion of procedure, no agreement can be reached on the arrangements for holidays, it is recognized that it then remains as the ultimate responsibility of management to determine when the holidays shall be taken.

3.05 To enable employees to make any necessary arrangements including the reservation of holiday accommodation, it is recommended that joint steps should be taken as early as possible to arrange the dates when the above holidays are taken. Workers who are engaged subsequent to these arrangements being made shall be notified accordingly at the time of their engagement.

Part C: Payment for Holidays

1. Annual Holidays

3.06 For the purpose of providing payment for 20 days of annual holiday, there shall be credited to every employee for each full week's work performed, a sum representing 1/12th of the appropriate time rate plus one-third or 1/12th of the rate at present used for calculating holiday credits, whichever is the greater.

3.07 These credits shall be accumulated in a special fund maintained by each firm and shall be payable to each employee up to the amount due in respect of the impending holiday at the time that holiday is taken, or at such other time as may be mutually agreed.

3.08 A full week's work shall be a week of 40 hours but workpeople employed on approved short-time provided they have worked a full short week shall be entitled to the full week's holiday credit.

3.09 Employees absent due to certified sickness or any accident incurred during the normal course of employment shall be credited with the appropriate holiday allowance for working hours so lost for a period up to 12 weeks in any one year.

3.10 When, in any week less than 40 hours have been worked, the appropriate proportion of the full week's allowance shall be credited. This provision is subject to the following exceptions:

(*a*) The other eight days of paid holiday provided by this Agreement shall be treated as working days for the purpose of calculating holiday credits.

(*b*) Workpeople engaged on public work (e.g. local or county councillors, etc.) shall be allowed holiday credits for working hours thus spent.

2. Payment for certain Bank or other holidays

3.11 In addition to the reservation of holiday credits as defined above, separate payment shall also be made to every worker for eight Bank or other holidays each year at a sum equivalent to the appropriate time rate plus one-third (or the rate at present used for calculating his holiday pay if this is greater). This payment shall be made on a uniform basis of eight hours for all site workers.

3.12 Workpeople shall not qualify for payment for the eight 'Bank or other' holidays who fail to work the full normal working day immediately preceding and the full working day immediately following the holiday, unless they can produce evidence to the satisfaction of the employer that their absence was due to causes beyond their control.

AGREEMENTS (E.E.F.)

3.13 Notwithstanding the foregoing, it is agreed that where any of the above eight 'Bank or other' holidays are included in or attached to the annual holiday period, the qualifying condition in respect of the working day preceding the holiday will be waived. The qualifying day following the holiday period will be retained.

3.14 Workers who would otherwise be eligible for holiday pay who are absent on either or both of the qualifying days will not be disqualified from holiday pay provided that:

(1) they are in the employment of a federated firm at the time of the holiday and

(2) they are absent with leave or their absence is due to accident or sickness duly certified.

3.15 Travelling men who have long journeys to make to their homes and who have been given special permission to leave site early will not forfeit holiday payments, fares or travelling time for the return journey to site. On the other hand, if circumstances do not permit any general curtailment of site working hours on the days immediately preceding or following a holiday, any other men who leave early or fail to report for work and exceed their permitted leave without permission or reasonable excuse will not qualify for their holiday payment (and fares and travelling time) for the holiday in question.

Part D: Payment For Work Done on Holidays

3.16 Work done on the eight Bank or other days provided in Part A of this Section shall be paid at premium rates. Two days in each year shall be paid at the rate of double-time and the other six days at the rate of time-and-a-half. These days shall be computed from midnight to midnight.

3.17 In England, Wales and Northern Ireland the two 'double-time' days shall be:

(1) Christmas Day (or the following Monday if Christmas Day falls on a Sunday).

(2) Either Good Friday (in districts where this day is recognized as a holiday) or Easter Monday (in districts where that day is recognized as a holiday and Good Friday is not).

3.18 In Scotland the two 'double-time' days shall be:

(1) New Year's Day, or the following day if New Year's Day falls on a Sunday.

(2) One other day, being the day recognized by the Engineering Industry in the district concerned as the second 'double-time' day.

3.19 The remaining six 'time-and-a-half' days shall be arranged locally having due regard to the holidays observed by the Engineering Industry in the district concerned.

3.20 Work done on days of annual holiday will not be subject to premium payment.

SECTION IV: GUARANTEE OF EMPLOYMENT

4.01 All hourly rated manual workers who have been continuously employed by the employer for not less than four weeks shall be guaranteed employment for five days in each normal pay week. In the event of work not being available for the whole or part of the five days, employees covered by the guarantee will be assured earnings equivalent to their time rate for 40 hours.

4.02 This guarantee is subject to the following conditions:

(*a*) That the employees are capable of, available for, and willing to perform satisfactorily, during the period of the guarantee, the work associated with their usual occupation, or reasonable alternative work where their usual work is not available.

AGREEMENTS (E.E.F.)

(*b*) Where approved short time is worked as an alternative to redundancy, or in the case of a holiday recognized by agreement, custom or practice, the period of guarantee shall be reduced proportionately.

(*c*) In the event of dislocation of production as a result of an industrial dispute on the site of a federated contractor, or in any federated establishment or site, the operation of the period of the guarantee shall be automatically suspended.

(*d*) In computing the assured earnings referred to above, premium payments due for overtime worked on weekdays, and premium payments for work done on Sundays and holidays, shall be ignored.

SECTION V: INCLEMENT WEATHER

5.01 In cases where men have to cease their normal duties owing to interruption by inclement weather, every effort shall be made to find them reasonable alternative work. Where work has to continue in adverse weather conditions, such as in the interests of emergency, safety, or for other accepted essential reasons, the employers shall provide weatherproof clothing where necessary.

5.02 All men who can be found work under cover must remain at work without interference from other men.

5.03 Any men for whom no work can be found, and who are required to remain on site ready to resume their normal tasks when circumstances permit, shall be guaranteed payment at their time rate for time thus spent.

5.04 Payment for inclement weather is made subject to the provisions of Clause (a) of the Guarantee of Employment section of this Agreement.

5.05 The subject of inclement weather (wet time) payments is reserved for determination at national level and firms shall not have latitude to negotiate higher payments for cabined up time than the national rate.

SECTION VI: CONDITION PAYMENTS

(a) **Height Money**

6.01 All men working at a height of 50 feet or more on any type of framework or structure shall receive an allowance as provided by the following scales:

At 50 feet but under 75 feet	£0·0150
„ 75 „ „ „ 100 „	£0·0300
„ 100 „ „ „ 150 „	£0·0450
„ 150 „ „ „ 200 „	£0·0900
„ 200 „ „ „ 250 „	£0·1350
„ 250 „ „ „ 300 „	£0·1800

with an addition of £0·0450 per hour for each additional 50 feet.

6.02 When the framework or structure concerned is open to basement level, the height shall be calculated from basement level.

6.03 When steel framework is covered in at ground floor level (except for staircase or hoist openings), the height shall be calculated from ground floor level. In the case of a boiler structure the height shall be calculated from the nearest adjacent filled-in floor levels.

AGREEMENTS (E.E.F.)

6.04 In the case of work which is being carried out on or from temporary or permanent staging or flooring, having adequate safeguards constructed in accordance with existing regulations, height money shall be calculated from that level (unless the work is over the side of the floor or over an opening in the floor).

6.05 It is understood that the allowances shall also be paid:

(*a*) To a man working adjacent to or above an opening in a filled-in floor, where there is a possibility of his falling through the opening 50 feet or more.

(*b*) To a man called upon to handle materials over a parapet or barrier preparatory to landing at levels of 50 feet or more.

6.06 In the case of conveyor structures based on ground floor level, the height shall be calculated from the ground adjacent to each structure.

6.07 Cranedrivers and banksmen operating a crane on open steelwork at heights of 50 feet or more shall be paid the allowance.

(b) Conditions of Dirt and Heat

6.08 Workers may raise with their foreman any case in which it is considered that conditions are abnormal, e.g., excessive dirt, heat or fumes or working in confined space. In respect of each condition established as abnormal, a maximum allowance of £0·0300 per hour worked per condition shall be paid to each worker who is required to work under the condition. This allowance shall also apply to workers brought into contact with flue deposits, and in cases where flue deposits are established as corrosive then the allowance shall be increased to £0·0600 per hour for that condition.

6.09 In the event of obnoxious or dangerous fumes or acids being encountered, a special allowance shall be mutually agreed.

(c) Muddy Conditions

6.10 Where it is agreed that site conditions are such that rubber boots are required, the employer shall be responsible for providing same and in addition a payment of £0·0100 per hour shall be made. Rubber boots will not be used by men working aloft.

SECTION VII: ALLOWANCES

Part A: Radius Allowance

7.01 Radius allowance shall be paid to any man who travels daily between his home and a site over 2 miles away. This allowance is compensation for travelling time and fares and will, therefore, only be paid in respect of days worked.

7.02 The scales of radius allowance payable are as follows:

(*a*) **Scale 1**

		Per day
Over 2 miles and not exceeding 5 miles each way	. . . . .	£1·13
,, 5 ,, ,, ,, ,, 8 ,, ,, ,,	. . . . .	£1·56
,, 8 ,, ,, ,, ,, 11 ,, ,, ,,	. . . . .	£2·07
,, 11 ,, ,, ,, ,, 14 ,, ,, ,,	. . . . .	£2·59
,, 14 ,, ,, ,, ,, 17 ,, ,, ,,	. . . . .	£3·20
,, 17 ,, ,, ,, ,, 20 ,, ,, ,,	. . . . .	£3·73
,, 20 ,, ,, ,, ,, 25 ,, ,, ,,	. . . . .	£4·19
,, 25 ,, ,, ,, ,, 30 ,, ,, ,,	. . . . .	£4·55
,, 30 ,, ,, ,, ,, 35 ,, ,, ,,	. . . . .	£4·86
,, 35 ,,	. . . .	£5·16

AGREEMENTS (E.E.F.)

7.03 Payable in respect of distances for which suitable free transport is provided.

(b) **Scale 2**

					Per day
Over	2 miles and not exceeding	5 miles each way			0·76
,,	5 ,, ,, ,, ,,	8 ,, ,, ,,			1·04
,,	8 ,, ,, ,, ,,	11 ,, ,, ,,			1·38
,,	11 ,, ,, ,, ,,	14 ,, ,, ,,			1·72
,,	14 ,, ,, ,, ,,	17 ,, ,, ,,			2·14
,,	17 ,, ,, ,, ,,	20 ,, ,, ,,			2·48
,,	20 ,, ,, ,, ,,	25 ,, ,, ,,			2·79
,,	25 ,, ,, ,, ,,	30 ,, ,, ,,			3·02
,,	30 ,, ,, ,, ,,	35 ,, ,, ,,			3·24
,,	35 ,,				3·45

7.04 All distances in each scale shall be measured on a straight line basis, but exceptional cases (e.g., circuitous routes due to natural barriers) where this is considered to be inappropriate may be raised with the employer in accordance with the agreed procedure.

7.05 Any man who, by arrangement with the employer, uses both provided and other transport and travels a total distance of over two miles each way daily shall be entitled to an extra payment of £0·100 per day in addition to any allowance(s) payable under Scale 1 or 2, or both, provided the total allowance does not exceed £5·25 per day.

7.06 Radius allowance is not payable to any man who is in receipt of lodging allowance, but special consideration shall be given where lodgings are not available close to the site.

7.07 Radius allowance is only payable on a daily basis, viz. those days of the week on which the worker is actually travelling to and from the site.

7.08 Any man who is absent during working hours, or leaves the site early or returns late at holiday times without good cause or without permission, is liable to forfeit the radius allowance to which he would otherwise be entitled for the day or days concerned.

Part B: Lodging Allowance

7.09 A lodging allowance of £5·25 per day, i.e., £36·75 per week, shall be paid to any man who, by agreement with his employer, lives away from home.

7.10 The company will determine with each employee a personal base for 'lodging allowance'. This will normally be his permanent home address and will be recorded and revised as necessary.

7.11 Lodging allowance shall be paid on a seven day basis except in the case of a broken week, e.g., at the beginning or end of a contract. A man who is absent during working hours on a normal working day (Monday to Friday) without good cause or without permission, shall forfeit the lodging allowance payable for that day.

7.12 Any case of sickness or accident arising out of, or in the course of, a worker's employment which necessitates a man entitled to lodging allowance remaining in his lodgings shall be dealt with sympathetically by the employer on its merits.

7.13 A man who is in receipt of lodging allowance shall not be paid radius allowance, but special consideration shall be given where lodgings are not available close to the site.

7.14 As regards holidays other than the annual holiday period, a retainer of £1·65 per day shall be paid in respect of those days of holiday when the lodgings are not occupied

AGREEMENTS (E.E.F.)

and lodging allowance not paid. For the purpose of this clause 'days of holiday' shall be understood as including any day on the 12 occasions referred to in Clause 7.23 when a man is released from site and does not occupy his lodgings.

7.15 As regards the annual holiday period, any man who is required to pay a retention fee for his lodgings shall be reimbursed to the amount actually paid to a maximum of £6·60 per week upon production of proof of payment to the employer's satisfaction.

7.16 In exceptional cases such as when lodgings are not available and men are required to travel daily from home, special arrangements may be made between the firm and the workers.

Lodging allowance at holiday periods

7.17 Where men return home at the recognized holiday periods, lodging allowance is paid on the days on which the men are travelling home from the site or returning to the site. In respect of the intervening days of the holiday the lodging retainer will be paid.

7.18 For example, if a job closes down on a Friday night due to the following Monday being a holiday, and work is resumed on the Tuesday, lodging allowance would be paid in respect of the Friday if the men travel home on that day, and in respect of the Tuesday, if the men travel back to the site on that day. The retainer would be payable in respect of the Saturday, Sunday and Monday.

7.19 If, however, a worker elected to remain over the holiday period in the site district and so would occupy his lodgings, he should receive lodging allowance for the week-end period (Saturday, Sunday and Monday) but would not receive travelling expenses.

Part C: Transport Expenses

7.20 In the case of men paid lodging allowance, transport expenses consisting of fares or vouchers for second class travel, together with travelling time at single time rates shall be allowed at the start of an 'away' contract, and in addition at the finish of an 'away' contract, except in a case in which a worker can reasonably be transferred to another contract on which he qualified for lodging allowance. When so transferred, a worker shall receive fare or voucher and be paid travelling time at single time rate for travelling from the original site to the contract.

7.21 The employer may direct the route by which the men travel and in the event of the men electing to travel by route other than that directed by the employer, the employer will be liable to pay only transport expenses which would have been payable on the directed route.

Note

7.22 In cases where men terminate their employment with the company during the course of a contract, such men are not entitled to transport expenses for the journey to their home base. Nevertheless where management is satisfied that the termination arises from genuine circumstances which require a man's return to his home area (as defined in Clause 7.10), the application for such expenses will be given proper consideration.

Part D: Transport Expenses and Allowances at Holiday Periods

7.23 Transport expenses consisting of travel vouchers or fares for second-class travel or equivalent value for the return journey between the site and the normal place of residence, shall be allowed to men employed on 'away' contracts and in receipt of a lodging allowance, on twelve occasions during the year. These occasions shall normally be once per month and so arranged to coincide with recognized holiday periods.

AGREEMENTS (E.E.F.)

7.24 In addition, payment at single time rates shall be made for the time spent in returning from the holiday to the site on the occasion of each holiday period. Lodging allowance shall be paid in respect of the days on which the men are travelling to and from the site.

7.25 The payment of transport expenses, travelling time for the journey back to the site and the retainer shall be conditional on the punctual return of the men to the site on the next working day after the holiday period, unless they can produce evidence to the satisfaction of the employer that their absence was due to causes beyond their control or are absent or return late with the permission of the employer, such permission not to be unreasonably withheld.

7.26 Where a worker on an 'away' contract is required by the employer to remain at the site to continue to work during a recognized holiday period, he shall receive an allowance equal to the transport expenses he would have received had he returned home. In addition to payment at appropriate rates for work done during the holiday, he shall receive lodging allowance in respect of the days he is required to remain at the site, though no more than one day's lodging allowance shall be payable in respect of any one day.

7.27 In exceptional cases, the foregoing provisions may be varied by mutual arrangement between the employer and the individual worker.

7.28 Except in very exceptional cases, workers are expected to return to the site during the course of the morning of the day following the holiday period.

SECTION VIII: PROCEDURE

Part A: Provisions for Avoidance of Disputes

Procedure for dealing with questions arising concerning manual workers on outside construction sites

8.01 This section of the Agreement is designed to provide broad principles within which procedures for settling disputes at all levels can operate between each federated contractor and his workpeople.

8.02 There shall be effective joint consultation on all appropriate matters.

8.03 The Federation confirms its recognition of the rights of the Unions to negotiate on behalf of their members and that it is in the mutual interests of federated contractors and their workpeople that the latter should be members of an appropriate signatory trade union in order that negotiations can be conducted on a fully representative and authoritative basis.

8.04 Any questions arising on a site, including the interpretation or application of national agreements, shall be raised first of all between the workers or their representatives with their immediate supervisor.

8.05 Thereafter there may be other stages in the domestic procedure, through which the question should be progressed, if it is not resolved at the first stage. These stages should preferably be set out in written form and shall be agreed between the management and the stewards on site, assisted where necessary by the full-time local official(s) from both sides.

8.06 The domestic procedure, whilst being a matter for agreement at site level, shall be based on the following principles:

(*a*) There shall be an agreed number of stages, providing for specified time limits, within which any question or type of question raised by the employees or the management shall be discussed.

AGREEMENTS (E.E.F.)

(b) Clear provision shall be made regarding the appointment, functions, rights and responsibilities of stewards.

(c) The particular levels of management which should be approached and involved at each stage of the domestic procedure shall be clarified.

8.07 Failing settlement at the final stage of domestic procedure, the matter may be reported by either party to the local Employers' Association Sites Group and/or the local full-time official(s) of the union(s) concerned for consideration at a Site Conference which shall be held within seven working days of receipt of application or as otherwise mutually agreed.

8.08 Failing settlement at Site Conference the matter may be referred directly for discussion at a Statutory Conference in terms of Clause 8.09 below. Notwithstanding this, the party pursuing the reference shall have the option of referring the question first of all for further consideration at a Local Conference between representatives of the local Employers' Association Sites Group and the local official(s) of the union(s) concerned thereafter, if the matter is to be pursued further it should be referred to Statutory Conference.

8.09 Statutory Conference shall be held on the first Thursday of each month when questions of which full particulars have been referred not less than 10 days beforehand, shall be discussed at national level between representatives of the Federation and the executive official(s) of the union(s) concerned.

8.10 Failing settlement at Statutory Conference the Procedure shall be regarded as exhausted on the question concerned.

8.11 Until the above procedure has been exhausted, there shall be no stoppage of work either of a partial or general character including a go slow, a work to rule, a strike, a lock out or any other kind of restriction in output or departure from normal working.

8.12 Negotiations under the above procedure may be instituted by either management or the workers concerned.

Part B: Appointment and Function of Stewards on Site

8.13 Stewards shall be elected in accordance with the rules of their union from among the members of the unions signatory to this Agreement. A steward shall represent the workers of his own employer in accordance with the above Procedure, and shall act in accordance with the terms of all relevant Agreements.

8.14 Upon appointment, the name of each steward and the group of workers he represents shall be officially notified in writing by the Union to management.

8.15 Stewards shall be afforded reasonable facilities to deal with questions raised by their constituent members. The scope of such facilities shall be for mutual agreement between stewards and their employer, assisted where necessary by the full-time local official(s) of both sides. In all other respects a steward shall conform to the same working conditions as his fellow employees.

SECTION IX: DISCIPLINARY RECOMMENDATIONS

9.01 The parties to this agreement recommend to their members that procedures should be introduced at site level, to provide fair and effective arrangements for dealing with disciplinary matters. The details of these procedures shall be determined by discussion between each employer and the Union(s) representing his employees, but shall have regard to the following provisions and principles.

AGREEMENTS (E.E.F.)

The procedure should clarify

9.02 (a) The kind of acts (or omissions) which constitute serious industrial misconduct and will normally result in summary dismissal, without prior cautioning or warning. (Examples of these may be listed for clarification purposes and may include fighting, stealing, wilful damage, false booking of work, clocking offences, serious breaches of safety regulations or company rules, and/or other matters which may be determined at site level.)

(b) Other acts (or omissions) of a less serious nature which if they recur within a given time will, after proper warning(s) has (have) been given, also lead to eventual dismissal (after notice or payment in lieu). Such instances could include poor attendance and timekeeping, absence unsatisfactorily explained, lack of capability, bad workmanship or performance, failure to comply with rules and regulations and/or other matters to be determined at site level.

(c) (i) The number of warnings relative to each offence in (b) above which should be given before final dismissal occurs should be clearly stated. The procedure may also provide for other penalties, such as reprimands or suspension to be administered, as an alternative or preliminary to dismissal.

(ii) Warnings to employees shall normally be given in the presence of a shop steward (or other representative) and shall be recorded in a book kept for that purpose by site supervision. Employees may also receive a written copy of their warning.

(d) The levels of management with responsibility for discipline and dismissal should be indicated.

(e) After formal warning has been given, due account shall be taken of cases where a man subsequently improves his performance and also of the period of time which has elapsed since a previous warning was given.

(f) A worker being cautioned, warned, or reprimanded shall be entitled to representation by his shop steward.

(g) There shall be provisions for appeal to a stipulated level of higher management (if the employee wishes) before a dismissal is enacted.

(h) If the fairness of a given dismissal, or warning, or reprimand is challenged, the Union may refer the matter for discussion under the normal avoidance of disputes procedure.

(i) Appeals or matters of discipline and dismissal shall be given priority both in the domestic and external stages of procedure.

SECTION X: TERMINATION OR TRANSFER OF EMPLOYMENT

Period of notice

10.01 During the first four weeks of continuous employment the notice required to be given by either the employee or a company to terminate employment shall be one hour on either side or upon payment of one hour's wages in lieu of notice by the company.

10.02 After four weeks' continuous employment the notice to be given by employee or company shall be one week or one week's wages in lieu of notice on the part of the company.

10.03 After thirteen weeks' (or more) continuous employment with the company, the length of notice required to be given by either side to terminate the employment shall be in accordance with the provisions of Section 1 of the Contracts of Employment Act 1972.

AGREEMENTS (E.E.F.)

10.04 The above provisions shall not necessarily apply in situations where misconduct is involved (see Section IX of the Agreement, entitled Disciplinary Recommendations).

Transfer of employees

10.05 In cases where a firm wishes to move an employee to another site (as opposed to terminating his employment) it is recommended that the maximum possible advance notice of intended transfer should be given on every occasion. This should not normally be less than one week's notice, other than in exceptional circumstances or where otherwise mutually agreed.

Travelling men

10.06 This National Agreement is not in itself intended to create an obligation to travel from site to site. Where such obligations exist these are separate personal matters between employer and employee.

SECTION XI, PART A: GRADES OF LABOUR AND DEFINITIONS OF WORK

11.01 There shall be three classifications of labour under this Agreement, namely, skilled workmen, semi-skilled, and unskilled. Their skills, qualifications and general duties are defined in the following paragraphs.

Skilled workman

11.02 A skilled workman is one who:

(*a*) has completed an apprenticeship or other recognized training period in an appropriate trade;
and
(*b*) is able and required to carry out appropriate skilled work without supervision or skilled assistance.

11.03 Notwithstanding (*a*) above, and in the absence of any unified training arrangements for the mechanical construction industry at present, it is accepted that provided that skilled labour as above is not available, a worker who has not received formal trade training may also be recognized as skilled provided that

(*c*) he has undergone a period of experience of the type of work in question which is agreed as being not less than the recognized training period for his trade, and he is also able to conform with the requirements of (*b*) above.

Semi-skilled workman

11.04 A semi-skilled workman is one who is able and required to carry out work of a semi-skilled nature, including the use of relevant equipment. A semi-skilled workman may work under the direction of but not necessarily under the immediate control of a skilled workman.

Unskilled workman

11.05 An unskilled workman is one who assists generally in the performance of unskilled tasks such as simple loading and unloading, fetching and carrying and in site housekeeping.

AGREEMENTS (E.E.F.)

11.06 A full list of the grades of labour covered by this Agreement and their appropriate rates of pay is contained in the wages section of this Agreement.

Job Definitions of Some of The Principal Craftsmen

11.07 The following definitions should be read in conjunction with the foregoing definition of a skilled workman set out in Clauses 11.02/11.03 of this Agreement.

Erector

11.08 An erector is a man who can work aground or aloft and who is from experience capable of positioning, fixing and bolting up the compartments of the plant, including minor fabrication modifications (using necessary equipment), and also providing where necessary his own temporary working platforms subject to the overall direction of the engineer in charge or the foreman.

Rigger

11.09 A rigger is a man who is able to rig and erect all tackle and lift units of any weight or description forming parts of the unit and auxiliary plant.

Fitter and Machinist

11.10 A skilled fitter or machinist is a man who is able and required to prepare, assemble, install, line up and/or test mechanical and/or electrical equipment (including work on turbine generators, reactors and ancillary plant, and also switchgear and control equipment) in a construction site environment and without supervision or skilled assistance.

Pipefitter

11.11 A skilled pipefitter is a man who is able and required to carry out the full normal range of pipefitting work encountered on construction sites without supervision or skilled assistance.

Crane driver

11.12 A crane driver who is able to familiarize himself with the workings of and who can satisfactorily operate the range of cranes normally used in construction work and who also has the knowledge and ability to implement statutory regulations concerning crane operations shall be regarded as a skilled man.

Constructional scaffolder

11.13 A constructional scaffolder is a man who from experience is capable of handling, erecting, dismantling, servicing and maintaining tubular scaffolding and any other temporary working platform, including where necessary the access thereto.

Plater

11.14 A plater is a man who is able and required to carry out appropriate skilled work without supervision or skilled assistance and who can be expected, in the course of his duties, to carry out welding which is normally associated with platers work – i.e. tack and stud welding.

AGREEMENTS (E.E.F.)

Welders

11.15 For the purpose of this agreement, a tube is defined as being up to and including 4½" O.D. irrespective of thickness.
A pipe is defined as being above 4½" O.D. irrespective of thickness.
The term 'Approved Welder' means a welder who is certified by an approved authority.

11.16 Welder qualification is in accordance with B.S.S. 2633 entitled 'Specification for Class 1 Arc Welding of Ferritic Steel Pipework for carrying fluids', or any other approved equivalent standard.

11.17 This standard applies to carbon and ferritic alloy steels as defined in Table 1 of the above British Standard (B.S.S. 2633) and to austenitic steels. The four grades of welding referred to as Classes A1, A2, A3 and B, shall be deemed to have the following meanings:

11.18 (A1) Approved welders who are engaged on the welding of carbon and ferritic alloy steel tubes only.

11.19 (A2) Approved welders who are engaged on any one of the following:

(a) Welding of carbon and ferritic alloy tubes and pipes.
(b) Welding of austenitic tubes.
(c) Semi-automatic and/or automatic welding of carbon and ferritic alloy pressure parts.
(d) Manual welding of carbon or ferritic alloy steel pressure vessels.

11.20 (A3) Approved welders who are engaged on:

(a) Any combination of the grades (a) to (d) listed in (A2) above.
(b) Austenitic pipes.
(c) Manual, semi-automatic or automatic welding of austenitic pressure vessels.

11.21 (B) Welders qualified to the satisfaction of the employer engaged on the manual and semi-automatic welding of non-pressure parts, e.g. casings, structural weldings, stud welding and the welding of attachments to pressure parts.

SECTION XI, PART B: SPECIAL PAYMENTS FOR SKILL OR PROFICIENCY

Qualification allowances

11.22 In addition to the time rates set out in the wages section to this Agreement, certain skilled workers shall be entitled to receive qualification allowances in accordance with the following provisions:

Erector/riggers

11.23 An erector/rigger who had demonstrated his competence in the terms of the schedule at the end of this section shall receive qualification allowance of £0·0250 per hour.

Fitters and machinists

11.24 A mechanical fitter, electrical fitter, pipefitter or machinist who is fully skilled within the definitions set out in Clauses 11.02/11.03 of this Agreement (see also Job Definitions) shall receive an allowance of £0·0250 per hour.

AGREEMENTS (E.E.F.)

Platers

11.25 A plater who is fully skilled within the definition set out in Clauses 11.02/11.03 of this Agreement (see also Job Definitions) shall receive an allowance of £0·0250 per hour.

Welders

11.26 By reference to the classification set out in Clauses 11.15–11.21 of this Agreement (see Job Definitions) welders shall receive the following allowances, according to the nature of the work on which they are engaged:

Grade A1 – £0·0521 per hour
Grade A2 – £0·0521 per hour
Grade A3 – £0·0604 per hour

11.27 The above allowances are payable for hours worked and shall not enter into the calculation of overtime and nightshift premiums or holiday payments.

PART C: CRANEDRIVERS

11.28 In the case of Diesel and electric cranes, one hour's pay per day at single rate to be paid in addition to the ordinary workings hours, such to include time for oiling round, cleaning and attention to the machine, with due regard to safety.

11.29 Where a driver is asked by the employer to do any repairs, at other times than ordinary working hours, the actual time worked to be paid for with the usual rates for overtime.

11.30 The time for the examination and greasing of all bonds and chains shall be left to the discretion of the employer, but if done at other times than during working hours, the time to be paid shall be three hours with usual rates for overtime, once every fortnight being recognized as normal.

Qualified Erector/Rigger

11.31 The term 'Erector/Rigger' shall be applied to a man who has served for at least 5 years as an erector or rigger and who had demonstrated that he has the requisite knowledge and skill described in the Training Manual for Erectors and who is able and willing to carry out the work of an erector/rigger at any height. The term may also be applied to a workman who has satisfactorily completed an improvership course (see Section XII) and who has subsequently worked as an erector/rigger for a period of two years in the same class of work or who has worked as a junior for three years on outside steelwork erection with a subsequent three years as an erector/rigger on the same class of work.

11.32 The erector/rigger shall be required to demonstrate his competence to his employer in one or other of the following ways:

(a) Where he has carried out relevant work during the whole of the previous five years in the employment of his present employer he must have clearly demonstrated to his employer that he possesses all the requisite skill and knowledge. Where an erector/rigger has not served the whole of the previous five years with his present employer, the service provision above will be deemed to be complied with where he can satisfactorily demonstrate that the balance of the period has been spent on relevant work with other federated firms covered by this National Agreement. (Reference shall be made to other employers to establish the facts if necessary).

AGREEMENTS (E.E.F.)

(b) In all cases where he cannot fulfil the requirement in (a) above he shall demonstrate his skill and knowledge by the passing of the appropriate erector qualification tests as laid down in the Training Manual for Erectors.

11.33 His knowledge, skill and duties shall cover the following areas:

(1) A relevant appreciation of statutory regulations.
(2) Wire ropes, slings and chains, etc.
(3) Ability to understand and work to sketches and working drawings.
(4) Ability to mark off, drill and cut steel plate and RS sections.
(5) Knowledge and use of knots for construction purposes.
(6) Rigging and lifting: knowledge of procedures, ability to understand and use all relevant equipment.
(7) Construction of stagings for erection purposes, including use of ladders.
(8) Setting, levelling and plumbing of steel structures.
(9) Erection of heavy steelwork components.
(10) Assembly and rigging of major items of lifting equipment.

SECTION XII: TRAINING AND IMPROVERSHIP

12.01 Suitable workpeople between the ages of 18 and 32 may be recruited for training as improvers in the occupations of erector, rigger, crane driver and scaffolder.

12.02 Not less than 3 years' instructive experience in the particular classification of work is to elapse before the full rate for the occupation in question shall be paid.

12.03 During training the following rates shall apply according to the age at which training is commenced.

Training commenced at age 18
18 – $67\frac{1}{2}$%
19 – 80%
20 – 90%
21 – skilled rate

Training commenced at age 19
19 – 80%
20 – 90%
21 – $92\frac{1}{2}$%
22 – skilled rate

Training commenced at age 20
20 – 90%
21 – $92\frac{1}{2}$%
22 – 95%
23 – skilled rate

Training commenced at age 21 or over
1st year – $92\frac{1}{2}$%
2nd year – 95%
3rd year – $97\frac{1}{2}$%

AGREEMENTS (E.E.F.)

12.04 As regards erector rigger training, the employer will supply each improver with a Registration Book in which will be entered the employee's name, permanent address and National Insurance number. The book will allow for entries to be made by his employer or employers giving dates of employment and details of experience and training during the three years of the Improvership Course. Registration of Improvers will be recorded by the Engineering Employers' Federation and the Amalgamated Union of Engineering Workers (Constructional Section).

12.05 The production of the Registration Book showing length of service as an improver shall be accepted as sufficient entitlement to the hourly rates provided by this Agreement.

12.06 It is agreed that no unconstitutional action will be supported as a result of a dispute arising from the operation of improverships and any questions arising out of the interpretation, length of service, experience or any other matter relating to improverships shall be dealt with by a joint panel set up for this purpose. The panel shall examine such questions on the date of the Statutory Conference arranged for the last month of the Quarter.

12.07 During the period of improvership the improvers shall be given, insofar as facilities at site are available, practical experience in the full range of work normally carried out by an erector or rigger employed on mechanical construction sitework. He should be instructed at site in:

(a) the application, fixing, rigging and operation of conventional erection equipment, i.e. wire rope, blocks, pull lifts, chain blocks, hoists and winches and other lifting tackle, and in the proper care and maintenance of such equipment;

(b) the safe application of other types of equipment including small tools, hand and power driven tools and oxygen and acetylene cutting equipment;

(c) the construction and dismantling of working platforms and stagings (both fixed and suspended) as well as means of access thereto, and of tubular scaffolding and special purpose equipment, and in the care of proper maintenance of same;

(d) safe working practices and safety discipline, with special emphasis on the statutory obligations imposed on the workmen by existing construction regulations.

12.08 Instruction at the site should, where practicable, be supplemented by other forms of instruction complementary to the work of erectors and riggers, and when called upon to do so improvers will be required to attend lectures, demonstrations, instruction films and discussions on subjects related to their training.

12.09 It is mutually agreed that improvers must be prepared to travel during their period of training, and that they should be encouraged to supplement their training wherever and whenever facilities are available by attending evening classes.

Apprentices, boys and youths

12.10 These shall be paid at the following percentages of the adult skilled rate:

16 years – $42\frac{1}{2}\%$
17 years – $57\frac{1}{2}\%$
18 years – $67\frac{1}{2}\%$
19 years – 80%
20 years – adult rate, subject to satisfactory completion of training.

AGREEMENTS (E.E.F.)

SECTION XIII: SAFETY REPRESENTATIVES

Introductory note

13.01 The purpose of this Agreement is to ensure that the above Regulations are applied in a manner which is suitable to the special needs of engineering construction sites and also to assist in promoting a safer working environment.

The parties accord with the view of the Health and Safety Commission that the degree of agreement necessary to operate the Regulations must be achieved through the existing industrial relations machinery. However the ensuing provisions, including the functions of safety representatives and safety committees should be regarded as an agreed expression of the intentions of the Regulations, the Guidance Notes and the Code of Practice and not as an extension of the industrial bargaining machinery.

It is jointly accepted that it is in the mutual interests of employers and union members that the following provisions should be observed on sitework contracts covered by the MCE Agreement.

General

13.02 The trade unions signatory to the MCE Agreement shall appoint safety representatives according to the following provisions:

(a) Similar arrangements shall apply as with the appointment of stewards under the Procedure Agreement and the workplace shall be each site work contract of an employer.

(b) In the event of any unresolved questions arising on a sitework contract concerning the operation of the Regulations these shall be channelled by either party through the external stages of the Avoidance of Disputes Procedure (see clause 11(a)).

Safety representatives – appointment

13.03 The number of safety representatives to be appointed on any sitework contract shall have regard to the needs of that contract including:

(a) the total numbers employed;
(b) the variety of different occupations;
(c) the size of the workplace and the variety of workplace locations;
(d) the operation of shift systems;
(e) the type of work activity and the degree and character of the inherent dangers.

However, it is not envisaged that in normal circumstances the number of safety representatives to be appointed on any contract under these Regulations should exceed the existing number of stewards.

13.04 Where a union wishes to appoint a safety representative on any sitework contract, the full-time union official(s) shall contact the Employers' Association to discuss the number of safety representatives and the names of person(s) to be appointed as may be appropriate for that workplace, with regard to the numbers and groups of operatives who are to be represented.

13.05 By mutual agreement a safety representative may represent members of another trade union on his employer's contract, particularly when the total number of workers employed is small.

AGREEMENTS (E.E.F.)

13.06 Safety representatives shall, upon appointment, be issued with appropriate credentials by their union and the union shall notify the employer in writing of each appointment made under the Regulations, indicating the group or groups of operatives on the contract who are covered by the arrangement.

13.07 With regard to Regulation 3(4) it is recognized that in the construction industry it will not always be feasible for a union to nominate employees who have had two years continuous service with their employer. Nevertheless every effort will be made to appoint workers whose overall experience makes them most suitable for the task of a safety representative, and the union will discuss with the employer what is the best arrangement for the contract in question. At least two years' adult experience in the industry would normally be the minimum acceptable length of service.

13.08 (a) When appointing a safety representative the union(s) shall indicate to the employer details of any relevant training already received by their nominee.

(b) The unions shall also indicate to the employer as many weeks as possible in advance of the event, in each case, their intentions to send nominated safety representatives on approved safety training courses. They shall at the same time indicate the location, duration and content of the training course which is proposed for each safety representative with regard to clauses 3, 4 and 5 of the Health and Safety Commission Code of Practice.

(c) The number of safety representatives attending a course at any one time shall be a matter for mutual agreement with regard to the availability of relevant courses and the operational requirements of the employer.

(d) With regard to the Code of Practice, employers are encouraged to participate in the training programme and the parties approve the development of a joint approach to training on all items essential to fulfil the purposes of the Regulations.

13.09 The employer shall normally confirm his acceptance (in writing) of the appointment of a safety representative within seven days of receipt of the union's notification. When so doing the employer shall also indicate the name(s) and designation(s) of the appropriate member(s) of management to whom the safety representative should normally refer in the exercising of his functions under these Regulations.

13.10 A person shall cease to be a safety representative when:

(a) the trade union which appointed him notifies the employer in writing that his appointment has been terminated; or

(b) he ceases to be employed at the workplace (contract) in respect of which he was first appointed; or

(c) he is transferred to another contract by his employer, or to another work group outside of the terms of his appointment (and which already has another safety representative); or

(d) he resigns.

13.11 In accordance with the Regulations and the Code of Practice the employer shall allow safety representatives time off with pay to carry out their functions in the workplace and to attend necessary training courses of an approved nature.

13.12 (a) Any differences between the parties at site level *arising out of the operation of the Regulations* shall be referred by either party to the full-time officials of the Trade Unions or the Employers' and shall not give rise to any industrial action unless the Procedure has been exhausted.

(b) Any other differences such as might arise over safe systems of work (i.e. safety

AGREEMENTS (E.E.F.)

matters of a technical nature), or on the interpretation and application of the Health and Safety at Work Act and other supplementary Regulations may be referred to an Inspector of the Health and Safety Executive or some other suitably qualified expert for advice and assistance.

Functions of safety representatives

13.13 The functions of a safety representative are defined as follows:

(a) In consultation with his employer, to represent other employees on his contract, as defined in his appointment, on the making of arrangements for co-operation in promoting and developing measures to ensure the health and safety at work of employees and checking the effectiveness of such measures;
(b) To investigate hazards and dangerous occurrences at his employer's work place and to examine the causes of accidents;
(c) To investigate complaints by any employee in his constituency relating to that employee's health and safety or welfare at work;
(d) To make representations to his employer on matters arising out of (b) and (c) above and general matters regarding the health, safety or welfare at work, of the employees whom he represents.
(e) To carry out inspections subject to the provisions of paragraphs 13.14/17.
(f) To represent employees in his constituency in consultations at the workplace with inspectors of the Health and Safety Executive and of any other enforcing authority and to receive information from inspectors in accordance with section 28(8) of the 1974 Act. To this end, the employer will, so far as he is able and aware notify the safety representative(s) of any intended visit of an inspector.
(g) To attend safety committee meetings where he attends in his capacity as a safety representative in connection with any of the above functions.
(h) To encourage safe practices among the employees he represents.

Inspections of the workplace

13.14 Safety representatives may carry out inspections of the workplace to which their appointment refers at such intervals and after giving such notice (in writing) as shall normally be agreed with his employer, and providing that any safety representative has not inspected the workplace or relevant part of it in the previous three months, subject to the following:

An additional inspection may be made where it is jointly agreed in advance to be necessary because

(a) there has been a substantial change in the conditions of work since the time of the last inspection; or
(b) the Health and Safety Commission or Health and Safety Executive have published new significant information relevant to the hazards of the workplace since the last inspection; or
(c) there has been a notifiable accident or dangerous occurrence or a notifiable disease has been contracted (i.e. when notice of which is required to be given by virtue of any of the relevant statutory provisions within the meaning of section 53(1) of the Health and Safety at Work etc. Act) and the interests of the employees represented by the safety representative are directly involved.

AGREEMENTS (E.E.F.)

(d) emergency inspections would normally only follow a serious accident resulting in e.g. a fracture, an amputation, the loss of sight of an eye or serious damage thereto, a dislocation, or any other injury which is likely to cause prolonged incapacity for work or permanent damage to the employee's health e.g. internal haemorrhage, severe burns or scalds etc.

13.15 Safety representatives may carry out an inspection only where it is safe to do so, and in accordance with any safety instruction issued by the client and any 'permit to work' system which may apply.

13.16 The employer shall provide all necessary facilities and assistance for inspections and he or his representative shall be entitled to be present. Facilities shall include opportunities for private discussion between the safety representative and the employees he represents. It is recommended that inspections should be carried out on a joint basis.

13.17 The safety representative shall report in writing to the employer on the result of each inspection whether or not anything unsafe has been observed. The format of such reports will normally be jointly agreed and a predetermined checklist is recommended.

Inspection of documents and provision of information

13.18 Subject to reasonable notice, an employer shall permit his safety representatives to inspect and take copies of any document which:

(a) the employer is required to keep by any relevant statutory provisions within the meaning of section 53(1) of the Health and Safety at Work etc. Act 1974, other than a document concerning the health record of an identifiable individual; and

(b) is relevant to the workplace or to the employees they represent.

13.19 The employer shall make available to his safety representatives such information within the employer's knowledge as would enable the safety representatives to carry out their functions except information which:

(a) it would be against the interests of national security to disclose; or

(b) he cannot disclose without contravening a prohibition imposed by or under an enactment, or

(c) relates specifically to an individual unless the individual has agreed to disclosure, or

(d) cannot be disclosed without substantially injuring the employers' undertaking or that of the supplier of the information, for reasons other than the effect on health, safety and welfare at work; or

(e) the employer has obtained for the purpose of bringing, prosecuting or defending any legal proceedings.

13.20 An employer is not required to produce or allow inspection of any document or part of a document which is not related to health, safety or welfare.

Safety committees

13.21 An employer shall establish a safety committee if such a request is made in writing by at least two of his safety representatives subject to the following:

(a) the employer shall consult with his safety representatives who made the request and

AGREEMENTS (E.E.F.)

with the representatives of recognized trade unions whose members work in any workplace in respect of which the employer proposes that the committee should function;
(b) the employer shall display a notice of the composition of the safety committee and the employees covered by it;
(c) in reaching agreement on the functions, objectives, conduct of the safety committee, the employer, union representatives, safety representatives shall, where appropriate, have regard to the Guidance Notes for Safety Committees issued by the Health and Safety Commission.

N.B. The introduction of any sitewide safety committee is not a statutory requirement and this subject shall be a matter for separate discussion between the national executives.

SECTION XIII: WAGES AND TIME RATES

Supplementary National Agreement of 30 May 1979 on wages and other matters Amending The Provisions of The Mechanical Construction Engineering Agreement

Parties

1 This agreement is made between the Engineering Employers' Federation and the following Trade Unions:

A.U.E.W. (Constructional Section);
A.U.E.W. (Engineering Section);
Amalgamated Society of Boilermakers, Shipwrights, Blacksmiths and Structural Workers;
Electrical, Electronic, Telecommunication and Plumbing Union;
National Union of Sheet Metal Workers, Coppersmiths, Heating and Domestic Engineers;
Transport and General Workers' Union;
General and Municipal Workers' Union.

Part A: Items effective from Monday, 11 June 1979

General Wage Increases and New Time Rates

2 As from Monday, 11 June 1979, all workers covered by the MCE Agreement will receive the following general increases in wages:

Skilled workers	24 pence per hour
Semi-skilled workers	21·5 pence per hour
Unskilled workers	19 pence per hour

These amounts shall be added to the existing time rates contained in the MCE Agreement.

Minimum Pay Levels

3 Also from the same date, the time rates shall be further improved by the following minimum rate increments:

Skilled workers	7·5 pence per hour
Semi-skilled workers	6·75 pence per hour
Unskilled workers	6·0 pence per hour

The resultant new time rates are set out in the schedule to this agreement.

AGREEMENTS (E.E.F.)

4 The above minimum rate increments shall not be applied as general wage increases except that any worker whose hourly remuneration, after receipt of the appropriate amount of general increase in para 2 above, still falls below the level of the new time rate for his grade, shall receive a sufficient minimum rate increment to raise his hourly earnings to that level. In all other cases, an amount equivalent to the minimum rate increment in 3 above shall be transferred from existing second tier (e.g. bonus and lieu bonus payments) currently paid in excess of the rate.

5 This agreement also provides corresponding pay improvements for Improvers and Apprentices and the increases in their cases are set out in the schedule.

6 The new time rates shall be used for the calculation of overtime and shift premiums and holiday pay, and for authorized cabined-up time during inclement weather. It is not the intention of this agreement that the new time rates should be used as bonus calculators or give rise to any increase in bonus payments. Where bonus earnings are expressed as a percentage of the existing time rate(s), the percentage relationship(s) shall be altered so as to give effect to the principle that all workers receive neither more nor less than the general wage increases provided by this settlement, and bonus earnings remain unchanged for unchanged effort or output.

Performance, Pay and Productivity at Site Level

7 (*a*) The improvements provided by this agreement represent the limit of concessions to be made by federated firms at national, local or site level during its currency.
Any claims for additional general improvements in wages by means of further increases in bonus, lieu bonus or other second tier payments will not be entertained or supported by the parties.
(*b*) The Executives of the unions and the Federation are gravely concerned at the harmful effects on the industry of any pay increases which are not linked to commensurate improvements in productivity. It is also agreed that to resort to a practice of supplementary pay bargaining at site or domestic level following this national award is not in the best long-term interests of the industry and its future propects.
(*c*) The parties jointly support the principle that once on any contract the level and system of any second tier payment arrangement has been properly established, it shall not be subject to re-negotiation during the currency of this National Agreement, other than in circumstances which are approved by the Federation and the signatory trade unions.
(*d*) The parties also recognize that if any of their members continue to attempt to secure further increases in second tier payments which are not linked to productivity, or which are contrary to the above provisions, then this will only undermine the capacity of their union Executives to conduct meaningful negotiations with the employers at national level, and the stability which all parties seek from current and future national agreements will be seriously undermined.
(*e*) The unions and the Federation intend to take joint steps very shortly to decide what mechanisms should be introduced to ensure that all their members, contractors and clients adhere fully to the above provisions, and the responsibilities which this entails for the industry's future.

AGREEMENTS (E.E.F.)

Paid Bereavement Leave

8 In the event of the death of a man's wife or child he shall be granted up to a maximum of three days special leave for which payment will be made at eight hours of his basic time rate for each day.

Severance Payments for Employees with under Two Years Service

9 (a) For each week of continuous employment up to a maximum of 103 weeks service, certain payments will be made by his employer to any employee who is dismissed as redundant before he has completed two years unbroken service with the employer. These payments will be calculated at the rate of £2·00 for each complete week of service for skilled employees and pro-rata (to time rate) for other grades.

(b) These new arrangements shall be brought into effect immediately on existing contracts where there are currently no arrangements of this nature or similar already in operation. They shall also be introduced immediately on all new contracts.

(c) The above arrangements will be waived upon the completion of the necessary length of service by the employee to qualify for a statutory redundancy payment under the provisions of the Employment Protection (Consolidation) Act 1978.

(d) On any existing contracts where short term severance payments (or other arrangements having similar intent or effect) have already been negotiated for workers with under two years service, and these arrangements are more favourable than or broadly compatible with the provisions in sub paragraphs (a) and (b) above, then the existing arrangements may continue, by agreement between the parties concerned, for the duration of that contract only.

(e) It is agreed that all future contracts shall be covered uniformly by the new arrangements, and that any different practices on terminal payments on existing contracts shall be phased out. Any questions arising at site level about the implementation of these provisions shall be raised under the Avoidance of Disputes Procedure.

Note for Guidance

The computation of a worker's period of continuous employment for the purpose of applying the above arrangements shall be made along similar lines to those applicable to statutory redundancy payments in the appropriate legislation. Days of holiday, sickness and temporary absence from work in any period covered by the worker's contract of employment will be reckonable as qualifying service. On the other hand no week shall count during the whole or part of which the worker has participated in a stoppage of work amounting to a strike or similar industrial action. Such weeks will not be reckoned as weeks of service in computing the number of weeks for which he has been continuously employed. They will not however break the continuity of employment.

Part B: Effective from 1 September 1979

Overtime Premium – New Arrangements

10 (a) With effect from the above date, all overtime hours worked before midnight on weekdays (Monday to Friday) shall be paid at the rate of time plus one half.

(b) On Saturdays, time and one half will be paid until 4.00 p.m. or until eight hours have been worked, whichever is the sooner. Thereafter the remaining hours on Saturday will be paid at double time.

AGREEMENTS (E.E.F.)

(c) All hours worked on Sundays will (continue to) be paid at double time.

(d) An employee required to work extended overtime after midnight shall be paid at double time for the hours worked after midnight until his task is completed. If this is before 7.00 a.m. then he shall receive an allowance of time and one half from the time he completes his task until 7.00 a.m.
If his task continues up to or beyond 7.00 a.m. and (for example, transport reasons) he is unable to leave site before that time, then if the next day is a normal working day he shall not be required to attend further for work on that day, and he shall receive an additional allowance of eight hours pay at plain time rate in respect of that day.

(e) All premium payments are calculated on the national basic time rate for the grade concerned.

Part C: Effective from 1 January 1980

Work done on Statutory Holidays from 1 January 1980

11 As from the above date, payment for work done on *all* of the eight Bank or other days (stipulated in Clauses 3.01 and 3.16 of the MCE Agreement) shall be made at the rate of double time.

Part D: General

Sick Pay Scheme

12 (a) The Federation has agreed, on behalf of its Sites Group member firms, that those of their employees who regularly work under the terms of the MCE Agreement should be covered by a Sick Pay Scheme (similar to the 'Crusader Scheme' which operates on the OCPCA Site Agreement Sites).

(b) The precise date of implementing new arrangements of this nature by all member firms cannot be pre-determined at this stage. However, the Federation will undertake early enquiries about the most effective method of implementing such an arrangement and once the necessary information is at hand will take prompt steps to enrol and inform its member firms. Thereafter, the matter will be further discussed with the unions and an appropriate time for implementation will be agreed.

(c) It is understood that any member firms whose existing sick pay schemes are more favourable than or broadly comparable with the eventual national scheme may by arrangement with their employees opt out of the national scheme.

(d) It is also understood that if a member firm enrols in the national sick pay scheme in respect of its construction employees under the MCE Agreement, this confers no obligation on them to enrol any other of their employees who work in different parts of their organization.

Joint Working Party

13 It is agreed that a joint working party shall be set up consisting of representatives of the Federation and the signatory trade unions to consider and make recommendations on other matters on which it was not possible to reach a final position in the negotiations on 30 May 1979, with the intention of reaching a satisfactory conclusion by the end of 1979.

These include:

(a) The introduction of *status quo* or similar provisions into the MCE Agreement to

AGREEMENTS (E.E.F)

those contained in Clauses 5 and 6 of the EEF/CSEU Agreement of 1 March 1976 (Procedure for the Avoidance of Disputes).

(*b*) The formulation of guidelines for any necessary overtime work on construction sites, including the question of reasonable limits to overtime hours and also the categories of work and situations where it would be impracticable or contrary to the interests of all parties to attempt to restrict overtime working.

(*c*) The unions' claim for an initial free issue of protective clothing such as safety boots, overalls or donkey jackets.

(*d*) The unions' claims for an increase in proficiency payments for coded welders and the employers desire that the job definitions and classifications of work should be brought up to date before this claim can be properly considered.

New National Agreement for the Engineering Construction Industry

14 The Executives of the Federation and the signatory trade unions re-affirm their commitment to work together over the next few months in the tripartite working party with the OCPA (and under the auspices of the NEDO) towards the earliest possible conclusion of a new unifying national agreement for all engineering construction sitework. One of the principal objectives of this agreement would be to introduce better regulated systems of payment for the industry especially on large multi-contractor sites.

Duration

15 The provisions of this agreement shall apply for not less than 12 months from 11 June 1979 unless superseded by the terms of any new industry-wide national agreement. Except where altered to take into account the foregoing terms of this settlement, the existing terms of the MCE Agreement shall continue to apply in full.

Large Industrial Projects – Labour
AGREEMENTS (E.E.F.)

Schedule of New Hourly Rates
Adult Operatives

Changes in rate structure on 11 June 1979

Grade	Former Rate as from 17.7.78	General increase	Intermediate position	Minimum Rate Increment transferred from second tier payment	Resultant Time Rate w.e.f. 11.6.79
Erectors Riggers Constructional Scaffolders Chippers Caulkers Gougers Platers Tubers Expanders Fitters Machinists Pipe Fitters Sheeters Burners Riveters Crane Drivers Welders (all grades)	£1·46	24p	£1·70	7·5p	£1·7750
Semi-skilled	£1·31	21·5p	£1·525	6·75p	£1·5925
Unskilled	£1·17	19p	£1·36	6·0p	£1·42

AGREEMENTS (E.E.F.)

Improvers Schedule of Hourly Rates

Age and agreed percentage of adult skilled rate	Former Rate as from 17.7.78 £	General increase p	Intermediate position £	Minimum Rate Increment p	Resultant Time Rate w.e.f. 11.6.79 £
Training Commenced Age 18					
18 (67½%)	0·986	16·2	1·148	5·1	1·199
19 (80%)	1·168	19·2	1·360	6·0	1·420
20 (90%)	1·314	21·6	1·530	6·8	1·598
21 skilled rate	1·460	24·0	1·700	7·5	1·775
Training Commenced Age 19					
19 (80%)	1·168	19·2	1·360	6·0	1·420
20 (90%)	1·314	21·6	1·530	6·8	1·598
21 (92½%)	1·351	22·2	1·573	6·9	1·642
22 skilled rate	1·460	24·0	1·700	7·5	1·775
Training Commenced Age 20					
20 (90%)	1·314	21·6	1·530	6·8	1·598
21 (92½%)	1·351	22·2	1·573	6·9	1·642
22 (95%)	1·387	22·8	1·615	7·1	1·686
23 skilled rate	1·460	24·0	1·700	7·5	1·775
Training Commenced Age 21					
1st year (92½%)	1·351	22·2	1·573	6·9	1·642
2nd year (95%)	1·387	22·8	1·615	7·1	1·686
3rd year (97½%)	1·424	23·4	1·657	7·3	1·731

Juniors and apprentices Schedule of Hourly Rates

Age and agreed percentage of adult skilled rate	Former Rate as from 17.7.78 £	General increase p	Intermediate position £	Minimum rate Increment p	Resultant Time Rate w.e.f. 11.6.79 £
16 (42½%)	0·6205	10·20	0·7225	3·19	0·7544
17 (57½%)	0·8395	13·80	0·9775	4·13	1·0206
18 (67½%)	0·9855	16·20	1·1475	5·06	1·1981
19 (80%)	1·1680	19·20	1·3600	6·00	1·4200

AGREEMENTS (E.E.F.)

Re: Radius And Lodging Allowances: 1979

With effect from Monday, 14 May 1979 radius and lodging allowances payable under the Mechanical Construction Engineering Agreement shall be increased as follows:

1 Lodging Allowances to be increased from £5·25 to £5·80 per day (£40·60 per week).

2 Scales I and II Radius Allowances to be increased by 11% to give new daily values as follows:

					Per Day Scale I £	Scale II £
Over	2 miles and not exceeding	5 miles each way			1·25	0·84
,,	5	,,	,,	8 ,, ,,	1·73	1·15
,,	8	,,	,,	11 ,, ,,	2·30	1·53
,,	11	,,	,,	14 ,, ,,	2·87	1·91
,,	14	,,	,,	17 ,, ,,	3·55	2·37
,,	17	,,	,,	20 ,, ,,	4·14	2·75
,,	20	,,	,,	25 ,, ,,	4·65	3·10
,,	25	,,	,,	30 ,, ,,	5·05	3·35
,,	30	,,	,,	35 ,, ,,	5·39	3·60
,,	35	,,			5·72	3·83

3 The daily holiday retainer for Lodgings (in Clause 7.14) to be increased from £1·65 to £1·80 per day.

4 The weekly holiday retainer for Lodgings (in Clause 7.15) to be increased from £6·60 to £7·25 per week.

5 This agreement is in full and final settlement of the claim presented by the trade unions signatory to the MCE Agreement to the EEF on 13 March 1979 and shall apply for a period of not less than 12 months from Monday, 14 May 1979.

LARGE INDUSTRIAL PROJECTS

Estimating

INTRODUCTION

The limiting factors which dictate the type of estimating procedure to be adopted are usually (*i*) amount of time available and/or (*ii*) the amount of information to hand. The general types of estimate used and their respective accuracies can be classified as follows:

(*a*) Order of Magnitude (Rough) Estimate $\pm 30\%$ accuracy.
(*b*) Budget Estimate $\pm 15\%$ accuracy.
(*c*) Definitive Estimate $\pm 5\%$ accuracy.

ORDER OF MAGNITUDE ESTIMATES

These are prepared in response to requests for information to enable a customer to consider the viability of an outline scheme. The method of estimating is usually restricted to scaling up or down the cost of a similar plant of different capacity. This method can be known as the 'six tenths rule' after the exponential or logarithmic factor that is commonly applied as follows:

$$C_2 = \left(\frac{P_2}{P_1}\right)^{0.6} \times C_1$$

where C_1 = the cost of the previous plant (updated to present-day costs),
C_2 = the cost of the proposed plant being sought,
P_1 = the capacity of the previous plant,
and P_2 = the capacity of the proposed plant.

For example:
C_1 = £30,000
P_1 = 3,000 M.T.P.A. (metric tonnes per annum)
and P_2 = 6,000 M.T.P.A.

Then $C_2 = \left(\frac{6,000}{3,000}\right)^{0.6} \times £30,000$

$\frac{6,000}{3,000} = 2.00$

The log of 2·00 is 0·30103,
therefore $\log 2.00^{0.6} = 0.30103 \times 0.6 = 0.18062$
The antilog of 0·18062 is 1·5158
therefore $C_2 = 1.5158 \times £30,000 = £45,474$ (say £45,500)

However, the 0·6 factor should be used with caution, as experience will show that the factor can vary between 0·5 and 0·75 depending upon the type of plant or process, the proportions of equipment, etc.

It should be borne in mind that the six tenths rule can only be applied to process units (not off-sites for instance) and then only when comparing one single-stream process with another of different capacity.

INTRODUCTION

In using a published exponential or logarithmic factor, its origin should be known and suitable allowance made. For instance, American factors should not be used for work in this country without adjustment, because (*inter alia*) the American industry has a proportionately higher labour cost and a proportionately lower material cost (owing to high automation) than its British counterpart.

BUDGET ESTIMATES

These usually form part of preliminary proposals which are prepared for appropriation purposes. The method of estimating normally consists of determining the costs of items of equipment from statistics, built up from previous plants, based on preliminary process data sheets and flow sheets which should be available at this stage. These statistics are in graphic or tabular form and are further explained in a later sub-section. The remaining costs are then expressed as a factor of the total equipment cost. This factor, which is often referred to as a 'Lang Factor', includes for not only bulk materials but for design and construction costs as well. The magnitude of this 'Lang Factor' will obviously depend on (*i*) the type of plant, (*ii*) the size of plant, and (*iii*) the pre-dominant materials of construction. Thus, the range can extend from 2·5 to 5·0 and whilst a factor of 3·75 is a reasonable average figure, it is a figure that should be treated cautiously. A budget estimate can sometimes be refined by using individual 'Lang factors' for each category of bulk materials, construction, design etc.

DEFINITIVE ESTIMATES

These are prepared in response to a customer's formal invitation to tender; generally on a 'lump sum' basis. This involves quite a considerable amount of pre-tender engineering design work. The estimator may elect to obtain quotations from vendors for the major items of equipment to sharpen the accuracy of the estimate, depending upon the amount of previous experience of the particular type of plant concerned and the amount of confidence generated. In addition, full take off of bulk materials may be carried out and priced to improve the accuracy of the estimate still further, if time permits. Definitive estimates are sometimes produced for cost control purposes or even process evaluation. Depending on the degree of refinement required, the preparation of a definitive estimate can cost about 1% of the total cost of a plant.

Large Industrial Projects – Estimating   439

COST CODE LIST

Most contractors and some customers have their own systems of cost coding and a list is published by the Association of Cost Engineers. As no one list enjoys wide acceptance in the industry, the list that follows may be considered as typical of many.

The use of a cost code list is a useful contribution to cost control and cost prediction. The code references are included to illustrate the sort of system that could be adopted. The two digit codes may be expanded as required to identify individual units within an item and gaps between codes may be filled by additional items.

The list gives good coverage of most contractors' prime cost items likely to be encountered in normal contracts within the industry but it should be borne in mind that general overheads and profit are not included.

COMMODITY CODES

A – Tanks, sub-divided as follows:
AA	Bulk storage tanks	AR	Rectangular tanks
AB	Bins	AS	Silos, bunkers
AC	Cyclones	AV	Vats
AH	Hoppers	AX	Other tanks
AL	Lining	AZ	Spares

B – Vessels, sub-divided as follows:
BC	Columns	BR	Reactors
BG	Gas holders	BS	Spheres
BI	Trays (internals)	BX	Other vessels
BN	Non-pressure vessels	BZ	Spares
BP	Pressure vessels		

C – Heat exchangers, sub-divided as follows:
CB	Block type	CK	Kettle type reboilers
CC	Tank coils (where supplied separately)	CP	Plate type reboilers
		CS	Shell and tube type reboilers
CD	Double tube type	CT	Cooling Towers
CF	Air-cooled type	CX	Other types
CJ	Jet type coolers	CZ	Spares

D – Mechanical equipment, sub-divided as follows:
DB	Blowers, fans	DP	Pumps
DC	Compressors	DR	Refrigeration plants
DD	Driers	DS	Desuperheaters
DE	Internal combustion engines	DT	Turbines
DF	Filters, centrifuges	DW	Water treatment plants
DG	Gas generators	DX	Other mechanical equipment
DJ	Ejectors, eductors	DZ	Spares
DM	Mixers, agitators		

COST CODE LIST

E – Heaters, sub-divided as follows:
    EB   Steam boilers
    ED   Drum driers
    EF   Furnaces
    EH   Fired heaters
    EI   Incinerators
    EJ   Ducting
    EK   Kilns, ovens
    ES   Stacks, chimneys
    EW   Waste heat boilers
    EX   Other heaters
    EZ   Spares

F – Materials Handling Equipment, sub-divided as follows:
    FC   Conveyors
    FE   Elevators
    FG   Grinders, crushers
    FL   Lifts, cranes, hoists
    FM   Magnetic separators
    FP   Packaging machinery
    FR   Rotary valves
    FS   Screens
    FT   Trucks, trolleys
    FW   Weighing machinery
    FX   Other materials handling equipment
    FZ   Spares

G – Miscellaneous Equipment not falling within any of the above categories

J – Civil Works, sub-divided as follows:
    JA   Demolition
    JB   Dewatering
    JC   Culverts and tunnels
    JD   Drainage and sewers
    JF   Foundations
    JG   Grading
    JH   Paving and hardstanding
    JL   Landscaping
    JM   Marine installations
    JP   Piling
    JQ   Rail sidings
    JR   Roads, paths and fences
    JS   Surveys and tests
    JT   Trenching and backfilling
    JV   Reservoirs, sumps and ponds
    JW   Bund walls and dykes
    JX   Other civil items
    JZ   Spares

K – Buildings, sub-divided as follows:
    KA   Administration
    KB   Boiler house
    KC   Compressor house
    KD   Pump house
    KE   Workshops
    KG   Gatehouse
    KL   Laboratories and medical
    KM   Canteen
    KP   Process/production
    KS   Control and switch rooms
    KV   Garage
    KW   Warehouse
    KX   Other buildings

L – Structures, sub-divided as follows:
    LB   Vehicle and foot bridges
    LE   Equipment structures
    LP   Piping structures
    LX   Other structural items
    LY   Yard structures

M – Fire Protection, sub-divided as follows:
    MC   Clothing and portable equipment
    ME   Fireproofing equipment
    MF   Foam generators
    MG   Carbon dioxide generators
    MH   Hydrants and extinguishers
    MP   Fireproofing pipework
    MS   Fireproofing steelwork
    MV   Fire fighting vehicles
    MW   Fire water mains system
    MX   Other fire protection items
    MZ   Spares

Large Industrial Projects – Estimating

COST CODE LIST

P – Pipework, sub-divided as follows:

PA	Pipes		PS	Pipe supports
PB	Fittings		PT	Steam tracing
PF	Flanges		PV	Valves and specials
PJ	Gaskets and bolting		PW	Wrapping/coating
PL	Lining		PX	Other piping items
PP	Prefabrication		PZ	Spares

Q – Electrical, sub-divided as follows:

QA	Communications		QP	Power
QC	Cathodic protection		QS	Switchgear including starters
QE	Earthing		QT	Transformers and extra high tension cables
QH	Trace heating			
QI	Instrument cabling		QX	Other electrical items
QJ	Cable jointing		QZ	Spares
QL	Lighting			

R – Instrumentation, sub-divided as follows:

RA	Analysers etc.		RP	Pressure instruments
RC	Control and reducing valves		RR	Relief valves
RD	Panels and control desks		RT	Temperature instruments
RF	Flow instruments		RX	Other instrumentation units
RI	Installation materials		RZ	Spares
RL	Level instruments			

S – Insulation, sub-divided as follows:

SE	Cold insulation of equipment		SQ	Hot insulation of pipework
SF	Hot insulation of equipment		SX	Other cold insulation
SP	Cold insulation of pipework		SY	Other hot insulation

T – Mechanical painting, sub-divided as follows:

TE	Painting equipment		TX	Other mechanical painting
TP	Painting pipework			

W – Catalysts and Chemicals

X – Export Packing and Shipment

Y – Import Duties and Taxes

Z – Other Charges not falling within any of the above categories

HEAD OFFICE CODES

02	Project		24	Mechanical equipment design
04	Planning		26	Materials handling design
06	Process design		28	Civil design
08	Layout and model		30	Architectural design
10	Design co-ordination		32	Structural design
12	Head office construction		34	Piping design
14	Procurement		36	Electrical design
16	Cost control		38	Instrumentation design
18	Estimating		40	Licensing fees
20	Vessel design		42	Overseas Head Office expenses
22	Heat exchanger design		48	Other Head Office items

COST CODE LIST

FIELD CONSTRUCTION CODES

- 50 Direct labour
- 54 Direct civil labour
- 58 Indirect labour
- 60 Trade supervision
- 64 Civil trade supervision
- 68 Field management and administration
- 69 Commissioning engineers
- 70 Site buildings
- 73 Temporary services
- 77 Site transport
- 79 Scaffolding
- 80 Construction plant
- 82 Civil construction plant
- 84 Cranes and rigging tackle
- 85 Testing equipment and services
- 86 Consumables
- 87 Small tools
- 89 Other construction items

GENERAL CODES

- 92 Contract insurance
- 94 Finance costs
- 96 Export credit guarantee department costs
- 98 Other general costs not falling within any of the above categories

COST ANALYSIS

Upon completion of a contract, it is customary for a contractor for his own use to produce a final cost report in which to set out in detail all costs incurred on the contract broken down into their appropriate cost codes. Such a final report is of immense value to the estimator, for it not only records the actual cost of various items, but it analyses these costs with regard to the quantity, size and type of commodity they represent. The relationship between various cost codes can also be given. For example, there could be a large chapter on piping to set out the different materials of construction segregated into units, showing the range of sizes, average sizes, ratio of fittings, valves, flanges and supports to pipes, average man hours per metre run and so on.

What follows is an overall summary of the cost analysis of a typical petro-chemical plant showing the main cost categories and their relationship with one another. It should be emphasized, however, that while the following example is fairly representative, wide variations in costs and their relationships to one another do occur in the many types of plant that are encountered (for instance, the Finance Cost figure (Code 94) could be considered high if compared with projects on which the terms of payment are more favourable).

The total of the typical cost analysis when compared with the cost of the procurement of the major equipment (total (1) columns (*a*) and (*b*)) represents a 'Lang factor' of 3·36.

Contractor's profit is excluded from the cost analysis.

COST ANALYSIS

	Item and Code	a Commodity Material £	b Sub- Contract £	c % of Total 1, Columns a and b	d Code	e Field Con- Man Hours
A	Tanks	10,698	48,122	1·70	50	125
B	Vessels	1,146,374	—	33·18	50	12,665
C	Heat exchangers	1,141,199	—	33·03	50	7,445
D	Mechanical equipment	686,031	—	19·86	50	3,170
E	Heaters	—	415,583	12·03		
F	Materials handling	6,902	—	0·20	50	60
	Total (1)	2,991,204	463,705	100·00		23,465
J	Civil works	124,576	372,677	14·39	54	—
K	Buildings	50,560	116,152	4·83	54	—
L	Structures	—	223,728	6·47		—
M	Fire protection	44,170	—	1·28	54	
P	Pipework	944,781	15,942	27·81	50	114,535
Q	Electrical	555,895	87,037	18·61	50	26,900
R	Instrumentation	934,429	—	27·05	50	28,600
S	Insulation	—	247,936	7·18		—
T	Mechanical painting	—	25,136	0·73		—
W	Catalysts	125,253	—	3·63	50	3,500
X	Shipping	107,309	—	3·11		—
Y	Import duties	181,848	—	5·27		—
	Total (2)	6,060,025	1,552,313			197,000

Indirect labour 58
Trade supervision 60/64
Field management and Admin. 68
Amenities (Scaffold, £90,000) 70/79
Construction plant 80/85
Consumables/small tools 86/87
Project planning, etc.
Process design
Procurement
Cost control
Prints, expenses, etc.

Total (3)   6,060,025   1,552,313

COST ANALYSIS

	f struction Cost £	g % of Column a	h Total of columns a, b and f £	Code	Head Office Cost £	l Total of columns h and k £
	629	5·88	59,449	20 ⎱	26,599	
	65,788	5·74	1,212,162	20 ⎰		
	38,658	3·39	1,179,857	22	9,816	
	16,217	2·36	702,248	24 ⎱	31,032	
	—	—	415,583	24 ⎰		
	314	4·55	7,216	26	950	
	121,606	4·07	3,576,515		68,397	3,644,912
	140,537	112·81	637,790	28 ⎫		
	60,050	118·77	226,762	30 ⎪	17,099	
	—	—	223,728	32 ⎬		
	50,933	115·31	95,103	28 ⎭		
	534,792	56·60	1,495,515	08, 34	152,942	
	145,252	26·13	788,184	36	37,364	
	154,998	16·59	1,089,427	38	53,514	
	—	—	247,936			
	—	—	25,136			
	18,864	15·06	144,117			
	—	—	107,309			
	—	—	181,848			
	1,227,032	20·25	8,839,370		329,316	9,168,686
		% of Total 2 Column f				
	61,825	5·04	61,825			
	275,532	22·46	275,532			
	170,129	13·87	170,129			
	277,296	22·60	277,296			
	200,764	16·36	200,764			
	59,202	4·82	59,202			
				02, 04, 12 ⎫		
				06 ⎪		
				14 ⎬	206,918	
				16 ⎪		
				48 ⎭		
	2,271,780		9,884,118		536,234	10,420,352
	Overheads		395,364		616,677	1,012,041
	Total (4)		10,279,482		1,152,911	11,432,393

General: Contract insurances	(Code 92)	72,302
Finance costs	(Code 94)	171,179
Other charges	(Code 98)	12,887
	Total (5)	11,688,761

COST GUIDE

ABBREVIATIONS

A.C.	Alternating current
A.N.S.I.	American National Standards Institute
A.P.I.	American Petroleum Institute
A.S.T.M.	American Society for Testing Materials
A.W.G.	American wire gauge
B.S.	British Standard
D.C.	Direct current
D.S.	Direct switching
H.B.C.	High breaking capacity
O.C.M.A.	Oil Companies Materials Association
O.S. & Y.	Outside screw and yoke
PVC	Polyvinyl chloride
T.E.M.A.	Tubular Exchanger Manufacturers Association

COST GUIDE

BULK STORAGE TANKS (SUBCONTRACT PRICES) (CODE AA)

Vertical cylindrical bulk storage tanks, fixed conical roof, atmospheric working pressure, carbon steel construction, complete with handrailing and stairways. Supplied in plate small condition and completely erected at site, tested and painted one coat of primer.

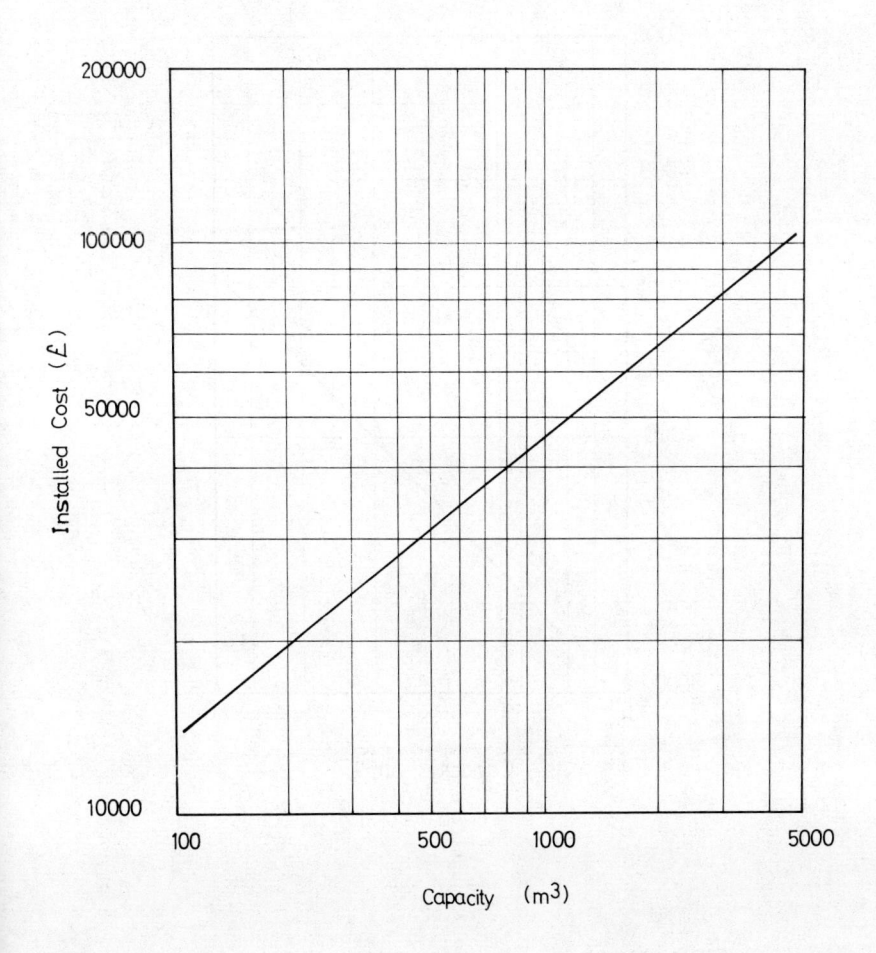

COST GUIDE

SILOS (MARKET PRICES OF MATERIALS) (CODE AS)

Carbon steel silos complete with access ladder, access hatch, 100 mm diameter inlet pipe and 455 mm diameter flanged outlet.

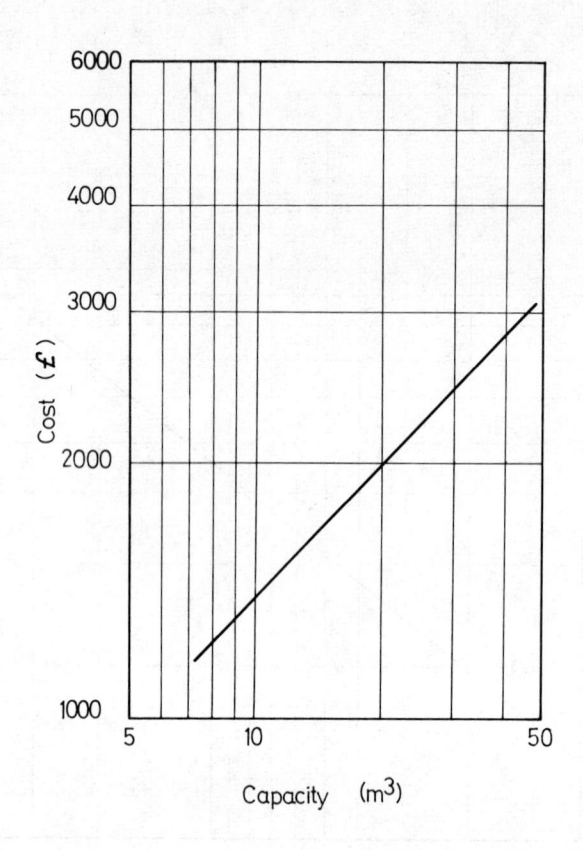

COST GUIDE

COLUMNS (MARKET PRICES OF MATERIALS) (CODE BC)

Vertical cylindrical columns, dished and flanged heads, carbon steel welded construction, spot X-rayed, supplied complete with tray supports, manways, nozzles, support skirt, etc., finished in one coat of primer.

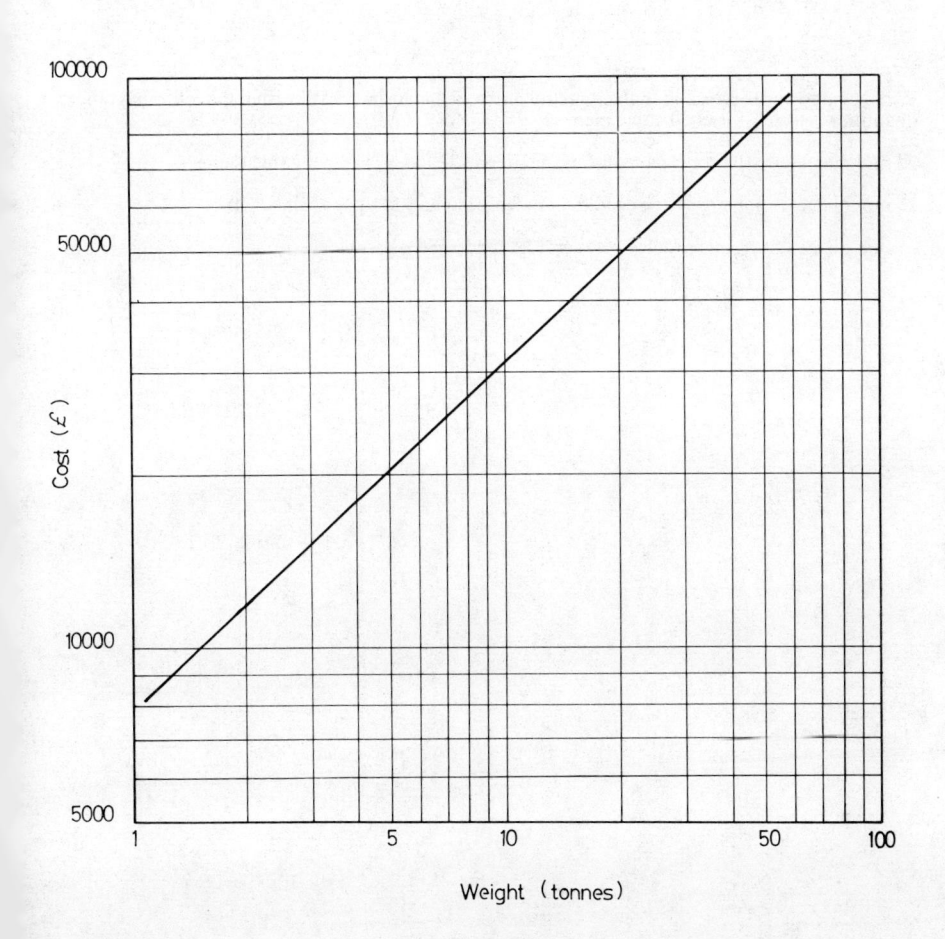

COST GUIDE

TRAYS (MARKET PRICES OF MATERIALS) (CODE BI)

Carbon steel trays for the above columns, sieve type, supplied complete with downcomers, seal pan, etc.

Price tends to vary between £50 and £175 per square metre depending upon the number, size and degree of perforation.

Packings

Prices tend to vary according to the quantity purchased and should fall within the following ranges for quantities between 1 and 50 cubic metres.

25 mm diameter polypropylene pall rings between £320 and £585 per cubic metre

25 mm diameter polypropylene saddles between £280 and £510 per cubic metre

25 mm diameter ceramic saddles between £290 and £685 per cubic metre.

Large Industrial Projects – Estimating

COST GUIDE

PRESSURE VESSELS (MARKET PRICES OF MATERIALS) (CODE BP)

Horizontal or vertical cylindrical pressure vessels, dished and flanged heads, carbon steel welded construction, spot X-rayed supplied complete with manways, nozzles, supports, etc., finished in one coat of primer.

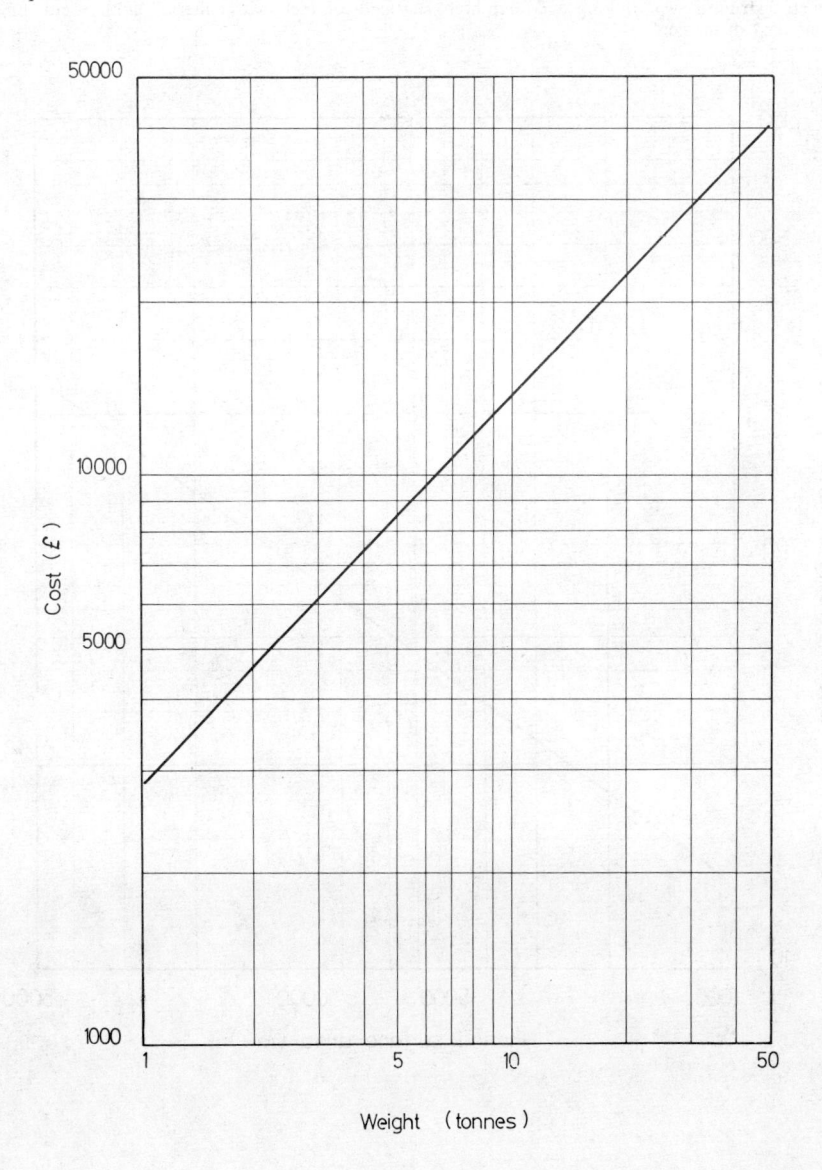

[Graph: Cost (£) vs Weight (tonnes), log-log plot. Cost axis from 1000 to 50000; Weight axis from 1 to 50. Linear relationship on log-log scale.]

COST GUIDE

AIR-COOLED HEAT EXCHANGERS (MARKET PRICES OF MATERIALS) (CODE CF)

Horizontal forced draught air-cooled type heat exchangers, comprising carbon steel tubes 25 mm o/d. × 9·15 m long, aluminium finned, manifolding, plenum chambers, vee-rope driven fans, mounted on a steel structure 9·75 m long × 7·00 m high, supplied complete with galleries, ladders, etc., finished in one coat of primer.

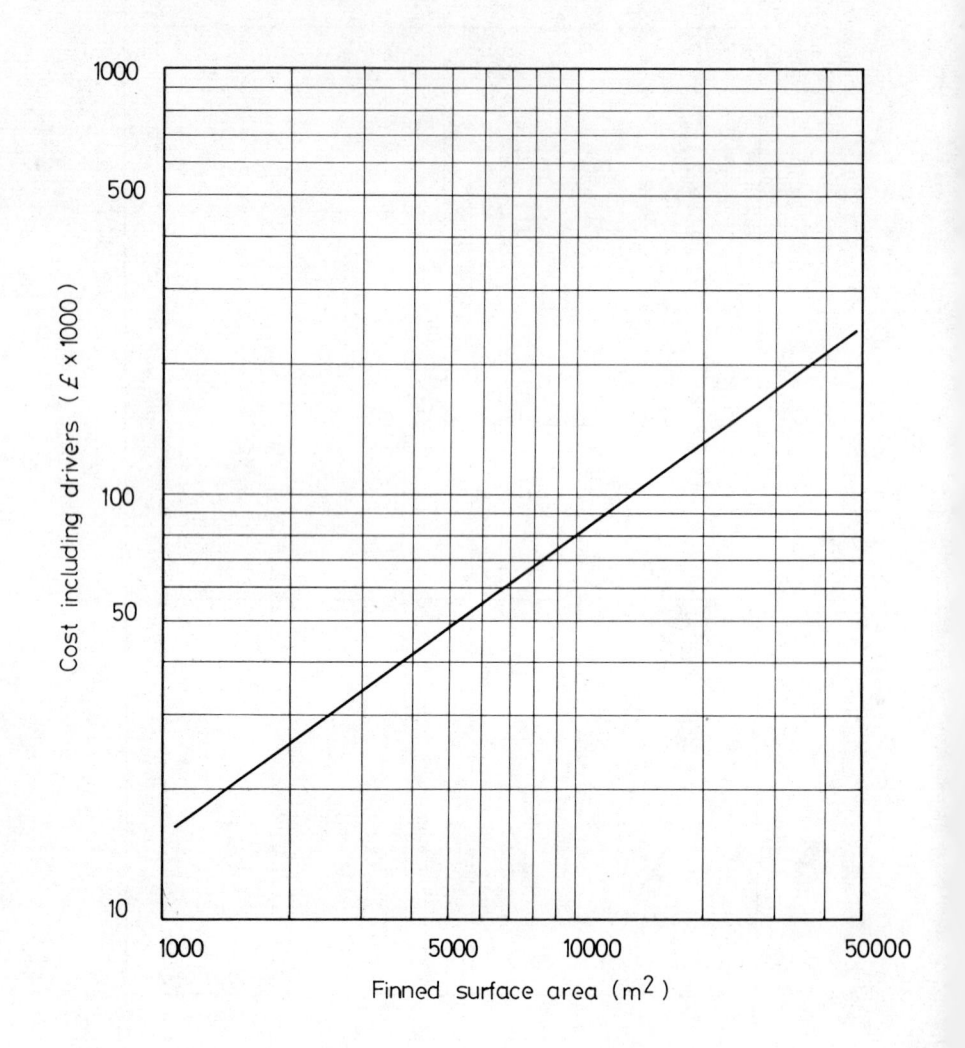

Cost including drivers (£ × 1000)

Finned surface area (m^2)

Large Industrial Projects – Estimating 453

COST GUIDE

SHELL AND TUBE HEAT EXCHANGERS (MARKET PRICES OF MATERIALS) (CODE CS)

Horizontal floating head shell and tube heat exchangers, T.E.M.A. type A.E.T., 25 mm o/d. tubes × 5·00 m long, design pressure 11½-bars, spot X-rayed, carbon steel construction throughout, supplied complete with baffles, nozzles, supports, etc., finished in one coat of primer.

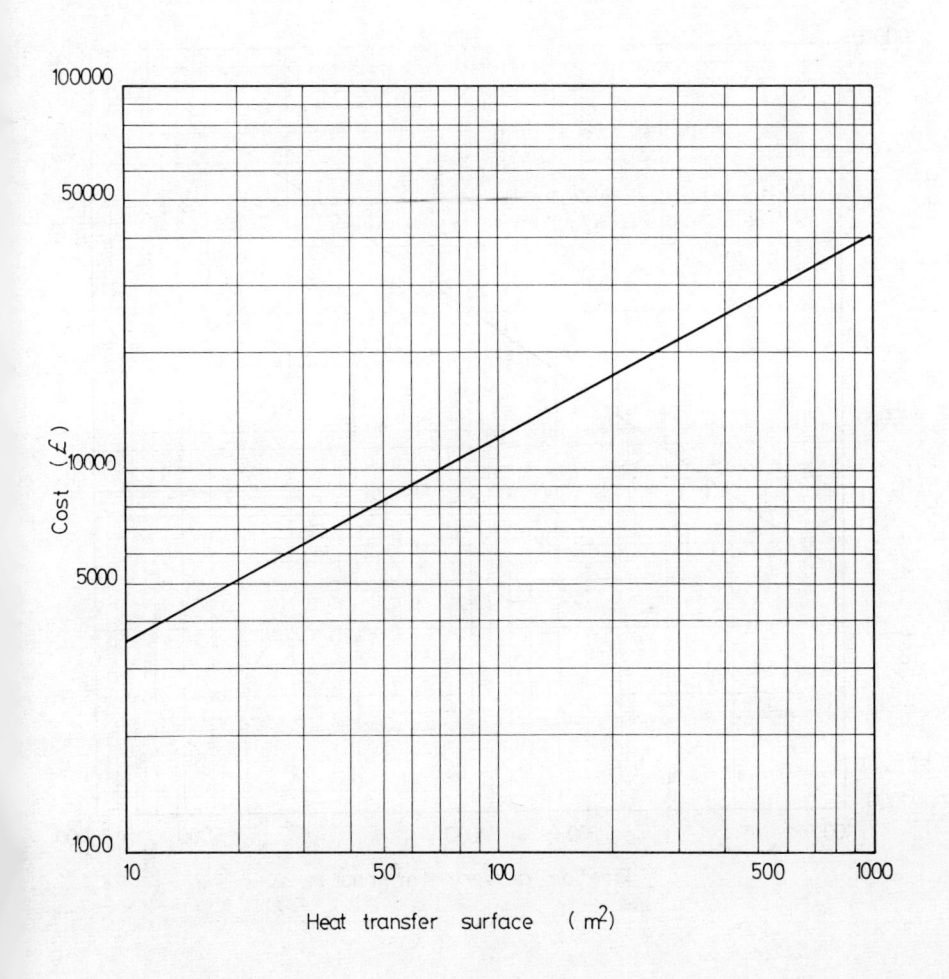

Cost (£) vs Heat transfer surface (m²)

COST GUIDE

AIR COMPRESSORS (MARKET PRICES OF MATERIALS) (CODE DC)

Oil-free reciprocating type air compressors, two stage, to produce air at 8 bars, supplied complete with electric motor, flywheel, combined suction filter and silencer, non-return valve, aftercooler, air receiver, control panel, etc.

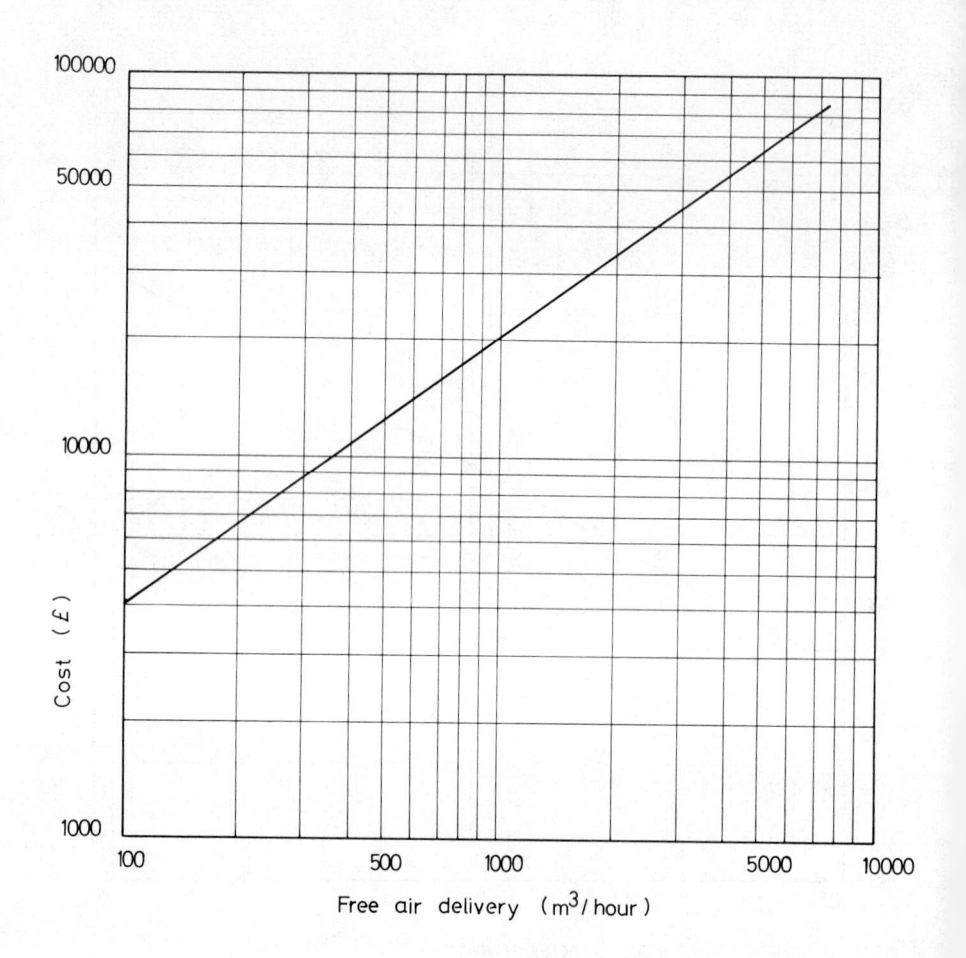

Large Industrial Projects – Estimating

COST GUIDE

PUMPS (MARKET PRICES OF MATERIALS) (CODE DP)

Horizontal centrifugal chemical process type pumps, carbon steel casing, mechanical seal, supplied complete with combined baseplate, coupling and guard. Cost of electric motor is excluded but cost of fitting driver to baseplate and coupling up to pump is included.

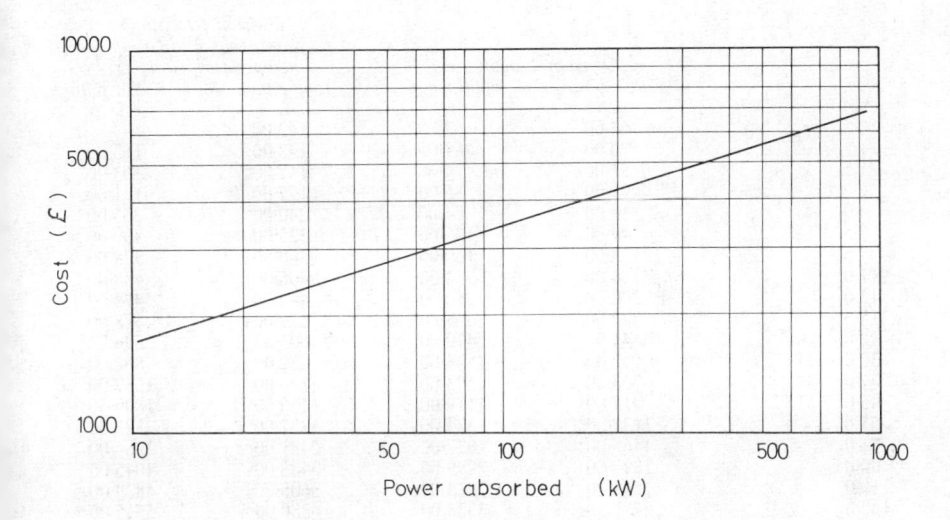

Power absorbed (kW) = hydraulic power (kW) $\div$ pump efficiency $\left(\dfrac{\%}{100}\right)$

Hydraulic power (kW) = $\dfrac{H_e Q \varrho}{367 \cdot 8}$

where H_e = effective total head (m)
  Q = flow (m^3/hr.)
and ϱ = specific gravity
Pump efficiencies can vary widely between 20% and 80%.

Large Industrial Projects – Estimating

COST GUIDE

ELECTRIC MOTORS FOR PUMPS (MARKET PRICES OF MATERIALS) (CODE DP)

Pump motors supplied, totally enclosed fan-cooled, weatherproof, non-sparking outdoor squirrel cage induction motors, dimensions to B.S. 3979 (metric), continuously rated to B.S. 2613, horizontal foot mounted, suitable for 400/440 volts 3-phase, 50-Hz supply for operation as defined in O.C.M.A. Elec. 1.

Size (kW)	Unit	Division 2 locations 1500 r.p.m. £	3000 r.p.m. £	Division 1 locations (flameproof Exd Group II temperature class T5) 1500 r.p.m. £	3000 r.p.m. £
0·75	No.	64·00	67·00	143·00	—
1·1	,,	70·00	69·00	143·00	163·00
1·5	,,	81·00	74·00	157·00	163·00
2·2	,,	91·00	85·00	172·00	191·00
4·0	,,	119·00	124·00	240·00	255·00
5·5	,,	149·00	136·00	323·00	300·00
7·5	,,	177·00	160·00	360·00	378·00
11·0	,,	254·00	254·00	406·00	438·00
15·0	,,	300·00	284·00	449·00	474·00
18·5	,,	379·00	358·00	529·00	578·00
22·0	,,	441·00	439·00	603·00	694·00
30·0	,,	552·00	546·00	852·00	862·00
37·0	,,	733·00	593·00	1126·00	1057·00
45·0	,,	913·00	1119·00	1430·00	1809·00
55·0	,,	1131·00	1384·00	1832·00	2385·00
75·0	,,	1465·00	1877·00	2431·00	3264·00
90·0	,,	1899·00	2299·00	3483·00	4045·00
110·0	,,	2399·00	2915·00	3608·00	4829·00
132·0	,,	2771·00	3338·00	4380·00	6515·00
150·0	,,	3199·00	3815·00	4782·00	6515·00

Extra for providing a set of wiped type cable gland, sealing box and armour clamp including jointing sleeves, sealing compound etc.

	Size	Unit	£
Up to	110 kW	No.	36·50
,,	150 kW	,,	54·00

FURNACES (SUB-CONTRACT PRICES) (CODE F)

There are several types of process furnaces, selection largely depending on the thermal duty requirement and the process application. Consequently, estimating the cost of furnaces can be a complex undertaking requiring a specialist knowledge. Typical installed costs for vertical cylindrical process furnaces supplied and constructed complete in carbon steel, tend to range from about £12·00 to £35·00 per kW of heat absorbed.

Large Industrial Projects – Estimating

COST GUIDE

STEEL TUBING AND FITTINGS (MARKET PRICES OF MATERIALS)
STEEL TUBE (CODE PA)

Carbon Steel Tubes to A.P.I. 5L grade 'B' supplied in random lengths and quantities of under 6 tonnes to 1 tonne of each size and type.

Cold drawn seamless, plain ends.

Nominal bore	Thickness mm (A.P.I. Classification)	Unit	£
15 mm	2·8 (st.w.)	100 m	119·00
	3·7 (x st.)	,,	246·00
	7·5 (xx st.)	,,	397·00
20 mm	2·9 (st.w.)	,,	183·00
	3·9 (x st.)	,,	246·00
	5·6	,,	365·00
	7·8 (xx st.)	,,	521·00
25 mm	3·4 (st.w.)	,,	205·00
	4·5 (x st.)	,,	273·00
	6·4	,,	410·00
	9·1 (xx st.)	,,	645·00

Hot finished seamless, bevelled ends.

Nominal bore	Thickness mm (A.P.I. Classification)	Unit	£
32 mm	3·56 (st.w.)	100 m	236·24
	4·85 (x st.)	,,	297·49
40 mm	3·68 (st.w.)	,,	283·05
	5·08 (x st.)	,,	360·08
50 mm	3·91 (st.w.)	,,	367·08
	5·54 (x st.)	,,	480·70
65 mm	5·16 (st.w.)	,,	553·96
	7·01 (x st.)	,,	747·90
80 mm	5·49 (st.w.)	,,	673·73
	7·62 (x st.)	,,	911·18
	11·13	,,	1298·27
100 mm	4·78	,,	768·79
	6·02 (st.w.)	,,	958·95
	8·56 (x st.)	,,	1357·94
150 mm	5·56	,,	1344·13
	7·11 (st.w.)	,,	1702·57
	8·74	,,	2071·84
	9·52	,,	2139·05
	10·97	,,	2490·79
200 mm	6·35	,,	1946·67
	7·04	,,	2150·06
	8·18 (st.w.)	,,	2485·49
	9·52	,,	2738·94
	10·31	,,	2954·87
	12·70 (x st.)	,,	3669·07
250 mm	6·35	,,	2447·55
	7·80	,,	2988·66
	8·74	,,	3336·70
	9·27 (st.w.)	,,	3533·02
	9·52	,,	3626·80
	10·31	,,	3914·52
	12·70 (x st.)	,,	4873·76

COST GUIDE

STEEL TUBING AND FITTINGS (MARKET PRICES OF MATERIALS)
STEEL TUBE (CODE PA) *continued*

Carbon steel Tubes to A.P.I. 5L grade 'B' supplied in random lengths and quantities of under 6 tonnes to 1 tonne of each size and type.
Hot finished seamless, bevelled ends *continued*

Nominal bore	*Thickness mm* (A.P.I. Classification)	Unit	£
300 mm	6·35	100 m	2913·76
	7·14	,,	3266·41
	8·38	,,	3820·76
	9·52 (st.w.)	,,	4325·93
	10·31	,,	4671·66
	12·70	,,	5824·18
350 mm	7·92	,,	4064·22
	9·52 (st.w.)	,,	4862·25
	10·31	,,	5251·67
	12·70 (x st.)	,,	6423·59
400 mm	9·52 (st.w.)	,,	5689·65
	10·31	,,	6148·07
	12·70 (x st.)	,,	7525·78
450 mm	9·52 (st.w.)	,,	6417·90
	10·31	,,	6936·11
	11·91	,,	7984·53
	12·70 (x st.)	,,	8496·29

Large sizes.

Outside diameter mm	Unit	7·14 £	7·92 £	8·74 £	*Thickness (mm)* 9·52 £	10·31 £	11·13 £	11·91 £	12·70 £
508·0	100 m	5689·88	6306·97	6942·75	7556·61	8168·51	8797·86	9405·89	10011·35
558·8	,,	6266·94	6948·67	7648·91	8327·31	9001·84	9697·00	10368·95	11038·31
609·6	,,	6843·36	7588·88	8355·71	9097·36	9836·44	10591·81	11331·35	12065·25
660·4	,,	7420·41	8229·84	9063·61	9867·41	10649·74	11495·95	12295·06	13092·22
711·2	,,	8005·56	8870·16	9769·30	10637·47	11503·70	12895·56	13252·10	13088·81
762·0	,,	8263·65	9166·22	10094·28	10993·12	11888·84	12812·56	13705·16	14595·29
812·8	,,	—	9783·27	10775·40	11735·19	12682·40	13673·04	14688·83	15584·95
863·6	,,	—	10400·33	11456·00	12447·27	13496·78	14546·14	15560·68	16574·59
914·4	,,	—	10601·77	11678·45	12719·04	13759·52	14230·20	15866·35	16900·01
965·2	,,	—	11195·44	12333·83	13433·93	14533·23	15664·63	16759·34	17852·25
1016·0	,,	—	—	—	14447·93	15305·53	16498·39	17662·94	18804·51
1066·8	,,	—	—	—	14862·02	16078·22	17332·73	18545·95	19756·77
1117·6	,,	—	—	—	—	18174·06	19592·12	20963·16	22833·52
1168·4	,,	—	—	—	—	19007·37	20491·32	21826·87	23359·83
1219·2	,,	—	—	—	—	19841·33	21391·11	22889·91	24386·77

Large Industrial Projects – Estimating

COST GUIDE

STEEL TUBING AND FITTINGS (MARKET PRICES OF MATERIALS)
STEEL TUBE (CODE PA) continued

Carbon Steel Tubes to A.S.T.M. A-106 grade 'B' supplied in random lengths and quantities of under 6 tonnes to 1 tonne of each size and type.

Cold drawn seamless, plain ends.

Nominal bore	Thickness mm (A.P.I. Classification)	Unit	£
15 mm	2·8 (st.w.)	100 m	119·00
	3·7 (x st.)	,,	246·00
	7·5 (xx st.)	,,	397·00
20 mm	2·9 (st.w.)	,,	183·00
	3·9 (x st.)	,,	246·00
	5·6	,,	365·00
	7·8 (xx st.)	,,	521·00
25 mm	3·4 (st.w.)	,,	205·00
	4·5 (x st.)	,,	273·00
	6·4	,,	410·00
	9·1 (xx st.)	,,	645·00

Hot finished seamless, bevelled ends.

Nominal bore			
32 mm	3·56 (st.w.)	,,	243·00
	4·85 (x st.)	,,	306·00
40 mm	3·68 (st.w.)	,,	286·00
	5·08 (x st.)	,,	341·00
	7·10	,,	596·00
	10·20 (xx st.)	,,	745·00
50 mm	3·9 (st.w.)	,,	323·00
	5·5 (x st.)	,,	534·00
	8·7	,,	655·00
	11·1 (xx st.)	,,	1142·00
65 mm	5·2 (st.w.)	,,	546·00
	7·0 (x st.)	,,	769·00
	9·5	,,	1216·00
80 mm	5·5 (st.w.)	,,	685·00
	7·6 (x st.)	,,	794·00
	11·1	,,	1353·00
	15·2 (xx st.)	,,	2010·00
100 mm	6·0 (st.w.)	,,	1030·00
	8·6 (x st.)	,,	1142·00
	13·5	,,	2308·00
	17·1 (xx st.)	,,	3062·00
150 mm	7·1 (st.w.)	,,	1601·00
	11·1 (x st.)	,,	2147·00
	18·2	,,	4219·00
	21·9 (xx st.)	,,	5584·00

COST GUIDE

STEEL TUBING AND FITTINGS (MARKET PRICES OF MATERIALS)
STEEL TUBE (CODE PA) continued

Carbon Steel Tubes to A.S.T.M. A-106 grade 'B' supplied in random lengths and quantities of under 6 tonnes to 1 tonne of each size and type.

Hot finished seamless, bevelled ends *continued*

Nominal bore	Thickness mm (A.P.I. Classification)	Unit	£
200 mm	8·2 (st.w.)	100 m	2628·00
	12·7 (x st.)	,,	3474·00
	23·0	,,	6701·00
250 mm	9·3 (st.w.)	,,	3896·00
	15·1	,,	6601·00
	28·6	,,	11788·00
300 mm	9·5 (st.w.)	,,	4033·00
	12·7 (x st.)	,,	6279·00
	17·5	,,	9679·00
	33·3	,,	18365·00
350 mm	9·5 (st.w.)	,,	4566·00
	12·7 (x st.)	,,	6744·00
400 mm	9·5 (st.w.)	,,	6179·00
	12·7 (x st.)	,,	7048·00
450 mm	9·5 (st.w.)	,,	6664·00

STEEL TUBING AND FITTINGS (MARKET PRICES OF MATERIALS)
SCREWED FITTINGS (CODE PB)

Forged Carbon Steel Fittings, 3000 lb (210 bars) rating, in accordance with A.N.S.I. B16·11 (B.S. 3799) material to A.S.T.M. A105 (B.S. 1503–161 grade B) ends screwed to A.P.I.

	Unit	15 mm £	20 mm £	25 mm £	Nominal bore 40 mm £	50 mm £	65 mm £	80 mm £	100 mm £
45 degree elbow	No.	1·45	1·82	2·40	5·23	6·78	15·86	24·53	46·99
90 degree elbow	,,	0·96	1·21	1·79	3·96	5·09	13·56	20·57	42·71
Tee	,,	1·27	1·65	2·29	5·07	6·98	21·04	43·58	55·35
Cross	,,	3·46	4·19	6·00	13·01	20·15	—	—	—
Coupling	,,	0·46	0·59	0·88	2·06	2·68	5·85	9·50	15·55
Half-coupling	,,	0·41	0·45	0·66	1·53	2·00	4·28	—	—
Reducing coupling	,,	0·59	0·77	1·16	2·67	3·49	—	—	—
Cap	,,	0·41	0·61	0·79	1·85	2·43	5·22	8·54	13·44
Union	,,	1·63	1·94	2·30	3·96	5·79	18·15	23·59	50·61
Hexagon nipple	,,	0·40	0·51	0·77	1·40	2·17	6·17	9·40	14·70
Reducing hexagon nipple	,,	0·45	0·64	0·96	2·21	2·71	7·40	11·28	17·78
Hexagon bush	,,	0·37	0·50	0·68	1·33	2·05	5·99	7·15	12·47
Hexagon plug	,,	0·29	0·32	0·47	1·09	1·65	4·52	6·98	10·97
Square head plug	,,	0·13	0·21	0·28	0·40	0·64	1·93	3·00	4·62

For screwed fittings to B.S. 1740 refer to page 25

COST GUIDE

STEEL TUBING AND FITTINGS (MARKET PRICES OF MATERIALS)
WELDED FITTINGS (CODE PB)

Forged Carbon Steel Fittings, butt-welded type, material to A.S.T.M. A 234 grade 'B' dimensions to A.N.S.I. B16·9 (B.S. 1640).

		Unit	50 mm std £	50 mm xs £	65 mm std £	65 mm xs £	80 mm std £	80 mm xs £	100 mm std £	100 mm xs £
45 degree elbow, long radius		No.	1·36	1·96	1·67	2·24	2·19	2·89	3·46	4·49
90 degree elbow, long radius		,,	1·36	1·96	2·09	2·80	2·74	3·61	4·32	5·61
Equal tee		,,	4·59	5·76	6·98	8·13	7·98	10·40	11·00	15·24
Reducing tee		,,	5·88	7·37	8·93	10·41	10·21	13·31	14·08	19·51
Cap		,,	1·65	2·03	1·83	2·19	2·46	3·16	3·61	4·84

	150 mm std £	150 mm xs £	200 mm std £	200 mm xs £	250 mm std £	250 mm xs £	300 mm std £	300 mm xs £
45 degree elbow, long radius	8·34	11·29	16·01	21·26	28·68	36·40	41·70	54·54
90 degree elbow, long radius	10·43	14·11	20·11	26·58	35·85	45·50	52·13	68·18
Equal tee	23·88	30·48	41·50	53·39	64·64	88·02	94·14	124·47
Reducing tee	30·57	39·01	53·12	68·34	82·74	112·66	120·50	159·32
Cap	7·13	8·97	9·57	13·12	12·86	17·22	21·59	23·79

	350 mm std £	350 mm xs £	400 mm std £	400 mm xs £	450 mm std £	450 mm xs £	500 mm std £	500 mm xs £
45 degree elbow, long radius	66·46	86·15	84·66	109·09	114·65	145·10	147·26	170·11
90 degree elbow, long radius	83·08	107·69	105·83	136·36	148·31	181·38	184·08	212·64
Equal tee	139·45	172·22	213·42	296·96	346·51	440·20	452·00	589·53
Reducing tee	178·50	220·44	218·26	303·70	354·37	450·19	461·26	602·91
Cap	26·40	33·21	31·00	38·07	35·38	42·35	46·20	61·57

	Unit	600 mm std £	600 mm xs £
45 degree elbow, long radius	No.	208·26	258·25
90 degree elbow, long radius	,,	260·32	322·81
Equal tee	,,	625·24	746·74
Reducing tee	,,	639·43	763·69
Cap	,,	60·85	79·12

For Welded Fittings to B.S. 1965 refer to page 33

Abbreviations:
std = fittings to suit standard wall pipe
xs = fittings to suit extra strong wall pipe

Large Industrial Projects – Estimating

COST GUIDE

STEEL TUBING AND FITTINGS (MARKET PRICES OF MATERIALS)
WELDED FITTINGS (CODE PB) continued

Forged Carbon Steel Fittings, butt-welded type, material to A.S.T.M. A 234 grade 'B' dimensions to A.N.S.I. B16·9 (B.S. 1640).

Nominal bore

Reducer Concentric	Unit	25 mm std £	25 mm xs £	40 mm std £	40 mm xs £	50 mm std £	50 mm xs £	65 mm std £	65 mm xs £
50 mm × .	No.	1·77	2·42	1·77	2·42	—	—	—	—
65 mm × .	,,	2·35	3·27	1·98	2·82	1·98	2·82	—	—
80 mm × .	,,	2·46	3·58	2·22	3·23	2·22	3·28	2·22	3·23
100 mm × .	,,	—	—	3·99	5·49	3·44	4·66	3·44	4·66
150 mm × .	,,	—	—	—	—	9·58	20·24	6·65	10·91
Eccentric									
50 mm × .	,,	2·75	4·38	2·75	4·38	—	—	—	—
65 mm × .	,,	3·15	4·97	3·03	4·76	3·03	4·76	—	—
80 mm × .	,,	3·53	5·59	3·18	5·05	3·18	5·05	3·18	5·05
100 mm × .	,,	—	—	5·58	8·36	4·84	7·29	4·84	7·29
150 mm × .	,,	—	—	—	—	13·88	28·67	11·47	20·93

Reducer Concentric		80 mm std £	80 mm xs £	100 mm std £	100 mm xs £	150 mm std £	150 mm xs £	200 mm std £	200 mm xs £
100 mm × .	,,	3·44	4·66	—	—	—	—	—	—
150 mm × .	,,	6·06	9·93	5·39	8·90	—	—	—	—
200 mm × .	,,	12·16	23·12	10·08	19·17	8·06	16·74	—	—
250 mm × .	,,	—	—	19·89	34·33	13·33	30·70	11·94	28·71
Eccentric									
100 mm × .	,,	4·84	7·29	—	—	—	—	—	—
150 mm × .	,,	10·19	18·52	9·11	16·58	—	—	—	—
200 mm × .	,,	19·20	40·45	15·58	31·48	13·62	27·53	—	—
250 mm × .	,,	—	—	33·14	65·20	22·60	44·76	20·28	41·91

Abbreviations: std = fittings to suit standard wall pipe
xs = fittings to suit extra strong wall pipe
For Welded Fittings to B.S. 1965 refer to page 34

Large Industrial Projects – Estimating 463

COST GUIDE

STEEL TUBING AND FITTINGS (MARKET PRICES OF MATERIALS)
WELDED FITTINGS (CODE PB) continued

Forged Carbon Steel Fittings, butt-welded type material to A.S.T.M. A-234 grade 'B' dimensions to A.N.S.I. B16·9 (B.S. 1640)

		Nominal bore							
		150 mm		200 mm		250 mm		300 mm	
Reducer		std	xs	std	xs	std	xs	std	xs
Concentric	Unit	£	£	£	£	£	£	£	£
300 mm ×. . .	No.	22·04	40·45	22·04	40·45	19·59	36·11	—	—
350 mm ×. . .	„	43·59	74·68	39·27	67·13	34·83	63·80	34·83	63·80
400 mm ×. . .	„	—	—	70·01	117·52	68·15	115·45	59·63	91·99
450 mm ×. . .	„	—	—	—	—	—	—	100·50	148·48
500 mm ×. . .	„	—	—	—	—	—	—	151·90	229·80
Eccentric									
300 mm ×. . .	„	41·76	74·26	41·76	74·26	37·11	66·25	—	—
350 mm ×. . .	„	73·26	120·54	65·85	—	59·04	96·55	59·04	96·55
400 mm ×. . .	„	—	—	114·31	176·30	107·55	173·16	102·14	137·84
450 mm ×. . .	„	—	—	—	—	—	—	120·18	185·54
500 mm ×. . .	„	—	—	—	—	—	—	201·64	298·77

		350 mm		400 mm		450 mm		500 mm	
Reducer		std	xs	std	xs	std	xs	std	xs
Concentric	Unit	£	£	£	£	£	£	£	£
400 mm ×. . .	No.	59·63	91·90	—	—	—	—	—	—
450 mm ×. . .	„	91·64	135·60	91·64	135·34	—	—	—	—
500 mm ×. . .	„	114·57	203·69	104·37	165·83	104·37	165·83	—	—
600 mm ×. . .	„	172·64	341·08	159·14	325·62	146·40	289·80	127·31	258·30
Eccentric									
400 mm ×. . .	„	102·14	137·84	—	—	—	—	—	—
450 mm ×. . .	„	106·28	169·18	106·28	169·18	—	—	—	—
500 mm ×. . .	„	163·39	264·80	146·26	215·58	146·26	215·58	—	—
600 mm ×. . .	„	249·69	530·00	230·17	488·45	213·81	434·69	192·63	388·34

COST GUIDE

STEEL TUBING AND FITTINGS (MARKET PRICES OF MATERIALS)
FLANGES (CODE PF)

Forged Carbon Steel Flanges, material to A.S.T.M. A-105 grade I raised or flat face, drilled, dimensions to A.N.S.I. B16·5 (B.S. 1560)

150 lb (11¼ bars) *rating*

	Unit	*Slip on* £	*Standard* £	*Weld neck Extra strength* £	*Blind* £
50 mm	No.	2·53	3·25	3·41	4·54
65 ,,	,,	3·90	4·41	4·65	5·69
80 ,,	,,	3·90	4·79	5·19	6·50
100 ,,	,,	5·45	6·80	7·26	8·60
150 ,,	,,	8·20	10·21	10·83	12·36
200 ,,	,,	12·53	15·61	17·01	19·18
250 ,,	,,	19·33	24·58	26·14	26·44
300 ,,	,,	28·45	36·33	37·89	43·91
350 ,,	,,	38·19	49·48	51·03	—
400 ,,	,,	49·64	63·40	64·94	—
450 ,,	,,	63·40	81·95	83·49	—
500 ,,	,,	74·68	103·59	105·14	—
600 ,,	,,	110·40	146·89	148·43	—

For flanges to B.S. 10 refer to page 46.

300 lb (21¼ bars) *rating*

	Unit	*Slip on* £	*Standard.* £	*Weld neck Extra strength* £	*Blind* £
50 mm	No.	3·66	4·18	4·33	4·09
65 ,,	,,	6·01	6·43	6·65	6·73
80 ,,	,,	6·01	6·56	6·80	6·73
100 ,,	,,	8·69	10·51	10·38	10·21
150 ,,	,,	14·69	16·70	17·78	17·64
200 ,,	,,	23·20	26·91	28·45	30·93
250 ,,	,,	33·85	40·20	41·74	47·94
300 ,,	,,	51·95	58·75	60·30	66·49
350 ,,	,,	77·15	88·90	90·45	97·41
400 ,,	,,	105·59	115·95	117·50	128·3
450 ,,	,,	138·06	157·70	159·25	153·06
500 ,,	,,	167·29	179·34	180·90	208·73
600 ,,	,,	243·50	293·75	295·30	324·68

Large Industrial Projects – Estimating 465

COST GUIDE

STEEL TUBING AND FITTINGS (MARKET PRICES OF MATERIALS)
FLANGES (CODE PF) continued

Forged Carbon Steel flanges, material to A.S.T.M. A-105 grade I'raised or flat face, drilled, dimensions to A.N.S.I. B16·5 (B.S. 1560).

600 lb (42 bars) *rating*

	Unit	Slip on £	Standard £	Weld neck Extra Strength £	Blind £
50 mm	No.	6·70	6·70	6·70	6·81
65 ,,	,,	10·26	10·26	10·26	10·71
80 ,,	,,	10·26	10·26	10·26	10·71
100 ,,	,,	17·31	17·31	17·31	20·23
150 ,,	,,	35·25	35·25	35·25	37·26
200 ,,	,,	54·58	54·58	54·58	55·81
250 ,,	,,	84·56	84·56	84·56	99·70
300 ,,	,,	117·81	117·81	117·81	153·83
350 ,,	,,	181·50	181·50	181·50	184·14
400 ,,	,,	219·85	219·85	219·85	259·58
450 ,,	,,	312·72	312·76	312·76	319·10
500 ,,	,,	—	356·05	356·05	—

FLANGED JOINTS (CODE PJ)

Compressed Asbestos Fibre Gaskets 1·59 mm ($\frac{1}{16}''$) thick, complete with studbolts, nuts and washers to B.S. 1750 to suit raised face flanges of the diameter and rating shown (for flange prices see page 415) to A.N.S.I. B16·5 (B.S. 1560).

Flange rating

Nominal bore	Unit	150 lb (11½ bars) £	300 lb (21½ bars) £
50 mm	No.	1·32	2·63
65 ,,	,,	2·46	4·14
80 ,,	,,	2·88	4·28
100 ,,	,,	2·96	4·58
150 ,,	,,	4·28	9·29
200 ,,	,,	4·35	9·97
250 ,,	,,	9·51	17·67
300 ,,	,,	9·74	20·30
350 ,,	,,	12·65	26·88
400 ,,	,,	19·64	29·20
450 ,,	,,	20·06	34·99
500 ,,	,,	26·79	35·45
600 ,,	,,	27·20	43·91

PIPE SUPPORTS (MARKET PRICES OF MATERIALS) (CODE PS)

There are many types of pipe supports in existence. Indeed, most contractors and some customers have their own standard supports. These are sometimes priced definitively, but are more likely to be estimated as a percentage of the total pipe cost. Clearly, the percentage will vary according to the type and complexity of the pipework, but will range between 3½% and 7½% for a typical petrochemical plant.

COST GUIDE

GLASS TUBING (CODE PA) (MARKET PRICES OF MATERIALS)

Borosilicate glass tubing to I.S.O. 3585 and I.S.O. 4704 with flat buttress ends, delivered in lots exceeding £250·00 of mixed or single sizes.

Length mm	Unit	15 mm £	25 mm £	40 mm £	*Nominal bore* 50 mm £	80 mm £	100 mm £	150 mm £	225 mm £	300 mm £
75	No.	1·44	3·49	—	—	—	—	—	—	—
100	,,	1·45	2·67	3·97	5·17	—	—	—	—	—
125	,,	1·46	2·67	3·97	5·25	8·37	—	—	—	—
150	,,	1·56	2·75	4·06	5·42	8·58	13·42	24·01	—	—
175	,,	1·59	2·83	4·06	5·50	8·89	14·06	25·55	—	—
200	,,	1·63	2·84	4·20	5·66	9·14	14·39	25·55	—	—
300	,,	1·70	2·91	4·45	6·15	10·03	16·43	29·51	54·02	101·06
400	,,	1·87	3·16	4·53	6·72	10·83	18·44	33·57	—	—
500	,,	2·01	3·55	4·94	7·29	11·68	21·02	37·51	68·56	121·28
700	,,	2·43	3·88	5·83	7·92	13·42	23·04	44·04	—	—
1000	,,	2·50	4·94	6·95	9·86	16·00	27·50	54·58	89·59	173·99
1500	,,	3·16	6·07	8·65	11·97	20·05	34·53	73·59	111·09	197·73
2000	,,	3·72	8·08	10·83	14·24	24·57	46·53	90·66	141·08	297·18
3000	,,	—	10·19	14·06	18·84	32·50	56·03	123·00	—	—

Large Industrial Projects – Estimating

COST GUIDE

VALVES (MARKET PRICES OF MATERIALS) (CODE PV)

GATE VALVES

Bronze gate valves 200 lb (14½-bars) rating in accordance with B.S. 1952, material to A.S.T.M. B-62, rising stem inside screw, solid nickel alloy wedge discs, integral seats, union bonnet, ends screwed to A.P.I. . . . . . Unit No.

Nominal bore

	15 mm £	20 mm £	25 mm £	32 mm £	40 mm £	50 mm £
	9·09	11·17	14·74	23·77	35·19	49·77

Cast iron gate valves, 125 lb (10-bars) rating, material to A.S.T.M. A-126 grade 'B' or 'C' rising stem, O.S. and Y, wedge disc, bronze trimmed, flanged and drilled to A.N.S.I. B16.1 (B.S. 1575) . . . „

50 mm £	65 mm £	80 mm £	100 mm £	150 mm £	200 mm £	250 mm £
49·14	72·03	79·78	112·47	172·51	307·05	466·59

Ditto . . . . . „

300 mm £	350 mm £	400 mm £	450 mm £
608·73	1386·89	1832·76	2140·16

Cast iron gate valves 150 lb (11½-bars) rating, material to B.S. 1452, rising stem, O.S. and Y, wedge disc, bronze trimmed, flanged and drilled to B.S. 10, table 'F'. . . . . . „

50 mm £	65 mm £	80 mm £	100 mm £	150 mm £
47·33	53·66	60·60	80·12	133·94

Ditto . . . . . „

200 mm £	250 mm £	300 mm £	350 mm £	400 mm £
234·85	348·75	436·95	739·43	1033·31

COST GUIDE

VALVES (MARKET PRICES OF MATERIALS) (CODE PV) continued

GATE VALVES continued

Cast carbon steel gate valves, 150 lb (11½-bars) rating, material to A.S.T.M. A-216 grade W.C.B. rising stem, O.S. and Y, solid wedge disc, 13% chrome trim, flanged and drilled to A.N.S.I. B-16.5 (B.S. 1560) . . Unit No.

	\multicolumn{7}{c}{Nominal bore}
	50 mm £	65 mm £	80 mm £	100 mm £	150 mm £	200 mm £	250 mm £
	67·57	107·91	116·32	137·50	187·75	288·15	—

Cast carbon steel gate valves, 300 lb (21½-bars) rating, material to A.S.T.M. A-216 grade W.C.B. rising stem, O.S. and Y, solid wedge disc, 13% chrome trim, flanged and drilled to A.N.S.I. B-16.5 (B.S. 1560) . . . . . ,,

50 mm £	65 mm £	80 mm £	100 mm £	150 mm £	200 mm £
121·75	162·16	172·39	222·14	373·23	554·11

Ditto . . . . . ,,

250 mm £	300 mm £	350 mm £	400 mm £	450 mm £
936·34	1334·01	1744·57	2518·83	2904·72

Forged carbon steel gate valves, 800 lb (56 bars) rating, material to A.S.T.M. A-105 grade II, O.S. and Y, bolted bonnet, solid wedge disc, 13% chrome trim, ends screwed to A.P.I. . . . . . ,,

15 mm £	20 mm £	25 mm £	32 mm £	40 mm £
12·66	13·92	17·73	32·00	40·48

Large Industrial Projects – Estimating

COST GUIDE

VALVES (MARKET PRICES OF MATERIALS) (CODE PV) continued

CHECK VALVES

Cast carbon steel check valves, 150 lb (11½-bars) rating, material to A.S.T.M. A-216 grade W.C.B., swing pattern, 13% chrome trim, flanged and drilled to A.N.S.I. B.16.5 (B.S. 1560)

	Unit	\multicolumn{5}{c}{Nominal bore}
		50 mm £	80 mm £	100 mm £	150 mm £	200 mm £	250 mm £
	No.	57·80	87·50	114·12	173·14	303·05	480·48

Ditto	„	300 mm £ 663·08	350 mm £ 903·87	400 mm £ 1289·86	450 mm £ 1769·02	500 mm £ 2180·42

Cast carbon steel check valves, 300 lb (21¼-bars) rating, material to A.S.T.M. A-216 grade, W.C.B. swing pattern, 13% chrome trim, flanged and drilled to A.N.S.I. B.16.5 (B.S. 1560)

		50 mm £ 75·07	80 mm £ 118·85	100 mm £ 139·70	150 mm £ 238·92	200 mm £ 399·30

Ditto	„	250 mm £ 675·85	300 mm £ 943·80	350 mm £ 1909·93	400 mm £ 3052·65

Forged carbon steel check valves, 800 lb (56-bars) rating, material to A.S.T.M. A-105 grade II, piston type, bolted cover, 13% chrome trim, ends screwed to A.P.I.

	„	15 mm £ 8·90	20 mm £ 11·50	25 mm £ 15·30	32 mm £ 24·10	40 mm £ 54·10

Bronze check valves 200 lb (14¼-bars) rating in accordance with B.S. 1953, material to A.S.T.M. B-62, lift type, renewable nickel alloy disc, integral seat, union bonnet, ends screwed to A.P.I.

	„	15 mm £ 4·39	20 mm £ 6·49	25 mm £ 9·30	32 mm £ 12·44	40 mm £ 14·53	65 mm £ 38·08	80 mm £ 63·85	100 mm £ 110·68

Cast iron check valves 125 lb (10 bars) rating, material to A.S.T.M. A-126 grade 'B' or 'C', swing pattern, bronze trimmed, flanged and drilled to A.N.S.I. B-16.1 (B.S. 1575)

	„	50 mm £ 29·74	65 mm £ 33·40	80 mm £ 37·42	100 mm £ 48·87	150 mm £ 80·96	200 mm £ 179·83	250 mm £ 259·99	300 mm £ 349·79

Cast iron check valves, 150 lb (11½-bars) rating, material to B.S. 1452, swing pattern, bronze trimmed, flanged and drilled to B.S. 10, table 'F'

	„	48·76	57·79	69·48	103·73	204·67	344·20	578·56	650·75

COST GUIDE

VALVES (MARKET PRICES OF MATERIALS) (CODE PV) *continued*

GLOBE VALVES

Bronze globe valves, 200 lb (14½-bars) rating, in accordance with B.S. 2060, material to A.S.T.M. B-62, rising stem, inside screw, nickel alloy renewable plug discs and seats, union bonnet, ends screwed to A.P.I.

Nominal bore

	Unit	15 mm £	20 mm £	25 mm £	32 mm £	40 mm £	50 mm £	65 mm £	80 mm £	100 mm £
A.P.I.	No.	2·33	4·01	6·09	8·59	10·74	13·14	26·20	37·93	82·16

Cast iron globe valves, 125 lb (10-bars) rating material to A.S.T.M. A-126 grade 'B' or 'C', rising stem, O.S. and Y, renewable disc and seat ring, bronze trimmed, flanged and drilled to A.N.S.I. B.16.1 (B.S. 1575)

		40 mm £	50 mm £	65 mm £	80 mm £	100 mm £	150 mm £	200 mm £	250 mm £	300 mm £
	,,	54·87	73·11	78·96	93·07	127·12	225·28	508·28	806·08	1226·52

Cast iron globe valves 150 lb (11½-bars) rating material to B.S. 1452, rising stem, O.S. and Y, renewable disc and seat ring, bronze trimmed flanged and drilled to B.S. 10, table 'F'   ,,   —   73·11   78·96   93·07   127·12   225·28   —   —   —

Cast carbon steel globe valves, 150 lb (11½-bars) rating, material to A.S.T.M. A-216 grade W.C.B., rising stem, O.S. and Y, plug type disc, 13% chrome trim, flanged and drilled to A.N.S.I. B.16.5 (B.S. 1560)   ,,   —   92·34   —   150·72   175·93   271·94   632·21   —   —

Ditto but 300 lb (21½-bars) rating   ,,   —   111·84   —   183·29   236·81   570·71   797·81   —   —

Forged carbon steel globe valves, 800 lb (56-bars) rating, material to A.S.T.M. A-105 grade II, rising stem, O.S. and Y, bolted bonnet, 13% chrome trim, ends screwed to A.P.I.

	Unit	15 mm £	20 mm £	25 mm £	32 mm £	40 mm £
A.P.I.	No.	16·32	19·42	30·30	44·46	87·28

COST GUIDE

VALVES (MARKET PRICES OF MATERIALS) (CODE PV) continued

GLASS VALVES

Borosilicate glass valves to I.S.O. 3587 with flat buttress ends, prices delivered in mixed or individual sizes	Unit	15 mm £	25 mm £	40 mm £	50 mm £	80 mm £	100 mm £	150 mm £
				Nominal bore				
Straight through	No.	32·26	47·54	76·17	97·80	245·70	479·12	810·81
Angle pattern	,,	32·26	47·54	76·17	97·80	245·70	479·12	810·81
Regulating – straight through pattern	,,	63·68	100·21	137·40	177·64	—	—	—
Flap type non-return valve	,,	—	78·00	101·00	128·00	220·00	349·00	—

Large Industrial Projects – Estimating

COST GUIDE

EARTHING (MARKET PRICES OF MATERIALS) (CODE QE)

Whilst earthing requirements tend to vary according to the type of plant, a rough approximation of $\frac{3}{4}$% to 1$\frac{1}{2}$% of the total value of the electrical materials can be applied to a typical petro-chemical plant.

ELECTRICAL-INSTRUMENT CABLING (MARKET PRICES OF MATERIALS) (CODE QI)

Multicore, polyethylene insulated and served, single wire armoured and PVC sheathed, 120 volt grade A.C. or D.C. in accordance with O.C.M.A. INP. 4.

			Pairs individually screened			Pairs connectively screened		
			1/0·80 mm		1/1·13 mm	1/0·80 mm		1/1·13 mm
				gland		gland	gland		gland
	Unit	£	size	£	size	£	size	£	size
			mm		mm		mm		mm
2 pair	100	204·36	20	253·41	20	146·35	20	174·02	20
5 ,,	metres	307·80	20	408·72	25	230·14	20	312·67	25
10 ,,	,,	523·00	25	665·27	32	328·23	25	495·81	32
20 ,,	,,	919·31	32	1093·96	40	601·91	32	770·75	40
30 ,,	,,	1276·31	40	1593·85	50	708·03	40	1025·57	50
50 ,,	,,	2083·53	50	2620·84	64	1008·13	50	1516·35	64

Multicore, polyethylene insulated PVC served, lead alloy sheathed, served single wire armoured and PVC sheathed, 120 volt grade A.C. or D.C. in accordance with O.C.M.A. INP. 4.

			Pairs individually screened			Pairs collectively screened		
			1/0·80 mm		1/1·13 mm	1/0·80 mm		1/1·13 mm
				gland		gland	gland		gland
	Unit	£	size	£	size	£	size	£	size
			mm		mm		mm		mm
2 pair	100	313·77	20	394·26	25	241·82	20	273·21	25
5 ,,	metres	455·88	25	592·01	25	358·10	25	475·05	25
10 ,,	,,	716·98	32	908·62	40	484·65	32	681·78	40
20 ,,	,,	1168·94	40	1358·52	50	717·15	40	1022·43	50
30 ,,	,,	1583·79	50	2043·60	50	926·54	40	1331·01	50
50 ,,	,,	2550·88	50	3231·09	64	1309·79	50	1968·62	64

Polyethylene insulated, screened and PVC sheathed, 120 volt grade A.C. or D.C. in accordance with O.C.M.A. INP 4.

		1/0·80 mm		1/1·13 mm	
	Unit	£	gland size	£	gland size
Single pair	100 metres	34·34	—	50·80	—

Large Industrial Projects – Estimating 473

COST GUIDE

ELECTRICAL – THERMO-COUPLE EXTENSION CABLES (MARKET PRICES OF MATERIALS) (CODE QI)

Multicore, copper/constantan conductors, PVC insulated and served single wire armoured and PVC sheathed, in accordance with O.C.M.A.

	Unit	Pairs individually screened £	gland size mm	Pairs collectively screened £	gland size mm
2 pair 20 A.W.G. (1/0·80 mm)	100 metres	272·90	20	202·47	20
5 ,, ,, ,,	,,	433·40	20	334·36	20
10 ,, ,, ,,	,,	748·43	25	495·34	25
20 ,, ,, ,,	,,	1334·47	32	839·45	32
30 ,, ,, ,,	,,	1867·38	40	1123·35	40
50 ,, ,, ,,	,,	2978·78	50	1649·03	50

Multicore, copper/constantan conductors, PVC insulated, served lead alloy sheathed, served, single wire armoured and PVC sheathed in accordance with O.C.M.A.

	Unit	Pairs individually screened £	gland size mm	Pairs collectively screened £	gland size mm
2 pair 20 A.W.G. (1/0·80 mm)	100 metres	415·00	20	322·73	20
5 ,, ,, ,,	,,	619·37	25	497·06	25
10 ,, ,, ,,	,,	995·07	32	693·72	32
20 ,, ,, ,,	,,	1661·60	40	1063·93	40
30 ,, ,, ,,	,,	2262·11	40	1408·51	40
50 ,, ,, ,,	,,	3665·43	50	2030·71	50

Copper/constantan conductors, PVC insulated, screened and PVC sheathed, in accordance with O.C.M.A., complete with draw wire.

	Unit	£	gland size
Single pair 16 A.W.G. (1/1·29 mm)	100 metres	89·29	—

ELECTRICAL – JUNCTION BOXES (MARKET PRICES OF MATERIALS) (CODE QI)

Junction and Distribution Boxes as defined in O.C.M.A. Elect. 1, supplied complete with rail mounted plastic terminal blocks, gland plates and gaskets and side hung door, galvanized finished.

	Unit	Sheet steel weatherproof enclosure, gland plate undrilled suitable for operation in division 2 locations £	Cast iron weatherproof enclosure, gland plate drilled suitable for operation in division 1 locations (flameproof) £
100 terminals	No.	77·00	—
60 ,,	,,	53·79	—
40 ,,	,,	47·73	33·46
20 ,,	,,	34·52	19·71
10 terminals and under	,,	31·54	34·70

COST GUIDE

ELECTRICAL – ELECTRIC POWER CABLING (MARKET PRICES OF MATERIALS) (CODE QP)

PVC insulated and served, single wire armoured and PVC sheathed, 600–1000 volt grade, in accordance with B.S. 6346/1969.

Cross sectional area mm²	Unit	£	Single core gland size mm	£	3 core gland size mm	£	4 core gland size mm
2·5	100	—	—	80·55	20	—	—
4	metres	—	—	118·96	20	—	—
6	,,	—	—	155·47	20	—	—
10	,,	—	—	235·14	25	—	—
16	,,	—	—	341·11	25	—	—
25	,,	—	—	327·43	25	392·34	32
35	,,	—	—	422·18	32	525·72	32
50	,,	—	—	562·90	32	675·17	40
70	,,	—	—	772·76	40	—	—
95	,,	—	—	1024·39	40	—	—
120	,,	—	—	1281·99	50	—	—
150	,,	—	—	1586·60	50	—	—
185	,,	—	—	1943·84	50	—	—
240	,,	—	—	2516·92	64	—	—
300	,,	1159·86	40	3087·92	64	—	—
400	,,	1461·65	50	—	—	—	—
500	,,	1783·83	50	—	—	—	—

PVC insulated, lead alloy sheathed, served, single wire armoured and PVC sheathed, 600–1000 volt grade, in accordance with B.S. 6346/1969 with lead alloy sheath to O.C.M.A. Elec. 4.

Cross sectional area mm²	Unit	£	Single core gland size mm	£	3 core gland size mm	£	4 core gland size mm
2·5	100	—	—	254·66	20	—	—
4	metres	—	—	316·18	20	—	—
6	,,	—	—	394·39	20	—	—
10	,,	—	—	544·92	25	—	—
16	,,	—	—	688·27	25	—	—
25	,,	—	—	476·54	25	548·63	32
35	,,	—	—	592·40	32	735·70	32
50	,,	—	—	788·59	32	944·77	40
70	,,	—	—	1057·71	40	—	—
95	,,	—	—	1416·37	40	—	—
120	,,	—	—	1753·70	50	—	—
150	,,	—	—	2147·16	50	—	—
185	,,	—	—	2655·48	50	—	—
240	,,	—	—	3547·20	64	—	—
300	,,	1689·59	40	4192·21	64	—	—
400	,,	2122·61	50	—	—	—	—
500	,,	2600·64	50	—	—	—	—

Large Industrial Projects – Estimating

COST GUIDE

ELECTRICAL – ELECTRIC POWER CABLING (MARKET PRICES OF MATERIALS) (CODE QP) continued

Paper insulated, lead alloy sheathed, served, single wire armoured and PVC sheathed, 600–1000 volt grade in accordance with B.S. 480/1966 and B.S. 6480/1969.

Cross sectional area mm^2	Unit	Single core £	3 core £	4 core £
16	100 metres	—	484·41	—
25	,,	—	610·07	724·64
35	,,	—	774·12	930·85
50	,,	—	974·82	1184·72
70	,,	—	1256·00	
95	,,	—	1619·03	
120	,,	—	1984·40	
150	,,	—	2161·39	
185	,,	—	2906·09	
240	,,	—	3706·61	
300	,,	1809·90	4520·02	
400	,,	2222·23	—	
500	,,	2707·08	—	

ELECTRICAL – CABLE GLANDS (MARKET PRICES OF MATERIALS) (CODE QP)

Compression type cable glands to B.S. 4121, weatherpoof and watertight, to suit PVC or polyethylene insulated, served, single wire armoured and PVC sheathed cable complete with lock nuts and shrouds.

Conduit size mm	Unit	Non lead sheathed Non-flameproof £	Flameproof £	Lead sheathed Non-flameproof £	Flameproof £
20	No.	1·68	1·77	1·93	2·02
20	,,	1·99	2·09	2·30	2·42
25	,,	2·77	3·09	3·12	3·45
32	,,	3·94	4·11	4·37	4·55
40	,,	6·89	7·19	7·49	7·78
50	,,	9·73	10·16	10·39	10·82
64	,,	15·86	16·59	16·91	17·64
76	,,	19·81	21·96	26·70	27·93

Wiped joint type cable glands for use with paper insulated, lead alloy sheathed, served, single wire armoured and PVC sheathed cable are normally supplied with item of equipment, i.e., electric motor, etc.

Large Industrial Projects – Estimating

COST GUIDE

ELECTRICAL – CABLE TRAYS (MARKET PRICES OF MATERIALS) (CODE QP)

Perforated Mild Steel Cable Trays, Standard Admiralty pattern, galvanized finish.

	Per 2·44 m £	Bends each £	Toes each £
50 × 13 × 1·5 mm	2·67	1·47	2·21
75 × 13 × 1·5 mm	2·92	1·51	2·27
100 × 13 × 1·5 mm	3·41	1·62	2·43
150 × 13 × 1·5 mm	4·80	2·11	3·17
200 × 13 × 1·5 mm	5·22	2·57	3·86
225 × 13 × 1·5 mm	6·68	3·09	4·64
300 × 19 × 1·5 mm	8·81	4·57	6·86
375 × 19 × 1·5 mm	10·70	6·70	10·05
450 × 19 × 1·5 mm	12·35	7·87	11·96
600 × 19 × 2·0 mm	20·93	12·43	18·65
750 × 19 × 2·0 mm	27·70	19·04	28·56
900 × 19 × 2·0 mm	33·27	27·62	41·43

Add approximately 10% to the above prices to cover for fixings, etc.

ELECTRICAL – UNDERGROUND CABLE CONDUIT (MARKET PRICES OF MATERIALS) (CODE QP)

Vitrified clay cable conduit, glazed ceramic finish internally with plain ends and integral polythene jointing sleeves in 1·50 m lengths.

	Unit	90 mm £	100 mm £	*Internal diameter* 125 mm £	150 mm £	225 mm £	300 mm £
Straight lengths	m	1·16	1·20	1·92	2·35	4·31	8·10
Bends	No.	1·16	1·20	1·92	2·30	5·47	9·05
Bellmouth	,,	1·16	1·20	1·92	2·35	5·47	9·05

Above prices are based on a quantity of 1000 m and over of each size.

ELECTRICAL – CABLE-MARKERS (MARKET PRICES OF MATERIALS) (CODE QP)

	Unit	£
Reinforced concrete cable marker posts 760 mm high, supplied complete with stamped lead plate.	No.	5·46
Reinforced concrete block type cable markers.	,,	3·84
Hard burned clay interlocking cable protection covers, apex pattern to B.S. 2484 size 229 mm long × 114 mm wide × 51/32 mm deep.	,,	0·11

Price based on 1200 lot load.

Large Industrial Projects – Estimating

COST GUIDE

ELECTRICAL – MOTOR CONTROL CENTRES (MARKET PRICES OF MATERIALS) (CODE QS)

Grouped Motor Control Boards, non-withdrawable cubicle type, floor mounted with sheet steel enclosure to B.S. 587, comprising incoming fuse switch, ammeter, current transformers, volt meter, phase selector switches, bus bar arrangement, earth bar, glanding chamber, various starters and possibly outgoing fuse switches, and fuse distribution boards.
  Incoming fuse switch, triple pole neutral arranged as incoming feeder.
  Starters, air-break contactor type, complete with magnetic overload relay with start/stop push buttons, ammeter, pilot light, re-set button, air-break motor circuit switch inter-locked with door and H.B.C. fuse-links on both motor and control circuits.
  Outgoing fuse switches, triple pole, H.B.C. fuse-links mounted in insulated carriers, arranged as outgoing feeder.
  Fuse distribution boards, 12-way single or triple pole and neutral.

Type of Unit	Rating
Incoming fuse switch	500 A
Starters direct switching	13 kW
	26 kW
	37 kW
D.S. reversing	13 kW
	26 kW
	37 kW
Star-delta	22 kW
	37 kW
	55 kW
Outgoing fuse switch	45 A
	75 A
	100 A
Distribution boards. S.P. and N.	30 A
T.P. and N.	30 A

Approximate price per complete tier of board £2591·18.

Approximate price for an additional star delta motor starting panel up to 55 kW comprising cubicle, start/stop buttons, ammeter, pilot light and re-set button.   Unit No.   £ 370·73

Large Industrial Projects – Estimating
COST GUIDE

ELECTRICAL – MOTOR CONTROL CENTRES (MARKET PRICES OF MATERIALS) (CODE QS) continued

Grouped Motor Control Boards, withdrawable cubicle type, floor mounted with sheet steel enclosure to B.S. 587, comprising incoming fuse switch, ammeter, current transformer, volt meter, phase selector switches, bus bar arrangement, earth bar, glanding chamber, various starters and possibly outgoing fuse switches and fuse distribution boards.
  Incoming fuse switch, triple pole and neutral, arranged as incoming feeder.
  Starters, air-break contactor type, complete with magnetic overload relay with start/stop push button, ammeter, pilot light, re-set button, air-break motor circuit switch inter-locked with door and HBC fuse-links on both motor and control circuits.
  Outgoing fuse switches, triple pole, HBC fuse-links mounted in insulated carriers, arranged as outgoing feeder.
  Fuse distribution boards, 12-way single or triple pole and neutral.

Type of unit	Rating
Incoming fuse switch	500 A
Starters direct switching	7½ kW
	26 kW
	37 kW
D.S. reversing	7½ kW
	26 kW
	37 kW
Star-delta	11 kW
	37 kW
	55 kW
Outgoing fuse switch	45 A
	75 A
	100 A

Approximate price per complete tier of board £3552·72.

Approximate price for additional star-delta motor starting panel up to 55 kW comprising cubicle, start/stop buttons, ammeter, pilot light and re-set button . . Unit No. £ 503·04

ELECTRICAL – PUSH-BUTTON STATIONS (MARKET PRICES OF MATERIALS) (CODE QS)

Push-button switches, double break type with start and stop buttons, weatherproof enclosure as defined in O.C.M.A. Elec. 1 using 400/440-volts 3-phase 50 hz supply, complete with duty label 1000 mm high post and mounting plate.

Suitable for operation in division 1 or 2 locations (flameproof) . . . . Unit No. £ 77·29

TRANSFORMERS (MARKET PRICES OF MATERIALS) (CODE QT)

Distribution transformers, 11000/440 volts or 3300/440 volts, delta-star oil immersed, naturally cooled, double wound, core type standard transformers complete with first filling of oil.

Rating kVA	Approximate price per kVA £
500	7·86
1000	5·57
2000	8·52

Large Industrial Projects – Estimating

COST GUIDE

INSTRUMENTS (MARKET PRICES OF MATERIALS) CONTROL VALVES (CODE RC)

Cast carbon steel globe control valves, 300 lb (21¼-bars) rating, single or double seat, material to A.S.T.M. A-216 grade WCB, equal percentage contour plug, stainless steel type 316 trim, flanged and drilled to A.N.S.I. B.16. 5 (B.S. 1560), supplied complete with direct acting diaphragm actuator.

Nominal bore mm	Unit	£
25	No.	329·00
40	,,	434·00
50	,,	527·00
65	,,	620·00
80	,,	714·00
100	,,	1039·00
150	,,	1722·00
200	,,	2340·00
250	,,	3288·00
300	,,	5277·00

Cast carbon steel butterfly control valves, 150 lb (11¼-bars) rating, heavy duty pattern, material to A.S.T.M. A-216 grade WCB, fishtail or equal design, stainless steel type 316 trim, flanged and drilled to A.N.S.I. B.16·5 (B.S. 1560), supplied complete with direct acting diaphragm actuator.

Nominal bore mm	Unit	£
100	No.	728·00
150	,,	769·00
200	,,	935·00
250	,,	1010·00
300	,,	1089·00
350	,,	1102·00
400	,,	2322·00
450	,,	2503·00
500	,,	2706·00

Extra for supplying the above valves with pneumatic positioners £108·00 each.

ORIFICE PLATES AND FLANGES (CODE RF)

Stainless steel orifice plate, material to A.S.T.M. A-240 grade 316, supplied complete with a pair of forged carbon steel orifice flanges weldneck type, 300 lb (21¼-bars) rating, material to A.S.T.M. A-105 grade I, raised face, each drilled and tapped two 15 mm A.P.I. connections, compressed asbestos fibre gaskets, stud bolts, nuts, washers and jackscrews.

Nominal bore mm	Unit	£
25	No.	76·00
40	,,	78·00
50	,,	81·00
80	,,	98·00
100	,,	111·00
150	,,	116·00
200	,,	138·00
250	,,	198·00
300	,,	274·00
350	,,	401·00
400	,,	501·00

COST GUIDE

INSTRUMENTS (MARKET PRICES OF MATERIALS) FLOW INDICATORS (CODE RF)

Variable area operation flowmeter, full view type, stainless steel float and guides within glass tube scale 250 mm long, ends flanged 150 lb (11½-bars) rating.

Nominal bore mm	Unit	£
25	No.	228·33
40	,,	310·89
50	,,	367·65

PRESSURE INDICATORS (CODE RP)

Bourdon tube type pressure gauges 100 mm diameter dial, calibrated 0–5 bar, stainless steel measuring element, process connection screwed to ½" A.P.I.   Unit No.   £ 23·74

RELIEF VALVES (CODE RR)

Cast Carbon Steel Safety-Relief Valves, 300 lb (21½-bars) (inlet) and 150 lb (11½-bars) (outlet) ratings material to A.S.T.M. A216 grade WCB, conventional spring-loaded type with plain screwed cap, stainless steel trim, flanged and drilled to A.N.S.I. B-16·5 (B.S. 1560).

Pipe size	Unit	Orifice	£
25 × 50 mm	No.	D 0·710 cm²	267·00
25 × 50 ,,	,,	E 1·265 ,,	269·00
40 × 50 ,,	,,	F 1·980 ,,	328·00
40 × 65 ,,	,,	G 3·245 ,,	332·00
40 × 80 ,,	,,	H 5·065 ,,	358·00
50 × 80 ,,	,,	J 8·300 ,,	398·00
80 × 100 ,,	,,	K 11·860 ,,	548·00
80 × 100 ,,	,,	L 18·405 ,,	566·00
100 × 150 ,,	,,	M 23·225 ,,	779·00
100 × 150 ,,	,,	N 28·000 ,,	835·00
100 × 150 ,,	,,	P 41·165 ,,	971·00
150 × 200 ,,	,,	Q 71·295 ,,	1264·00
150 × 200 ,,	,,	R 103·230 ,,	1615·00
250 × 250 ,,	,,	T 167·750 ,,	2668·00

TEMPERATURE INDICATORS (CODE RT)

Mercury-in-steel type thermometer, 150 mm diameter dial, calibrated 0–150 °C, rigid stem, 190 mm insertion length, supplied complete with stainless steel thermowell screwed 1" API   Unit No.   £ 115·46
As above but Bi-metallic type.   ,,   54·12
As above but capillary type, supplied complete with 5 m length of capillary tubing   ,,   146·42

Large Industrial Projects – Estimating

COST GUIDE

INSULATION (SUB-CONTRACT PRICES) (CODE SQ) HOT SERVICE

Mineral wool sections, secured in place by means of galvanized steel wire and finally cladded overall with 22 or 24 S.W.G. galvanized steel sheeting and secured by self-tapping screws or galvanized steel straps suitably sealed.

Nominal pipe size mm	Thickness insulation mm	Pipe per metre untraced £	Pipe per metre traced £	LR 90° bends No. £	Flange boxes No. £	Valve boxes No. £
20	25	4·02	4·42	4·42	5·89	11·76
25	25	4·14	4·80	4·80	6·37	12·74
40	40	5·38	6·14	6·14	8·15	16·32
50	40	5·70	6·51	6·51	8·65	17·29
65	40	6·13	6·77	6·77	9·01	18·03
80	40	6·49	7·09	7·09	9·43	18·87
100	40	7·09	7·93	7·93	10·57	21·12
150	50	10·33	11·61	11·61	15·43	30·88
200	50	11·99	12·80	12·80	17·05	34·08
250	50	14·37	15·10	15·10	20·10	40·20
300	50	15·98	17·18	17·18	22·85	45·69
350	50	17·12	18·71	18·71	24·87	49·43
400	65	24·62	26·69	26·69	35·59	71·18
450	65	27·16	29·10	29·10	38·70	77·41
500	65	29·75	32·30	32·30	40·84	82·43
600	65	34·12	38·18	38·18	46·79	93·62

Flat surfaces 65 mm thickness of insulation at £14·27 per square metre.
Above prices exclude scaffolding and painting.

MECHANICAL PAINTING (SUBCONTRACT PRICES) (CODE TE/TP)

Wire brushed to remove rust and loose scale, one coat Calcium Plumbate, one undercoat and one coat gloss paint.

Nominal pipe size mm	Per metre £	LR 90° bends metre £	Flanges pair £	Valves No. £
20	0·64			
25	0·70			
40	0·76			
50	0·82			
65	0·87			
80	0·91	Priced as	Priced as	Priced as
100	0·96	pipe	one third	one
150	1·22	measured	metre	metre
200	1·54	along	of pipe	of pipe
250	1·90	centre line		
300	2·25			
350	2·53			
400	2·81			
450	3·09			
500	3·37			
600	3·93			

Flat surfaces at £3·35 per square metre.
Above prices exclude scaffolding.

COST GUIDE

HEAD OFFICE COSTS (CODES 02 TO 48)

This section predominantly covers the process and engineering design aspects of a project, but also includes project management, planning, head office construction, procurement, cost control and sometimes estimating involvement. In the preparation of definitive type estimates, it is usual for a contractor to assess the number of man-hours for each of these disciplines using norms that have been developed in the light of past experience. The resultant man-hour totals are then priced at the appropriate cost per man-hour.

When considering budget estimates, head office costs (including overheads) can be expressed as a percentage of the total installed cost of the plant (also including contractor's overheads). The percentage will vary according to the size and complexity of the plant, the amount of design work that has already been carried out by the customer and the predominant materials of construction. A typical Refinery/Petro-chemical type plant manufactured in carbon steel tends to range between $7\frac{1}{2}$ and 25%.

The term procurement (code 14) usually embraces enquiry, bid analysis (excluding engineering) purchase, progress, inspection and expedition.

Large Industrial Projects – Estimating

COST GUIDE

FIELD CONSTRUCTION COSTS (CODES 50 TO 89)

The basic cost element of field construction is the direct labour cost, to which all other field costs relate in some way. By definition, direct labour is labour that is employed by the contractor and engaged in the actual construction of the plant. Grades of men covered in this definition include riggers, pipe fitters, electricians, etc., together with attendant mates (if any).

Indirect labour, is labour that is also employed by the contractor (not sub-contracted) but is not engaged in the actual construction of the plant. These men include maintenance fitters and electricians, drivers, yard gangs, stores assistants, janitors, etc.

DIRECT LABOUR COSTS (CODE 50)

These are generally assessed in terms of man-hours initially and then priced at a cost per man-hour appropriate to the grade of labour involved. The method of assessing the man-hour content of a project can vary with various norms being applied in different ways. Size of project, difficulty in construction and geographical location are amongst factors to be taken into account when determining the man-hour requirements. For example, productivity varies in different parts of the country. Given below are ranges of typical man-hour rates for various types of construction work.

Item		Principal variable factors
Vessels and columns	3–25 man-hours per tonne	Size, method of lift.
Column trays	12–25 man-hours per m²	Type, height of tray within column, number of pieces.
Heat exchangers:		
shell & tube type	5–15 man-hours per tonne	Size, complexity.
air-cooled type	25–75 man-hours per tonne	Size, complexity.
Pumps, fans and compressors	60–150 man-hours per tonne	Size, number of pieces.
Pipework	150–600 man-hours per tonne	Nominal bore, complexity height of installation, on site or off site.
Electrical power distribution	5–15 man-hours per kW	
Instruments (pneumatic)	20%–30% of materials cost (excluding panels and high priced analysers, etc.)	Size of drive, length of cable run, complexity.

A cost per man-hour for each grade of labour can be calculated by taking the man's basic wage rate and adding to it the various costs required to convey him to and from site in addition to maintaining him while he is at site during the construction period. Shown below is a method for arriving at a total cost per man-hour for two grades of direct labour working at a typical refinery or petrochemical plant site where a site agreement and a basic 40 hour working week (without overtime) is in operation. (The basic rates used are not related to an actual site.)

	Cost per man-hour	
	Imported pipe fitter £	Local attendant mate £
Basic rate	2·2000	1·7100
Premium time (applicable to sites with regularly worked overtime)	—	—
Payroll burden @ 25% of basic rate	0·5500	0·4275
Subsistence @ £40·60 per week	1·0150	
Periodic fares @ £36·40 every 4 weeks	0·2275	
Periodic travelling time 4 hours @ basic rate every 4 weeks .	0·0550	
Daily travelling allowance, say £1·53 per day . . .		0·1912
Productivity payment (fixed allowance)	0·6500	0·4875
Abnormal conditions money	0·0750	0·0750
Total	4·7725	2·8912

Large Industrial Projects – Estimating

COST GUIDE

Payroll burden is normally interpreted as comprising annual holidays, public holidays, national insurance and graduated pension contributions, sick pay scheme contributions, employer's liability insurance and provision for labour recruitment and redundancy.

INDIRECT LABOUR COSTS (CODE 58)

These are either priced definitively or expressed as a percentage of the direct labour cost. The percentage will usually vary between 4% and 6% depending upon the size and complexity of the project.

FIELD SUPERVISION (CODES 60 TO 69)

The extent of supervision will vary according to the size and complexity of the project and the calibre of the direct labour employed, but given below is a typical Field Organization Chart for a medium sized Refinery/Petrochemical Project.

Generally, field supervision can be considered in two parts, namely trade supervision (codes 60 to 64) and field management and administration (code 68). These are either priced definitively or expressed as a percentage of the direct labour cost. While trade supervision tends to vary between $17\frac{1}{2}\%$ and $27\frac{1}{2}\%$ of the direct labour cost, the field management and administration costs range between 5% and 25%, depending on the size and complexity of the construction aspect of the project.

AMENITIES (CODES 70 TO 79)

This section covers the costs involved in providing the labour and supervision with all the facilities at site necessary for them to carry out the construction work using the appropriate plant and tools. Such facilities include site offices, changing rooms, canteen and stores together with temporary electric and water services, telephones, fencing, roads, transport both to and from site and on site, scaffolding, etc.

Pricing amenities can be carried out definitively using the contractor's own hire rates and estimating the cost of temporary materials and labour for erection and dismantling, or they can be expressed as a percentage of the sum total cost of all labour (both direct and indirect), trade supervision and site administration. The percentage will vary according to the size of the work force employed in relation to the duration of the construction period, the location of the site and the amount of amenities already made available by the customer, but a range between $7\frac{1}{2}\%$ and 15% would be fairly representative (not including scaffolding).

Scaffolding (code 79) would have to be considered in isolation. It is dangerous to attempt to determine scaffolding costs by applying percentages. Each project should be treated on its merits.

CONSTRUCTION PLANT AND TOOLS (CODES 80 TO 89)

This section can be considered as comprising four main parts as follows:

(a) *Consumables* – These are generally interpreted as consisting of welding rods, inert shielding gas, solders, fluxes, acetylene and oxygen gases, etc., used in the actual construction of the plant. There are various methods of assessing the costs of consumables, but the end result usually represents from $1\frac{1}{2}\%$ to $2\frac{1}{2}\%$ of the direct labour cost of the mechanical bulk materials (codes P to T). Petrol, oil, diesel fuel and electricity (if chargeable) are generally considered as being part of the cost of the machine that is fuelled by them. Similarly, electric light bulbs, fuses, etc., are considered part of the temporary electrical services.

(b) *Small Tools* – These are usually interpreted as comprising craftsmen's tool kits, protective clothing, clamps, rollers, small jacks, tarpaulins, powered hand tools, other portable tools, etc. Whilst these many items of small tools can be priced individually, it is more common to express the total

COST GUIDE

cost as a percentage of the total direct and indirect labour costs. This percentage tends to vary between $3\frac{1}{2}\%$ and 5% dependent upon the predominant type of construction work, anticipated losses, etc.

(c) *Cranes and Rigging Tackle* – It is difficult to make a general assessment of the craneage and rigging aspects of any construction work due to the fact that there are so many factors which will affect this cost. Therefore, it is advisable for the estimator to carry out a study of the lifting requirements of the particular project, taking into account the rationalization of equipment, delivery dates, access difficulties, etc., establishing which types of crane would be necessary and the duration they would be required on site.

(d) *Other Construction Plant* – This can broadly be described as fabrication and testing equipment, including such items as welding machines, pipe cutters and pipe bending machines. These items can be priced definitively using the contractor's own hire rates but, as a rough guide, the cost, excluding welding machines tends to vary between $1\frac{1}{2}\%$ and $2\frac{1}{4}\%$ of the total direct labour cost.

OVERHEADS AND PROFIT

These items are excluded from the cost code list.

The calculation of overheads varies from contractor to contractor depending upon such factors as head office organization and structure, size and type of projects undertaken and accounting and costing policy. Overheads for the purposes of budget estimating for a typical petro-chemical project in carbon steel will probably fall in the range of 5% to 15% of the preceding project costs, providing costs are allocated in a similar manner to the cost code list and typical analysis.

Profit varies from project to project depending on the contractor's policy, market conditions, size and content of the project and the degree of competition and is frequently not disclosed to the customer. For the purposes of budget estimating, profit can be taken to fall in the range of 5% and 10% of the preceding project costs including overheads.

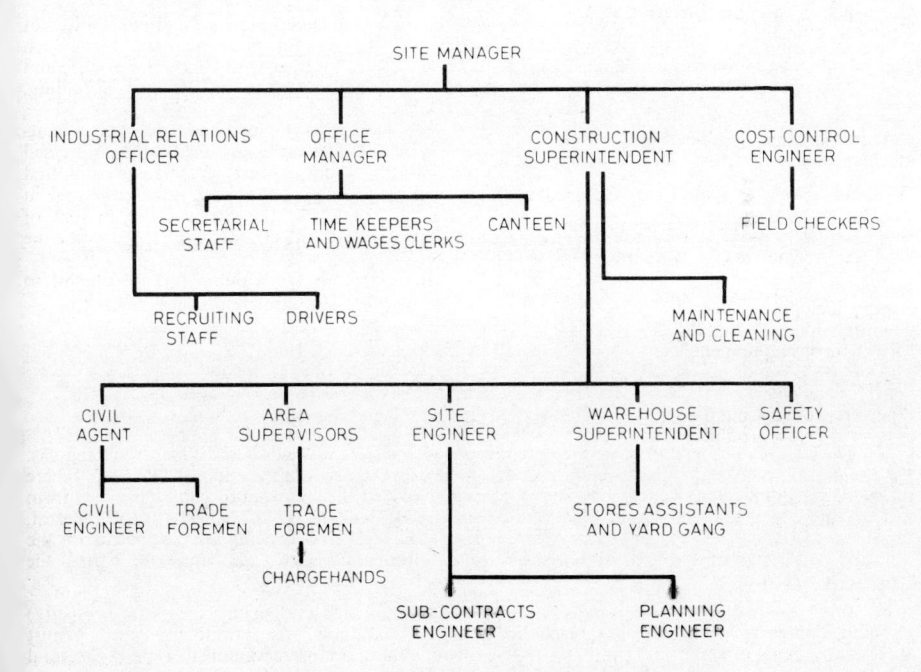

```
                          SITE MANAGER
        ┌──────────────┬──────────────┬──────────────┐
INDUSTRIAL RELATIONS   OFFICE      CONSTRUCTION    COST CONTROL
    OFFICER           MANAGER      SUPERINTENDENT    ENGINEER
        │                │                │              │
   SECRETARIAL      TIME KEEPERS      CANTEEN       FIELD CHECKERS
      STAFF      AND WAGES CLERKS
        │
   RECRUITING       DRIVERS                        MAINTENANCE
     STAFF                                         AND CLEANING

    CIVIL        AREA              SITE         WAREHOUSE         SAFETY
   AGENT     SUPERVISORS         ENGINEER    SUPERINTENDENT      OFFICER
     │            │                 │              │
   CIVIL       TRADE              TRADE                      STORES ASSISTANTS
 ENGINEER    FOREMEN            FOREMEN                       AND YARD GANG
                                   │
                              CHARGEHANDS
                                          SUB-CONTRACTS      PLANNING
                                            ENGINEER         ENGINEER
```

COST INDICES

The following indices of erected costs of process plants are reproduced by courtesy of The Association of Cost Engineers. The basis of calculation was explained in an article in 'The Cost Engineer' in March 1964 and updated index figures have been included in subsequent editions of that journal which are obtainable from the Secretary of the Association, 33, Ovington Square, London, SW3 1LJ.

The indices are related to chemical, petrochemical and petroleum projects in the United Kingdom and reflect the major changes in cost to contractors but not changes in tender levels. The implications of this are similar to those applying to the cost indices for mechanical and electrical services and the reader is referred to the first paragraph of page 324.

A recent revision of the Department of Trade and Industry's Productivity Index is taken into account in the cost indices from the first quarter of 1969 onwards.

The indices are based on four typical plants lettered A, B, C and D. The essential contents of the four plants can be gauged from the following two tables that show the cost proportions of the major elements as they were during the base year (1958).

Cost proportions of major elements in 1958

Category of Cost	Plant A %	Plant B %	Plant C %	Plant D %
Mechanical/electrical material and equipment	54·3	57·8	54·1	62·1
Erection labour	17·9	16·2	18·4	13·0
Foundations, materials and labour	8·2	8·7	8·1	9·3
Engineering and site supervision	10·9	9·2	10·8	8·1
Construction equipment/transport	6·0	5·2	5·9	4·4
Temporary facilities	2·7	2·9	2·7	3·1
	100·0	100·0	100·0	100·0

Cost proportions of sub-elements within mechanical/electrical material and equipment in 1958

Type of material and equipment	Plant A %	Plant B %	Plant C %	Plant D %
Towers, vessels and drums	15	25	20	30
Exchangers	10	15	23	—
Furnaces	15	10	—	—
Pumps, including drivers	8	5	5	10
Compressors, including drivers	4	—	2	—
Miscellaneous equipment	6	10	—	30
Major equipment	58	65	50	70
Pipework above ground	21	16	22	9
Pipework below ground	2	1	1	1
Structures	6	4	5	8
Electrical	7	4	8	7
Fireproofing and insulation	1	2	2	1
Instruments	5	8	12	4
	100	100	100	100

Large Industrial Projects – Estimating

COST INDICES

Mean 1958 = 100

PLANT A

Year	First quarter	Second quarter	Third quarter	Fourth quarter
1967	132·3	131·9	133·1	133·6
1968	136·9	139·2	141·0	140·8
1969	143·1	144·6	149·0	151·1
1970	159·0	163·4	166·5	169·3
1971	174·4	177·9	181·4	181·9
1972	183·1	185·2	190·5	204·8
1973	204·2	207·9	224·2	232·3
1974	241·5	261·8	277·3	288·9
1975	312·8	330·3	353·9	363·5
1976	371·6	383·1	404·9	421·3
1977	437·0			

PLANT B

Year	First quarter	Second quarter	Third quarter	Fourth quarter
1967	132·6	132·3	133·2	133·7
1968	136·9	138·9	140·4	139·8
1969	142·1	143·6	148·8	150·9
1970	159·4	165·1	168·0	170·5
1971	175·1	179·1	182·7	182·6
1972	183·3	185·6	190·9	204·3
1973	203·8	208·2	224·3	232·2
1974	241·7	263·3	278·2	279·8
1975	317·2	335·3	358·0	366·5
1976	375·4	387·0	408·2	425·0
1977	441·0			

PLANT C

Year	First quarter	Second quarter	Third quarter	Fourth quarter
1967	134·1	133·8	134·9	135·4
1968	139·0	141·0	142·4	141·8
1969	144·2	145·9	151·3	153·8
1970	163·0	169·0	172·1	174·7
1971	180·0	184·1	187·5	187·6
1972	187·9	189·8	195·2	209·6
1973	209·0	213·6	231·0	239·5
1974	245·6	268·1	283·6	296·5
1975	330·1	350·6	375·5	382·5
1976	391·4	403·9	426·3	444·2
1977	461·0			

PLANT D

Year	First quarter	Second quarter	Third quarter	Fourth quarter
1967	132·0	131·9	133·1	133·7
1968	136·9	138·8	140·1	140·0
1969	142·5	144·0	148·7	150·4
1970	157·9	161·5	165·5	168·6
1971	172·9	176·3	179·8	180·3
1972	182·2	184·6	190·1	202·5
1973	202·2	206·5	221·5	228·7
1974	239·3	258·3	272·6	282·4
1975	304·2	318·7	339·3	350·3
1976	360·7	371·2	392·0	407·1
1977	422·0			

COST INDICES

Mean 1970 = 100

From the first quarter of 1978 the Association of Cost Engineers intend to continue the above series of indices but related to a base date of 1970 which will bring them into line with other similar publications. To cover the transitional period indices are given below for 1977 and the first quarter of 1978. These are basically mathematical conversions of the '1958' figures with certain minor exceptions.

PLANT A

Year	First quarter	Second quarter	Third quarter	Fourth quarter
1977	287·8	290·4	298·6	302·9
1978	311·9	319·9	332·4	336·5
1979	359·8	375·3	385·3	396·6*

PLANT B

Year				
1977	288·1	290·5	299·0	303·6
1978	312·4	320·3	332·0	334·2
1979	356·3	370·8	380·7	392·5*

PLANT C

Year				
1977	293·4	295·6	304·8	308·7
1978	317·7	326·2	337·2	337·9
1979	361·4	376·3	384·8	394·7*

PLANT D

Year				
1977	281·2	284·4	291·0	297·2
1978	306·4	313·2	324·2	331·7
1979	348·3	363·6	373·5	388·1*

* Estimated

Note

Figures in all the indices on these pages relate to the last month in each quarter.